ORBIS TERRARUM

Band 12

ORBIS TERRARUM

Internationale Zeitschrift für historische Geographie der Alten Welt
Journal of Historical Geography of the Ancient World
Revue internationale de géographie historique du monde antique
Rivista internazionale della geografia storica del mondo antico

OT 12 (2014)

Franz Steiner Verlag

ORBIS TERRARUM

Internationale Zeitschrift für historische Geographie der Alten Welt
Journal of Historical Geography of the Ancient World
Revue internationale de géographie historique du monde antique
Rivista internazionale della geografia storica del mondo antico

Publikationsorgan der *Ernst Kirsten Gesellschaft: Internationale Gesellschaft für Historische Geographie der Alten Welt*

begründet von Prof. Dr. Eckart Olshausen
herausgegeben von Prof. Dr. Klaus Geus unter Mitarbeit von Prof. Dr. Tønnes Bekker-Nielsen

www.steiner-verlag.de/orbis

Das Emblem auf dem Cover der Zeitschrift ist abgeleitet von einem Relief in der Galleria Estense, Módena, Inv. Nr. 2627 (vgl. Vera Sauer, OT 1, 1995, 9–23).

Bibliografische Information der Deutschen Nationalbibliothek:
Die Deutsche Nationalbibliothek verzeichnet diese Publikation in der Deutschen Nationalbibliografie; detaillierte bibliografische Daten sind im Internet über <http://dnb.d-nb.de> abrufbar.

Druck: Laupp & Göbel, Nehrewn
Gedruckt auf säurefreiem, alterungsbeständigem Papier.
Printed in Germany.
ISSN 1385-285X
ISBN 978-3-515-11205-5 (Print)
ISBN 978-3-515-11206-2 (E-Book)

INHALTSVERZEICHNIS

VORWORT DES HERAUSGEBERS

Im Juli 2014 hat die Ernst Kirsten Gesellschaft Herrn Prof. Dr. Michael Rathmann (Eichstätt) zu ihrem 1. Vorsitzenden gewählt. Damit fällt auch die Herausgabe des *Orbis Terrarum*, des publizistischen Organs der Gesellschaft, in seine Verantwortung. Aus redaktionellen Gründen ist der vorliegende Band aber noch in Berlin entstanden.

Um die vorliegende Publikation haben sich viele Kolleginnen und Kollegen, Freudinnen und Freunde verdient gemacht. Stellvertretend nenne ich hier Tønnes Bekker-Nielsen, Martin Fröhling, Niklaas Görsch, Eckart Olshausen, Vera Sauer, Søren Lund Sørensen und Rainer Streng.

Meinem Nachfolger darf ich an dieser Stelle viel Erfolg wünschen!

Berlin, im Juli 2015 Klaus Geus

WATER AND PARADOXOGRAPHY: POLEMON'S WORK
Περὶ τῶν ἐν Σικελίᾳ θαυμαζομένων ποταμῶν

Mariachiara Angelucci

1. INTRODUCTION

The wondrous world in the field of myth, art, ethnography and especially of nature had aroused a great interest in the Greeks from very ancient times.[1] Archilochus, in a famous passage,[2] used for the first time the term θαυμάσιος to indicate the solar eclipse in 648 BC, a phenomenon that was perceived as belonging to the divine sphere, according to a typical way of thinking of the ancient world, in which wonderful aspects of nature had long been seen as a manifestation of the supernatural.[3]

Curiosity about the wonders of nature and fabled peoples is already to be found in Homer and in particular in the *Odyssey*, where the link between the marvellous facts and the divine sphere is clearly evident:[4] the term τέρας always implies a reference to the divine, often present even where it is written θαῦμα, a word used to

1 See the paradoxographic studies by K. Ziegler, s. v. *Paradoxographoi*, in RE XVIII 3, 1949, pp. 1137–1166; A. Giannini, *Studi sulla paradossografia greca. I. Da Omero a Callimaco: motivi e forme del meraviglioso,* "RIL" 97 (1963), pp. 247–66; Id., *Studi sulla paradossografia greca. II. Da Callimaco all'età imperiale,* "Acme" 17 (1964), pp. 99–138; C. Jacob, *De l'Art de compiler à la fabrication du merveilleux. Sur la paradoxographie grecque*, "LALIES" 2 (1980), pp. 121–40; G. Schepens – K. Delcroix, *Ancient Paradoxography: Origin, Evolution, Production and Reception*, in O. Pecere (ed.), *La letteratura di consumo nel mondo greco-latino*, Cassino 1996, pp. 373–460; M. M. Sassi, "*Mirabilia*", in C. Cambiano, L. Canfora, D. Lanza (edd.), *Lo spazio letterario della Grecia antica*, vol. 1, *La produzione e la circolazione del testo,* tome 2, *L'Ellenismo*, Roma 1993, pp. 449–68; O. Wenskus, s. v. *Paradoxographoi*, in *Der Neue Pauly* 9, 2000, pp. 309–312; Á. Ibáñez Chacón, *Poesia y paradoxographia*, "Maia" 60 (2008), pp. 393–404; I. Pajon Leyra, *Paradoxografia griega: estudio de un género literario*, Diss. Madrid 2008, Madrid 2009; Ead., *Entre ciencia y maravilla: El género literario de la paradoxographia griega*, Zaragoza 2011. For the texts of the paradoxographic authors see the collections by A. Westermann (ed.), Παραδοξόγραφοι, *Scriptores rerum mirabilium Graeci*, Brunsvigae – Londini 1839 (= Amsterdam 1963); O. Keller (ed.), *Rerum naturalium scriptores graeci minores*, Leipzig 1877; A. Giannini (ed.), *Paradoxographorum Graecorum Reliquiae*, Milano 1967.
I would like to thank Prof. Dr. Klaus Geus who gave me the opportunity to read the essay *Paradoxography* ed. by K. Geus – C. G. King, in P. Keyser – J. Scarborough, *Oxford Handbook of Science and Medicine in the Classical World*, currently being printed.

2 Archilochus, fr. 122 West.

3 Giannini, *Studi* cit., I, pp. 250, 253 with note 39.

4 See G. Nenci, *La concezione del meraviglioso nei poemi omerici*, "AAT" 92 (1957–8), pp. 275–311, part. pp. 281–93. Homer had already been attributed by the Stoà with the creation of the paradoxographic literature (Ps.-Plut. *Vit. Hom.* 6.618), see G. Vanotti (ed.), *De mirabilibus auscultationibus*, Padova 1997, p. VII.

indicate a miraculous event or a marvel from an aesthetic point of view.[5] However, it is in the episode of Polyphemus that the interest for the wondrous, intended in the way which will be typical of the paradoxographic literature, emerges. He is referred to as a "wondrous monster" who "was not like a man that lives by bread, but like a wooded peak of lofty mountains, which stands out to view alone, apart from the rest"[6]. Polyphemus was therefore a creature who astonishes because he goes beyond the horizon of what is known, contrary to all expectations. The attention to what is out of the ordinary and exceeds the bounds of human experience will effectively be one of the elements which will cross the centuries up to the Hellenistic period and beyond, becoming one of the main features of the paradoxographic writings.

The Homeric Odysseus, so curious towards the lands and the peoples he comes into contact with, embodies the spirit of the Ionic world. The colonisation and the commercial development that took place in the VIII–VII century B.C. spurred on the Ionians of Asia Minor to compete with new human and geographical realities, which both attracted and repelled. Fascination with the marvellous developed in conjunction with the rise of geographical and ethnographic interests which led to the birth of the *periploi* literature and the subsequent literary experiences of Herodotus and the logographers. The *mirabilia* in Herodotus, especially in the naturalistic and ethnographic fields, represent a frequent τόπος.[7] Alongside the attention with which climatic and zoological peculiarities as well as customs of people such as the Egyptians[8] and the Arabs[9] are described, a particular interest for the world of water is observed in Herodotus, with a view to highlighting wonderful and extraordinary aspects. Paradoxographic aura can be perceived in the description of the Nile and the Meride lake in Egypt: the historian notes that the Nile behaves in a totally different way from all other rivers and he is therefore led to delve more deeply into the reason for this peculiarity, all the more remarkable for the fact that none of the Egyptians, when questioned about it, could provide a scientific and acceptable explanation;[10] he also describes the Meride lake as a *thauma* because of the size and human ingenuity with which it had been excavated and realized.[11] The amplitude of the Pontus Euxinus is observed with a similar degree of astonishment.[12]

The lively curiosity for whatever is strange and abnormal, which could be found in the Ionic world that Polemon himself came from, was inherited from the

5 Vanotti, *De mirabilibus* cit., pp. 251–52.

6 Hom. *Od.* 9.190–191 (translation by A.T. Murray, Loeb edition).

7 Cf. K. Trüdinger, *Studien zur Geschichte der griechisch-römischen Ethnographie*, Diss. Basel 1918, pp. 34–37; A. Momigliano, *Alien Wisdom: the Limits of Hellenization*, Cambridge 1975, p. 25; W.M. Bloomer, *The Superlative Nomoi of Herodotus's Histories*, "ClAnt" 12 (1993), pp. 30–50.

8 Hdt. II.35–36. The presence of numerous *thaumasia* in Egypt is indicated by Herodotus as the reason why he devotes considerable time to this region where natural phenomena and customs are very different from those the Greeks are accustomed to.

9 On Arabia and the Arabs see Hdt. II.75; III.107–113.

10 Hdt. II.19–22.

11 Hdt. II.149.1.

12 Hdt. IV.85.

rhetorical historiography of the Isocrates school with Ephorus[13] and especially with Theopompus,[14] in whose writings we find it combined with the awareness of offering the readers a welcome distraction from the historical description. In the Alexandrine age the wondrous as a reason for ἡδονή will be the basis for the development of paradoxography as an independent genre. Among the many authors who in the fourth century B.C. were interested in *paradoxa*, a prominent role should be attributed to the historians of Alexander, who were particularly interested in the fabulous customs of the East, in the flora and fauna of the lands conquered by the Macedonian king.

2. POLEMON AND THE RISE OF THE PARADOXOGRAPHIC GENRE

The line of studies that made the most significant contribution to the rise of a specific literary εἶδος was the Peripatetic school. Among the exponents of the Peripathos the taste for everything that is curious and unusual resulted, at least initially, in scientific research. In this sense Aristotle's critical investigation, devoted especially to the animal kingdom and meteorological phenomena, was crucial and aimed not at a mere collection of material but at a detailed analysis which would make it possible to comprehend the laws of nature. He promoted an ambitious gathering of data, intended to cover all branches of knowledge, from literature and natural phenomena to science, seen from every possible angle. In the late Peripathos research which had originally been directed towards the acquisition of new knowledge, gradually gave way to a taste for the gathering of data for its own sake. Aristotle's successor in the direction of the Peripathos was Theophrastus, who turned his attention primarily to the field of botany but did not neglect other interests dear to paradoxography which are evident in his writings περὶ ὑδάτων, περὶ λίθων and περὶ ζῴων. He had access to a huge amount of data, deriving from the studies by Aristotle, some of which could be rationalised, while other items lacked an immediate scientific explanation. Theophrastus, followed by members of the Peripathos, devoted his attention to the latter and over time the impetus to scientific inquiry that had been typical of their master Aristotle was lost.[15]

Polemon's erudition is clearly indebted to the Peripatetic model and was undoubtedly influenced by the Aristotelian school, although he never explicitly de-

13 Giannini, *Reliquiae* cit., p. 364.

14 The *mirabilia,* which are extensively present in the historical work of Theopompus, have been widely used by the paradoxographers of later periods. A writing entitled Θαυμάσια has been ascribed to him. The presence in the eighteenth book of his *Philippika* (*FGrHist* 115, F 64–77) of a large number of *mirabilia,* however, has suggested two hypotheses: the work Θαυμάσια could be a section of the *Philippika* (Jacoby, *FGrHist* II B, *Komm.* p. 365), detached some time later from the rest of his work, or it could be a subsequent work made up of extracts taken from the author's total production (cf. Ziegler, s. v. *Paradoxographoi* cit., pp. 1144–45). Cf. Giannini, *Studi* cit., II, pp. 102–104; Id., *Reliquiae* cit., pp. 365–68; W. R. Connor, *Theopompus and fifth-Century Athens*, Washington 1968, pp. 12–13; G. S. Shrimpton, *Theopompus the Historian*, Montreal – London 1991, pp. 15–20; Schepens, *Paradoxography* cit., p. 380 with note 18.

15 Sassi, *Mirabilia* cit., pp. 454–57.

clares his indebtedness or makes direct references in any of his fragments. We do not know for certain whether Polemon joined one specific philosophical school, but the interest he showed for the *thaumasia* in the naturalistic field as well as the information that he provides on topics of botany and zoology and the almost encyclopedic collection of data does recall the Aristotelian school as it had developed over time.[16] His learned approach to research with this insistence on details and his attention to the precise organization of the information according to criteria among which that of geography, which charaterised the classification of knowledge of this school, lead to a similar reflection.[17]

The previous examples of the paradoxographic literature were resumed by Callimachus, who was undoubtedly known to Polemon and who devoted particular attention to the theme of the wondrous for its own sake. All the elements present in the authors, who had already made the marvellous the object of their interests, merged in him and from this point onwards we can refer to paradoxography as an independent literary tradition.[18]

The work Θαυμάτων τῶν εἰς ἅπασαν τὴν γῆν κατὰ τόπους ὄντων συναγωγή, handed down to us by the Suida[19] and known also by the shorter titles Ἐκλογὴ τῶν παραδόξων[20] and Θαυμάσια, which are likely to have been of common use, must be assigned to Callimachus, who is generally recognized as the founder of the paradoxographic genre. If the title is by no means sure, because of textual problems which cannot be easily solved,[21] it can be said with a reasonable degree of certainty that his writing was organized by geographical sections, each of which in turn is thought to have been subdivided into different topics. The theme regarding *paradoxa* of water received particular attention as is shown by the surviving fragments, thirty-nine out of forty-eight of which were related to hydrographic material.[22] Philostephanus[23] can be placed in the same line as Callimachus. He too was a native of Cyrene and may also have been his student, the author of περὶ παραδόξων ποταμῶν, which was very close to the interests of his probable master, as can be seen in the work on the specific and sectorial subject of the wonderful rivers.

16 M. Angelucci, *Polemone di Ilio: fra ricostruzione biografica e interessi antiquari*, "SCO" 49 (2003), p. 170; Ead., *Polemon's Contribution to the Hellenistic Literature of the Second Century B. C.*, "Hormos" 3 n. s. (2011), p. 329.

17 R. Capel Badino (ed.), *Filostefano di Cirene. Testimonianze e frammenti*, Milano 2010, pp. 35–36.

18 Giannini, *Studi* cit., I, pp. 264–65. See also R. Pfeiffer, *Storia della filologia classica, dalle origini alla fine dell'età ellenistica*, ed. it. Napoli 1973, pp. 223–24; Giannini, *Reliquiae* cit., pp. 15–20.

19 Suid. κ 227.

20 On the problem of the connection between ἐκλογή and συναγωγή see Schepens, *Paradoxographoi* cit., p. 395 note 68 who rejects the idea that considers the ἐκλογή an epitome and gives an account of the different positions of modern scholars. See also Giannini, *Studi* cit., II, pp.105–106 with note 33.

21 On the textual problems of the title see Schepens, *Paradoxography* cit., p. 395 note 68.

22 Giannini, *Studi* cit., I, pp. 107–108.

23 Giannini, *ibid.*, pp. 21–23.

Polemon was also unquestionably familiar with the writings of Antigonus of Carystus, whom he attacked in the work πρὸς Ἀδαῖον καὶ Ἀντίγονον, which consisted of at least six books. The author, a contemporary of Philostephanus and slightly younger than Callimachus, may be identified with the sculptor and versatile writer who flourished at the school of Pergamon, particularly during the reign of Attalus I. Modern scholars are in some doubt as to whether he should be credited with the ἱστοριῶν παραδόξων συναγωγή on various paradoxographic topics, organised according to an organic plan where the thematic criterion can be found. The topic which was best structured and richest in details after the zoological theme, is unquestionably the one connected with water (ch. 129–65).[24] Dorandi,[25] however, has questioned the authenticity of this writing, claiming that it should be considered the work of a Byzantine compiler, who lived probably at the time of Constantine VII Porphyrogenitus. He argues that it is the result of *excerpta* of paradoxographic works including the tract Περὶ ζῴων by Antigonus, the Περὶ ζῴων ἱστορίας by Aristotle, the work Θαυμάτων τῶν εἰς ἅπασαν τὴν γῆν κατὰ τόπους ὄντων συναγωγή by Callimachus. Alongside the thematic criterion there is the bibliographical, which allows the subdivision of the material on the basis of the sources used, as was the intention of the author, who points out the beginning and the end of the Aristotelian sections and the beginning of the Callimachean ones.[26]

Irrespective of who the author really is, the collection is essential for the information conveyed and for the authors whose testimony is thus preserved. It enables us to understand more clearly the paradoxographic material, which is frequently passed on by fragments very brief and unsystematic.

In this context Polemon's paradoxographic work, with its particular emphasis on the *paradoxa* of water, according to a typical predilection of the Hellenistic age, which saw the flowering of paradoxographic writings,[27] can more easily be understood: the perieget, a native of Asia Minor, where the cultural stimulus of the Ionic literature originated, is part of a wide and varied tradition, which he helped to enrich, thus giving expression to a commonly felt interest on the part of his readers.

24 Jacob, *De l'art* cit., pp. 124–29; Giannini, *Studi* cit., II, p. 116; Schepens, *Paradoxography* cit., p. 396.

25 T. Dorandi (ed.), *Antigone de Caryste. Fragments, texte établi et traduit*, Paris 1999, pp. XIV–XVII; O. Musso, *Sulla Struttura del Cod. Pal. Gr. 398 e deduzioni storico-letterarie,* "Prometheus" 2 (1976), pp. 1–10 has confirmed this hypothesis by making reference to issues of codicology. He has shown that the *Palatinus gr.* 398, the only code that hands down to us the collection of Antigonus' fragments, is the work of a copyist of the tenth century A.D. who flourished at the court of Constantine VII Porphyrogenitus, who had gathered around him scholars whose task was to create a project of συλλογαί. There are also many affinities between the work ascribed to Antigonus and the *Excerpta de animalibus* composed as part of the studies sponsored by the emperor.

26 Schepens, *Paradoxography* cit., p. 396.

27 For a list of writers of paradoxography in Hellenistic times see Giannini, *Studi* cit., II, pp. 105–127, 139.

3. THE PROBLEM OF THE TITLE

Polemon's paradoxographic work has been handed down to us with four titles: Περὶ τῶν ἐν Σικελίᾳ θαυμαζομένων ποταμῶν (F 83), περὶ Θαυμασίων (F 84), περὶ τῶν ἐν Σικελίᾳ ποταμῶν (F 82) and περὶ ποταμῶν (F 81).[28] The first two reveal the paradoxographic nature of his work; the two latter confirm the attention he paid to rivers.

Despite the uncertainty that often surrounds the titles of ancient works, in this case they already give us some idea as to the problems underlying the writing of Polemon: the fragments we have do not seem to deal always with paradoxographic information; not all fragments concern rivers, nor was Sicily the only geographical area considered. And this is why it is difficult to attempt to establish the exact structure and content of Polemon's work, of which we can get only a partial idea.

The most significant fragment is undoubtedly F 83, taken from Macrobius, in which Polemon refers to two sulphur springs located in Sicily in the district of Mineo in what is now the province of Catania,[29] which were linked with the ancient deities of Palici and had the extraordinary power to unmask and punish perjurers. This passage testifies to the unquestionable interest of the perieget in natural phenomena related to the world of water in Sicily and provides us with an example of crucial importance – and one that is unique for its length – about his attitude to the world of *paradoxa*.

The first problem arises from the discovery that the perieget does not confine himself to the wonders related to waterways as shown by F 84: he mentions the exceptional thinness of the diviner Archestratus, who weighed no more than an obol, and of Panaretus,[30] who was a disciple of Arcesilaus of Pitane and flourished at the court of Ptolemy Euergetes, where he lived without ever falling ill despite his slight physique.

Two hypotheses have been therefore formulated, either that they are two independent writings, the one relating to wonderful facts in general, the other more specific and dealing with wonderful waterways, addressed by region, with particular attention to Sicily or that it is a single work concerning *thaumasia*, divided into sections according to a geographical or thematic criterion. In the case of a geographical criterion, it was argued that sections had been ordered by topic, according to the model of Callimachus; in the case of the thematic criterion, apparently used by Philostephanus, it was believed that the thematic sections had been divided into geographical subsections.

28 References to Polemon's fragments follow the numbering adopted by L. Preller *Polemonis Periegetae Fragmenta*, Leipzig 1838 (= Amsterdam 1964): F 83 (= F 2 Giannini) Macr. *Sat.* V.19.26–30; F 84 (= F 1 Giannini) Athen. XII 552 c; F 82 (= F 3 Giannini) Athen. VII 307 b–c; F 81 (= F 4 Giannini) *Schol. Eurip. Med.* 835.

29 See Vanotti, *De mirabilibus* cit., p. 100; N. Cusumano, *Siculi*, in P. Anello, G. Martorana, R. Sammartano (edd.), *Ethne e religioni nella Sicilia antica*, Atti del Convegno Internazionale (Palermo 6–7 dicembre 2000), Roma 2006, p. 122.

30 Athen. XII 552 c; Eusth. *ad. Il.* 1287 44 (*Il.* 23.72). Cf. Ael. *Var. Hist.* 10.6.

The hypothesis of the two independent writings is supported by Preller, who believes the περὶ Θαυμασίων to be separate from the one dealing with rivers, to which the three remaining titles would accordingly refer. He imagines a work with a title similar to the περὶ παραδόξων ποταμῶν by Philostephanus, who had dealt more with marvellous and special issues related to rivers than with a description of their origin or of the route they followed, a fact which would presuppose a more specific geographical approach.[31]

Likewise, Müller keeps the two writings separate and treats the *mirabilia* of the rivers of Sicily as one section of a wider περὶ παραδόξων ποταμῶν.[32] Deichgräber, while following Preller, recognizes the problematic nature of the issue and argues that rivers and unusual springs found in Sicily were just one of the topics covered by Polemon, who then moved on seamlessly to deal with other matters according to his customary mode of exposition.[33]

Giannini on the other hand agrees with the hypothesis of a single work entitled, according to the current use, περὶ Θαυμασίων, in which special attention was paid to what was wondrous in the world of water, a topic from which the perieget deviated to tackle subjects which had little to do with paradoxography.[34]

This is the second problem posed by Polemon's fragments, namely the presence of passages that do not seem at first glance to have *mirabilia* as their theme: in F 81 Polemon records the rivers named Cephisus to be found in Greece and writes "there is a Cephisus in Athens, in Sicyon and in Argos", while in F 82 he notes that mullets (κεστρεῖς) were also called πλῶτες. A third hypothesis on the organization of Polemon's work has therefore been put forward, in addition to the other two previously mentioned: the existence of a work entitled περὶ ποταμῶν, not paradoxographic in nature, has been envisaged, divided into sections on the basis of geographical criteria, each of which was in turn divided into a subsection on the *paradoxa*.[35] It is a hypothesis, however, that would appear to be superfluous inasmuch as these two cases do not deal with θαυμάσια, but fall within what can

31 Preller, *Fragmenta* cit., pp. 125, 131.

32 C. Müller, *FHG* III, Paris 1883, pp. 139–41.

33 K. Deichgräber, s. v. *Polemon*, in *RE* XXI 2, 1952, pp. 1315–16. On Polemon's tendency to resort to *excursus* see G. Pasquali, *Die Schrifstellerische Form des Pausanias*, "Hermes" 48 (1913), pp. 176–86; Angelucci, *Polemon's contribution* cit., pp. 334–35; Ead., *Polemone* cit., p. 176.

34 Giannini, *Studi* cit., II, pp. 120–21. F. Susemihl, *Geschichte der griechischen Literatur in der Alexandrinerzeit*, II, Leipzig 1891–92 (= Hildesheim 1965), p. 673 note 134 largely agrees with him, even though, when he quotes Polemon's writings, he keeps separate the two titles περὶ Θαυμασίων and περὶ ποταμῶν. The περὶ τῶν ἐν Σικελίᾳ ποταμῶν and the Περὶ τῶν ἐν Σικελίᾳ θαυμαζομένων ποταμῶν may represent a part of this latter work.

35 Giannini, *Studi* cit., I, p. 121 note 134, even though he is in favour of a single work περὶ Θαυμασίων and thinks that there is little doubt that the περὶ τῶν ἐν Σικελίᾳ ποταμῶν and the Περὶ τῶν ἐν Σικελίᾳ θαυμαζομένων ποταμῶν are one and the same work, does not rule out completely this hypotesis. W. Schmid – O. Stählin, *Wilhelm von Christs Geschichte der griechischen Literatur*, II 1, *Die nachklassische Periode der griechischen Litteratur. Von 320 vor Christus bis 100 nach Christus,* München 1920[6], p. 243, affirm that the περὶ ποταμῶν had as its theme focused on the homonyms of rivers and did not possess a paradoxographic nature.

be regarded as ἴδιον[36]: the river is cited for the particular fact that there are several rivers with the same name, just as it is a linguistic peculiarity to find two terms to describe the same species of fish.

Having discarded this third hypothesis, a fourth can be proposed, in my opinion, on the basis of considerations about Polemon's type of writing. He is remembered by the tradition above all for his periegetic writings, organized geographically, in which he describes not only regions and cities of ancient Greece but also Sicily, Magna Graecia and the areas of Greek culture in Asia Minor.[37] We do not know for sure if he was the author of separate writings or of one work in which he systematized the data in a περιήγησις κοσμική [38] with macro-sections dedicated to the different areas covered, including in particular Greece, organized in turn into subsections relating to different regions such as Attica, Argolis, Elis etc. Writings such as περὶ τῶν ἀναθημάτων τῶν ἐν 'Ακροπόλει would thus form part of far more extensive work. It is also possible to argue that the macro-sections were in actual fact self-contained and subdivided into chapters. Following this line of reasoning the controversial title Ἑλλαδικός has been on occasion interpreted as an indication of an alleged *periegesis* of Greece.[39]

On the contrary, there also exists a belief that the idea of a universal periegesis or of a *periegesis* of Greece has its origin in later times as a result of the weight of the authority of Pausanias, who is the lens through which periegetic writings have not infrequently been read. Furthermore, the lack of reliable data confirming the existence of such general works and the extensive presence of single and specific titles have induced scholars to think that Polemon's writings were composed and edited independently.[40] This thesis tallies with the encyclopedic nature of his re-

36 On the terminology connected with the marvellous see Schepens, *Paradoxography* cit., pp. 381–82 with note 24 and p. 398.

37 See Angelucci, *Polemon's Contribution* cit., pp. 331–41. Cf. Ead., *Polemone* cit., pp. 165–83.

38 According to Preller, *Fragmenta* cit., p. 23 the adjective κοσμική dates back to the Byzantine era so it is preferable to use the expression περιήγησις οἰκουμένης. At any rate, the authenticity of the περιήγησις κοσμική is a source of debate. The expression περιήγησις κοσμική is never quoted in Polemon's writings but appears only in the Suida and almost at the end after the incomplete list of some monographic writings of the perieget. Furthermore, there are almost no periegetical Hellenistic works bearing the title περιήγησις and the adjective κοσμική appears to date back to Byzantine times. See F. De Angelis, *Pausania e i periegeti. La guidistica antica sulla Grecia* in E. Vaiani (ed.), *Dell'antiquaria e dei suoi metodi*, Pisa 1998, p. 2.

39 This is the opinion of Preller, *Fragmenta* cit., pp. 23–25. See also Müller, *FHG* III, pp. 112–13. The work known as Ἑλλαδικός is actually controversial. Athenaeus himself doubts its authenticity and in both passages (XI 479; XIII 606 b), in which he specifically refers to the title, he writes: Πολέμων γοῦν ἢ ὅστις ἐστὶν ὁ ποιήσας τὸν ἐπιγραφόμενον Ἑλλαδικός. Deichgräber, s. v. *Polemon* cit., pp. 1302–1303 ascribes it to Polemon. On the contrary Susemihl, *Geschichte* cit., p. 669 affirms that the work is a sort of epitome, composed by another author. The theme of the two passages, which deal respectively with the votive offerings found in the Treasuries of Olympia and with an anecdote related to the Treasury of Spina in Delphi, actually conforms to the perieget's erudite approach and so it is possible to attribute them to him. It is more the title and the nature of the work that cause controversy.

40 Preller, *Fragmenta* cit., p. 23, believes that it is a single work made up of individual writings collected together by Polemon or by later grammarians. See also Müller, *FHG* III, p. 112. W.

search which produced works in which a very considerable amount of space was devoted to the description of limited portions of territory. Strabo informs us that the work περὶ τῶν ἀναθημάτων τῶν ἐν Ἀκροπόλει consisted of four books. For that matter the author closest to Polemon for his type of writings and undoubtedly known to him was Heliodorus of Athens,[41] who dedicated fifteen books to the Acropolis. This was the nature of the antiquarian research in Hellenistic times and it clearly shows the influence of the Peripatetic model. Polemon himself is the author of a work contesting Timaeus in fifteen books, in which he seems to focus on erudite issues. These reflections do not exclude that the perieget had a broader project in mind or may have systematized his writings at a later date, just as we cannot rule out that their organization in a single work was the fruit of subsequent efforts.

In the light of these considerations, the Περὶ τῶν ἐν Σικελίᾳ θαυμαζομένων ποταμῶν, abbreviated to περὶ τῶν ἐν Σικελίᾳ ποταμῶν, might arguably be a self-contained work perhaps thematically connected to Polemon's work Κτίσεις Ἰταλικῶν καὶ Σικελικῶν πόλεων of a periegetic and antiquarian nature. The fragment on the river Cephisus could similarly belong to a work on the marvellous rivers in Attica, abbreviated to περὶ ποταμῶν, which might have followed up his periegetic writings about this region. The content of the fragment concerning Archestratus and Panaretus lead us to think that the perieget had recourse to the *excursus* technique or it might suggest the existence of a further paradoxographic work. The περὶ Θαυμασίων, to which this fragment belongs, would be an abbreviation commonly used to refer to one or more of his paradoxographic works, in the same way as Pausanias defines the writers of paradoxography with the generic formula of οἱ ἐπὶ τοῖς θαύμασι.[42] According to the line of reasoning adopted for the περιήγησις κοσμική and for the Ἑλλαδικός the title περὶ Θαυμασίων may have been used by Polemon himself or by others in a later period to group together the paradoxographic writings which were first composed as self-contained works.

At any rate it is clear that the lack of a paradoxographic title on Attica or at least on another region apart from Sicily prevents us from doing more than formulate a simple hypothesis and that in the attempt to solve the problem of the organisation of the antiquarian data in Polemon's works the two authors who immediately come to mind are Callimachus with his geographical criteria (this would presuppose, however, an organisational approach and the systematic orderings of the material which we do not know if Polemon possessed) and Philostephanus with his thematic criteria.

Judeich, *Topographie von Athen*, München 1931, p. 11 is in favour of a single work, while Deichgräber, *Polemon* cit., p. 1292 and pp. 1303–1304, prefers the hypothesis of individual self-contained writings, arguing that had it been a single work the perieget would not have been given the title of *proxenos* of Delphi (p. 1303). Capel Badino, *Filostefano* cit., pp. 36–37 believes that Philostephanus' works, as is the case with Polemon, represented "the gathering together of heterogeneous monographs which were assembled at a later stage on the basis of the organisational criterion they shared".

41 *FGrHist* 373.

42 Paus. VIII.46.5.

4. THE FRAGMENT ON THE PALICI

Polemon's fragment on the Palici (F 83) is undoubtedly his most significant passage in terms of hydrographic *paradoxa*. Macrobius cites him in connection with his exegetical commentary on some verses of Virgil's *Aeneid* (9.581–85), in which the altar of Palicus, located on the banks of the river Symaethus in Sicily,[43] is mentioned. He relies on the perieget and on three other Greek authors, Callias,[44] Aeschylus[45] and Xenagoras,[46] to explain the myth of the Palici and the trial by ordeal related to them, accusing the Latin authors of ignorance, in that they confine themselves to defining Palicus as a god, without adding anything more precise. In point of fact, Silius Italicus[47] and Vibius Sequestres[48] had already dealt with the Palici and the oath, albeit briefly.

The Palici[49] were local deities of Sicily who were rapidly absorbed by the Greek mythology. Traditionally they were believed to be twin brothers and sons of the nymph Thalia who, fearing the wrath of Zeus, were granted the possibility of being swallowed up by the earth at the time of birth. The twins, at the moment they were born, sprang from the earth itself and were therefore called Palici ἀπὸ τοῦ πάλιν ἱκέσθαι, according to the etymology given by Aeschylus.[50] In the district of Mineo,[51] near the point where the two brothers are believed to have appeared, there were two sulphur springs, which according to the testimony of Polemon formed

43 Macr. *Sat.* V.19.15–19.

44 *FGrHist* 564. Callias, who lived between the IV and the III century B.C., is the author of a history of Agathocles in twenty-two books.

45 Aesch. fr. 6 *TGrF* 3, 127–28 (= Macr. *Sat.* V.19.24).

46 *FGrHist* 240. Xenagoras, who lived in the II century B.C., composed works entitled χρονική and *On the islands*.

47 Sil. It. *Pun.* 14.219–20. Silius Italicus, a Latin poet who lived between the I and the II century A.D. and who was a great admirer of Cicero and Virgil, is known above all for the epic-historical poem in seventeen books, in which he narrated the Second Punic War from the battle of Sagunto to the battle of Zama.

48 Vib. Seq. 177. Vibius Sequester, a scholar from the IV–V century A.D., was a contemporary of Macrobius. He is the author of a lexicon containing geographical place-names taken from poets.

49 K. Ziegler, s.v. *Palikoi*, in *RE* XVIII 3, 1949, pp. 99–123; J.H. Croon, *The Palici. An Autochthonous Cult in Ancient Sicily*, "Mnemosyne" s. IV, 5 (1952), pp. 116–29; G. Glotz, s.v. *Palici*, in *DA*, IV 1, pp. 284–85; L. Bello, *Ricerche sui Palici*, "Kokalos" 6 (1970), pp. 71–97; E. Manni, *Divagazioni sul culto dei Palici. Hommages à Robert Schilling*, Paris 1983, pp. 175–85; F.P. Rizzo, s.v. *Palico*, in *Enciclopedia Virgiliana*, directed by F. Della Corte, Roma 1987, pp. 935–36 with bibliography.

50 Aesch. fr. 6 *TGrF* 3, 127–28 (= Macr. *Sat.* V.19.24). Cf. Macr. *Sat.* V.19.18.

51 According to Callias's testimony (*FGrHist* 564, F 1) the place was located not far from the river Symaethus, which flows to the south of Catania. See Vanotti, *De mirabilibus* cit., p. 100; Cusumano, *Siculi* cit., p. 122. On the site of Palice see G. Di Stefano, s.v. *Palice*, in G. Nenci – G. Vallet (edd.), *Bibliografia topografica della colonizzazione greca in Italia e nelle isole tirreniche*, pp. 280–82; A. Messina, s.v. *Mineo*, in *Bibliografia topografica* cit., pp. 145–51.

very deep craters,[52] whose connection with those deities was established by the ancients. These are the words of the perieget:

> The so-called Palici are believed by the inhabitants to be gods native to the region. They have as brothers craters sunk in the ground: those who approach them must be free of pollution and abstain from sexual intercourse and also from certain foods. 27. The heavy odor of brimstone rises from them and produces a strange drowsiness in bystanders; their water is foul, with a color very like white soapsuds and rises in frothing waves, like the swirling and bubbling of water on a high boil. They say that these craters are unfathomably deep, so that cattle that fall into them and a mule-team driven into them disappear, as do grazing mares that jump in them.[53]

Here those who were under accusation could come to declare their innocence and undergo a singular rite, aimed at exposing perjurers. Polemon provides a detailed description of the oath-swearing ceremony, which the Sicilians considered to be the most powerful and whose sacredness is confirmed by the conditions and the acts laid down by the rite.[54] The swearer had to appear purified from all defilements, crowned, dressed only in his tunic and without a belt. From the rim of the crater, which he had to approach waving a branch, he uttered the formula of the oath that was communicated to him by the ὁρκωταί, who are to be seen as priests of the Palici' shrine, rather than as accusers. If the accused was guilty, he died instantly. Therefore, before swearing, he had to provide guarantors who had to undertake to cover the purification expenses in the event of perjury.[55] Polemon does not specify how perjurers died, but according to Ps.-Aristotle it was by combustion.[56] In the text that precedes Polemon's quotation Macrobius, perhaps citing the perieget,

52 The craters are called by the ancient sources with different names: *krateres, pegai, krenai, lacus, stagna*. See Cusumano, *Siculi* cit., pp. 122–23.

53 Macr. V.19.26–27 (translation by R.A. Kaster, Loeb edition).

54 For an in-depth study of the Palici and the trial by ordeal see N. Cusumano, *Ordalia e soteria nella Sicilia antica. I Palici,* in *Mythos* 2, 1990, pp. 9–186 with bibliography; Cusumano, *Siculi* cit., pp. 121–45 particularly pp. 122–128. See also on the Palici K. Ziegler, s. v. *Palikoi*, in *RE* XVIII 3, 1949, pp. 99–123; Croon, *The Palici* cit., pp. 116–39; G. Glotz, s. v. *Palici*, in *DA*, IV 1, pp. 284v85; L. Bello, *Ricerche sui Palici*, "Kokalos" 6 (1970), pp. 71–97; E. Manni, *Divagazioni sul culto dei Palici. Hommages à Robert Schilling*, Paris 1983, pp. 175–85; F.P. Rizzo, s. v. *Palico*, in *Enciclopedia Virgiliana*, Roma 1987, pp. 935–36 with bibliography; L. Maniscalco (ed.), *Il santuario dei Palici: un centro di culto nella Valle del Margi*, Palermo 2008.

55 Macr. *Sat.* V.19.28–29: "For the Sicilians the craters provide the most powerful oath, when opponents who have issued a challenge have been ritually purified. With a writing tablet in hand, the persons administering the oath address the parties to the oath about whatever matter it is for which the oath is being sought. Then one party to the oath, garlanded and waving a green bough, ungirt and wearing only a tunic, dips his hand in the crater and repeats the words of the oath after the person administering it. Should he make good the oath taken, he departs for home unscathed; but if he is shown to have transgressed against the gods, he dies on the spot. In the course of the ritual the parties promise that they will provide bondsmen for the priests, who are liable for purifying the shrine should anything untoward happen. Near this spot the Palikênoi founded the city Palikê, named after the gods" (translation by R.A. Kaster, Loeb edition).

56 Ps.-Arist. *Mir. Ausc.* 57. According to Diodorus (XI.89) the perjurer didn't die but became blind.

whose testimony probably was not limited to the passage referred to above, states that they were not general allegations, but accusations of theft or similar crimes.

Polemon's testimony is essential because it reveals his interest towards *mirabilia* in the field of nature but it is also interesting for the reconstruction of the rite[57] and because it focuses on local gods, later acknowledged by the Greeks, that is to say on a cult that was a point of contact between Greeks and natives.[58] Even though the passage has not been seen as directly relevant to the sphere of wonders,[59] we cannot deny the existence of a description of the natural phenomena considered a marvellous fact in reference to the depth of the craters and to the water that "rises in frothing waves, like the swirling and bubbling of water on a high boil". Likewise the power given to them in the trial by ordeal intended to unmask perjurers can rightly be considered to be something extraordinary.

Polemon can justly be included in the authors who came under the influence of the Peripathos, which no longer sets out to uncover scientific explanation of phenomena in accordance with Aristotle's approach, but extrapolates especially those facts most lacking in a possible rational interpretation from the wide range of Aristotelian data – and not just from them.[60] After all "θαῦμα is no match for explanation; the sense of the marvellous cannot survive on a rational basis. It is imperative for the paradoxographer to concentrate on *historia*, the establishment and the recording of facts without explaining them." [61] The amazement at a phenomenon that goes beyond what man is accustomed to observe in nature, without a rational explanation being provided, emerges from the passage taken from the perieget referred to above and we can also find the aspect of the divine on which the ordeal rite is based and is revealed in the immediate death of the perjurer.

The extraordinary springs of the Palici were a topic dear to the Aristotelian school, as is evident in an extract from the περὶ θαυμασίων ἀκουσμάτων, preserved in the *corpus aristotelicum* which should be considered as belonging to the Peripatetic tradition and dating back perhaps to the third century B. C.[62] The text, which is provided below and which is useful for a comparison with that of the perieget, shows the same sense of wonder for the miracle related to water:

> There is a spring among the Palici in Sicily, covering the space of ten couches; this throws up water to the height of six cubits, so that the whole place is thought to be inundated; and it falls back again to the same spot. There is an oath which is regarded as very sacred there; for a man writes down the oath he takes on a small tablet and casts it into the water. If he swears truly, the

57 Polemon provides the most complete and reliable evidence for the trial by ordeal. In addition to him we have only Diod. IX.89.

58 See E. Manni, *Culti greci e culti indigeni nella Sicilia antica. Problemi di metodo e spunti di ricerca*, "ASS" (1980), pp. 5–17.

59 Giannini, *Studi* cit., II, p. 121 note 135.

60 P. M. Fraser, *Ptolemaic Alexandria*, Oxford 1972, I, p. 770.

61 Schepens, *Paradoxography* cit., p. 391.

62 It is difficult to establish with certainty the chronological context of the work whose authenticity had already been questioned by Erasmus in the XVI century. If certain clues point to the era of the emperor Adrianus, the placing of the work in the III century B. C. nevertheless would seem to be preferable. See Sassi, *Mirabilia* cit., pp. 457–59 and the analysis of the two chronological hypotheses in Vanotti, *De Mirabilibus* cit., pp. XI–XIV.

tablet floats. If he swears falsely, the tablet is said to grow heavy and disappear and the man is burned. So the priest takes security from him that someone shall purify the temple.[63]

The text of Ps.-Aristotle, though short, provides some information regarding the trial by ordeal that cannot be found in Polemon: the height to which the geyser spouts, the size of the area covered by water and the cause of death of the perjurer. Given the length of Polemon's passage, so rich in details and given his sophisticated erudition, the fact that this information is missing suggests that the perieget did not use the περὶ θαυμασίων ἀκουσμάτων as a source – always assuming that it is to be attributed, as would appear to be the case, to the third century B. C. – and that at any rate the two authors relied on different sources.[64]

In none of the paradoxographic fragments are we given the name of the authors on whom Polemon based his writings. The indication of the names serves, for the paradoxographers, as proof of the credibility of what they say, as is already clear from Callimachus himself:[65] in fact the paradoxical phenomenon, despite exceeding all expectations and human experience, is considered a marvel which can be believed in precisely because it belongs to the real world and not to the world of imagination.[66] The author of paradoxography reveals a critical spirit in his selection of the documentary sources in support of the reported facts which belies the opinion of those who believe that the works of *mirabilia* are devoid of any critical sense. Similarly, the credibility of what is narrated is reinforced by the indication of the places to which the wonders are related. This is a feature which can clearly be found in the paradoxographers: "Such geographical precision had in fact been paramount from the beginnings of the genre in Callimachus".[67]

While the indication as to location is present in Polemon, the fragmentary nature of his writings makes it impossible for us to know if he omitted the name of the source or if it has simply not been passed on to us. It is, however, possible to infer from the expression παρὰ τοῖς ἐγχωρίοις that he consulted local sources.[68] We can moreover assume that Polemon, in addition to first-hand autopsy, which certainly cannot be excluded, also used literary evidence including Timaeus, although we do not possess any passages in support of this hypothesis. Similarly, he

63 Ps.-Arist. *Mir. Ausc.* 57 (translation by W. S. Hett, Loeb edition).

64 The problem of Ps.-Aristotle's source for the passage about the Palici is still open. See Vanotti, *De Mirabilibus* cit., p. 101.

65 Callimachus cites among his sources Theophrastus, Megasthenes, Timaeus, Polycritus or Polycleitus, Theopompus, Aristotle, Ctesias, Amometus, Xenophilus, Heraclides Ponticus, Pha(i)nias, Nicagoras. See Schepens, *Paradoxography* cit., p. 383.

66 Schepens, *Paradoxography* cit., p. 382: "An astonishing item can only be termed θαυμαστόν if, indeed, it belongs to the real world, if it is witnessed or reported to have happened or to have been observed". Cf. anche Geus – King, *Paradoxography* cit., currently being printed.

67 Cf. J. Stern, *Paradoxographus Vaticanus*, in S. Heilen, R. Kirstein, R. Scott Smith, S.M. Trzaskoma, R. van der Wal, M. Vorwerk (edd.), *In Pursuit of Wissenschaft: Festschrift für William M. Calder III zum 75. Geburtstag*, Hildesheim and New York 2008, pp. 439–440; Geus – King, *Paradoxography* cit., currently being printed.

68 Cf. D. Ambaglio, Ἐπιχώριος: *un termine tecnico storiografico?*, in *Storiografia locale e storiografia universale. Forme di acquisizione del sapere storico nella cultura antica,* Atti del Congresso (Bologna 16–18 dicembre 1999), Como 2001, pp. 7–21.

may have known the work περὶ τῶν ἐν Σικελίᾳ θαυμαζομένων by Nymphodorus of Syracuse,[69] which was possibly intended to follow up a *periegesis* of Sicily, known only in the title of a scholium to *Odyssey*.[70] Nymphodorus's fragments do not make any mention of the Palici but we cannot rule out that he describes the natural phenomenon in a passage that has since been lost. Even though we have very little information, the perieget's familiarity with Nymphodorus and maybe even specific reference to his work can be deduced from the fact that both authors indicate the city of Iccara as the origin of the hetaera Laide, information that they probably derive from Timaeus.[71] Even Callimachus himself, who exerted so much influence on the way the material and data were organized by both contemporary and subsequent writers, may well have been consulted by Polemon. The belief that Ps.-Aristotle draws on the Callimachean *mirabilia*, a view which has long been upheld,[72] has on occasion been questioned and the two works may have an independent origin: "there is nothing to prevent the pseudo-Aristotelian work from being independent from Callimachus (to whom, effectively no references are made): also because there is no reason why the paradoxographic genre has at all costs to be attributed to a sole line of descent. "[73]

It is difficult to identify with certainty the sources Polemon may have used. Indeed, the natural phenomenon he describes attracted the attention of ancient writers[74] and was dear to the historians and antiquarians of Hellenistic times who wrote about Sicily and of whom we still have some records:[75] the Παλικίνη κρήνη is mentioned by Theophilus, who quotes it in his Περιήγησις Σικελίας, as Stephanus from Byzantium reports;[76] the Byzantine lexicographer is likewise the source for the reference to the same spring by Silenus of Calacte;[77] similarly, Isigonus of Nicea cannot hide his astonishment at the remarkable existence of the waters linked to the Palici as handed down by the *Paradoxographus Florentinus*, which is dedicated to hydrographic *mirabilia* and contain much of his work Ἄπιστα.[78]

69 *FGrHist* 572 = Giannini, *Reliquiae* cit., pp. 112–15.

70 *FGrHist* 572, T 2.

71 Jacoby, *FGrHist* 570 *Komm.* p. 603; S. Spada, *La storiografia occidentale di età ellenistica*, in R. Vattuone (ed.), *Storici greci d'Occidente*, Bologna 2002, pp. 254–55.

72 Sassi, *Mirabilia* cit., p. 458 with note 13.

73 Sassi, *ibid.* p. 459.

74 See also Plut. *Timol.* 12; Diod. XI.89.2; Strab. VI.2.9; Hesych. π 176; Theogn. *Can.* 323 (Cramer, *An. Ox.* II, p. 60).

75 As regards Antigonus of Charystus, leaving aside the problems of chronological dating, we possess a fragment (F 121 Giannini) relating to the miraculous phenomenon connected with the temple of the Palici and not with the two craters: φησὶν [...] τῆς Σικελίας ἐν Παλικίοις οἰκοδομηθῆναι τόπον, εἰς ὃν ὅστις ἂν εἰσέλθῃ, εἰ μὲν κατακλιθείη, ἀποθνήσκειν, εἰ δὲ περιπατοίη, οὐδὲν πάσχειν.

76 *FGrHist* 573, F 1 (= Steph. Byz. s. v. Παλίκη).

77 *FGrHist* 175, F 3 (= Steph. Byz. s. v. Παλίκη).

78 Isig. F 3 Giannini = *Parad. Flor.* F 8 Giannini. Isigonus passes on the same information we find in Ps.-Aristotle in connection with the height of six cubits allegedly reached by the jet of water. See also Spada, *La storiografia* cit., p. 262.

5. CONCLUSIONS

Unfortunately, we do not possess many fragments that enable us to understand more clearly how Polemon's work Περὶ τῶν ἐν Σικελίᾳ θαυμαζομένων ποταμῶν was organized and what other natural phenomena the perieget dealt with. His attention to whatever is strange and unusual is also evident in his approach to nature, whose most extraordinary aspects form one of the topics of his writings. This kind of information, so delightful and fascinating, has been greatly appreciated by a wide but at the same time cultured public.[79] We are on a very different level from that regarding the political historiography of Thucydides and Polybius, who could be understood only by a very restricted circle of readers. The spread of a light literature was favoured by the cultural atmosphere which had arisen after the conquests of Alexander the Great and the widening of the *oikumene* with the resulting encounter between the Greeks and completely new realities. In addition, we have to consider the important role played by the Hellenistic courts.[80] They promoted the development of paradoxographic literature thanks to the well-stocked libraries necessary to find the texts from which the paradoxographers extracted their wonders.[81] In the case of Polemon it is worth remembering the close ties he probably had with the Attalids who, by surrounding themselves with scholars and artists, transformed Pergamon into a stimulating intellectual environment.[82] In this context it is therefore hardly surprising that Polemon, who prided himself on his extensive knowledge in the antiquarian field, should try his hand at the paradoxographic genre, so popular and widely practised in the Hellenistic period.

79 On the readers of the paradoxographic literature and on the social and cultural environment in which it develops see E. Gabba, *True History and False History in Classical Antiquity*, "JRS" 71 (1981), pp. 50–62; Jacob, *De l'art* cit., pp. 135–39; Schepens, *Paradoxography* cit., pp. 399–409.

80 It is worth remembering in relation to this point the interest shown by Ptolemy II Philadelphus for exotic animals which were then displayed in Alexandria (Agatharch. in Phot., *Bibl.*, Cod. 250.1, 441b = F 1 Burstein; Strab. XVII.1.5. Cf. Diod. III.36.3–4; 37.7 = Agatharch. F 80 Burstein). There is an interesting passage by Athenaeus (XIV 654 b–d) which contains the description of the zoo in Alexandria written by Ptolemy VIII Euergetes II (*FGrHist* 234, F 2). Cf. J. Trinquier, *Localisation et functions des animaux sauvages dans l'Alexandrie lagide: la question du "zoo d'Alexandrie"*, "MEFRA" 11, 2 (2002), pp. 861–919; Schepens, *Paradoxography* cit., pp. 404–407; M. Angelucci, *Le ricchezze africane in Agatarchide di Cnido e nel Periplus Maris Erythraei*', Atti del XVII convegno internazionale "L'Africa Romana. Le ricchezze dell'Africa. Risorse, produzioni, scambi", Siviglia 14–17 dicembre 2006, Roma 2008, pp. 115–24.

81 Cf. Schepens, *Paradoxography* cit., p. 389: "Paradoxography is derivative literature: a form of writing which presupposes the existence of other written works." Cf. Geus – King, *Paradoxography* cit., currently being printed.

82 See Angelucci, *Polemone* cit., pp. 170–73; Ead., *Polemon's contribution* cit., pp. 329–330.

BIBLIOGRAPHY

Ambaglio, D., Ἐπιχώριος: *un termine tecnico storiografico?*, in *Storiografia locale e storiografia universale. Forme di acquisizione del sapere storico nella cultura antica,* Atti del Congresso (Bologna 16–18 dicembre 1999), Como 2000, pp. 7–21.

Angelucci, M., *Polemone di Ilio: fra ricostruzione biografica e interessi antiquari,* "SCO" 49 (2003), pp. 165–84.

Angelucci, M., *Le ricchezze africane in Agatarchide di Cnido e nel "Periplus Maris Erythraei"*, Atti del XVII convegno internazionale "L'Africa Romana. Le ricchezze dell'Africa. Risorse, produzioni, scambi", Siviglia 14–17 dicembre 2006, Roma 2008, pp. 115–24.

Angelucci, M., *Polemon's Contribution to the Hellenistic Literature of the Second Century B. C.*, "Hormos" 3 n. s. (2011), pp. 326–41.

Bello, L., *Ricerche sui Palici*, "Kokalos" 6 (1970), pp. 71–97.

Bloomer, W. M., *The Superlative Nomoi of Herodotus's Histories,* "ClAnt" 12 (1993), pp. 30–50.

Burstein, S. M. (ed.), *Agatharchides of Cnidus: On the Erythraean Sea,* London 1989.

Capel Badino, R. (ed.), *Filostefano di Cirene. Testimonianze e frammenti,* Milano 2010.

Connor, W. R., *Theopompus and Fifth-Century Athens,* Washington 1968.

Cramer, J. A. (ed.), *Anecdota graeca e codd. manuscriptis bibliothecarum oxoniensium (An. Ox.)*, 2 vols., Amsterdam 1963 (= Oxford 1835–37).

Croon, J. H., *The Palici. An Autochtonous Cult in Ancient Sicily*, "Mnemosyne" s. IV, 5 (1952), pp. 116–29.

Cusumano, N., *Ordalia e soteria nella Sicilia antica. I Palici,* in *Mythos* 2, 1990, pp. 9–186.

Cusumano, N., *Siculi*, in P. Anello, G. Martorana, R. Sammartano (edd.), *Ethne e religioni nella Sicilia antica*, Atti del Convegno Internazionale (Palermo 6–7 dicembre 2000), Roma 2006, pp. 121–45.

De Angelis, F., *Pausania e i periegeti. La guidistica antica sulla Grecia* in E. Vaiani (ed.), *Dell'antiquaria e dei suoi metodi*, Pisa 1998, pp. 1–14.

Deichgräber, K., s. v. *Polemon*, in *RE* XXI 2, 1952, pp. 1287–1320.

Di Stefano, G., s. v. *Palice*, in G. Nenci – G. Vallet (edd.), *Bibliografia topografica della colonizzazione greca in Italia e nelle isole tirreniche*, pp. 280–82.

Dorandi, T. (ed.), *Antigone de Caryste. Fragments, texte établi et traduit*, Paris 1999.

Fraser. P. M., *Ptolemaic Alexandria*, 3 vols, Oxford 1972.

Gabba, E., *True History and False History in Classical Antiquity*, "JRS" 71 (1981), pp. 50–62.

Geus, K. – C. G. King, *Paradoxography*, in P. Keyser – J. Scarborough (edd.), *Oxford Handbook of Science and Medicine in the Classical World*, currently being printed.

Giannini, A. (ed.), *Paradoxographorum Graecorum Reliquiae*, Milano 1967.

Giannini, A., *Studi sulla paradossografia greca. I. Da Omero a Callimaco: motivi e forme del meraviglioso,* "RIL" 97 (1963), pp. 247–66.

Giannini, A., *Studi sulla paradossografia greca. II. Da Callimaco all'età imperiale,* "Acme" 17 (1964), pp. 99–138.

Glotz, G., s. v. *Palici*, in *Dictionnaire des antiquités grecques et romaines d'après les textes et les monuments (DA)*, edd. by Ch. Daremberg – E. Saglio, 5 vols, Paris 1877–1919.

Hett, W. S. (ed.), Aristotle. *Minor Works*: *On Colours, On Things heard, On Plants, On Marvellous things heard, Mechanical Problems, On indivisible Lines, Situations and Names of Winds, On Melissus, Xenophanes and Gorgias*, Cambridge Mass. – London 1980.

Ibáñez Chacón, Á., *Poesia y paradoxographia*, "Maia" 60 (2008), pp. 393–404.

Jacob, C., *De l'Art de compiler à la fabrication du merveilleux. Sur la paradoxographie grecque*, "LALIES" 2 (1980), pp. 121–40.

Judeich, W., *Topographie von Athen*, München 1931.

Trüdinger, K., *Studien zur Geschichte der griechisch-römischen Ethnographie*, Diss. Basel 1918.

Kaster, R. A. (ed.), Macrobius. *Saturnalia*, books 3–5, Cambridge Mass. – London 2011.

Keller, O. (ed.), *Rerum naturalium scriptores graeci minores*, Leipzig 1877.

MANISCALCO, L. (ed.), *Il santuario dei Palici: un centro di culto nella Valle del Margi*, Palermo 2008.
MANNI, E., *Culti greci e culti indigeni nella Sicilia antica. Problemi di metodo e spunti di ricerca*, "ASS" (1980), pp. 5–17.
MANNI, E., *Divagazioni sul culto dei Palici. Hommages à Robert Schilling*, Paris 1983, pp. 175–85.
MESSINA, A., s. v. *Mineo*, in G. Nenci – G. Vallet (edd.), *Bibliografia topografica della colonizzazione greca in Italia e nelle isole tirreniche*, pp. 145–51.
MOMIGLIANO, A., *Alien Wisdom: the Limits of Hellenization*, Cambridge 1975.
MÜLLER, C., *Fragmenta Historicorum Graecorum (FHG)*, 4 vols., Paris 1841–70.
MURRAY, A. T. (ed.), Homer. *The Odyssey*, Cambridge Mass. – London 1976.
MUSSO, O., *Sulla struttura del Cod. Pal. Gr. 398 e deduzioni storico-letterarie,* "Prometheus" 2 (1976), pp. 1–10.
NENCI, G., *La concezione del meraviglioso nei poemi omerici*, "AAT" 92 (1957–8), pp. 275–311.
PAJÓN LEYRA, I., *Entre ciencia y maravilla: El género literario de la paradoxographia griega*, Zaragoza 2011.
PAJÓN LEYRA, I., *Paradoxografia griega: estudio de un género literario*, Diss. Madrid 2008, Madrid 2009.
PASQUALI, G., *Die schrifstellerische Form des Pausanias*, "Hermes" 48 (1913), pp. 161–223.
PFEIFFER, R., *Storia della filologia classica, dalle origini alla fine dell'età ellenistica*, ed. it. Napoli 1973.
PRELLER, L., *Polemonis Periegetae Fragmenta*, Leipzig 1838 (= Amsterdam 1964).
RADT, S. (ed.), Aeschylus, *Tragicorum graecorum fragmenta (TGrF),* vol. 3, Göttingen 1985.
RIZZO, F. P., s. v. *Palico*, in *Enciclopedia Virgiliana*, directed by F. Della Corte, Roma 1987, pp. 935–36.
SASSI, M. M., "*Mirabilia*", in C. Cambiano, L. Canfora, D. Lanza (edd.), *Lo spazio letterario della Grecia antica*, vol. 1, *La produzione e la circolazione del testo,* tome 2, *L'Ellenismo*, Roma 1993, pp. 449–68.
SCHEPENS, G. – K . DELCROIX, *Ancient Paradossography: Origin, Evolution, Production and Reception*, in O. Pecere (ed.), *La letteratura di consumo nel mondo greco-latino*, Cassino 1996, pp. 373–460.
SCHMID, W. – O. STÄHLIN, *Wilhelm von Christs Geschichte der griechischen Literatur*, vol. II, 1, *Die nachklassische Periode der griechischen Litteratur. Von 320 vor Christus bis 100 nach Christus,* München 19206.
SHRIMPTON, G. S., *Theopompus the Historian*, Montreal – London 1991.
SPADA, S., *La storiografia occidentale di età ellenistica*, in R. Vattuone (ed.), *Storici greci d'Occidente*, Bologna 2002, pp. 254–55.
STERN, J., *Paradoxographus Vaticanus*, in S. Heilen, R. Kirstein, R. Scott Smith, S. M. Trzaskoma, R. Van der Wal, M. Vorwerk (edd.), *In Pursuit of Wissenschaft: Festschrift für William M. Calder III zum 75. Geburtstag*, Hildesheim and New York 2008, pp. 439–440.
SUSEMIHL, F., *Geschichte der griechischen Litteratur in der Alexandrinerzeit*, II, Leipzig 1891–92 (= Hildesheim 1965).
TRINQUIER, J., *Localisation et functions des animaux sauvages dans l'Alexandrie lagide: la question du "zoo d'Alexandrie"*, "MEFRA" 114, 2 (2002), pp. 861–919.
VANOTTI, G. (ed.), *De mirabilibus auscultationibus*, Padova 1997.
WENSKUS, O., s. v. *Paradoxographoi*, in *Der Neue Pauly: Enzyklopädie der Antike*, 9, 2000, pp. 309–312.
WEST, M. L. (ed.), *Iambi et elegi graeci ante Alexandrum cantati*, I–II, Oxford 1971–72.
WESTERMANN, A., Παραδοξόγραφοι, *Scriptores rerum mirabilium Graeci*, Brunsvigae – Londini 1839 (= Amsterdam 1963).
ZIEGLER, K., s. v. *Palikoi*, in *Paulys Realencyclopädie der classischen Altertumswissenschaft (RE)*, XVIII 3, 1949, pp. 99–123.
ZIEGLER, K., s. v. *Paradoxographoi*, in *Paulys Realencyclopädie der classischen Altertumswissenschaft* (RE), XVIII 3, 1949, pp. 1137–1166.

A NOTE ON STRABO, *GEOGRAPHY* 14.6

Tønnes Bekker-Nielsen

In her 1995 article on 'Strabo, Polybios, and the Stade', Sarah Pothecary argues, contrary to the views of Aubrey Diller (1934) and Germaine Aujac (1966), that Strabo used a consistent value for the Greek stade (*stadion*) throughout his *Geography*. Strabo was neither 'a happy ignoramus, who simply did not know that these distances were based on varying stades', nor 'a careless scholar who realised the situation but did not care'.[1] When giving distances, Strabo consistently employed a conversion rate of one stade to one-eighth of a Roman mile (185 metres). According to Pothecary, the only exception found in Strabo's work is the conversion ratio of 8 1/3 miles to the stade used by Polybios, a discrepancy to which Strabo himself draws attention.[2]

While Pothecary's conclusion remains valid on general grounds, it does not take account of some apparent exceptions to the rule in book 14 of the *Geography*, where Strabo is describing Cyprus. The relevant passages, all of which deal with overland distances,[3] will be discussed below.

[1] μῆκος δὲ ἀπὸ Κλειδῶν ἐπὶ τὸν Ἀκάμαντα πεζῇ σταδίων χιλίων τετρακοσίων ὁδεύοντι ἀπ᾽ ἀνατολῆς ἐπὶ δύσιν. (Strabo, *Geography* 14.6.2 C 682 Radt)
The distance from the Kleides [islands] to the Akamas [promontory] on foot, travelling from east to west, is one thousand four hundred stades.

The distance of 1400 stades (259 km) by road from Akamas to the Kleides islands is close to the distance in real space measured on a modern map and to the estimate of Artemidoros, quoted by Pliny the Elder,[4] of 162½ Roman miles =1300 stades for the distance from Akamas to cape St. Andreas (not including the sailing distance to the islands). In this case, Strabo is evidently applying the 'standard' stade of 1/8 Roman mile.

[2] Εἶθ᾽ ἡ Πάφος, κτίσμα Ἀγαπήνορος καὶ λιμένα ἔχουσα καὶ ἱερὰ εὖ κατεσκευασμένα· διέχει δὲ πεζῇ σταδίους ἑξήκοντα τῆς Παλαιπάφου. (Strabo, *Geography* 14.6.3 C 682 Radt)
Next Paphos, a foundation of Agapenor, having a harbour and a well-built sanctuary. The distance by road to Palaipaphos is sixty stades.

1 Pothecary 1995, 49. For more recent assessments of Strabo as a scholar and geographer, see Koelsch 2004, 507–13.

2 Strabo 7.7.4, C 322; Pothecary 1995, 51; cf. Hultsch 1882, 85–86.

3 Sailing distances have been omitted, as sea distances are generally too imprecise for the purpose of the present analysis.

4 Artemidoros ap. Plin. *nat.* 5.129. Like Strabo, Pliny reckons a stade at 1/8 of a Roman mile, equivalent to 185 metres, cf. *nat.* 5.63.

The *Tabula Peutingeriana* gives the distance separating New Paphos and Old Paphos as 11 Roman miles = 88 stades and the twelfth milestone from Nea Paphos[5] was found c. 2.5 kilometres east of Old Paphos. Measured on a modern map, the distance from the site of the northeast gate[6] of New Paphos to the ruins of the sanctuary in Old Paphos (Kouklia) is 15.2 km = 10.3 Roman miles = 82 stades. Thus Strabo's figure is slightly less than three quarters of the actual distance.

> [3] εἶτα Καρπασία πόλις λιμένα ἔχουσα, κεῖται δὲ κατὰ τὴν ἄκραν τὴν Σαρπηδόνα· ἐκ δὲ τῆς Καρπασίας ὑπέρβασίς ἐστιν ἰσθμοῦ τριάκοντα σταδίων πρὸς τὰς νήσους τὰς Καρπασίας καὶ τὸ νότιον πέλαγος· (Strabo, *Geography* 14.6.3 Radt C 683)
> Next Karpasia, a city having a port, and nearby, the Sarpedon promontory. From Karpasia, the passage across the isthmus is thirty stades, to the Karpasia islands and the sea on the southern side.

The ὑπέρβασίς ἰσθμοῦ, 'passage across the isthmus' is stated by Strabo to be thirty stades long measured from Karpasia (Agios Philon). On a modern map, the distance is at least 7.5 km = 41 stades from shoreline to shoreline; perhaps slightly less if Strabo's distance was reckoned from the landward side of Karpasia's *pomerium*.[7] Again, Strabo under-estimates the distance by a factor of c. 0.75.

> [4] εἶτα Τρῆτα καὶ Βοόσουρα καὶ Παλαίπαφος, ὅσον ἐν δέκα σταδίοις ὑπὲρ τῆς θαλάττης ἱδρυμένη, ὕφορμον ἔχουσα καὶ ἱερὸν ἀρχαῖον τῆς Παφίας Ἀφροδίτης· εἶτ΄ ἄκρα Ζεφυρία πρόσορμον ἔχουσα, καὶ ἄλλη Ἀρσινόη ὁμοίως πρόσορμον ἔχουσα καὶ ἱερὸν καὶ ἄλσος· μικρὸν δ΄ ἀπὸ τῆς θαλάττης καὶ ἡ Ἱεροκηπία. (Strabo, *Geography* 14.6.3 Radt C 683)
> Next Treta, Boösoura and then Palaipaphos, which is situated ten stades inland from the sea, and has an anchorage and an ancient sanctuary of the Paphian Aphrodite. Then cape Zephyria which has a landing-place, and another Arsinoë which likewise has a landing-place, a temple and a sacred grove; and a short distance from the sea is Hierokepis.

Strabo tells us that Old Paphos is located 'ten stades' from the sea.[8] This, like the other figures, is a round number and Strabo does not tell us from which point the distance was reckoned: from the shoreline or from the anchorage mentioned in the following sentence. Thus we cannot tell whether he under-estimated the distance in this case as well.

> [5] εἶτ΄ Ἀφροδίσιον, καθ΄ ὃ στενὴ ἡ νῆσος· εἰς γὰρ Σαλαμῖνα ὑπέρβασις σταδίων ἑβδομήκοντα· εἶτ΄ Ἀχαιῶν ἀκτή, ὅπου Τεῦκρος προσωρμίσθη πρῶτον ὁ κτίσας Σαλαμῖνα τὴν ἐν Κύπρῳ, ἐκβληθείς, ὥς φασιν, ὑπὸ τοῦ πατρὸς Τελαμῶνος· (Strabo, *Geography* 14.6.3 Radt C 682)
> Next Aphrodision, where the island is narrow, for the passage across to Salamis is seventy stades; then the beach of the Achaians, where Teukros, the founder of Salamis in Cyprus, made his first landing: exiled, as they tell, by his father Telamon.

According to Strabo, the 'passage across' (*hyperbasis*) from Aphrodision on the northern coast to the city of Salamis is seventy stades in length.[9] This figure is

5 Bekker-Nielsen 2004, 273–74 no. 12 = Mitford 1980, 1334 no. 14.

6 Inter-city distances were normally measured from the gates or from the *pomerium*, not from the centre of a city; Bekker-Nielsen 2004, 170–71 n. 20; Chevallier 1997, 64.

7 Bekker-Nielsen 2004, 170.

8 Strabo 14.6.4 C 683.

9 Strabo 14.6.3 C 682.

clearly in error: as pointed out by D. G. Hogarth more than a century ago,[10] no location on the north coast lies within 70 stades of Salamis. From the site of ancient Salamis, a short distance north of modern Famagusta, the nearest point on the north coast is 28 kilometres or 151 stades distant as the crow flies. Hogarth himself offered one possible explanation, but went on to reject it:

> It is just possible that Strabo either stated, or intended to state, the distance from Aphrodisium to the *bay* of Salamis, and not to Salamis itself … But as this involves an emendation of Strabo's text, it must not be pressed; and on other grounds I cannot feel satisfied that the alteration is worth the making.[11]

A further argument against Hogarth's emendation is that the toponym 'bay of Salamis' is not attested in ancient literature; Strabo himself simply refers to 'the sea on the southern side'. The problem is compounded by the absence of unequivocal evidence for the location of Aphrodision. The place-name is found only in one other ancient author, Ptolemy,[12] who locates the city between the 'beach of the Achaians' mentioned by Strabo and an otherwise unknown settlement, Makaria.[13]

In 1852, the German traveller Ludwig Ross visited the north coast around the village of Akanthou and tentatively proposed that an *Aphrodision* might well have been the predecessor of the church of the 'Panagia Pergamou',[14] on the map accompanying Hogarth 1889 (fig. 1) marked as 'Pergamon'. An examination of the ruins seen by Ross led Hogarth to conclude that they dated from the Byzantine, not the Roman period.[15] In their place, Hogarth proposed another site, some distance further westward, known as Iastriká, as the ruins of ancient Aphrodision.[16] In his view, Akanthou village was the successor of ancient Aphrodision; here, as elsewhere in Cyprus, settlements had moved inland where they were less exposed to raids from the sea.

Some distance inland from Akanthou/Tatlısu, at the northern approach to the Mallıdağ pass, a steep, narrow stretch of roadway cut into the living rock can be seen running parallel to the modern highway (fig. 2).[17] For part of the ascent, a set of steps runs parallel to the roadway, perhaps a later addition for the benefit of pedestrians and pack-animals. From its resemblance to other rock-cut roadways on the island,[18] the ascent appears to be of Hellenistic or Roman date, and since the Mallıdağ pass offers the only convenient route through this part of the Northern Range, there is little doubt that the road formed part of the *hyperbasis* connecting

10 Hogarth 1889, 94.
11 Hogarth 1889, 95–96.
12 *Geogr*. 5.14.4. The *dekate Kyprou* mentioned by Stephen of Byzantium s. v. 'Aphrodisias' probably refers to the sanctuary of Aphrodite at Old Paphos.
13 Bekker-Nielsen 2010, 422.
14 Ross 1852, 134–35.
15 Hogarth 1889, 97–98.
16 Hogarth 1889, 99.
17 The road was studied in detail by Marit Jensen and the author in October 2012. To the best of our knowledge, it had not previously been recorded or published.
18 E. g., Bekker-Nielsen 2004, 129, 134.

Aphrodision and Salamis.[19] As such, it provides a strong argument in support of Hogarth's identification of Iastriká with ancient Aphrodision. The problem remains, however, that the shortest possible route from the presumed site of Aphrodision over the Mallıdağ pass to the site of ancient Salamis is far longer than the 70 stades cited by Strabo.

A possible conjecture may be offered here. Since there is no evidence that Strabo ever visited Cyprus,[20] he must have combined information from different sources to produce his description of the island. One of these sources was Eratosthenes;[21] another will have been a description of the coast intended for navigators (*periplous*), from which he took his description of the north coast; a third an itinerary containing overland distances. If this itinerary, as was customary in the Greek-speaking world, rendered distances in the Milesian numeral notation, *ρο* (170) could easily have been misread or miscopied as *ο* (70).

A route from Aphrodision over the Mallıdağ pass and passing through Lefkoniko/Geçitkale and Peristerona/Alanıçı to Salamis would be c. 42 kilometres or 230 stades. Three quarters of this figure comes to c. 170 stades. It would seems that as was the case at Nea Paphos and Karpasia, Strabo's figure for the journey from Aphrodision to Salamis may have been underestimated by a factor 0.75. But how did these errors arise?

All that can be said with certainty about the source from which Strabo, directly or indirectly, derived his overland distances is that it postdates the foundation of New Paphos c. 315 BC.[22] If compiled during the period of Ptolemaic rule in Cyprus, it is quite possible that its distances were given in the Egyptian unit of measurement known as the *schoinos*. Strabo complains that he could find no single uniform value for the schoinos: in some parts of Egypt it was equal to 30 stades, in other areas 40 or 60. [23]

On this theory, the original source gave the distance across the Karpas peninsula as one *schoinos* (40 stades), that between New and Old Paphos as 2 *schoinoi* (80 stades) and the 'passage across' from Aphrodision to Salamis as 5 3/4 *schoinoi* (230 stades[24]). At a later date, these values were converted using the ratio of 30 stades to the *schoinos*, producing the figures 30, 60 and 170 stades, which were recopied as 30, 60 and 70 into the document that served Strabo as his source.[25]

19 On an earlier occasion (Bekker-Nielsen 2004, 158) the present author suggested that the *hyperbasis* might have been a direct route across the mountains behind Akanthou/Tatlısu, in which case the distance to Salamis would not be far longer than 170 standard stades of one-eighth Roman mile. Following the discovery of the ancient road in the Mallıdağ pass, this hypothesis is no longer tenable.

20 For a full discussion of the extent of Strabo's travels, see Dueck 2000, 15–28.

21 Cf. Strabo 14.6.4 C 684.

22 For the date, Bekker-Nielsen 2000.

23 Geogr. 17.1.24 C 804.

24 Units were more often divided into halves, quarters and eights than into thirds; thus 5 ¾ or 5 5/8 are more likely than 5 2/3. Any one of the three would, after multiplication by 30, be rounded to 170.

25 An alternative possibility is that Strabo's source used 'stades' longer than the standard stade. Antoine Hirsch has argued for the co-existence of two metrological systems in Egypt, one

REFERENCES

AUJAC 1966 = Aujac, G. *Strabon et la science de son temps*. Paris.
BEKKER-NIELSEN 2000 = Bekker-Nielsen, T. 'The Foundation of Nea Paphos', *Proceedings of the Danish Institute at Athens* 3, 195–207.
BEKKER-NIELSEN 2004 = Bekker-Nielsen, T. *The Roads of Ancient Cyprus*. Copenhagen.
BEKKER-NIELSEN 2010 = Bekker-Nielsen, T. 'A new edition of Ptolemy's *Geography*', *Epetirida tou Kentrou Epistimonikon Ereunon* 35, 417–27.
CHEVALLIER 1997 = Chevallier, R. *Les voies romaines* (second edition). Paris.
DILLER 1934 = Diller, A. 'Geographical latitudes in Eratosthenes, Hipparchus and Posidonius', *Klio* 27, 258–69.
DUECK 2000 = Dueck, D. *Strabo of Amasia: A Greek Man of Letters in Augustan Rome*. London.
HIRSCH 2013 = Hirsch, A.P. Ancient Egyptian Cubits: Origin and Evolution (unpublished Ph.D. thesis, University of Toronto).
HOGARTH 1889 = Hogarth, D.G. *Devia Cypria: Notes of an Archaeological Journey in Cyprus in 1888*. London.
HULTSCH 1882 = Hultsch, F. *Griechische und römische Metrologie*, second edition. Berlin.
KOELSCH 2004 = Koelsch, W.A. 'Squinting back at Strabo', *Geographical Review* 94, 502–18.
MITFORD 1980 = Mitford, T.B. 'Roman Cyprus', *ANRW* II.2.7, 1285–1384.
POTHECARY 1995 = Pothecary, S. 'Strabo, Polybios, and the Stade', *Phoenix* 49, 49–67.
ROSS 1852 = Ross, L. *Reisen nach Kos, Halikarnassos, Rhodos und der Insel Cypern*. Halle.

based on the 'royal' cubit of c. 52.5 cm and an older system based on a 'great' cubit of 60 cm, which remained in use into the Ptolemaic period (HIRSCH 2013, 120; 149, table 28). As an Egyptian stade equalled 400 cubits, a hypothetical 'great stade' would have been c. 240 metres. At present, however, there is no textual evidence for the use of a 240-metre 'great stade' in the Hellenistic period.

FIGURES

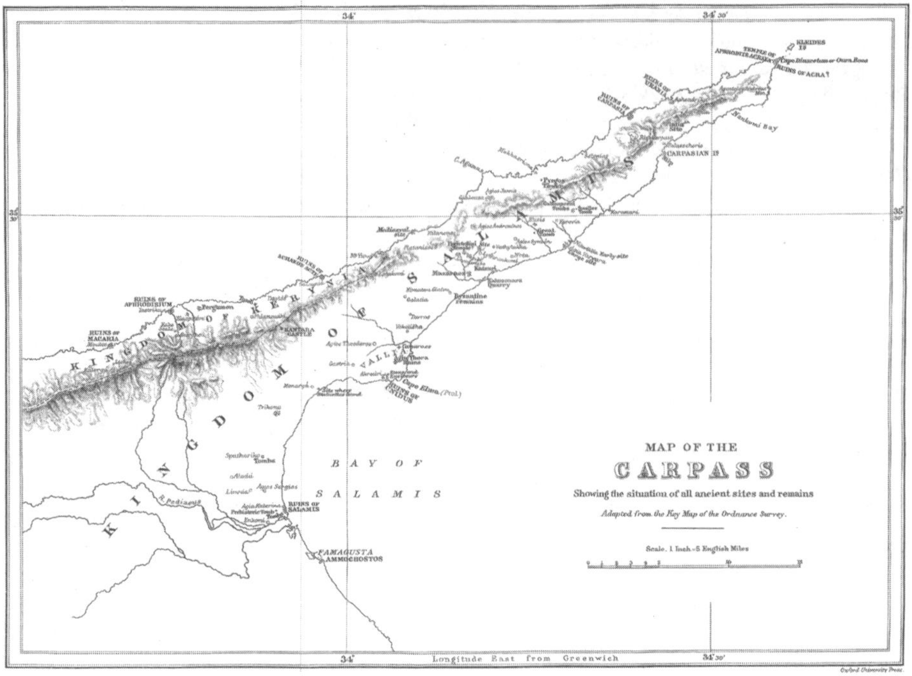

Fig. 1: Map of the Karpas peninsula (HOGARTH 1889).

Fig. 2: Rock-cut roadway north-east of Mallıdağ (author's photo, October 2012).

EINE LANDKARTE DES FAYUM AUS DER FRÜHEN RÖMISCHEN KAISERZEIT

Horst Beinlich

Es ist üblich geworden, historische Untersuchungen mit den Worten einzuleiten „schon die alten Ägypter …“. Tatsächlich haben schon die alten Ägypter die Fähigkeiten gehabt, ihre Umwelt aus der Vogelperspektive darzustellen. Eine große Anzahl von Architekturzeichnungen – Grundrisse und Aufrisse von Gebäuden – kann dies belegen.[1] Auch Grundrisse von Grabanlagen gehören zu diesem Bereich. Hin und wieder findet man unter dem Begriff von „Landkarten“ in populärer Literatur auch Darstellungen der Umwelt einer Person. In Ägypten betrifft das z. B. die Darstellungen auf prähistorischen Naqada-Gefäßen, auf denen man Elemente der Landschaft finden kann. Das sind dann z. B. Berge. Seltener sind in der ägyptischen Kunst Darstellungen von identifizierbaren Landschaftsteilen. Bekanntestes Beispiel ist dafür die Wiedergabe des Gebel Barkal beim 4. Katarakt des Nils mit seiner charakteristischen Form.[2] Landkarten sind dies alles aber nicht. Um eine Landkarte zu entwerfen, braucht man mehr, vor allem einen Grund, dies zu tun. Wenn man beispielsweise jemandem einen Weg beschreiben muß, der nicht zugegen ist, ist eine beschriftete Karte ein geeignetes Mittel dazu.[3] Es ist die Situation der sattsam bekannten Pläne von Piratenschätzen. Die ägyptische Goldminen-Karte aus der Zeit der XIX. Dynastie (ca. 1250 v. Chr.) zeigt, daß solche Pläne gebräuchlich waren. Sie ähnelt eher den Grundrisszeichnungen, die uns von Pharaonengräbern überliefert sind, als Landkarten. Zumindest im ägyptischen Kernland, dem engen Niltal und dem Nildelta, gab es keinen Grund dafür, ganze Landschaften aus der Vogelperspektive darzustellen.

Anders ist das „im Feindesland“. In Darstellungen der Kadesch-Schlacht werden einzelne Bereiche der Landschaft, wie die vom Orontes umflossene Stadt und das Kriegslager der Ägypter, aus der Vogelperspektive gezeigt, um den für die Ägypter positiven Schlachtverlauf zu illustrieren. Die Stadt Kadesch selbst wird allerdings wieder in einer Sicht von vorn dargestellt.

Auch für die hier vorgestellte Landkarte des Fayum ist die Frage zu stellen, aus welchem Grund sie angefertigt wurde, und dann, was wir von ihr erwarten dürfen: Wie bereits an anderer Stelle ausführlich dargestellt, ist das „Buch vom Fayum“

1 Grundlegend dazu mit vielen Beispielen: Badawy: Le dessin architectural. S. a. Lexikon der Ägyptologie, s. v. Architekturdarstellung.

2 S. dazu Kendall: Gebel Barkal. In: Bard (Hrsg.): Encyclopedia of the Archaeology of Ancient Egypt. Routledge, London 1999, S. 325–28. Zu Landschaftsdarstellungen im Alten Ägypten allgemein s. Lexikon der Ägyptologie, s. v. Landschaft (-Beschreibung und -Darstellung).

3 Aus diesem Grund sind „Landkarten“ aus dem Bereich der jenseitswissenschaftlichen Topographie im Alten Ägypten durchaus verbreitet.

Abb. 1: Das Fayum (Baedeckers Ägypten, 1928, nach S. 194). Die Bereiche, auf die sich die Papyrusabschnitte 1 und 3 beziehen, sind schematisch markiert. Dabei wird beim See des Fayum eine Höhe des Wasserspiegels bei ca. 20m unter NN angenommen

ein Text, der ein Ritual beschreibt, das auf der Grundlage einer ortsspezifischen Mythologie durchgeführt wird. Ortsspezifisch heißt dabei, daß für die geographische Ausnahmesituation der Depression el-Fayum zur Zeit ihrer expandierenden Besiedlung durch Ägypter und Griechen in der Zeit der Ptolemäer und Römer eine Mythologie aus dem Niltal an die andersartigen geographischen Gegebenheiten des Fayum angepaßt werden mußte. Dies war auch sicher nötig, um den Tempelneubauten im Fayum eine religiöse Grundlage zu geben. Um es mit zwei Sätzen kurz zu beschreiben: 1.) Im religiösen Weltbild der Ägypter geht die Sonne in den Bergen, die das Niltal begrenzen, im Osten auf und im Westen wieder unter. Für Oasen ist hier keine Nische vorgesehen. 2.) Der Nil fließt von Süden geradewegs nach Norden bis zum Mittelmeer. Ein Abzweigen auch nur eines Teils widerspricht der religiösen Vorstellung.

Im Fayum ist selbstverständlich die Sonne auch sichtbar, aber anders als im Niltal kann sie – aus bestimmten Orten gesehen – im Wasser des Fayum-Sees auf- und wieder untergehen. Um wieder an ihren Aufgangsort zurück zu gelangen, durchschwimmt die Sonne – als Krokodil – den See in der Nacht in West-Ost-Richtung, während sie in der Vorstellung des übrigen Ägyptens einen unterirdischen Weg durch Höhlen zurücklegt.

Die vom Niltal abweichende Situation des Fayum ist also sehr stark an die geographische Lage im Westen des Niltals gebunden. Besonders die Bewegungsrichtung des zufließenden Nilwassers von Osten nach Westen durch den Seitenarm des Nils, den Bahr Yussuf, und die Bewegungsrichtung der Sonnenbahn von Osten nach Westen bewegten den Verfasser des Papyrus, entgegen den üblichen

Gepflogenheiten auch die Leserichtung des Papyrus von links – der Ausdruck ist gleichbedeutend mit „Osten" – nach rechts – gleichbedeutend mit Westen – zu ändern.[4]

Die Darstellung dieser neuen, fayumspezifischen Mythologie ließ sich nur auf der Grundlage einer geographischen Karte des Fayumgebietes angemessen bewerkstelligen. Es ist dabei aber auch nachvollziehbar, daß dabei z. B. Wege und Siedlungen nicht berücksichtigt werden konnten, falls sie nicht innerhalb der Mythologie eine bestimmte Rolle spielten. Andererseits ist es aber auch verständlich, warum mythologisch gebundene Darstellungen auf dem Papyrus Platz gefunden haben. Die nachfolgenden Abschnitte werden sich nun auf die Darstellungen und Aussagen beschränken, die für eine Interpretation des „Buches vom Fayum" als Landkarte relevant sind.

1. ABSCHNITT: DER DURCHBRUCH DES BAHR YUSSUF DURCH DEN RAND DER LIBYSCHEN WÜSTE INS FAYUM

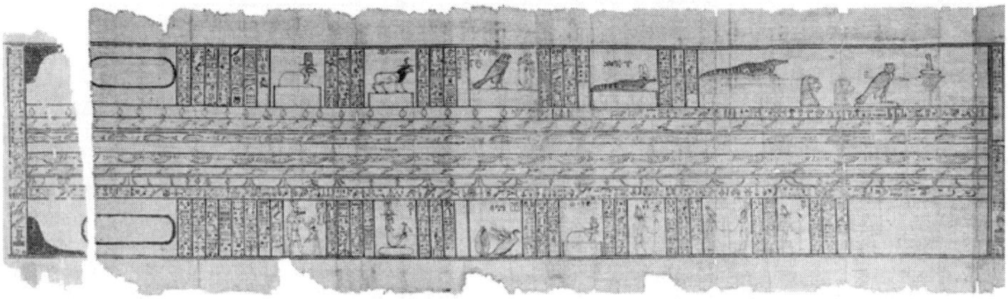

Abb. 2

Wie schon gesagt wurde und noch weiter zu zeigen sein wird, ist der Papyrus gesüdet, zur Herkunft des Nils ausgerichtet. Entsprechend ist der linke Rand des Papyrus der Osten, d. h. von hier kommt das Nilwasser durch den Bahr Yussuf in das Fayum. Der Nil ist selbst nicht dargestellt. Eine senkrechte Hieroglyphenzeile links (östlich) der beiden Darstellungen der Uferbegrenzungen weist aber auf den Nil hin. Sie besagt, daß das Nilwasser, das aus den mythischen Nilquellen bei Assuan kommt, teilweise in das Fayum hineinfließt, um dieses Gebiet zu bewässern, teilweise aber weiterfließt, um das Nildelta zu bewässern.

Der eigentliche Durchbruch durch die Libysche Wüste wird durch zwei Zeichen dargestellt, die in der Ägyptologie als Hieroglyphe mit dem Lautwert „q"verwendet werden und als Bild das Steilufer am Rand des Niltals darstellen („sandy hill-slopes"). Die beiden Zeichen orientieren sich an der Längsachse des Papyrus, signalisieren also schon, daß die Fläche des Papyrus der Fläche einer Landschaft gleichgesetzt wird und aus der Vogelperspektive zu betrachten ist. Jeweils rechts von diesen

4 Die Anpassung der Leserichtung an eine Sonnen-Bewegung ist in Papyri im Zusammenhang mit dem Sonnenlauf in der Unterwelt durchaus belegt, etwa im Amduat oder im Buch vom Ba.

Zeichen sieht man ein oblonges Zeichen – die Form einer Hieroglyphenkartusche. Dieses Zeichen bedeutet in der Hieroglyphenschrift Insel bzw. Hügel. Da im Überschwemmungsland Hügel schnell zu Inseln werden können, ist die Gleichsetzung der beiden Bedeutungen mit ähnlicher Aussprache sehr häufig.

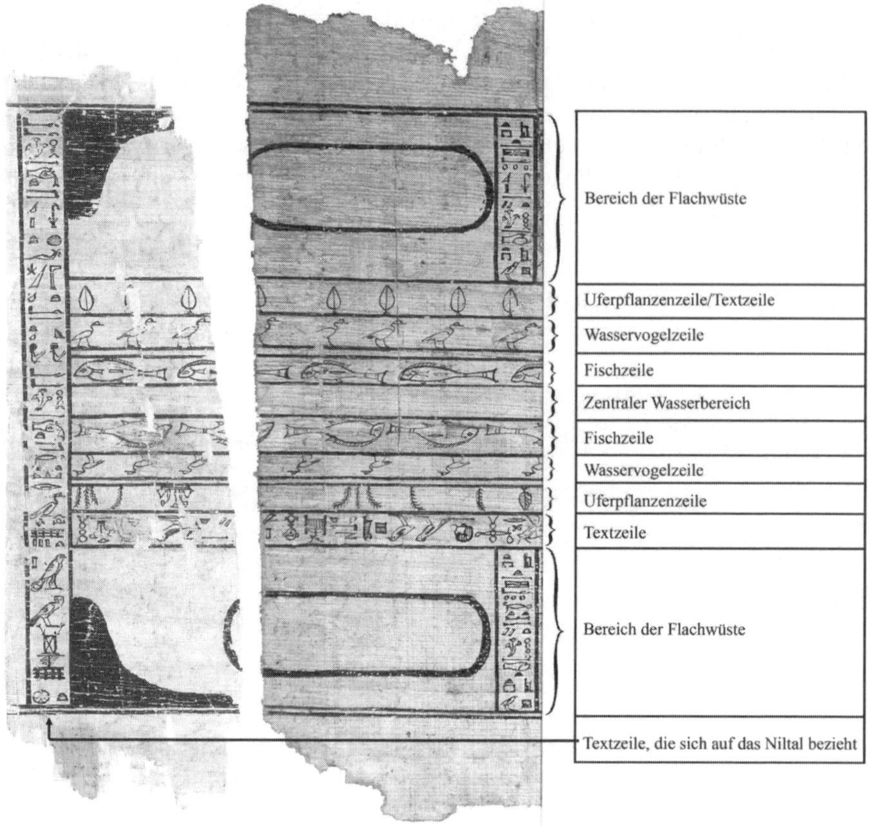

Abb. 3

Die hieroglyphische Beischrift bezeichnet den Hügel am unteren Rand des Papyrus als den „Nördlichen Sand", den am oberen Rand als den „Südlichen Sand". Sie werden als zusammengehöriges Paar betrachtet, das die Aufgabe hat, das Fayum vor üblen Dingen (dem Gott Seth), die aus dem Niltal kommen könnten, zu schützen. Der „Nördliche Sand" hat noch einen Eigennamen, nämlich „Menmenet", ein Name, der gleich noch eine Rolle spielen wird. Zusammen mit den Zeichen, die die Steilufer darstellen, sind diese „Sände" anscheinend die Flachwüstenzungen, die den Durchbruch des Bahr Yussuf von Norden und Süden bei der Lahun-Schwelle einengen.

Zwischen den Flachwüstenzungen sieht man auf dem Papyrus den Bahr Yussuf. Er ist in der Mitte durch ein nicht in jedem Papyrus[5] dargestelltes Zickzackmuster (Wasserlinien) gekennzeichnet. Über und unter dieser Fläche sind als Kennzeichen des Bahr Yussuf zwei Streifen mit Fischen und zwei mit Wasservögeln dargestellt. Dieser Wasserbereich wird wiederum von zwei Streifen mit Bäumen eingerahmt, die das Ufer mit seinen Pflanzen darstellen. Alle diese Streifen orientieren sich an der Längsachse des Papyrus und geben so die Vogelperspektive wieder. Auch die begleitenden Hieroglyphentexte ordnen sich diesem Blickwinkel unter, so daß man, um die Texte lesen zu können, den Papyrus teilweise um 180° drehen muß. Der Fischreichtum, den der Bahr Yussuf ins Fayum bringt, ist bei Herodot eingehend beschrieben, allerdings scheint er einen wichtigen Gesichtspunkt dabei übersehen zu haben.[6] Die breiten Bereiche oben und unten im Papyrus sind die Flachwüstenbereiche, die einem Hochwasser im Normalfall nicht ausgesetzt sind und deshalb als Bereich für Ansiedlungen dienen.

Rechts (westlich) des „Nördlichen Sandes" mit Namen Menmenet liegt der Ort „Kampfplatz". Er wird beschrieben als „westlich von Menmenet". Damit ist die Ausrichtung des Papyrus nach Himmelsrichtungen bestätigt. Der Ort wird dargestellt und beschrieben als „Platz, an dem Horus und Seth miteinander kämpfen". Gemeint ist hier der Kampf zwischen dem Positiven (hier wohl die Fruchtbarkeit des Überschwemmungswassers) und dem Negativen (hier wohl die Unfruchtbarkeit der Dürre oder der Wüste). Man kann dies als rein mythologische Aussage werten. Interessant wird die Aussage aber, wenn man sich die Bodenbeschaffenheit dieses Durchbruches ansieht. Hier ragt nämlich beim Ort Hawarat el-Maqta ein massiver Fels in das Bett des Bahr Yussuf und bildet damit eine natürliche Sperre, die das Wasser erst ab einer bestimmten Höhe überwinden kann.[7] Im Laufe der Geschichte wurden viele Versuche gemacht, diese Schwelle gangbarer zu machen. Setzt man den Ort des Papyrus mit Hawarat el-Maqta gleich, so hat man einen Fixpunkt für die Bestimmung weiterer Orte des Papyrus.

Der nächste Ort westlich läßt sich nicht genau festlegen. Es ist der Ort „der See von Memphis". Der Text bringt ihn mit den Nil-Kugelfischen zusammen, die mit dem Wasser der Nilüberschwemmung von Süden her ins Fayum gespült wurden. Wahrscheinlich handelt es sich bei dem See um ein Wasser-Rückhaltebecken, das zur Bewässerung während der Trockenzeit diente und das auch für längere Zeit als Fischreservoir diente.[8]

In die gleiche Richtung zielt die Interpretation des nächsten Ortes am Bahr Yussuf. Es ist der Platz „Pyramidenland". Zur Identifizierung kommt nur die Pyramide von Hawara in Frage, die schon am Ausgang des Durchbruches des Bahr Yussuf liegt. Damit haben wir einen sicheren Ort auf unserer Karte und zwar wie

5 Zu sehen im BF (W/P), s. Beinlich, (Mythos Tafelband), Taf. 19.

6 S. dazu unten.

7 S. dazu Brown, bes. S. 36 und S. 100 und Beinlich, Mythos 1, S. 157.

8 Das System der Beckenbewässerung war in Ägypten auch im Niltal verbreitet. Zum System s. Ibrahim, Ägypten (1996), S. 55. Ein großes Rückhaltebecken ist vom Leichtweiss-Institut erforscht worden, s. Garbrecht, Wasserspeicherung.

gefordert im Westen der vorher genannten Plätze.[9] Im Bild zu diesem Ort wird ein Kormoran gezeigt, der den Namen „Herr des Fischfanges“ trägt und nach dem Text im Geäst einer Sykomore sitzt. Wenn man sich an das eben erwähnte Fischreservoir erinnert, kann man sich den Sachverhalt leicht vorstellen. Auch hier müssen wir mit einem Wasser-Rückhaltebecken rechnen, in dem sich, nachdem das Überschwemmungswasser hineingeflossen ist, auch Fische tummeln.[10] Einen Kormoran direkt bei einem Wasser mit Fischen kann man sich gut vorstellen. Ob er an der Stelle sehr beliebt war, sei dahingestellt. Es ist aber hier an der Zeit, ein paar Bemerkungen zu Herodots Bericht über den Fischfang im Fayum zu machen.

Herodot (*Hist.* II, 149–150) berichtet, nachdem er sich darüber ausläßt, daß das (Nil)wasser sechs Monate ins Fayum hineinfließt und sechs Monate wieder hinausfließt, daß die Fischerei beim Hineinfließen 20 Minen einbringt, beim Hinausfließen aber jeden Tag ein Talent Silber. Da nun ein Hinausfließen des Wassers mit Fischen bei der Depression des Fayum nicht möglich ist, kann Herodots Aussage nur auf einem Mißverständnis beruhen. In der Zeit, wenn das Überschwemmungswasser ins Fayum strömt, gibt es einen Überfluß an Fischen für alle. Nach dieser Zeit sind frische Fische nur noch in den Wasser-Rückhaltebecken zu haben. Man kann sie leicht fangen, wenn man das Wasser daraus langsam abläßt und vor den Abfluß Reusen aufstellt. Der Fischfang im Fayum-See ist dagegen problematisch wegen des hohen Salzgehaltes. Dieser erlaubt nicht allen Süßwasserfischen das Überleben. Das Hinausfließen des Wassers bei Herodot beruht wohl darauf, daß ihm gesagt wurde, daß das Wasser zum Ozean bzw. zum Nil fließt. Nach ägyptischem Weltbild fließt das Wasser aus dem See des Fayum in den Urozean, der sich unter der Welt befindet, und daraus werden dann die Nilquellen gespeist. Dieser Geschichte wird im „Buch vom Fayum“ breiter Raum gewidmet.[11]

Es seien zwei Orte übersprungen. Der vorletzte Ort bezieht sich auf die beiden Göttinnen Satet und Anuket, die an dieser Stelle gemeinsam abgebildet sind und die eigentlich mit den mythischen Nilquellen bei Assuan in Verbindung stehen. Möglicherweise haben wir hier die Stelle, an der sich das Wasser des Bahr Yussuf verzweigt, um das Fayum zu bewässern. Bemerkenswerterweise haben wir – zumindest in der Antike – keinen direkten Abfluß zum See des Fayum, sondern das Wasser gelangt erst, nachdem es zur Bewässerung auf die Felder verteilt wurde, langsam und zum Teil als Grundwasser in den See. Eine Direktverbindung in der Antike anzunehmen, wäre höchst unsinnig, denn das Wasser bringt, sobald es im See angekommen ist, der Bevölkerung keinen Nutzen mehr.

9 Da die Ägypter nur die Kardinalhimmelsrichtungen benutzten und der Bahr Yussuf an dieser Stelle rechtwinklig vom Nil fließt, ergibt sich, daß er nach Westen fließt.

10 Eine besondere Rolle spielt hier immer der Nil-Kugelfisch (Tetraodon fahaka), der von Süden kommend in das Fayum geschwemmt wird. Die Entfernung von Assuan bis zum Fayum wird in diesem Zusammenhang mit 61 Iteru angegeben, das sind ca. 779,5 km bei einer Iteru-Länge von 12,64 km, was der tatsächlichen Strecke ziemlich genau entspricht. S. Beinlich, Iterumaß.

11 Auch das alte Rätsel um den angeblich „von Menschenhand gegrabenen See im Fayum“ wird im Buch vom Fayum gelöst. Es handelt sich um die Ätiologie des alten Ortsnamens Schedet (Krokodilopolis), s. BEINLICH, in: Göttinger Miszellen 100, 1987, S. 15–18.

Schließlich haben wir in dem letzten Ort am Bahr Yussuf mit dem Ort „Merwer“ den Hafen von Krokodilopolis, der Hauptstadt des Fayum. Auch dieser Ort liegt an der Linie, die vom „Nördlichen Sand“ über „Kampfplatz“ und „Pyramidenland“ zu ziehen ist, und bildet den westlichen Schlußpunkt.[12]

Bei den Orten, die am oberen Papyrusrand genannt werden, ist bisher keine Identifizierung möglich gewesen. Auch diese Orte halten sich aber an die Verteilung auf die Himmelsrichtungen. Bei einem der Orte wird gesagt, daß er im Süden liegt.

2. ABSCHNITT: DER DURCHBRUCH DES BAHR YUSSUF DURCH DEN RAND DER LIBYSCHEN WÜSTE INS FAYUM UND DER OSTEN DES SEES DES FAYUM IN MYTHOLOGISCHER AUSGESTALTUNG

Abb. 4

Die beiden Landzungen der libyschen Wüste

Die Göttin Mehet-weret („die große Flut“).

Der östliche Bereich des Fayum-Sees mit der Geburt des Sonnengottes.

Nördlicher See Südlicher See

12 S. dazu Beinlich, Mythos 1, S. 53, und Vandorpe, in: AfP 50, 2004, S. 61–78.

Der zweite Abschnitt des Papyrus ist für eine Frage nach einer kartographisch interpretierbaren Grundlage weniger ergiebig. Hier wird großenteils der 1. Abschnitt noch einmal wiedergegeben, wenn auch mythologisch verklärt. Die beiden Zungen der Libyschen Wüste werden noch einmal aufgegriffen und in einen Ritualkontext gestellt. Danach wird auf der nördlichen Zunge das Anschwellen der Nilüberschwemmung beobachtet und bei ihrem Nahen auf der südlichen Zunge ein Leuchtfeuer entzündet, damit Osiris, der dieses Wasser verkörpert, weiß, wo er ins Fayum abzubiegen hat. Das Wasser, das zwischen den beiden Orten „Haus der Fackel" im Süden und „Platz des Beobachtens" im Norden dargestellt sein müßte, ist aufgrund der fehlerhaften Aufteilung des Papyrusblattes nur als ein Paar horizontaler Linien zu sehen. Wasserlinien wären zwischen diesen Linien nicht darstellbar gewesen.

Der Bereich, den der Bahr Yussuf bewässert, ist als lebensspendende Göttin Mehet-weret („die große Flut") dargestellt. Dieses Gebiet wird im Ägyptischen als „Henet" bezeichnet, was wohl den Bereich meint, der vom Wasser erreicht werden kann, das Feuchtgebiet im Gegensatz zum stets trockenen Bereich des Fayum.

Es soll aber hier nicht wiederholt werden, was an anderer Stelle ausführlicher gesagt werden konnte. Unter den Füßen dieser Göttin – die scheinbar liegt – ist der östliche Bereich des Fayumsees dargestellt. Man sieht hier am östlichen Ufer die Geburt des Sonnengottes aus dem Wasser, nachdem er die nächtliche Fahrt von Westen nach Osten zurückgelegt hat.[13] Weitere ikonographische Elemente unterstreichen, daß es sich um den Osten handelt, etwa die Position der beiden Barken, der Tag- und der Nachtbarke, Bug an Bug. Dies ist der Barkenwechsel des Sonnengottes, der jeden Morgen im Osten des Himmels bei Tagesanbruch vollzogen wird.

Für die Fragestellung nach einer kartographischen Grundlage kann herangezogen werden, daß der obere Teil des dargestellten Sees als der Südliche See bezeichnet wird, der untere als der Nördliche See. Nach unserer Vorstellung der Fayum-Landschaft ist zwischen dem ersten und dem dritten Abschnitt des Papyrus ein Bruch, denn man muß vom Ende des Bahr Yussuf zum östlichen Ufer des Sees einen Sprung nach Nordosten machen. Dieser Sprung wird im Papyrus durch die Darstellung der Göttin ausgeglichen.

3. ABSCHNITT: DER SEE DES FAYUM

Der 3. Abschnitt des Papyrus wird von einer großen kartuschenförmigen Fläche beherrscht, die den See des Fayum darstellt. Das Innere dieser Fläche ist mit Zickzacklinien als Wasserfläche vorzustellen, wie das in einer Parallele zu dem ausführlichsten Papyrus auch zu sehen ist. Die Figuren innerhalb dieser Fläche sind im Zusammenhang mit dem Zurückschwimmen des Sonnengottes in Krokodilform von Westen nach Osten zu verstehen und sollen hier nicht weiter erläutert werden. Auch in diesem Abschnitt wird auf die Himmelsrichtungen geachtet. Auf der linken

13 Der Sonnengott steigt dabei aus dem Wasser, seine Beine sind noch nicht zu sehen. Der Text daneben bezeichnet ihn als „(Sonnengott) Re, der sich anschickt zu schwimmen". Darüber und darunter sieht man je zwei Urgottpaare.

Seite, also im Osten, ist der Geburts- bzw. Wiedergeburtsort des Sonnengottes nach erfolgreicher Verjüngung in der Nacht zu sehen. Auf der rechten Seite ist Westen, wo der Sonnengott am Abend in das Wasser eintaucht, vorher aber seine Feinde massakriert, so daß sich der Himmel rot färbt. Der Ort dieses Gemetzels wird mit einem Begriff des Westgebirges (Manu) gleichgesetzt. So wie im ersten Abschnitt des Papyrus ist der innere Bereich von einem Streifen mit Fischen und weiter außen mit einem Streifen von Wasservögeln umgeben.[14] Der See ist damit eindeutig charakterisiert. Auch Dinge und Geschehnisse, die mit bestimmten Himmelsrichtungen gleichzusetzen sind, festigen die Vorstellung vom Landkartencharakter des Papyrus, wie z. B. die Bewegungsrichtung der Sonne und die Tätigkeiten, die sie bei ihrem Lauf durchzuführen hat.

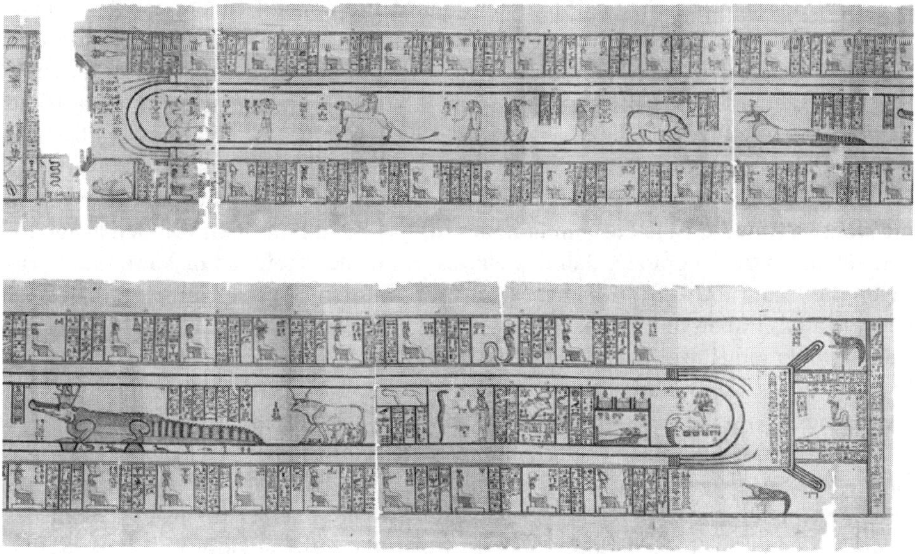

Abb. 5

Einzelne Orte lassen sich am Rand des Sees nicht identifizieren. Das hat mehrere Gründe. Zum einen ist das Geschehen, die Verjüngung der Sonne, ein Vorgang, der nicht mit realen Zuschauern verbunden werden kann. Zum anderen gehört es auch zum Charakter des Sees, daß hier keine, oder nur sehr wenig Siedlungen gewesen sind. Mit dem jährlichen Überschwemmungswasser steigt der Wasserspiegel um 160cm. Dementsprechend ist die jährliche Verdunstung. Das ist der Wert für die neueste Zeit. Im Altertum kann dieser Wert sehr viel höher, aber auch niedriger sein. Mit dem Steigen des Wassers vergrößert der See seinen Umfang schon unter

14 Im hier gezeigten Papyrus sind nur jeweils ein Fisch und ein Wasservogel ausgeführt. Nach dieser Vorlage hätte der Papyrus weiter ausgestaltet werden müssen. Dies ist aus uns nicht bekannten Gründen unterblieben. In parallelen Papyri ist aber die Ausarbeitung abgeschlossen worden.

normalen Umständen außerordentlich. Bei „Jahrhundertfluten“ kann dies zu Katastrophen führen.

Es ist also ungünstig, Siedlungen zu nahe am Wasserrand zu bauen. Weiter muß man sich den See als sehr dicht mit Schilf und Rohr bewachsen vorstellen. Er ist ziemlich flach und insgesamt heute im Mittel ca. 4 m tief. Das bedeutet, daß der Uferbewuchs sehr weit ins Wasser hineinreichen dürfte. Ein unmittelbarer Zugang zum offenen Wasser ist also die Ausnahme gewesen. Die Fischerei dürfte wegen des sehr salzhaltigen Wassers sehr eingeschränkt gewesen sein. Der Salzgehalt wird in verschiedenen Publikationen nicht einheitlich wiedergegeben. Er beträgt aber eindeutig das Mehrfache des Salzgehaltes, den man in den Weltmeeren findet. Heute hat man richtige Salzwasserfische dort ausgesetzt. Von den Fischen, die im Nil zu finden sind, haben allein die Tilapia im See des Fayum gute Überlebenschancen. Die Tiefe der Depression, ca. 45m unter dem Meeresspiegel, macht den Aufenthalt direkt am See nicht angenehmer. Dort tummelten sich vor allem Krokodile und vielleicht Nilpferde. Auch Bilharzia dürfte schon in pharaonischer Zeit im Fayum-See verbreitet gewesen sein. Es gab also m. E. keine Gründe, direkt am Wasserrand Siedlungen aufzubauen. Wichtig ist der See wegen seines Charakters als Platz der Götter und so wird er im Buch des Fayum auch verstanden, als Rückzugsort der Götter, als Ort der Regeneration, als Ort, in dem die Schöpfung begann. Hier haben die Acht Urgötter, als sie die Fundamente für das Fayum mit den Händen ausgehoben haben, so tief gegraben, daß das Urwasser aus der Tiefe herauskam, und sozusagen die Baugrube auffüllte. Dieses „Graben mit den Händen“ hat Herodot (Hist. II, 149) als „Graben mit (Menschen)händen“ mißverstanden[15] und mit seinem Bericht darüber ein Rätsel in die Welt gesetzt, das nicht unerheblich dazu beigetragen hat, daß er als Lügner angesehen wurde. Mithilfe des „Buches vom Fayum“ ließ sich Vieles von seinen Berichten als authentisch nachweisen.

Es seien noch ein paar Worte zum zweiten Teil des „Buches vom Fayum“ gestattet, der an dieser Stelle nicht besprochen wird. Hier werden ausführlich drei Orte besprochen, die in den Ritualen, die sich vor allem am Ostufer des Fayumsees abspielen, eine große Bedeutung haben. Außerdem wird in Bild und Wort die Erschaffung (Graben) des Fayum-Sees durch die Urgötter besprochen. Es gibt zwar auch hier bei den Besprechungen der Orte Angaben von Himmelsrichtungen, also eine geographisch fundierte Vorstellung, wo welcher Ort ist, es fehlt aber der genaue Bezugspunkt, an die diese Angaben anknüpfen können, so daß man auf Vermutungen angewiesen ist.

Es sind noch einige Informationen nachzutragen, die den Umfang der Einleitung gesprengt hätten, und die ich deshalb an das Ende des Aufsatzes gesetzt habe: Das „Buch vom Fayum“, in dem die oben besprochene Landkarte eine so wichtige Rolle spielt, ist in seiner Zeit das gewesen, was man heute einen „Bestseller“ nennt. Man kennt heute über 30 verschiedene Handschriften und es gibt nur zwei Texte, die häufiger belegt sind. Beim „Buch vom Fayum“ handelt es sich um Papyri, wie der hier vorgestellte, die mit vielen Abbildungen versehen und mit Hieroglyphen

15 Herodot schreibt allerdings nicht „Menschenhände“, auch wenn man aus seinen weiteren Ausführungen schließen kann, daß er dies gemeint hat. „Menschenhände“ findet man meist in Übersetzungen, weil man sich „Götterhände“ nicht vorstellen konnte.

geschrieben sind. Dann gibt es Abschriften davon, die keine Darstellungen haben und in hieratischer Schrift abgefaßt sind, der kursiven Schreibschrift der Hieroglyphen. Eine Handschrift scheint rein demotisch geschrieben zu sein. „Demotisch" ist eine noch weiter abgekürzte Schreibschrift der ägyptischen Spätzeit, die Normalschrift in der griechisch-römischen Zeit Ägyptens. Schließlich gibt es auch einige Handschriften, die den Text hieratisch wiedergeben und ihn dann ins Demotische übertragen und kommentieren. Andere ägyptische Texte, die in so vielen verschiedenen Schriften wie das „Buch vom Fayum" abgefaßt sind, sind in der Ägyptologie bisher nicht bekannt.

Alle erhaltenen Texte kann man in die römische Kaiserzeit in Ägypten datieren. Ein erster Textauszug, der an einer Mauer des Tempels von Kom Ombo steht, ist sehr genau in die Zeit des Kaisers Augustus zu datieren. Die längste hieratische Abschrift ist im Kolophon sogar auf den Tag genau datiert, auf den 15. September 135 p. Chr. n., d. i. in die Zeit des Kaisers Hadrian. Allerdings sind die Texte, die im „Buch vom Fayum" niedergeschrieben sind, nicht notwendigerweise erst in der Zeit der Ptolemäer und Römer entstanden. Man hat vielmehr z. T. sehr alte lokal gebundene Überlieferungen für den Text übernommen, wie das Beispiel „Herodot" zeigt, in dessen Historien einige Details über das Fayum zu finden sind, lange bevor das Buch vom Fayum in seiner uns bekannten Form entstand.

BENUTZTE UND WEITERFÜHRENDE LITERATUR

Badawy, Alexandre: Le dessin architectural chez les anciens Egyptiens, Le Caire 1948.

Beinlich, Horst: The Book of the Faiyum, in: Horst Beinlich – Regine Schulz – Alfried Wieczorek (Eds.): Egypt's Mysterious Book of the Faiyum, Dettelbach 2013, S. 27–77.

Beinlich, Horst: Das Buch vom Fayum, in: Horst Beinlich – Regine Schulz – Alfried Wieczorek (Eds.): Die Entstehung der Welt: Schöpfungsmythen aus dem Alten Ägypten nach dem Buch vom Fayum, Dettelbach 2014, S. 27–77.

Beinlich, Horst: Das Buch vom Fayum: Zum religiösen Eigenverständnis einer ägyptischen Landschaft (Ägyptologische Abhandlungen 51), Wiesbaden 1991.

Beinlich, Horst: Die Erneuerung der königlichen/göttlichen Macht und das Lebenshaus von Ra-sehet (Fayum), in: 7. Symposion zur ägyptischen Königsideologie, Prag 26.–28. Juni 2013, Wiesbaden, im Druck.

Beinlich, Horst: Geografia, 1. Carte geografiche, elenchi topografici, processioni, testi di esecrazione, in: Storia della Scienza, Vol 1, Rom 2001, S. 127–33.

Beinlich, Horst: Das Iteru-Maß nach dem „Buch vom Fayum", in: Mitteilungen des Deutschen Archäologischen Instituts Abteilung Kairo 43, 1987, S. 1–5.

Beinlich, Horst: Leuchtfeuer/Signalfeuer, in: Göttinger Miszellen 244, 2015, i. Dr.

Beinlich, Horst: Der Moeris-See nach Herodot, in: Göttinger Miszellen 100, 1987, S. 15–18.

Beinlich, Horst: (Mythos 1) Der Mythos in seiner Landschaft: das ägyptische „Buch vom Fayum", Bd. 1: die hieroglyphischen Texte, Textband, (SRaT 11,1), Dettelbach 2013.

Beinlich, Horst: (Mythos Tafelband) Der Mythos in seiner Landschaft: das ägyptische „Buch vom Fayum", Tafelband, (SRaT 11,1 Tafelband), Dettelbach 2013.

Beinlich, Horst: (Mythos 2) Der Mythos in seiner Landschaft: das ägyptische „Buch vom Fayum", Bd. 2: die hieratischen Texte, Textband, (SRaT 11,2), Dettelbach 2014.

Botti, Giuseppe: La glorificazione di Sobk e del Fayyum in un papiro ieratico da Tebtynis (Analecta Aegyptiaca 8), Copenhagen 1959. Rez.: Derchain, Philippe, in: Bibliotheca Orientalis 18, 1961, S. 40–43; Mertens, Paul, in: Chronique d'Egypte 37, 1962, S. 123–126.

Brown, Robert H.: The Fayum and Lake Moeris, London 1892.

Demuth, Siegfried: Die Oase Faiyum – Wasserhaushalt und Wasserbeschaffenheit, Diss. Freiburg i. Br. 1985.

Garbrecht, Günther: Wasserspeicherung im Fayum (Möris-See) – Legende oder Wirklichkeit?, in: Leichtweiss-Institut für Wasserbau der Technischen Universität Braunschweig, Mitteilungen 89, 1986; = Water storage (Lake Moeris) in the Fayum Depression, legend or reality?, in: Irrigation and Drainage Systems I, Dordrecht 1987, S. 143–157.

Ibrahim, Fouad: Ägypten, Wissenschaftliche Länderkunden Band 42, Darmstadt 1996.

Kendall, Timothy: Gebel Barkal. In: Kathryn A. Bard (Hrsg.): Encyclopedia of the Archaeology of Ancient Egypt. Routledge, London 1999, S. 325–28.

Lexikon der Ägyptologie I bis VI: Wiesbaden 1975 ff.

O'Connor, David: From topography to cosmos: ancient Egypt's multiple maps, in: Richard J.A. Talbert (Ed.): Ancient perspectives: maps and their place in Mesopotamia, Egypt, Greece, and Rome, Chicago 2012, S. 47–79.

Tait, John: The 'Book of the Fayum': Mystery in a Known Landscape, in: O'Connor, David / Quirke, Stephen: Mysterious Lands, London 2003, S. 183–202.

Vandorpe, Katelijn: The Henet of Moeris and the ancient administrative division of the Fayum in two parts, in: Archiv für Papyrusforschung 50, 2004, S. 61–78.

GLEICHNAMIGE FLÜSSE UND STÄDTE: GEOGRAPHISCHE NAMENSERKLÄRUNG ODER TOPOGRAPHISCHE GESCHICHTLICHKEIT?

Das Beispiel von Zariaspa/Baktra

Veronica Bucciantini

Die Hydrographie einer bestimmten Region bildet eines der entscheidenden Elemente der Gliederung und Strukturierung von Räumen: Die Namen der Flüsse, Quellen, Flussverläufe, Nebenflüsse und Mündungsgebiete verweisen nicht nur auf eine hydrographische Darstellung, sondern haben eine historische Dimension.[1]

Im Vordergrund steht die geographische Namenserklärung, aber auch die historische Dimension topographischer Angaben, die Engels[2] als „detaillierte Beschreibung des Raumes und der geographischen Bedingtheiten der Geschichte" bezeichnete.

Borysthenes/Olbia, das Strabon (7, 3, 17 C 306) und Ps.-Skymnos (804–808) erwähnen, ist ein interessantes Beispiel: Der Anonymus *Per. Pont. Eux.* 60 denkt, dass Olbia der alte Name der Stadt Borysthenes ist.

Es lässt sich daher nicht feststellen, ob der Flussname dem Namen der Stadt vorausging oder folgte. Die möglichen Ursachen für die unterschiedlichen Namen für denselben Ort sind noch zu untersuchen.

Als weitere Beispiele können die Flussnamen und Stadtnamen Xanthos (Strab. 14, 3, 6 C 666), Limyra und Limyros (Strab. 14, 3, 7 C 666), Askania und Askanios (Strab. 14, 5, 29 C 680) erwähnt werden.

Im Folgenden werden meine Untersuchungen zu Zariaspa/Batria vorgestellt; durch diese Stadt verläuft der Fluss mit dem Namen Zariaspis/Bactrus.

Baktra[3], das mit dem heutigen Balkh im Norden Afghanistans identifiziert wird,[4] war sowohl im Reich der Achämeniden als auch später unter den Seleukiden

1 Mein Dank geht an Dr. Gunvor Lindström (DAI Berlin) für das Lesen der vorliegenden Arbeit und für die wertwollen Informationen über Balkh.

2 J. Engels, Die Geschichte des Alexanderzuges und das Bild Alexanders des Großen in Strabons Geographika – Zur Interpretation der augusteischen Kulturgeographie Strabons als Quelle seiner historischen Auffassungen, in: W. Will (Hrsg.), Alexander der Große. Eine Welteroberung und ihr Hintergrund (Antiquitas Serie I), Bonn 1998, 133.

3 Zu Baktra als „Mutter der Städte" vgl. F. L. Holt, Alexander the Great and Bactria, The formation of a greek Frontier in Central Asia, Supplements to Mnemosyne, Leiden 1988, 29. Vgl. auch P. Leriche, Das Baktrien der 1.000 Städte, in S. Hansen, A. Wieczorek, M. Tellenbach (Hrsg.), Alexander der Grosse und die Öffnung der Welt, Asiens Kulturen im Wandel, Begleitband zur Sonderausstel., Regensburg 2010, 155–167.

4 Zu den archäologischen Ausgrabungen in Balkh und ihren Funden vgl. R. S. Young, The South Wall of Balkh-Bactra, AJA 59, 1955, 267–276, 29. Seit ein paar Jahren führen die Franzosen archäologische Ausgrabungen in Balkh und im südlich davon gelegenen Cheshm-e Shafa,

das wichtigste Verwaltungszentrum in dieser Region: die Satrapie Baktrien erstreckte sich südlich vom Mittellauf des Oxos (dem heutigen Amudarja)[5] bis zu den Hängen des Paropamisos (Hindukusch) und sie beinhaltete wohl auch die Gebiete nördlich des Oxos bis zum Hissar Range (Gissar) – der natürlichen Barriere zwischen den Flusssystemen von Baktrien und Sogdien.[6]

Zariaspa wird in der Forschung[7] als der älteste Name von Baktra betrachtet: so vermuten einige,[8] dass sich der Name von den Avestischen Worten für Gold – *zar* – und Pferd – *asp* – ableitet. Der Beiname der Stadt würde somit auf die Aufzucht der berühmten goldfarbenen Pferde verweisen. Eine andere Theorie besagt, dass sich die Bezeichnung Zariaspa nur auf einen Teil der Stadt bezieht.[9] So verbindet Cunningham[10] Zariaspa mit dem iranischen Begriff Azar-i-Asp, das den zoroastrischen Feuertempel in der Stadt bezeichnen würde, und er vermutete, dass Adrapsa (Strab. 11, 11, 2 C 516; 15, 2, 10 C 725) nur eine Variante für Zariaspa sei: Adar-i-Asp wäre dann nur eine andere Aussprache von Azar-i-Asp[11].

einer Befestigugsanlage aus achämenidischer Zeit, durch: vgl. P. Marquis, R. Besenval, P. Bernard, Du "mirage bactrien" aux réalités archéologiques: nouvelles fouilles de la Délégation archéologique française en Afghanistan (DAFA) à Bactres (2004–2005), CRAI 2006, 150.2, 1175–1248; P. Marquis, R. Besenval, La rêve accompli d' Alfred Foucher à Bactres: nouvelles fouilles de la DAFA 2002–2007, CRAI 2007, 151.4, 1847–1874.

5 Der heutige Flusslauf Vakhsh – Amudarja beginnt erst am Zusammenfluss von Vakhsh und Pandzh (bei Takht-i Sangin), obwohl auf Karten vielfach – aber irrtümlich – der Pandzh (antiker Ochos) auch als Amudarja bezeichnet wird. Zum Oxos als Grenze zum Emirat von Bokhara vgl. P. M. Fraser, Cities of Alexander the Great, Oxford 1996, 176.

6 Das afghanische Flusssystem besteht aus vier großen Flüssen: der Amudarja (Oxos), Hari Rud, Helmand-Arghandab, Kabul. Im Norden des heutigen Afghanistan befinden sich noch zwei Flüsse: der Zeravshan (Polytimetos), der in der Wüste endet und der Syr Darya (Iaxartes), der parallel zum Amudarja verläuft und in den Aral-See fließt. Vgl. Holt (wie Anm. 3), 12. Zu den Handelswegen zwischen Hindukusch und Hissar Range vgl. Fraser (wie Anm. 5), 176.

7 W. Tomaschek, s. v. Baktros, RE II.2, 1896, col. 2814; H. Treidler, s. v. Zariaspa, RE IX.A2 1967, coll. 2326–2328; W. W. Tarn, The Greeks in Bactria and India, New York 1984², 114–115 und 231. Zur Differenzierung zwischen Baktria/Zariaspa bei Ptolemaios vgl. A. B. Bosworth, An Historical Commentary on Arrian's History of Alexander II, Commentary On Books IV–V, Oxford 1995, 19, demzufolge Zariaspa „the alternative name of the capital Bactra" war. Vgl. auch Fraser (wie Anm. 5), 156–7; J. Wiesehöfer, s. v. „Zariaspa", Brill's New Pauly. Antiquity volumes edited by: H. Cancik and, H. Schneider. Brill Online. http://brillonline.nl/entries/brill-s-new-pauly/zariaspa-e12214850; B. Brentjes, H. Treidler, s. v. „Bactria", Brill's New Pauly. Antiquity volumes edited by: H. Cancik and, H. Schneider. Brill Online. http://brillonline.nl/entries/brill-s-new-pauly/bactria-e211780; B. Brentjes, H. Treidler, s. v. „Baktros", Brill's New Pauly. Antiquity volumes edited by: H. Cancik and, H. Schneider. Brill Online. http://brillonline.nl/entries/brill-s-new-pauly/baktros-e211840.

8 H. Kiepert, Lehrbuch der alten Geographie, Berlin 1878, 57; W. Tomaschek, s. v. Baktriane, RE II.2, 1896, coll. 2805–2806.

9 Tomaschek (wie Anm. 8), col. 2806.

10 Für die schwierige Problemen über Baktra als Zentrum des Zoroastrismus und als der Ort, an dem Zoroaster als erstes wirkte und begann, seine Religion zu verbreiten, und an dem er ebenfalls starb, vgl. A. Cunningham, Coins of Alexander's Successors in the East, Numismatic Chronicle 8, 1868, 107.

11 Zum Tempel der Göttin des Wassers Anahita, die mit Artemis identifiziert wird, vgl. O. Coloru, Da Alessandro Magno a Menandro, Il regno greco di Battriana, Pisa – Roma 2009, 206, n. 5.

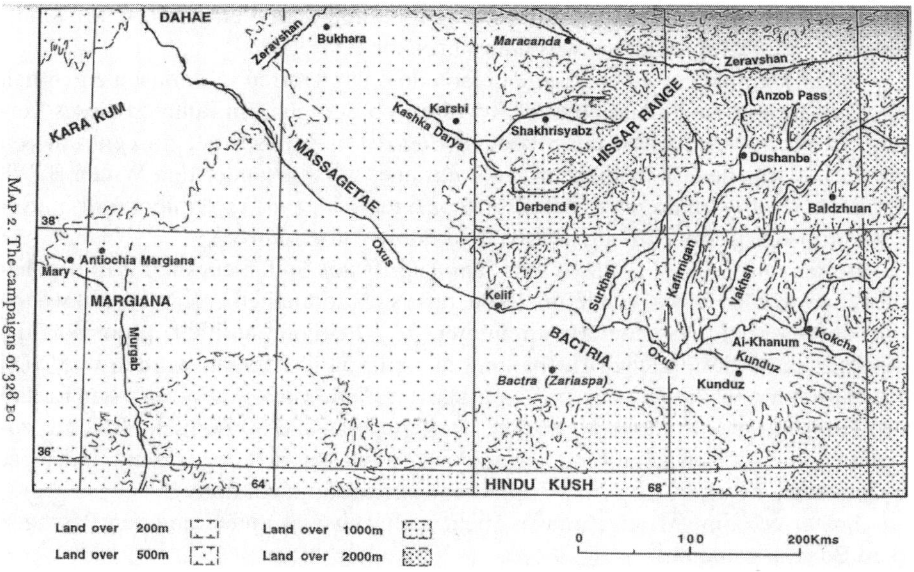

Abb. 1: A. B. Bosworth, *A Historical Commentary on Arrian's History of Alexander I,* Oxford 1980.

Seine Rolle als traditionelles Zentrum des Zoroastrismus ist für Baktra ausgesprochen bedeutsam: es ist der Ort, an dem Zoroaster als erstes wirkte und begann, seine Religion zu verbreiten, und an dem er ebenfalls starb.

In unseren Quellen finden wir bei Arrian sowohl den Begriff Zariaspa als auch Baktra, um die Hauptstadt von Baktrien zu bezeichnen.

Dabei fällt auf, dass die Verwendung des Begriffs Ζαρίασπα bei Arrian zwischen dem weiblichen Singular und dem Neutrum Plural variiert, was bereits von Carax von Pergamon[12] erwähnt wird, der eine Form im Neutrum vermutet – τὰ Ζαρίασπα – anstatt des üblichen Ζαρίασπα oder Ζαριάσπη.

Die Problemstellung der beiden Namen soll im Folgenden diskutiert werden, wobei ich von den Quellen ausgehe, die den Namen der Stadt Zariaspa und den des Flusses Zariaspis überliefern.

Ich werde mit einigen Passagen der *Anabasis* von Arrian beginnen: in diesen werden Zariaspa und Baktra im Einzelnen erwähnt, ohne dass angegeben wird, dass es sich um die gleiche Stadt handelt.

Als Spitamenes, der Bessos an Alexander ausgeliefert hatte, versuchte einen Aufstand der Bactrier und Sogdier zu organisieren, heißt es bei Arrian (*anab.* 4, 1, 5):[13] „Alexander nämlich hatte die Magnaten dieses Landes zu einer Versammlung

12 *FGrHist* 103, F 53 Et. Byz.: Ζαρίασπα ἢ Ζαριάσπη, πόλις Βακτριανή. Στράβων ενδεκάτη. Ἢ εκαλεῖτο καὶ Βάκτρα. Χάραξ δὲ ουδετέρως τὰ Ζαρίασπα.

13 Arr. *anab.* 4, 1, 5: ξυνεπελάβοντο δὲ αὐτοῖς τῆς ἀποστάσεως καὶ τῶν Σογδιανῶν οἱ πολλοί, ἐπαρθέντες πρὸς τῶν ξυλλαβόντων Βῆσσον, ὥστε καὶ τῶν Βακτριανῶν ἔστιν οὓς σφισιν οὗτοι ξυναπέστησαν, εἴτε δὴ καὶ δείσαντες Ἀλέξανδρον, εἴτε καὶ λόγον ἐπὶ τῇ ἀποστάσει διδόντες, ὅτι ἐς ἕνα ξύλλογον ἐπηγγέλκει Ἀλέξανδρος ξυνελθεῖν τοὺς ὑπάρχους τῆς χώρας ἐκείνης εἰς Ζαρίασπα, τὴν μεγίστην πόλιν, ὡς ἐπ' ἀγαθῷ οὐδενὶ τοῦ ξυλλόγου γιγνομένου.

nach Zariaspa, der größten Stadt des Gebietes, einberufen – eine solche Versammlung aber könne keinesfalls etwas Gutes bringen."[14]

In der Passage[15] wird die Sorge betont, dass das Treffen in Zariaspa eigentlich ein Vorwand sei, um alle Führer der Revolte zu beseitigen. Im Rahmen dieses Vortrags ist von Interesse, dass der Sitz des *syllogos*[16] in Zariaspa ist, ein Zentrum von großer historischer Bedeutung in der Baktriane, wo Alexander den Winter 329/8 verbrachte (vgl. Arr. *anab.* 4, 7, 1: Ταῦτα δὲ διαπραξάμενος ἐς Ζαρίασπα ἀφίκετο · καὶ αὐτοῦ κατέμενεν ἔστε παρελθεῖν τὸ ἀκμαῖον τοῦ χειμῶνος).

In Arr. *anab.* 4, 16, 5–6, wo von dem Angriff des Spitamenes auf Zariaspa berichtet wird, können wir wiederum lesen, dass sich dort neben einer „Anzahl wegen Krankheit zurückgelassener Hetairenreiter, [...] Peithon" aufhielt, „dem dort die Verwaltung der königlichen Hofhaltung in Zariaspa übertragen worden war" (ἐπὶ τῆς βασιλικῆς θεραπείας τῆς ἐν Ζαριάσποις τεταγμένος). Als diese vom Einfall der Truppen unter Spitamenes hörten, „holten sie an die 80 berittene Söldner zusammen, die als Bewachung in Zariaspa zurückgelassen worden waren" (ἱππέας ἐς ὀγδοήκοντα, οἳ ἐπὶ φυλακῇ τῶν Ζαριάσπων ὑπολελειμμένοι ἦσαν).[17]

Später erwähnt Arrian Zariaspa nicht mehr, aber in vielen anderen Passagen wird Βάκτρα genannt.[18]

Aus der Tatsache, dass der Name Ζαρίασπα in den beiden Passagen genannt wird, die auf Aristobulos zurückgehen (siehe *anab.* 4, 7, 1; 4, 16, 5–6), während sich die Bezeichnung Βάκτρα in einem Fragment des Ptolemaios befindet (*FGrHist* 138, F 14 = *anab.* 3, 29, 6–30, 5), schließt Bosworth, dass die verschiedenen Namen von der Quelle abhängen, die der Historiker aus Nikomedia jeweils verwendet hat.

Baktra[19] wird von Arrian in zweifachem Sinne verwendet: manchmal ist damit die gesamte Satrapie Baktrien[20] gemeint und manchmal auch nur die Stadt, wie im Falle von *anab.* 3, 25, 3.

Im Sinne der Satrapie wird der Begriff auch bei Herodot verwendet, wie in der Passage 6, 9, wo die Perser den Ioniern drohen, sie zu Sklaven und ihre Söhne zu Eunuchen zu machen, sowie ihre Töchter nach Baktrien – ἐς Βάκτρα – zu deportieren. So auch im neunten Buch (9, 113, 1), wenn Masistes, der Statthalter dieser Satrapie, nach einer Gewalttat des Xerxes gegen seine Frau und Tochter, nach Baktrien reist, um einen Aufstand auszulösen.

14 Vgl. Curt. 7, 6, 15. Übersetzung v. G. Wirth u. O. von Hinüber.

15 Vgl. für diese Passage C. Rapin, On the way to Roxane. The Route of Alexander the Great in Bactria and Sogdiana (328–327 BC), in G. Lindström, S. Hansen, A. Wieczorek, M. Tellenbach (Hrsg.), Zwischen Ost und West. Neue Forschungen zum antiken Zentralasien, Archäologie in Iran und Turan 14, Darmstadt, 2013, 46–47.

16 Zur Bedeutung von *syllogos* vgl. P. Briant, Histoire de l'empire perse de Cyrus à Alexandre, Paris 1996, 769.

17 Übersetzung v. G. Wirth u. O. von Hinüber.

18 Βάκτρα wird zwölfmal erwähnt, z. B. Arr. *anab.* 3, 25, 3: ἔχειν τε ἀμφ' αὐτὸν Περσῶν τε τοὺς ἐς Βάκτρα διαφυγόντας καὶ αὐτῶν Βακτριανῶν πολλούς· ... Ἀλέξανδρος δὲ ὁμοῦ ἤδη ἔχων τὴν πᾶσαν δύναμιν ᾖει ἐπὶ Βάκτρων.

19 A. B. Bosworth, A Historical Commentary on Arrian's History of Alexander I, Commentary on Books I–III, Oxford 1980, 356.

20 Βάκτριοι in Arist. *mir.* 833b14; Hdt. 3, 102; Βακτριανή in Strab. 11, 9, 2, C 515.

In einer ähnlichen Verwendung wie bei Arrian, die letztlich auf die Alexanderhistoriker zurückgeht, finden wir Zariaspa und Baktra bei Polybios erwähnt, der sie nicht als identisch begreift: Zariaspa erscheint in 10, 49, 15 γενομένης δὲ τῆς μάχης ταύτης ὁ μὲν Εὐθύδημος καταπλαγεὶς ἀνεχώρησε μετὰ τῆς δυνάμεως εἰς πόλιν Ζαριάσπαν τῆς Βακτριανῆς, in der Beschreibung des Angriffs von Antiochus dem Dritten gegen Euthydemus (206 v. Chr.), der sich nach der Schlacht in die Stadt Zariaspa zurückzieht. Polybios schreibt, dass es sich um Zariaspa in Baktrien handelt – Ζαριάσπαν τῆς Βακτριανῆς.

Polybios erwähnt Baktra[21] als Stadt nur einmal (29, 12, 8) im Rahmen der Beschreibung der Belagerungen, die *topoi* geworden sind.

Die beiden Orte werden von Strabon (Strab. 11, 8, 9 C 514[22]) wiederum als identisch erwähnt, der sagt, dass der Abstand zwischen Alexandria in Ariana[23] und der Stadt Baktra, die auch Zariaspa genannt wird, 3.870 Stadien beträgt; von dort zum Iaxartes sind es weitere 5.000 Stadien[24], von insgesamt 22.670 Stadien für die Länge vom Kaspischen Meer bis zum Punkt des Iaxartes, den Alexander erreicht hat.

In 11, 8, 9 lesen wir, dass die Stadt Baktra auch Zariaspa genannt wird – εἶτ' εἰς Βάκτραν τὴν πόλιν, ἣ καὶ Ζαριάσπα καλεῖται –, und noch wichtiger, was wir in einer anderen Passage von Strabon finden: 11, 11, 2 C 516 Πόλεις δ'εἶχον τά τε Βάκτρα ἥνπερ καὶ Ζαριάσπαν καλοῦσιν (ἣν διαρρεῖ ὁμώνυμος ποταμὸς ἐκβάλλων εἰς τὸν Ὦξον), καὶ Ἄδραψα[25] καὶ ἄλλας πλείους.

Beide Passagen gehen wahrscheinlich auf Eratosthenes zurück, der in dem Zusammenhang auch explizit genannt wird. Besonders interessant im Hinblick auf meine Forschung ist die zweite Passage, die den Fluss mit dem gleichen Namen nennt: „An Städten hatten sie Baktra, das auch Zariaspa genannt wird, hindurch strömt ein gleichnamiger Fluss, der in den Oxos mündet, Adrapsa und vielen anderen Städten".

21 Polybios erwähnt Βακτριανή an den folgenden Stellen 10, 48, 4; 10, 49, 6; 10, 49, 15; 11, 34, 3; Βάκτροι 10, 49, 8; Βακτριανοί 10, 49, 11.

22 Strab. 11, 8, 9 C 514: Τὰ δὲ διαστήματα οὕτω λέγει· ἀπὸ μὲν τοῦ Κασπίου ἐπὶ τὸν Κῦρον ὡς χιλίους ὀκτακοσίους σταδίους, ἔνθεν δ' ἐπὶ Κασπίας πύλας πεντακισχιλίους ἑξακοσίους, εἶτ' εἰς Ἀλεξάνδρειαν τὴν ἐν Ἀρίοις ἑξακισχιλίους τετρακοσίους, εἶτ' εἰς Βάκτραν τὴν πόλιν, ἣ καὶ Ζαριάσπα καλεῖται, τρισχιλίους ὀκτακοσίους ἑβδομήκοντα, εἶτ' ἐπὶ τὸν Ἰαξάρτην ποταμόν, ἐφ' ὃν Ἀλέξανδρος ἧκεν, ὡς πεντακισχιλίους· ὁμοῦ δισμύριοι δισχίλιοι ἑξακόσιοι ἑβδομήκοντα.

23 Vgl. Fraser (wie Anm. 5), 141 und 191–192; G. M. Cohen, The Hellenistic Settlements in the East from Armenia and Mesopotamia to Bactria and India, Berkeley – Los Angeles – London 2013, 260–261 identifiziert Alexandria in Ariana mit dem heutigen Herat.

24 Die gleiche Entfernung zwischen Iaxartes und Baktra findet sich bei Plin. *nat.* 6, 45: *hunc enim cardinem Alexandri Magni itinerum fecere ab <i>is Portis ad Indiae principium stadia <XV> DCLXXXX prodendo, ad Bactra oppidum, quod appellant Zariasta, MMM DCC, inde ad Iaxartem amnem <V>.*

25 H. L. Jones, The Geography of Strabo, V, London 1954, 280 und S. Radt (Hrsg.), Strabons Geographika: Band 3: Buch IX–XIII: Text und Ubersetzung, Göttingen 2004, 356 akzeptieren nicht Ἄδραψα von Meineke (*duce* Kramer), sondern Δάραψα. Cunningham (wie Anm. 10), 107 vertritt die These, dass Adrapsa eine Variante von Zariaspa ist und es mehr als eine Stadt mit dem Namen Zariaspa gab.

Ein Fluss mit dem gleichen Namen (Zariaspis?) der Stadt, der auch durch andere Städte fließt, bevor er in den Oxos mündet, steht in der Quelle von Arrian in erster Linie mit der Stadt in Verbindung, der er seinen Namen gibt.

Dies ist ein sehr bedeutsames Zeugnis, vor allem, wenn es, wie es scheint, auf Eratosthenes zurückgeht und sich auf einen Moment bezieht, wo das alte Zentrum des Zoroastrismus mit dem Namen des Flusses, der es durchquert, angegeben wurde.

Die Passage von Strabon wiederspricht dem, was wir bei Plinius lesen (*nat.* 6.48). Von Plinius erfahren wir in der Tat, dass die Stadt Zariastes später Baktra hieß, und zwar nach dem Fluss Bactrus. Die Bestätigung dafür finden wir wiederum bei Curtius Rufus: 7, 4, 3: *Sed qua mitior terra est, ingens hominum equorumque multitudo gignitur; itaque Bactriani equites XXX milia expleverant. Ipsa Bactra, regionis eius caput, sita sunt sub monte Parapamiso. Bactrus amnis praeterit moenia; is urbi et regioni dedit nomen.*

Nach Aristoteles (*meteor.* 1, 13,16) entspringt der Baktros aus dem Paropamisos, fließt durch Baktra und mündet in den Oxos (Amudarja) und nach *De fluviis* von Ps.-Plut. (23), muss er mit dem Araxes identifiziert werden: der Fluss war eine wichtige Handelsroute, da von dem indischen Fluss Kophen-Kabul die Waren durch den Baktrus in sieben Tagen zum Oxos kamen und von dort nach Hyrcania und darüber hinaus.

Plinius *nat.* 6, 52[26] informiert uns, dass sieben Tage benötigt wurden, um von Indien nach Baktra zu gelangen und diese Daten bestätigen die Bedeutung der Stadt als Drehscheibe an der Seidenstraße, wie wir bald sehen werden.

Es ist jedoch nicht mehr möglich, den Verlauf des Flusses Baktrus zu bestimmen, aus dem wahrscheinlich viele Bewässerungskanäle abgeleitet wurden, weil das Gewässer im Mittelalter in Richtung Mazar-i-Sherif umgeleitet wurde und dieser jetzt ein paar Meilen weiter östlich verläuft und in der Oase von Balkh endet[27].

Heute fließen in dieser Region rund um Balkh zwei Flüsse – Darrah und Band-i-Emir – aber keiner von beiden erreicht den Oxos, da sie entweder nur saisonal Wasser führen oder vorher unter der Erde verschwinden.

Um den scheinbaren Widerspruch zwischen den zitierten Quellen zu erklären, könnte man an verschiedene Ausführungen denken, die einen Aspekt der langen Geschichte zeigen: das alte Zentrum des Zoroastrismus übernimmt den Namen – wie Strabon berichtet – vom Fluss Zariaspis.

Die Bedeutung, die Baktrien in der Zeit der Achämeniden erhält, kann der Ursprung für den Namen der Stadt – Baktra – sein, der sich von der Region ableitet: dies zeigte die politische und institutionelle Konnotation der Hauptstadt der Satrapie.

26 Plin. *nat.* 6, 52: *adicit idem Pompei ductu exploratum, in Bactros septem diebus ex India perveniri ad Bactrum flumen quod in Oxum influat, et ex eo per Caspium in Cyrum subvectos, et V non amplius dierum terreno itinere ad Phasim in Pontum Indicas posse devehi merces.*

27 Nach den französischen Untersuchungen (DAFA vgl. Anm. 4) des Kanalsystems und der Siedlungen in der Oase von Balkh sieht es so aus, als hätte der Fluss zu keiner Zeit den Oxos erreicht: Vgl. É. De la Vassière 2010, Des Bactres a Balkh, par le Nowbahr, Journal Asiatique 2/298, 2010, 517–533; E. Fouache *et al.*, Palaechannels of the Balkh River (Nortern Afghanistan) and Human Occupation since the bronze Age Period, Journal of Archaeological Science 39, 2012, 3415–3427.

Die Koexistenz der beiden Namen, die an die Geschichte der zoroastrischen Zeit oder des Achämenidenreiches erinnern, kann das Nebeneinanderbestehen der beiden Namen bei den Alexanderhistorikern erklären, die später bei Arrian zusammengeführt werden.

In dem Licht einer Tradition, die den Namen der Stadt mit dem Fluss verbindet, der sie durchquert, und die um die Bedeutung von Baktra wusste, kann man davon ausgehen, dass Plinius und Curtius Rufus die Entstehung des Namens, die für Zariaspa galt, auf Baktra übertragen haben: der Fluss, der der Stadt ihren Namen gibt, ist bei Plinius ein ansonsten unbekannter Baktro (*nat.* 6, 48: *Oxyttagae, Moci, Bateni, Saraparae, Bactri, quorum oppidum Zariastes, quod postea Bactrum a flumine appellatum est*) und so auch bei Curtius Rufus (7, 4, 31: *Ipsa Bactra, regionis eius caput, sita sunt sub monte Parapamiso. Bactrus amnis praeterit moenia; is urbi et regioni dedit nomen*).

Bei der hier vorgeschlagenen Rekonstruktion soll schließlich berücksichtigt werden, dass Ptolemaios in einer Passage der *Geographie* (6, 11, 2)[28] Zariaspis erwähnt, nämlich in der Liste der Flüsse, die in Baktrien fließen und die im Oxos[29] zusammenfließen, deren Quellen auf dem 113. Längengrad und dem 39. Grad nördlicher Breite liegen.[30] Der Zusammenfluss ereignet sich bei einer Länge von 112° 30‘ und einer Breite von 44° N (Ptol. *geogr.* 6, 11, 3).[31]

In seiner Auflistung der Völkerschaften von Baktrien nördlich des Oxos, erwähnt Ptolemaios die Ζαριάσπαι (*geogr.* 6, 11, 6), eine Erwähnung der Baktrier findet sich hingegen nicht. In der Liste der Städte von Baktrien (*geogr.* 6, 11, 6–9) befindet sich Zariaspa jedoch bei einer Länge von 115° und 44° nördlicher Breite.[32] Es wird von Stückelberger und Grasshoff mit dem heutigen Balkh identifiziert.[33]

28 Ptol. *geogr.* 6, 11, 2: Διαῤῥέουσι δὲ τὴν Βακτριανὴν ποταμοὶ συμβάλλοντες τῷ Ὤξῳ ποταμῷ ὅ τε Ὦχος ποταμὸς, οὗ αἱ πηγαὶ ἐπέχουσι μοίρας <ρ>ι <λ>θ καὶ ὁ Δαργαμάνης … καὶ ὁ Ζαρίασπις, οὗ αἱ πηγαὶ ἐπέχουσι μοίρας ρι<γ> <λ>θ.

29 Zum Flusslauf des Oxos und Iaxartes, die in den Aral-See fließen, oder aber – nach Ptolemaios – in das Kaspische Meer vgl. H. Humbach, Hydrologische Probleme im Raum des Kaspischen Meeres und des Aralsee. a) Die Oxos Route und die Iaxartes Mündung, in A. Stückelberger, F. Mittenhuber, (Hrsg.), Ptolemaios – Handbuch der Geographie, Ergänzungsband, Basel 2009, 312–315; C. Sengör, Hydrologische Probleme im Raum des Kaspischen Meeres und des Aralsee. b) Zum Lauf des Oxos, ein Nachtrag, in Stückelberger, Mittenhuber (wie eben), 316–318.

30 G. Winkler, F. Mittenhuber, Asienkarte: Kaspisches Meer, Zentralasien, in Stückelberger, Mittenhuber (wie Anm. 29), 298–299: „Die Ausdehnung der Region in Länge und breite wir jedoch wieder deutlich überschätzt: Die breiten um en Oxos sind um durchschnittlich 3°–4° zu hoch, diejenigen um den Iaxartes sogar um 7°–8°. Die Längenausdehnung ist mit 25° statt 15° etwa um denselben Faktor zu gross wie in den anderen Regionen Zentralasiens“.

31 Ptol. *geogr.* 6, 11, 3: τῶν δὲ λοιπῶν ὁ μὲν Ἄρταμις καὶ ὁ Ζαρίασπις συμμίξαντες πρότερον ἀλλήλοις κατὰ θέσιν μοιρῶν.

32 Im Kanon bedeutender Städte sind z. B. die folgenden Städten erwähnt: Zariaspa 115° Länge und 43° 45’ nördliche Breite; Baktra 116° Länge und 41° nördliche Breite, vgl. Stückelberger, Mittenhuber (wie Anm. 29), 205.

33 A. Stückelberger, G. Grasshoff (Hrsg.), Ptolemaios – Handbuch der Geographie, 2. Teilband: Buch 5–8 und Indices, Basel 2006, 649.

Βάκτρα βασίλειον wird durch den Geographen auf 116° Länge und 41° Breite platziert und daher eher auf dem Fluss Dargoidos (*geogr*. 6, 11, 2), dessen Quellen bei 116°–39° N liegen, und nicht auf dem Fluss Zariaspis, dessen Quellen bei 113°–39° N liegen.

Es ist sehr bedeutsam, dass Baktra als „königlich“ definiert wird. Diese Bezeichnung tritt an keiner anderen Stelle auf.

Der Unterschied in den Koordinaten der beiden Städte wird auch in dem achten Buch wieder erwähnt: in *geogr*. 8, 23, 9[34], lesen wir, dass Zariaspa von Alexandria 3 Stunden und 2/3 entfernt ist, Baktra[35] hingegen 3 Stunden und 11/15. Zudem dauere in Zariaspa der längste Tag 15 Stunden und 3/8, in Baktra hingegen dauert er 15 Stunden und 11/15.

Das Problem, das sich stellt, ist, wie man den Unterschied von Zariaspa-Baktra bei Ptolemaios erklärt, weil der Geograph nur den Fluss Zariaspis kennt.

Droysen[36] dachte, dass es sich bei diesen um verschiedene Städte handelte, und dass zwischen Zariaspa und Baktra eine Verwechselung stattgefunden habe, die sehr schwierig zu erklären sei: der Gelehrte verglich *anab*. 4, 16, 1, wo wir lesen, dass Alexander in Baktra – ἐν Βακτροῖς – nur 4 Strategen zurücklässt, mit *anab*. 4, 16, 6, wo wir lesen, dass der König in Zariaspa Milizen mit Krankheitsproblemen und 80 Reiter zurücklässt, und er kam zu dem Schluss, dass dies verschiedene Städte waren.

Auch von Schwarz[37], der viele Jahre in Turkestan gelebt hatte, war der Meinung, dass man das antike Zariaspa nicht mit Baktra – Balkh identifizieren dürfe, sondern mit Tschardschui, und glaubte, dass es sich bei Zariaspa und Baktra nicht um die gleiche Stadt handelte.

Reuss[38] widersprach der Hypothese von Droysen und er dachte dass ἐν Βακτροῖς sich auf Baktrien beziehen müsse, unter Berufung auf die doppelte Bedeutung des Begriffs als Bezeichnung für die Stadt und die Region, die auch bei Arrian zu finden ist: Zariaspa wäre für den Gelehrten der älteste Name von Baktra.

Er tut die verschiedenen Breitengrade bei Ptolemaios mit der Beobachtung von Kiepert[39] ab, der glaubte, dass es sich um unterschiedliche Namen für dieselbe Stadt handelte. Dieser argumentierte für „die richtige Identification beider bei Strabon und Plinius, wogegen die Unterscheidung beider als verschiedener Orte in dem an ähnlichen Irrtümern reichen Kartenwerke des Ptolemaios nicht ins Gewicht fällt“.

34 Ptol. *geogr*. 8, 23, 9: Τῶν δὲ ἐν τῇ <Βακτριανῇ> … ἡ δὲ <Ζαρίσπα> τὴν μεγίστην ἡμέραν ἔχει ὡρῶν ι<ε> δη′, καὶ διέστηκεν Ἀλεξανδρείας πρὸς ἕω ὥραις τρισὶ καὶ διμοίρῳ· ἡ δὲ <Βάκτρα> τὴν μεγίστην ἡμέραν ἔχει ὡρῶν ι<ε>, καὶ διέστηκεν Ἀλεξανδρείας πρὸς ἕω ὥραις τρισὶ γο′ καὶ ιε′.

35 Ptol. *geogr*. 8, 23, 9 berechnet die Entfernung zwischen Alexandreia-Baktra auf 3 Stunden und 2/3, ebenso lang wie der Weg zwischen Alexandreia und Zariaspa. Zu den Vermessungsfehlern in codex X vgl. A. Stückelberger, G. Grasshoff (Hrsg.), Ptolemaios – Handbuch der Geographie, 1. Teilband: Einleitung und Buch 1–4, Basel 2006, 44.

36 J. G. Droysen, Geschichte Alexanders des Grossen, Hamburg 1833, 325–326, n. 57; *Id*., Geschichte des Hellenismus, Hamburg 1843, 617.

37 F. von Schwarz, Alexanders des Grossen Feldzüge in Turkestan, München 1893, 65.

38 F. Reuss, Hellenistische Beiträge, Bactra und Zariaspa, RhM 62, 1907, 591–595.

39 Kiepert, (wie Anm. 8), 57, n. 2.

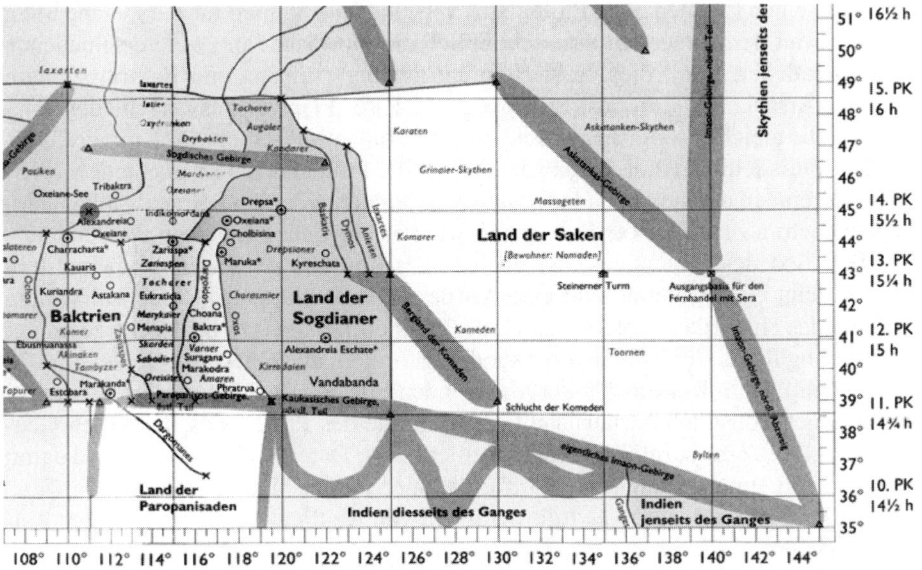

Abb. 2: A. Stückelberger, G. Graßhoff (Hrsg.), *Ptolemaios – Handbuch der Geographie, 2. Teilband: Buch 5–8 und Indices*, Basel 2006

Auch Tarn[40] dachte an einige Ungenauigkeiten vonseiten des Ptolemaios, den er als Geographen folgendermaßen charakterisiert: „extremely painstaking […] but devoid of critical and historical instinct; he collected or had at his disposal a vast material from every kind of source and many different dates."

Nach der Meinung von Tarn arbeitete Ptolemaios mit extrem heterogenen Quellen, meist bestehend aus Reiserouten von Kaufleuten oder anderen Reisenden, die aber selten exakt waren.

Geus[41] hat über Ptolemaios zu Recht festgestellt: „Im Einzelfall hat sich Ptolemaios einige, teilweise auch eklatante Irrtümer zuschulden kommen lassen. Doch sind die Ursachen für diese Irrtümer nicht in einer ideologischen Verbohrtheit oder einer fehlerhaften Methodik zu suchen. Sie liegen in erster Linie in den mangelnden technischen Möglichkeiten und den ungünstigen logistischen Voraussetzungen seiner Zeit."

40 Tarn (wie Anm. 7), 114.

41 K. Geus, Ptolemaios – Reaktionär, Theoretiker, Plagiator?, in T. Beck, M. d. S. Lopes, C. Rödel (Hrsg.), Barrieren und Zugänge: Die Geschichte der europäischen Expansion. Festschrift für Eberhard Schmitt zum 65. Geburtstag. Wiesbaden 2004, 50. Vgl. auch *Id.*, "… wie man auch beim Fehlen einer Kartenvorlage nur an Hand des Textes eine Weltkarte sehr leicht herstellen kann", Die Darstellung der Oikumene bei Klaudios Ptolemaios. In: Periplus: Jahrbuch für außereuropäische Geschichte 23, 2013, 76–91; *Id.*, Wie erstellt man eine Karte von der Welt? Die Lösung des Ptolemaios und ihre Probleme, in D. Boschung, T. Greub, J. Hammerstaedt (Hrsg.), Geographische Kenntnisse und ihre konkreten Ausformungen, München 2013, 119–136; K. Geus, I. Tupikova, The circumference of the Earth and Ptolemy's world map, Berlin 2013 (Max-Planck-Institut für Wissenschaftsgeschichte, Preprint 439), 1–26.

Aus seinen Quellen hätte Ptolemaios verschiedene Namen für die gleiche Stadt erhalten, mit dem Ergebnis, dass schließlich dieselbe Stadt an zwei verschiedenen Orten lokalisiert wird: dies passierte nicht nur bei Zariaspa und Baktra, sondern auch bei Arsacia und Europos (Ptol. *geogr.* 6, 2, 16–17), welches verschiedene Namen für die gleiche Stadt Rhagae sind, und bei Rabbath-Ammon (Ptol. *geogr.* 5, 17, 3) bzw. Philadelphia (Ptol. *geogr.* 5, 12–23), die Ptolemaios an verschiedenen Orten – das eine in der syrischen Dekapolis, das andere in Arabia Petraea – lokalisiert, was wiederum zeigt, dass er Dokumente aus verschiedenen Epochen verwendete.

Der Text des Ptolemaios, der etwa in der Zeit des Arrian zusammengestellt wurde, zeigt eine Methode und eine Art der Dokumentation, die sich vollständig von der des Historikers unterscheidet, der stolz erklärt, dass er beabsichtige, in der Darstellung der Geschichte den Alexanderhistorikern zu folgen.

In deutlichem Kontrast zu der mangelnden Prüfung und Aktualisierung der Informationen vonseiten Arrians steht die Methode des Ptolemaios. In dessen Geographie gibt es eine sorgfältige Prüfung aller neuen Daten und das Werk wird damit sozusagen zu einem Speicher von Informationen unterschiedlichster Art.

In der Geographie gibt es Informationen von Kaufleuten, die seit ältesten Zeiten die Seidenstraße[42] bereisten. Diese hatten ihre Handelsverbindungen stets ausgebaut und intensiviert, ausgelöst durch das Wachstum der Mittelmeermärkte, insbesondere Roms.

Die Seidenstraße führte von Xiang durch Lanzhou nach Westen und dann durch Turfan und Kucha, Dunhuang, und Cherchen Yarkande nach Kashgar, bis Samarkanda und dann weiter über den Nordosten des Iran bis nach Seleukia und dann nach Antiochia am Mittelmeer.

Baktrien hat auf dieser Route eine entscheidende Rolle gespielt: Neben den reichen natürlichen Ressourcen war die Region die einzige Landverbindung mit den Regionen Indiens: ohne Schwierigkeiten können sich Menschen und Tiere von Baktra nach Taxila bewegen, über den Hindukusch hinweg, bis nach Alexandria am Kaukasus (heute Kabul) und dann über den Khyber-Pass bis in die Region von Gandhara.

Im zweiten Jahrhundert v. Chr. wurde, dank der Herrscher des griechisch-baktrischen Reiches Demetrius und Menander, eine weitere Route durch Baktrien nach China hinzugefügt.

Baktra wurde ein entscheidender Punkt auf dem Routenweg, sowohl Richtung Süden und Indien als auch in Richtung Norden. Durch das Gebiet der Comedi und

42 Vgl. A. Herrmann, Die alten Seidenstraßen zwischen China und Syrien, Berlin 1910, 34; A. Foucher, La vieille route de l' Inde: de Bactre à Taxila, Vol. II, Paris 1947; L. Boulnois, The Silk Road, 1966; M. P. Charlesworth, Trade-routes and commerce of the Roman Empire, Chicago 1974, 102–104; H. W. Haussig, Die Geschichte Zentralasien und der Seidenstrasse, Darmstadt 1983; H. G. Franz (ed.), Kunst und Kultur entlang der Seidenstraße, Graz 1986; J. P. Drège, Viaggi lungo la via della seta, Milano 1986; J. P. Drège, Seidenstraße, Zürich 1993; B. Brentjes, s. v. „Silk Road". Brill's New Pauly. Antiquity volumes edited by: H. Cancik and H. Schneider. <http://brillonline.nl/entries/brill-s-new-pauly/silk-road-e1107020>; M. J. Olbrycht, Die Seidenstraße in der Geschichte der Antike, in Die Krim. Goldene Insel im Schwarzen Meer. Begleitbuch zur Ausstellung in Bonn, Darmstadt 2013, 67–87; J.-N. Robert, De Roma à la Chine. Sur les routes de la Soie au temps des Césars, Paris 2014.

der Sakai hindurch erreichte man den „Stone Tower"– wahrscheinlich das heutige Tashkurgan[43] – von wo aus man dann nach China und in die Stadt Singanfu gelangte.

Im Text der *Geographie*, die diesen Weg beschreibt, kommen die Informationen zum Tower of Stone von Marinus von Tyrus, der von Ptolemaios für übermäßige Maßnahmen kritisiert wurde, wie wir aus den Fragmenten des Maes Titianos wissen können. Er war ein Kaufmann von makedonischer Herkunft, der entlang der Seidenstraße gereist war: Maes Titianos beschreibt den Weg nach Osten mit den Etappen zwischen Hierapolis und dem Euphrat und dann in Richtung der Berge von Pamir und dem Tower of Stone: „Although nothing else can be known with certainty about his text, modern researchers tend to hold Maes Titianos, rightly or wrongly, as an ultimate source for several itineraries in Ptolemy's Book 6."[44]

Aus den Studien zur Seidenstraße ergibt sich, dass Baktra[45] ein Treffpunkt mehrerer Handelsrouten war und nach der Definition von Toynbee sei Baktra auch ein klassisches Beispiel für einen „round-about"-Kreisverkehr gewesen, wo alte und neue Strecken und Handelsrouten sich getroffen und gekreuzt haben.

Von Norden her führten nach Baktra die Handelswege aus Alexandria in Margiana, die dann weiter nach Alexandria in Ariana führten.

Diese Rolle von Baktra könnte die Bedeutung der Region in der hellenistischen Zeit rechtfertigen, da diese einen strategisch wichtigen Außenposten der Eroberungen Alexanders war und zugleich die sicherste Garantie für den Warenaustausch zwischen Ost und West darzustellen schien. Im Zuge dieses Bedeutungszuwachses von Baktra scheint Zariaspa in den Quellen zu verschwinden.

Wenn die hier vorgeschlagene Hypothese zutrifft, dann könnte die Passage des Ptolemaios über Zariaspa und Baktra die endgültige Bestätigung für eine nun aufgetretene Differenzierung sein: von der alten Stadt, die ihren Namen von dem Fluss erhalten hatte, ging die Hauptrolle in der Zeit der Achämeniden an Baktra über, dessen Namen sich von der Region und der gleichnamigen Satrapie ableitete.

Der unmittelbare Zugang zu den Orten und den einheimischen Menschen, die die lokalen Traditionen bewahrten, könnte die ursprüngliche Quelle der Berichte der Alexanderhistoriker sein, die die Namen Zariaspa und Baktra gekannt haben. Die Notwendigkeit, auch Baktra mit einem Fluss in Verbindung zu bringen, dürfte auch den Bactrus bei Plinius erzeugt haben, von dem wir auch das Zeugnis des Curtius Rufus haben.

Die komplexe, aber nie in Vergessenheit geratene Geschichte der Orte, die offenbar in Lokaltraditionen am Leben gehalten und dadurch von den Kaufleuten

43 Nach M. Mollat du Jourdin, J. Desanges, Les routes millénaires, Paris 1988, 118 befindet sich der Ort nicht in Afghanistan.

44 Vgl. A. Dan, *s. v.* „Maes Titianos (2213)". *Die Fragmente der Griechischen Historiker Part V.* Editor in Chief: H.-J.Gehrke. Brill online: http://referenceworks.brillonline.com/entries/fragmente-der-griechischen-historiker-v/maes-titianos-2213-a2213. First appeared online: 2013, mit Bibliographie.

45 A. Toynbee, Between Oxus and Jumna, New York – London 1961, 2; Holt (wie Anm. 3), 31–32.

in Erfahrung gebracht werden konnte, wurde schließlich im Text von Ptolemaios' *Geographie* zusammengeführt.

Damit schließt und erklärt sich gleichermaßen die gebrochene Geschichte eines Ortes, der durch die Dopplung des Namens schließlich in zwei Städte geteilt wird, die von geographischer Seite zu weit entfernt voneinander verortet wurden, um als derselbe Ort betrachtet werden zu können.

Die Gründe für diese Trennung können wir vielleicht in den Wegbeschreibungen von denjenigen finden, die verschiedene Wege mit verschiedenen Orten beschrieben, obwohl es sich doch eigentlich um die gleichen handelte.

THE THEME OF BATHING APHRODITES IN CLASSICAL GREECE: BIRTH OF AN ICONOGRAPHIC PATTERN, DEVELOPMENT, SUCCESS

Antonio Corso

The aim of this article is to consider ancient bathing Aphrodites[1] – especially the three most famous specimens of the bathing goddess: the Cnidia[2] (fig. 1), the Capitoline[3] (fig. 2) and the Crouching Aphrodite[4] (fig. 3) – and to suggest whether this visual evidence represents the bathing Cypris as a genre *topos* or in a specific mythological and / or narrative situation.

Of course the strong link of Aphrodite with water stems from her birth from the sea foam. She is born from the sea already in the 6th Homeric hymn as well as in the *Theogony* of Hesiod (vv. 188–206).[5] Her strong connection to the sea gave her the epikleses *Pontia*,[6] *Euploia*,[7] *Limenia*[8] and *Thalassaia*.[9] Moreover she guaranteed a fair sea – voyage already in the cyclic epos (Stasinus or Hegesias, *Cypria*, *frg.* 10 Evelyn White).

The rite of the bath of the goddess is equally important. She is known to have bathed in three occasions:

1. After her love with Ares. This episode is in Homer, *Odyssey* 8. 362–367:

> 'She, the laughter–loving Aphrodite, went to Cyprus, to Paphos, where is her demesne and fragrant altar. There the Graces bathed her and anointed her with immortal oil, such as gleams upon the gods that are forever. And they clothed her in lovely rairment, a wonder to behold' (transl. Loeb).

1 About the iconographic theme of the bathing female in classical Greece, see especially S. Pfisterer – Haas, 'Maedchen und Frauen am Wasser', *JdI* 117 (2002) 1–79 and A. Staehli, 'Women bathing', S.K. Lucore and M. Truemper (eds.), *Greek Baths and Bathing Culture*, Leuven (2013) 11–21.
I wish to thank Prof. Klaus Geus for his invitation to take part to the conference about Water in antiquity.

2 About the Cnidian Aphrodite, the bibliography is just huge. Here I cite only A. Corso, *The Art of Praxiteles ii*, Rome (2007) 9–187 and A.F. Stewart, 'Hellenistic Freestanding Sculpture from the Athenian Agora 1', *Hesperia* 81 (2012) 267–342.

3 About the Capitoline Aphrodite, see A.F. Stewart, 'A Tale of Seven Nudes', *Antichton* 44 (2010) 12–32.

4 About the Crouching Aphrodite, see E. Ghisellini, 'L'Afrodite Accovacciata tipo Doidalses'. C. Braidotti *et alii* (eds.), *Ou pan ephemeron*, Rome (2009) 663–685.

5 About Aphrodite and the sea, see D. Demetriou, 'Tes pases nautilis phylax. Aphrodite and the Sea', *Kernos* 23 (2010) 67–89.

6 See V. Pirenne – Delforge, *L'Aphrodite grecque*, Athens (1994) 434.

7 See Pirenne – Delforge (note 6) 433–434 and Corso (note 2) 23–24.

8 See Pirenne Delforge (note 6) 175.

9 See Pirenne Delforge (note 6) 434.

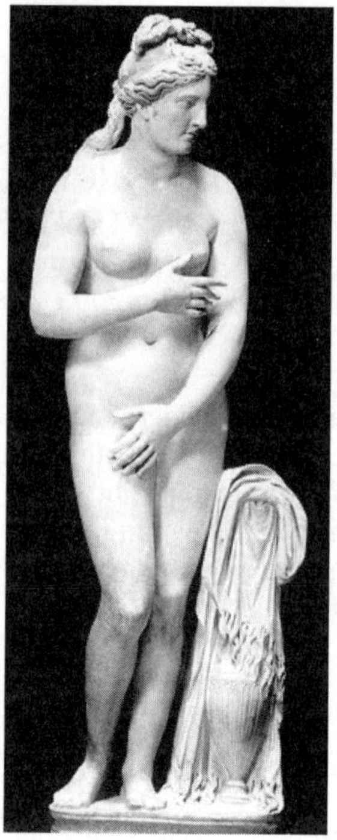

Fig. 1: Cnidian Aphrodite, Belvedere copy, Rome, Vatican Museums, no. 4260.

Fig. 2: Capitoline Aphrodite, Rome, Capitoline Museums, no. 409.

2. Before her love with Anchises. This episode is narrated in Homer, *Hymns* 5. 60–67:

> 'There she went in, and closed the gleaming doors, and there the Graces bathed her and rubbed her with olive oil, divine oil, as blooms upon the eternal gods, ambrosial bridal oil that she had ready perfumed. Her body well clad in all her fine garments, adorned with gold, smile–loving Aphrodite left fragrant Cyprus and sped towards Troy, rapidly making her way high among the clouds' (transl. Loeb).

3. Prior to the judgment of Paris. This bath of the goddess is known through the epigrams *Anthologia Graeca* 9. 619, 633 and 637.See *Anthologia Graeca* 9. 637 (by Anonymous):
'After bathing here divine limbs here, Cypris straightway showed herself to Alexander, and carried off the prize of the apple' (transl. Loeb).
See also *Anthologia Graeca* 9. 619 (by Agathias):
'Now I know, Cytherea, how thou didst conquer in the contest, stealing the vote of Alexander. It was here thou didst dip the body, and so didst find means to overcome Hera who had bathed in the streams of Inachus. It was the bath that won, and I fancy Pallas cried out thus: 'I was conquered by the water, not by the Paphian'' (transl. Loeb).
Finally see *Anthologia Graeca* 9. 633 (by Damocharis):
'This was the bath of Hera, Cypris and Pallas, when they were eager to get the golden apple. And perhaps now Paris will not be their judge, but their image reflected in the silver flood' (transl. Loeb).

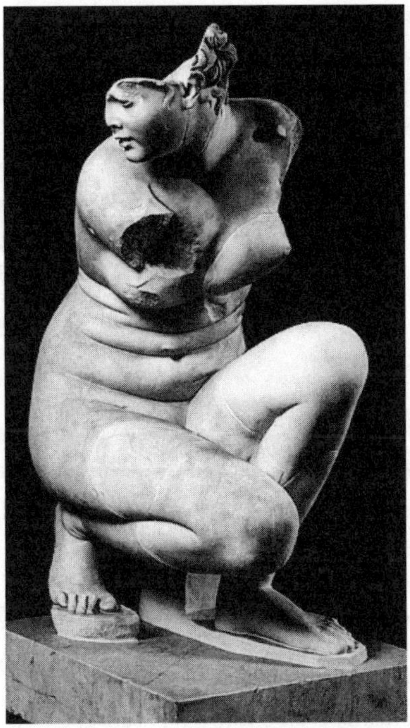

Fig. 3: Crouching Aphrodite from Tibur, Hadrian's villa, Rome, Roman National Museum, no. 108 597.

The occurrence of the bath in the love life of Aphrodite must be the mythological transfer of the ritual bath that maiden had among their preparation for the wedding.[10] This topos is already in Hesiod, *Works and Days*, vv. 519–524:

> 'And (the wind Boreas) does not blow through the tender maiden who stays indoors with her dear mother, unlearned as yet of the works of golden Aphrodite, and who washes her soft body and anoints herself with oil and lies down in an inner room within the house, on a winter day' (transl. Loeb).

References to the ritual bath of girls before their wedding became topical from the 5th c. BC onwards.[11] Here I quote only Euripides, *Iphigenia among the Taurians* 818–819:

> 'Orestes: And do you remember the bath you received from your mother to prepare you for Aulis? Iphigenia: I do. The goodness of the marriage has not effaced this memory' (transl. Loeb).

10 See R.F. Sutton, 'Female Bathers and the Emergence of the Female Nude in Greek Art', C. Kosso and A. Scott (eds.), *The Nature and Function of Water, Bath, Bathing and Hygiene*, Leiden (2009) 61–86.

11 See Stewart (note 3) 14 with note 6.

Fig. 4: Kneeling female bather, perhaps an echo of the Helen of Zeuxis, chous by the Eretria Painter, Sankt Petersburg, Ermitage, no. Pi 1872. 15.

Now let us turn to the Cnidian Aphrodite. This statue had been carved by Praxiteles around 360 BC and set up in Cnidus in the temple and sanctuary of Aphrodite Euploia, of the fair sea voyage.[12]

With the Cnidian Aphrodite the goddess is represented standing, her right hand shields her pubes as an act of *verecundia*, her left arm is picking up her garment from a small *kalpis* which must have contained the perfumed oil with which the Charites anointed the goddess, as it is argued from the above quoted Homeric passages.[13]

Praxiteles probably was inspired for the representation of a naked bathing mythical female by the Helen of Zeuxis (fig. 4) who was both naked as well as probably bathing.[14]

In any case, the representations of naked young females bathing was well rooted in the Athenian vase painting (figs. 5–7) and thus also in the visual culture of the Athenian Praxiteles.[15]

Now the following question arises: is the Cnidia a genre representation or a narrative one? In other words, does she illustrate a mythical episode?

A few considerations may suggest an answer to this question.

12 See note 2.

13 The copyist subtype representing a small *kalpis* near the goddess, named Cnidia Belvedere, is closer to the lost Praxitelean original than the subtype with a large *hydria*, named Cnidia Colonna, as it is revealed by the representation of the Cnidia on coin types of Cnidus (see Corso, cited in note 2).

14 See R. F. Sutton, 'The Invention of the Female Nude', J. H. Oakley and O. Palagia (eds.), *Athenian Potters and Painters*, Oxford (2009) 270–279.

15 See Stewart (note 2) 332–333.

First of all the epigram of the *Greek Anthology* 16. 161, attributed to Plato and presumably very early, explicitly asserts that the Cnidia was represented while she was about to be judged by Paris:

'Neither did Praxiteles nor the chisel work thee (*scil.*: the Cnidian Aphrodite), but so thou standest as of old when thou camest to judgment' (transl. Loeb).

This connection with the Judgment of Paris is repeated in *Anthologia Graeca* 16. 165–166 and 168–170.

See*Anthologia Graeca* 16. 165 (by Evenus):
'Pallas and the consort of Zeus said, when they saw the Cnidian, 'We are wrong in finding fault with Paris'' (transl. Loeb).

See also *Anthologia Graeca* 16. 166 (by Evenus):
'The neatherd alone saw of old on the mountains of Ida her who gained the prize of beauty, but Praxiteles has set her in full view of the Cnidians, having the vote of Paris to attest his skill.' (transl. Loeb).

Moreover the epigram *Anthologia Graeca* 16. 168 (by Anonymous) is also relevant:
'Paris, Anchises and Adonis saw me naked. Those are all I know of, but how did Praxiteles contrive it?' (transl. Loeb).

The poem *Anthologia Graeca* 16. 169 (also by Anonymous) should not be forgotten:
'Gaze from every side at the divine beauty of the foam–born Paphian and you will say, 'I applaud the Phrygian's judgment.' Again when you look at the Attic Pallas you will cry out, 'It was just like a neatherd for Paris to pass her by' ' (transl. Loeb).

Finally, the epigram *Anthologia Graeca* 16. 170 (by Hermodorus) confirms the close link of the Cnidia with the judgment of Paris:
'When you see, stranger, the Cnidian Cytherea, you would say this, 'Rule alone over mortals and immortals,' but when you look at Pallas in the city of Cecrops boldly brandishing her spear you will exclaim, 'Paris was really a bumpkin'' (transl. Loeb).

At the time when the Cnidia was created, Cnidus was in the satrapy of Caria, ruled by Mausolus.[16] The satraps of western Asia Minor regarded the kingdom of Troja the mythical ancestor of their own states. Thus, emphasis was given to the Amazons, allies of the Trojans, in the carved decoration of the Mausoleum.[17]

The satrap of Dascylium Artabazus commissioned a statue of Paris, set up in the *agora* of Parium, a town under his satrapy, where Paris was thought to have spent his youth.[18] The statue probably is the Paris of Euphranor, who represented the Phrygian hero as the slayer of Achilles, as Pliny 34. 77 observed.[19]

16 See A. Corso, 'The Cult and Political Background of the Knidian Aphrodite', *Proceedings of the Danish Institute at Athens* 5 (2007) 173–197.

17 See B. F. Cook, *Relief Sculpture of the Mausoleum at Halicarnassus*, Oxford (2005) 42–64.

18 See Athenagoras, *Legatio* 26. 3.

19 Considerations in favour of the identification of the statue of Paris at Parium with that made by Euphranor are suggested in A. Corso, *The Art of Praxiteles iv*, Rome (2013) 66–67.

The same satrap of Dascylium commissioned from Scopas the Apollo Smintheus at Chryse, commemorative of the pestilence sent by Apollo against the Greek army at Troja, an episode known from the 1st song of the Iliad.[20]

Thus it makes sense that the Cnidian Aphrodite illustrated the bath of the goddess prior to the judgment of Paris.

Let us now turn to the Capitoline Aphrodite. The goddess shields her breasts with her right arm and her pubes with her left hand. Her drapery is thrown on a *loutrophoros*, a typical Athenian vase for the wedding's bath.[21]

The original statue of the Capitoline copyist tradition is described by Ovid, *Ars Amatoria* 2. 613–614:

> 'Venus herself, as soon as she lays aside her robes, half stooping covers with her left hand her pubes' (transl. Loeb with amendments).

Ovid describes this statue in a poetic work in which he mentions repeatedly the *Appiades*,[22] one of the *monumenta Asini Pollionis*.[23] Thus this Aphrodite probably is the statue of the love goddess by Cephisodotus the Younger – Praxiteles' elder son – which was one of these *monumenta*.[24]

Is the Capitoline Aphrodite a genre representation of the bathing goddess or a narrative one? Since the *loutrophoros* is an Athenian vase used for the ritual baths of brides, we should search for a convincing interpretation inside the Athenian life.

The type of *loutrophoros* adopted for the Capitoline creation disappears around 300 BC,[25] thus this detail confirms the early date of the original statue, still in the Zeitgeist of the New Comedy.

Since the only marriage of Aphrodite was with Hephaestus but this married *status* was never celebrated in the literary or visual realms, a mythical context for the Capitoline Aphrodite is unlikely.

Thus the statue offers a transfer of the ritual baths of Athenian brides to the divine realm. Aphrodite as love goddess is here conceived as the bride par excellence. The shielding of her beauty reveals the embarrassment of the bride in the moment prior to her first love with her groom. The representation of a typical and often occurring episode of the Athenian society is trendy in the age of the New Comedy – a comedy made by common characters – and of the 'Characters' of Theophrastus, who are also permanent components of any society.[26]

20 See G. Calcani, *Skopas di Paros*, Rome (2009) 71–72. About the *Smintheion*, see C. Oezguenel, 'The Smintheion Sanctuary', F. Oezdem, *Canakkale*, Constantinople (2012) 274–293.

21 See Stewart (note 3) 15. See also F. Hildebrandt, 'Die Loutrophoros', K. Hitzl (ed.), *Kerameia*, Kiel (2011) 96–99.

22 See Ovid, *Ars amatoria* 1. 81–86 and 3. 451–545. See also Ovid, *remedia amoris* 659–660.

23 About these *monumenta*, see A. Bravi, *Griechische Kunstwerke im politischen Leben Roms und Konstantinopels*, Berlin (2014) 95–110.

24 See Pliny 36. 24.

25 See Stewart (note 3) 15.

26 This is argued especially from Attic late classical funerary representations: see *e. g. J.* Burnett Grossman, 'Funerary Sculpture', *The Athenian Agora* 35 (2013).

Fig. 5: Top: Bridal bath, pyxis, New York, The Metropolitan Museum of Art, no. 1972. 118. 148. Fig. 6: Left: Bathing nymph discovered by Silens. Bell-krater attributed to the Dinos Painter. Cracow, University, Institute of Archaeology 331. Fig. 7: Right: Peleus and Thetis. Pelike attributed to the Painter of Athens 1472. Toledo Museum of Art, no. 1993. 49.

It should not be forgotten that Cephisodotus the Younger together with his younger brother Timarchus fleshed out also the portrait of the prince of the poets of the New Comedy: Menander.[27]

Thus he and his patrons may have absorbed the tendency to transfer common moments of the society to the universal realm, from the *hic et nunc* to the *ubique et simper*.

Regarding the crouching Aphrodite, defined by Pliny *Venus lavans sese*,[28] I summarize here the vicissitudes which probably led to this creation. The king of Bithynia Nicomedes I around 260 BC asked the debt ridden city of Cnidus to buy the Cnidian Aphrodite in order to set it up in his new capital Nicomedia, offering in exchange to settle the huge debt of Cnidus. The Cnidians refused.[29] Thus he probably

27 See *e.g.* C. Vorster, 'Bildnis des Menander', K. Knoll (ed.), *Katalog der antiken Bildwerke* iii, Munich (2013) 74–76.
28 See Pliny 36. 35.
29 See Pliny 36. 20–21.

asked the local Bithynian sculptor Daedalsas to carve another bathing Aphrodite – the Crouching one – which was set up in Nicomedia and later moved to Rome.[30]

Thus the statue must have been endowed with the same interpretation which characterized the Cnidia: the bath is that which occurred prior to the judgment of Paris. The closeness of Nicomedia to the Trojan region strengthens this probability. The Romans may have brought the statue to Rome because Aphrodite was represented with this image as the protectress of Troja, thus of Rome, thus of the *gens Julia*, thus worthy to be settled in the *porticus Octaviae*.[31]

In conclusion, Aphrodite is first of all represented bathing in a mythological context, that of the *judicium Paridis*, in the context of the prevalence of Asia forwarded by the satrapic ideology.

Then, in keeping with the study of universal patterns of the society typical of the Aristotelism, the bathing goddess personified the bathing bride prior to her first love with her groom.

Finally, the bathing pattern is again understood to be a moment of the Trojan saga: emphasized in Rome and described in epigrams which stress that the water of the bath where Aphrodite merged her body was the real winner of the competition.[32]

30 See A. Corso, 'Nicomede I, Dedalsa e le Afroditi nude al bagno', *NumAntCl* 19 (1990) 135–160 and Ghisellini (note 4).

31 See Bravi (note 23) 149–163.

32 See the above quoted epigrams *Anthologia Graeca* 9. 619. 633 and 637.

DIE BAUINSCHRIFT AUF DEM DRUCKROHRAQUÄDUKT IN PATARA. EIN BEITRAG ZUR HISTORISCHEN KATASTROPHENFORSCHUNG*

Philipp Deeg

EIN BISLANG UNBEKANNTES ERDBEBEN IM JAHRE 69 N. CHR.

Vor einigen Jahren hat Sencer Şahin die Bauinschrift – genauer eine Doppelinschrift, deren Text auf West- und Ostseite identisch ist – vom Druckrohraquädukt in Patara vorgelegt.[1] Auch Werner Eck, Helmut Engelmann und Havva İşkan-Işik haben sich kurz darauf mit dieser Inschrift beschäftigt.[2] Sie ist aufschlussreich hinsichtlich der Wasserversorgung Pataras, mehr aber noch für die historische Katastrophenforschung.[3] Am Schluss der Inschrift wird in Kürze die Baugeschichte des Aquädukts beschrieben: Es wurde während der Herrschaft des Claudius unter der Statthalterschaft des Vilius Flaccus begonnen und ebenfalls zur Zeit des Claudius, aber vom Statthalter Eprius Marcellus fertiggestellt. Zuvor gibt der größere Teil der Inschrift,

* Den größten Anteil, dass aus einer Idee ein Aufsatz wurde, hatten Holger Sonnabend und Jörn Kobes. Der Dank des Verfassers gilt aber ebenso Frank Daubner, Jonas Scherr und insbesondere Werner Eck, die die Lektüre einer ersten – höchst unbefriedigenden – Fassung des vorliegenden Beitrags auf sich nahmen und wertvolle, kritische wie ermutigende Rückmeldungen gaben. Nicht zuletzt halfen auch die anonymen Gutachter durch konstruktive Hinweise, die so weit, aber auch behutsam als möglich eingearbeitet wurden. Klaus Geus schließlich gebührt Dank für die Aufnahme des Beitrags ins Peer-Review-Verfahren. Für etwaige Unstimmigkeiten tragen die Genannten freilich keine Verantwortung.

1 Şahin, Sencer: Die Bauinschrift auf dem Druckrohraquädukt von Delikkemer bei Patara. In: Schuler, Christof (Hrsg.): Griechische Epigraphik in Lykien. Eine Zwischenbilanz. Wien 2007, 99–109; vgl. auch Şahin, Sencer: Der neronische Leuchtturm und die vespasianischen Thermen von Patara. In: Gephyra 5 (2008), 1–32, hier 9–18. – Da der Stein auf der Westseite breiter ist als auf der Ostseite, erstreckt sich der Text auf der Westseite über zwölf, auf der Ostseite hingegen über 15 Zeilen.

2 Eck, Werner / Engelmann, Helmut / İşkan-Işik, Havva: Der Leuchtturm von Patara und Sex. Marcius Priscus als Statthalter der Provinz Lycia von Nero bis Vespasian. In: ZPE 164 (2008), 91–121, hier 115–118; vgl. İplikçioğlu, Bülent: Die Provinz *Lycia* unter Galba und die Gründung der Doppelprovinz *Lycia et Pamphylia* unter Vespasian. Veröffentlichungen der kleinasiatischen Kommission Nr. 25. In: AAWW 143 (2008), 5–23, hier 7, Nr. 05; SEG 57 (2007), 1673. – Des schnelleren Zugangs wegen findet sich der Text der Inschrift auch im Anhang dieses Aufsatzes.

3 Noch vor nicht allzu langer Zeit konstatierte Sonnabend, Holger: Epigraphische Zeugnisse für antike Naturkatastrophen. In: Bianchetti, Serena (Hrsg.): Poikilma. Bd. 2. La Spezia 2001, 1219–1224, hier 1221, dass in antiken Bauinschriften, die Naturkatastrophen erwähnen, „die Rede nur von einzelnen Gebäuden, nicht aber von Auswirkungen auf Privathäuser oder gar die städtische Infrastruktur“ sei. Die patarensische Aquäduktinschrift drückt uns hierzu immerhin ein Puzzlestück in die Hand.

mit einigen technischen Details versehen, Auskunft über Reparaturmaßnahmen, die der Statthalter Sex. Marcius Priscus unter der Schirmherrschaft Vespasians wegen Erdbebenschäden durchführen ließ. Es findet sich auch eine Zeitangabe, wieviele Monate Patara ohne die Leitung auskommen musste. An eben dieser Stelle ergaben sich unterschiedliche Lesarten. Şahin erkennt bei der Zeitangabe ein Lambda (μῆνας λ′), sodass er auf eine Reparaturzeit von 30 Monaten kommt. Eck u. a. erhalten durch ihre Lesung μῆνας δ′, also eines Delta, die Spanne von vier Monaten.[4] Während Şahin vorwiegend mit Fotographien und am noch stark versinterten Stein arbeitet, entfernten Eck u. a. den Sinter so weit als möglich für ihre Untersuchung.[5] Die daraus resultierende revidierte Lesart bezweifelt Şahin und fordert daher weitergehende „philologische [oder] historische Erklärungen“ für die Veränderung seiner Entzifferung.[6] Solche gleichsam äußeren Argumente, die gegen Şahins Lesung sprechen, lassen sich durchaus aufzeigen. Zunächst enthält die Titulatur Vespasians in der Inschrift das Gentiliz Flavius. Eck konnte bereits vor geraumer Zeit anhand der Thermeninschrift in Patara überzeugend nachweisen, dass eine solche Titulatur auf die früheste Regierungszeit Vespasians hindeutet.[7] Dies bestätigen einschlägige Arbeiten zur Kaisertitulatur.[8] Obwohl Şahin noch immer skeptisch ist,[9] kann seine Lesung von 30 Monaten kaum aufrechterhalten werden. Denn das Erdbeben, auf das er die in der Inschrift genannten Reparaturarbeiten zurückführt – und worin

4 Şahin: Bauinschrift (Anm. 1), 103; Eck u. a.: Leuchtturm (Anm. 2), 116; vgl. İplikçioğlu: Provinz (Anm. 2), 12, Anm. 39; SEG 57 (2007), 1673.

5 Şahin: Bauinschrift (Anm. 1), 102; Şahin: Leuchtturm (Anm. 1), 9 m. Anm. 28; Eck u. a.: Leuchtturm (Anm. 2), 116. – Dem Verf. war eine Betrachtung vor Ort nicht möglich.

6 Şahin: Leuchtturm (Anm. 1), 10. – Seiner Lesart folgt jüngst noch Jones, Christopher P.: Earthquakes and Emperors. In: Kolb, Anne (Hrsg.): Infrastruktur und Herrschaftsorganisation im Imperium Romanum. Berlin 2014, 52–65, hier 59: „It is tempting to prefer the higher number, […].“ Leider begründet Jones diese Ansicht nicht näher.

7 Eck, Werner: Die Legaten von Lykien und Pamphylien unter Vespasian. In: ZPE 6 (1970), 65–75.

8 Buttrey, Theodore V.: Documentary Evidence for the Chronology of the Flavian Titulature. Meisenheim am Glan 1980, 6, 8–10; Isaac, Benjamin: Vespasian's Titulature in A. D. 69. In: ZPE 55 (1984), 143–144; Kienast, Dietmar: Römische Kaisertabelle. Grundzüge einer römischen Kaiserchronologie. Darmstadt 52011, 108; vgl. İplikçioğlu: Provinz (Anm. 2), 11 mit epigraphischen Parallelbelegen. – Siehe auch Ruff, Christina: Ne quid popularitatis augendae praetermitteret. Studien zur Herrschaftsdarstellung der flavischen Kaiser. Marburg 2012, 81–87, die zwar keine Datierung anbietet, wann Vespasian das Gentiliz Flavius zugunsten des Gentilnomens Caesar fallen ließ, aber ebd., 82 (Hervorh. im Orig.; d. Verf.) in aller Deutlichkeit hervorhebt: „Die als *Flavische* bezeichneten Kaiser legen [sic] also Wert darauf, nicht als Flavier, sondern als Mitglieder des Herrscherhauses und legitime Nachfolger des Augustus gesehen zu werden.“

9 Şahin: Bauinschrift (Anm. 1), 107; Şahin: Leuchtturm (Anm. 1), passim; ferner Şahin, Sencer: Kaiserbauten und Kaiserehrungen in Patara. In: Winter, Engelbert (Hrsg.): Vom Euphrat zum Bosporus. Kleinasien in der Antike. Bd. 2. Bonn 2008, 597–610, bes. 598. – Gegen Şahin jetzt scharf, aber inhaltlich überzeugend Eck, Werner: Die Bauinschrift der neronischen Thermen in Patara. Zur methodischen Auswertung einer partiell eradierten Inschrift. In: ZPE 166 (2008), 269–275. – Dagegen Şahin: Leuchtturm (Anm. 1), 19–22.

Eck u. a. ihm zumindest nicht widersprechen[10] –, lässt sich, folgt man Cassius Dios Schilderung der Ereignisse, nur vage auf den Zeitraum zwischen der Erhebung des Vindex und dessen Tod, also auf März bis Mai 68, datieren.[11] Bei 30 Monaten Wiederaufbauzeit könnte die Wasserleitung also frühestens im September, spätestens im November 70 wiederhergestellt gewesen sein, mithin zu einer Zeit, als sich Vespasians offizielle Titulatur ohne Gentiliz längst durchgesetzt haben musste.[12] Indem Şahin aber für gesichert hält, „daß die Wiederinbetriebnahme der Wasserleitung zwischen Juli 69 (Beginn der Herrschaft Vespasians) und der ersten Hälfte des Jahres 70 erfolgte, weil Vespasian in der Inschrift noch als Flavius erscheint“,[13] schwächt er seine frühere Argumentation. Wenn neben dem epigraphischen Befund nach Eck u. a. somit auch ergänzende Überlegungen gegen Şahins Lesung sprechen, lassen die Zeitangabe von vier Monaten und die Nennung Vespasians (mit dem nur in seiner frühesten Regierungszeit belegten Gentiliz Flavius) als Initiator der Baumaßnahme nur einen Schluss zu: Nach dem Beben von 68, bei dem es sich übrigens wohl eher um ein Seebeben mit anschließendem Tsunami handelte,[14] ereignete sich schon um die Mitte des Jahres 69 ein weiteres Erdbeben. Diesen Schluss zieht auch Şahin.[15] Er verwirft ihn jedoch wieder, weil ein solches Beben nicht belegt sei.[16] Die antiken Quelleninformationen über Erdbeben oder vergleichbare Katastrophen sind jedoch meist sehr dürr, wenn sie überhaupt vorhanden sind. Es ist davon auszugehen, dass aus verschiedenen, nicht vollumfänglich erklärbaren Gründen nur ein Bruchteil aller antiken Katastrophen einschließlich Erdbeben überliefert ist.[17] Die Bauinschrift kann demnach gerade die bislang fehlende Information darstellen.[18] Nach diesem so belegten Beben leitete der Statthalter Priscus die Reparaturmaßnahmen ein. Zur Finanzierung wurden die angesparte

10 Şahin: Bauinschrift (Anm. 1), 105; Şahin: Kaiserbauten (Anm. 9), 598, Anm. 7; Şahin: Leuchtturm (Anm. 1), 9; vgl. eher vage Eck u. a.: Leuchtturm (Anm. 2), 115; indirekt zustimmend Eck: Bauinschrift (Anm. 9), 273; siehe auch İplikçioğlu: Provinz (Anm. 2), 12.

11 Cass. Dio 63,26,5; Kienast: Kaisertabelle (Anm. 8), 101.

12 İplikçioğlu: Provinz (Anm. 2), 12, Anm. 39.

13 Şahin: Leuchtturm (Anm. 1), 11; vgl. ebd., 10. – Der hier angenommene Zeitraum, in dem das Gentiliz als Teil der Titulatur plausibel sei, ist mit an Sicherheit grenzender Wahrscheinlichkeit zu großzügig veranschlagt wird, wie aus den Verweisen weiter oben hervorgeht.

14 Cass. Dio 63,26,5; Or. Sib. 4,109–113; Şahin: Bauinschrift (Anm. 1), 105; Şahin: Leuchtturm (Anm. 1), 12; İplikçioğlu: Provinz (Anm. 2), 12; Guidoboni, Emanuela/Comastri, Alberto/Traina, Giusto: Catalogue of ancient earthquakes in the Mediterranean area up to the 10th century. Rom 1994, 211.

15 Vgl. Şahin: Leuchtturm (Anm. 1), 11; vgl. Şahin: Bauinschrift (Anm. 1), 103. – Eck u. a.: Leuchtturm (Anm. 2), 116 und 118 stellen hierzu keine expliziten Überlegungen an.

16 Şahin: Leuchtturm (Anm. 1), 11; vgl. Şahin: Bauinschrift (Anm. 1), 103 und 105.

17 Vgl. Sonnabend, Holger: Naturkatastrophen in der Antike. Wahrnehmung, Deutung, Management. Stuttgart/Weimar 1999, 43 sowie ausführlich Waldherr, Gerhard H.: Antike Quellen zu Erdbeben und ihre Problematik. In: Waldherr, Gerhard H./Smolka, Anselm (Hrsg.): Antike Erdbeben im alpinen und zirkumalpinen Raum. Stuttgart 2007, 15–22.

18 Siehe auch Sonnabend: Zeugnisse (Anm. 3), 1220: „Inschriften sind Zeugnisse für den Eintritt von Naturkatastrophen. Auch wenn sie nicht über das Ereignis als solches informieren wollen, liefern sie doch einen Beleg dafür, daß eine Naturkatastrophe stattgefunden hat.“

Kopfsteuer[19] und ein Zuschuss des lykischen Bundes verwendet.[20] Da Priscus sich wahrscheinlich „Vespasian sofort angeschlossen hat[te], ja vielleicht [...] bereits vor der Ausrufung zum Kaiser durch Mucian etwa ins Vertrauen gezogen worden" war,[21] wurde folgerichtig Vespasian als Veranlasser und also Legitimationsinstanz der Maßnahme[22] vermerkt. Damit stützten sich Kaiser und Statthalter gegenseitig: Letzterer konnte sich auf eine höhere Autorität beziehen. Und indem er dies tat, half er mit, ebendiese Autorität des Usurpators Vespasian – zumindest vor Ort – zu etablieren bzw. zu stabilisieren.

An diesen Befund schließen sich einige Fragen an, die nachfolgend behandelt werden sollen.

EIN ERDBEBEN ODER EINE ERDBEBENSERIE?

Auf der Inschrift begegnet uns die Pluralformulierung σεισμοῖς. Sie dürfte weder Zufall noch Versehen sein, wie eine Inschrift aus Xanthos belegt.[23] Jüngst deutet Şahin dies als Hinweis auf eine Erdbebenserie.[24] Das ist eine durchaus mögliche

19 So ist die Formulierung συντηρηθέντα χρημάτα auf der Inschrift mit Eck u. a.: Leuchtturm (Anm. 2), 118 zu verstehen.

20 Warum die Höhe dieses Zuschusses nicht eingemeißelt, sondern „später mit Farbe nachgetragen" wurde (Eck u. a.: Leuchtturm [Anm. 2], 118; vgl. Şahin: Bauinschrift [Anm. 1], 108; İplikçioğlu: Provinz [Anm. 2], 7), kann nur vermutet werden. Vielleicht kommt hier die Hoffnung zum Ausdruck, Vespasian werde nach seinem – zumindest wohl von Priscus erhofften – Sieg noch eine Summe beisteuern. Falls diese Hypothese zutrifft, muss offen bleiben, ob besagte Hoffnung erfüllt wurde. Der Geldbedarf des Reiches spricht entschieden dagegen (Suet. Vesp. 16,3). Zudem stellte schon der Verzicht des Kaisers auf ihm zustehende Gelder einen Akt der Großzügigkeit dar (Veyne, Paul: Brot und Spiele. Gesellschaftliche Macht und politische Herrschaft in der Antike. Frankfurt/New York 1992, 544; vgl. Jones: Earthquakes [Anm. 6], 60); die Kopfsteuer war aber prinzipiell eine an Rom abzuführende Abgabe (DNP 11 [2001], s. v. Steuern, 982 [Galsterer]; Schwartz, Hertha: Soll oder Haben? Die Finanzwirtschaft kleinasiatischer Städte in der Römischen Kaiserzeit am Beispiel von Bithynien, Lykien und Ephesos [29 v. Chr. – 284 n. Chr.]. Bonn 2001, 280 und 281). Für einen weiteren Zuschuss spricht lediglich, dass Vespasian im Fortlauf seiner Herrschaft großes Augenmerk auf die Wasserversorgung legte (Andreu, Javier: Das Bild Vespasians und Titus' in den Städten des römischen Imperiums: Öffentliches Bauwesen und öffentliche Ehrungen. Epigraphische Beweise. In: Kramer, Norbert/Reitz, Christiane (Hrsg.): Tradition und Erneuerung. Mediale Strategien in der Zeit der Flavier. Berlin/New York 2010, 361–396, hier 377). Entschieden davon überzeugt, dass Vespasian keine Mittel nachschoss, ist Jones: Earthquakes (Anm. 6), 59.

21 Eck: Legaten (Anm. 7), 68.

22 Dazu allgemein Veyne: Brot und Spiele (Anm. 20), 543–545.

23 Balland, André: Fouilles de Xanthos VII. Inscriptions d'époque impériale du Letôon. Paris 1981, 29–31, Nr. 12; vgl. İplikçioğlu: Provinz (Anm. 2), 8, Nr. 11, und 13. – Die Rekonstruktion dieser Inschrift dürfte unstrittig sein und wird jedenfalls von Eck u. a.: Leuchtturm (Anm. 2), 116 sowie Şahin: Bauinschrift (Anm. 1), 106 m. Anm. 37 gleichermaßen akzeptiert. – Einzig die Frage, ob die Titulatur Vespasians mit dem Gentiliz Caesar oder Flavius zu rekonstruieren ist, ist nicht eindeutig geklärt; siehe İplikçioğlu: Provinz (Anm. 2), 8, Nr. 11 (Caesar), und 20 (Flavius).

24 Şahin: Leuchtturm (Anm. 1), 12.

Interpretation.[25] Diese Serie will Şahin im Jahre 67 einsetzen und bis 68 andauern lassen.[26] Dieser Vorschlag kann nicht überzeugen. Wenn er für ein Beben im Jahre 69 mehr Belege fordert als die Inschrift am Aquädukt, wären umgekehrt auch Beweise für seine These wünschenswert. Er versucht, diese beizubringen, indem er auf Dio verweist. Dabei entgeht ihm allerdings der vom antiken Historiographen gesetzte chronologische Rahmen.[27] Die mögliche Erdbebenserie also 68 einsetzen und bis 69 andauern zu lassen, erscheint plausibler. Aber auch diese Vermutung hängt an einer Zusatzannahme: dass die Katastrophe von 68 keine oder zumindest keine allzu schweren Schäden verursacht hatte, sodass die Stadt, wenn überhaupt, nur sehr kurze Zeit ohne Wasserversorgung auskommen musste und eine provisorische Reparatur reichte. Denn anderenfalls hätte Patara, das laut Şahin schon vor 68 knapp bei Kasse war,[28] aus finanzieller Not kaum eine andere Wahl gehabt, als einen Hilfsantrag beim Kaiser zu stellen, was aber gemäß Şahins Lesung der Inschrift gerade nicht passiert sein soll.[29]

Eck u. a. interpretieren die Pluralformulierung auf der Inschrift umstandslos als „Erdbeben, das mit mehreren Nachbeben einherging".[30] In seinen früheren Publikationen zum Thema geht Şahin in dieselbe Richtung, wenn er einmal meint, „daß das Bauwerk vom Erdbeben [sic!] mehrmals erschüttert wurde",[31] ein andermal Beschädigungen „durch ein Erdbeben (σεισμοῖς) [sic!]" erkennt[32] und zuletzt vermutet, „daß mit σεισμοῖς das verheerende Erdbeben [sic!] des Jahres 68 n. Chr. gemeint" sei.[33] In diesem Fall, der den Vorzug besitzt, keine Zusatzannahmen zu erfordern, besteht schlicht kein direkter Zusammenhang inhaltlicher oder kausaler Art zwischen dem Desaster von 68 und der Inschrift von 69.

25 Vgl. Balland: Xanthos (Anm. 23), 30 m. Anm. 134. Zudem ist auch hier die oben erwähnte Quellensituation bei Naturkatastrophen zu bedenken.

26 Şahin: Leuchtturm (Anm. 1), 12.

27 Şahin: Leuchtturm (Anm. 1), 12. – Zweifellos ist die Nichtbeachtung des dionischen Zeitrahmens ein Versehen, da Şahin triftige Gründe, am Historiographen zu zweifeln, sicher ausdrücklich genannt hätte. Ähnlich dürfte es sich bei Guidoboni u. a.: Catalogue (Anm. 14), 211, Nr. 97 verhalten, da sie keine näheren Angaben als das Jahr nennen. Dieses legen sie aber klar auf 68 fest. – Anders noch Şahin: Bauinschrift (Anm. 1), 105, wo ebenfalls auf Dio 63,26,5 Bezug genommen, dessen Schilderung jedoch eindeutig ins Jahr 68 gesetzt wird. – Şahin hätte vielleicht versuchen können, eine Verbindung zum Erdbeben auf Kreta, von dem Dictys *FGrHist* 49, T 4,3 für 66 berichtet, herzustellen. Möglicherweise schreckte er, was völlig nachvollziehbar wäre, vor den chronologischen Fallstricken zurück; vgl. dazu auch Panessa, Giangiacomo: Fonti Greche e Latine per la Storia dell'ambiente e del clima nel mondo Greco. Bd. 1. Pisa 1991, 347, T 53c–e, und 348–350; Guidoboni u. a.: Catalogue (Anm. 14), 193, Nr. 91.

28 Şahin: Leuchtturm (Anm. 1), 15.

29 Şahin: Leuchtturm (Anm. 1), 15; siehe dazu auch die Ausführungen weiter unten.

30 Eck u. a.: Leuchtturm (Anm. 2), 116.

31 Şahin: Bauinschrift (Anm. 1), 105; vgl. ebd., 106.

32 Şahin: Kaiserbauten (Anm. 9), 598.

33 Şahin, Sencer: Kragos Oros, Titanis Petra und der Apollontempel von Patara. Lokalisierungsversuche in der historischen Geographie Lykiens. In: Olshausen, Eckart/Sauer, Vera (Hrsg.): Die Landschaft und die Religion. Stuttgart 2009, 337–352, hier 342.

DIE LESART VON ZEILE 9 (WESTSEITE) BZW. ZEILE 11 (OSTSEITE)

Außer der Entzifferung der Zeitangabe ist ein weiterer Punkt zwischen Şahin und Eck u. a. strittig. In seiner Publikation belässt Şahin die betreffende Stelle in Majuskelschrift (ΚΑΤΑΝΛΡΑ), weil er die Entzifferung für zweifelhaft hält.[34] Eck u. a. lesen hier κατ' ἄνδρα, sodass der Sinn der Aussage wäre, dass keine Sondersteuer zum Wiederaufbau eingetrieben worden sei.[35] Şahin unterbreitet hingegen den Vorschlag, hier einen „Genitivus Absolutus mit temporaler oder kausaler Bedeutung" zu erkennen.[36] Die Inschrift würde demnach besagen, der Kaiser resp. Statthalter habe den Wiederaufbau aus der Kopfsteuer finanziert, weil die Stadt beim Kaiser keinen Antrag auf Wiederaufbau gestellt hatte.[37]

Welche dieser Lesarten korrekt ist, lässt sich hier nicht klären. Dies mag bedauerlich sein, ist an dieser Stelle jedoch zu verschmerzen, da, wie gleich zu sehen sein wird, inhaltlich beide plausibel sind und v. a. der Annahme eines Erdbebens im Jahre 69 nicht widersprechen.

Şahins Interpretation wirft die Frage auf, weshalb die Stadt keinen Hilfsantrag gestellt hatte und der Kaiser nicht seine Freigiebigkeit demonstrierte, wie es sonst in solchen Fällen durchaus üblich war.[38] Als um die Mitte des Jahres 69 das Beben stattfand, war eine solide Reparaturmaßnahme offenbar unerlässlich. Doch an wen sollte Patara sich mitten im Vierkaiserjahr um Finanzhilfe wenden? Vitellius oder Vespasian? Zudem hatten beide Thronaspiranten wohl Dringenderes zu tun, als sich um die Schäden in Lykien zu kümmern. Der Vespasian-Anhänger Marcius Priscus sah offensichtlich keine andere Möglichkeit, als mit der Kopfsteuer einen Wiederaufbau zu finanzieren.[39]

34 Şahin: Bauinschrift (Anm. 1), 103.

35 Eck u. a.: Leuchtturm (Anm. 2), 118; SEG 57 (2007), 1673 sowie İplikçioğlu: Provinz (Anm. 2), 7, Nr. 05 folgen dieser Lesart; vgl. auch die Herausgeberanmerkung von Christof Schuler bei Şahin: Bauinschrift (Anm. 1), 109; ferner Şahin (Anm. 1): Leuchtturm, 13; kritisch ebd., 12 sowie Şahin (Anm. 1): Bauinschrift, 103.

36 Şahin: Leuchtturm (Anm. 1), 15. – Zu betonen ist, dass Şahin diesen Vorschlag sehr zurückhaltend unterbreitet. Obwohl er ihn für sinnvoller hält als die Gegenposition, plädiert er schließlich für die ursprünglich von ihm gegebene, nicht klar aufgelöste Lesart; ebd., 17.

37 Şahin: Leuchtturm (Anm. 1), 16; vgl. bereits Şahin: Bauinschrift (Anm. 1), 103.

38 Kloft, Hans: Liberalitas principis. Herkunft und Bedeutung. Studien zur Prinzipatsideologie. Köln/Wien 1970, 118–120; Sonnabend: Naturkatastrophen (Anm. 17), 153 und 217; Sonnabend: Zeugnisse (Anm. 3), 1222; Meier, Mischa: Roman Emperors and ‚Natural Disasters' in the First Century AD. In: Janku, Andrea/Schenk, Gerrit Jasper/Mauelshagen, Franz (Hrsg.): Historical Disasters in Context. Science, Religion, and Politics. New York/London 2012, 15–30, bes. 15 und 16; Winter, Engelbert: Staatliche Baupolitik und Baufürsorge in den römischen Provinzen des kaiserzeitlichen Kleinasien. Bonn 1996, 94.

39 Dies könnte wiederum darauf hindeuten, dass Patara auch nicht im Stande war, selbst für die Reparatur aufzukommen. Wie das? Wir wissen nicht, welche Schäden das Desaster von 68 in Patara angerichtet hat. Dass Nero nicht eingriff, ist kein allzu zwingendes Indiz, da Katastrophenhilfe bei ihm offenbar keine hohe Priorität genoss. Das bekannteste Beispiel hierfür mag das Erdbeben in Laodikeia sein, von dem Tacitus berichtet, die Laodikeer hätten den Wiederaufbau selbst finanziert (Tac. ann. 14,27,1; vgl. Meier: Emperors [Anm. 38], 21; Jones: Earth-

Die Lesart von Eck u.a. verwirft Şahin, weil ihm nicht einleuchtet, warum eigens erwähnt worden sein sollte, dass bestimmte Mittel *nicht* herangezogen wurden.[40] Diesem Einwand kann mit einen Hinweis Suetons begegnet werden, wonach Lykien von Vespasian später die *libertas* entzogen worden sei.[41] Da Priscus aber nachweislich seit Nero Statthalter dieser Provinz war,[42] muss hinter dieser Freiheit eher ein Privileg, etwa in Form eines Steuernachlasses, stecken.[43] Damit wäre es durchaus denkbar und sinnvoll, dass der Statthalter eigens betonte, nur mit angesparten, d.h. längst eingezogenen Mitteln gearbeitet, mithin das Steuerprivileg nicht aufgehoben oder aufgeweicht zu haben.[44]

quakes [Anm. 6], 62). Dass dies anscheinend funktionierte, spricht für den Wohlstand der Stadt bzw. ihrer Honoratiorenschicht (Sonnabend: Naturkatastrophen [Anm. 17], 230; Winter: Baupolitik [Anm. 38], 95). Im Falle Pataras mag hinzukommen, dass Nero trotz seiner von Dio beschriebenen, geradezu demonstrativ wirkenden Sorglosigkeit angesichts des Vindex-Aufstandes andere Sorgen hatte als das Leid in Lykien (Cass. Dio 63,26). Möglicherweise musste Patara sich also wie Laodikeia selbst helfen. Das führte dann vielleicht erst zu der von Şahin festgestellten schwierigen Finanzlage der Stadt (Şahin: Leuchtturm [Anm. 1], 15), die 69 das Eingreifen des Statthalters nötig machte. – Vgl. zu Laodikeia auch Toner, Jerry: Roman Disasters. Malden 2013, 50 und 55, der die Selbsthilfe anders als Meier, Sonnabend und Winter für durchaus üblich hält. Dies ist auch die These von Jones, Earthquakes (Anm. 6). – Dass Nero sich nicht für Katastrophenhilfe interessierte, stellte zuerst Sonnabend: Naturkatastrophen (Anm. 17), 231 fest. Meier: Emperors (Anm. 38), 21–24 erhärtet diesen Eindruck. Weiter ausdifferenziert wird diese These demnächst in einem Aufsatz des Verfassers, der voraussichtlich in einem von Jonas Borsch und Laura Carrara herausgegebenen Sammelband publiziert wird und derzeit im Review-Verfahren ist. Besagter Sammelband knüpft an eine Tübinger Tagung an; zu dieser siehe Deeg, Philipp: Tagungsbericht Erdbeben in der Antike. Deutungen, Folgen, Repräsentationen. 28.03.2014–29.03.2014, Tübingen. In: H-Soz-Kult, 26.05.2014, http://www.hsozkult.de/conferencereport/id/tagungsberichte-5388, zuletzt aufgerufen am 27.06.2015.

40 Şahin: Leuchtturm (Anm. 1), 13

41 Suet. Vesp. 8,4.

42 Eck: Legaten (Anm. 6); Eck u.a.: Leuchtturm (Anm. 2); Eck: Bauinschrift (Anm. 9); Eck, Werner: Der Anschluss der kleinasiatischen Provinzen an Vespasian und ihre Restrukturierung unter den Flaviern. In: Capogrossi Colognesi, Luigi/Tassi Scandone, Elena (Hrsg.): Vespasiano e l'imperio dei Flavi. Rom 2012, 27–44, hier 36; İplikçioğlu: Provinz (Anm. 2), 6 und 8; Jones, Christopher P.: The Neronian Inscription on the Lighthouse of Patara. In: ZPE 166 (2008), 153–154.

43 Eck u.a.: Leuchtturm (Anm. 2), 114; Eck: Provinzen (Anm. 42), 37; Eck, Werner: Die politisch-administrative Struktur der kleinasiatischen Provinzen während der hohen Kaiserzeit. In: Urso, Gianpaolo (Hrsg.): Tra Oriente e Occidente. Indigeni, Greci e Romani in Asia Minore. Pisa 2007, 189–207, hier 198; vgl. Eck: Legaten (Anm. 6), 71; Stylow, Armin U.: LIBERTAS und LIBERALITAS. Untersuchungen zur innenpolitischen Propaganda der Römer. Diss. München 1972, 44, 51, 58; Behrwald, Ralf: Der Lykische Bund. Untersuchungen zu Geschichte und Verfassung. Bonn 2000, 143 m. Anm. 485. – Für ein bereits von Claudius erteiltes Privileg könnten einige lykische Münzen von 43 sprechen, doch gesichert ist dieser Befund nicht; siehe Behrwald: Bund (diese Anm.), 141–144; Imhoof-Blumer, Friedrich: Zur griechischen und römischen Münzkunde. Hildesheim/New York 1977 [ND der Ausgabe Genf 1908], 24, Nr. 13, 25.

44 Eck u.a.: Leuchtturm (Anm. 2), 118 lesen den Text aber als „eine Rechtfertigung der vermutlich eben erst neu eingeführten Kopfsteuer". Doch dürfte Vespasian in seinen ersten Regierungsmonaten angesichts der Auseinandersetzung mit Vitellius kaum seinen Fokus auf solche

DIE THERMENINSCHRIFT IN PATARA

Dass die Therme in Patara[45] neronisch ist, kann als gesichert gelten.[46] Warum wurde ausgerechnet auf dieser Inschrift die Titulatur Neros vollständig eradiert und durch die Vespasians ersetzt, auf dem prestigeträchtigeren Leuchtturm aber nicht?[47] Jedenfalls ist die Annahme, eine solche Komplettrasur sei beispiellos,[48] nicht zutreffend.[49] Vielleicht lässt sich der Befund durch den technischen Zusammenhang zwischen Wasserleitung und Therme erklären: ohne Wasserzufuhr kein Bad. Letzteres musste aufgrund des Schadens an der Wasserleitung wohl „außer Betrieb gesetzt" werden.[50] Der Zusammenhang zwischen beiden Konstruktionen sollte vielleicht auch bei der Nennung des Bauherrn zum Ausdruck gebracht werden. Am Aquädukt brachte man eine neue Inschrift mit der gesamten Baugeschichte an. Da das Bad offenbar keinen Schaden genommen hatte,[51] die Baugeschichte mit Eck aber lediglich auf Neros Schirmherrschaft zurückging, verzichtete man vielleicht der Einfachheit halber auf eine neue Inschrift, entfernte aber den Namen Neros vollständig und ersetzte ihn durch den des neuen Machthabers.

DAS APOLLON-HEILIGTUM

Abschließend einige Überlegungen zum Apollon-Heiligtum, die in Ermangelung hinreichender Belege eher spekulativer Natur sind und daher lediglich als Denk-

administrativen Reformen, zumal in den militärisch nicht relevanten Provinzen Kleinasiens, gelegt haben (Eck: Legaten [Anm. 7], 72). Ferner wäre diese Maßnahme kontraproduktiv gewesen, da die Kopfsteuer als besonders eindringliches Zeichen römischer Herrschaft verstanden wurde und der Flavier sich somit keine Freunde gemacht hätte (so überzeugend Eck: Provinzen [Anm. 42], 37, Anm. 45). Nicht ausgeschlossen ist damit, dass Vespasian die Kopfsteuer nach Wiedereinführung – wann immer diese erfolgte – für lokale Bauvorhaben zur Verfügung stellte (Eck: Struktur [Anm. 43], 198; vgl. Jones: Earthquakes [Anm. 6], 60). Bislang liegen hierfür keine Quellenbelege vor (vgl. Eck u.a: Leuchtturm [Anm. 2], 118).

45 TAM II 396; SEG 57 (2007), 1671.

46 Eck: Legaten (Anm. 7); Eck: Bauinschrift (Anm. 9); anderer Auffassung ist Şahin: Bauinschrift (Anm. 1), 107–109; Şahin: Kaiserbauten (Anm. 9), 598–600; Şahin: Leuchtturm (Anm. 1), 7–9.

47 Şahin: Leuchtturm (Anm. 1), 3–5; zur Leuchtturminschrift auch Eck u. a.: Leuchtturm (Anm. 2).

48 Şahin: Leuchtturm (Anm. 1), 3, 7.

49 Balland: Xanthos (Anm. 23), 3 m. Anm. 20 sowie 123–129, Nr. 48.

50 Şahin: Leuchtturm (Anm. 1), 5.

51 Die Annahme, auch das Bad hätte, wenn es aus neronischer Zeit stammte, bei einem Erdbeben Schaden nehmen müssen – so Şahin: Bauinschrift (Anm. 1), 107; zurückhaltender Şahin: Kaiserbauten (Anm. 9), 600 und Şahin: Leuchtturm (Anm. 1), 5; vgl. Eck: Bauinschrift (Anm. 9), 273 – , ist nicht zwingend, da gemäß dieser Logik bei einem Erdbeben stets jede betroffene Ortschaft vollständig in Schutt und Asche liegen müsste. Demgemäß hätte dann auch der Leuchtturm beschädigt werden müssen, worüber aber die zugehörige Inschrift schweigt (Eck u. a.: Leuchtturm [Anm. 2], 116). Ebenso findet sich kein Hinweis auf der Thermeninschrift (Eck: Bauinschrift [Anm. 9], 273). Warum sollte aber im Falle des Aquädukts über das Beben informiert werden, bei anderen beschädigten Gebäuden aber nicht?

anstöße für die weitere Forschung verstanden sein wollen. Wenn, wie weiter oben angenommen, die Rekonstruktion des Aquädukts eine herrschaftssichernde Funktion für Vespasian hatte, stellt sich die Frage, warum nicht auch das Heiligtum des Apollon mit seinem Orakel im Namen des Flaviers wiederhergestellt wurde, was sicherlich ebenfalls ein prestigeträchtiges Unternehmen gewesen wäre. Dass das Heiligtum aber erst vom Euergeten Opramoas nach einer Phase langen Schweigens restauriert wurde, überliefert das bekannte Opramoas-Dossier.[52] Über den Grund der Zerstörung informieren uns die Oracula Sibyllina: Vom Kragos herabstürzende Wassermassen werden für das Verstummen des Orakels verantwortlich gemacht.[53] Es ist damit wahrscheinlich, dass der Tempel durch die Katastrophe von 68 zerstört wurde.[54] Der Verzicht auf eine umgehende Rekonstruktion ließe sich vielleicht mit Geldmangel erklären[55]: der Stadt Patara[56] oder der Reichskassen,[57] vielleicht auch des wegen des Steuernachlasses[58] vorläufig nicht mehr nachfüllbaren Steuersäckels. Aber erscheint es nachvollziehbar, dass sich keine Finanzierungsmöglichkeit gefunden haben soll, wo der Tempel doch von hoher Bedeutung für den Ruf und das Selbstverständnis der Stadt gewesen sein muss?[59] Erst recht, wenn das Heiligtum 68 beschädigt, beim Beben 69 aber trotz keineswegs besserer Finanzlage umstandslos die Wasserleitung repariert wurde? Führt man sich die geographischen Hinweise von Şahin und die Bemerkungen der Oracula Sibyllina vor Augen,[60] wirkt ein anderer Gedanke plausibel. Die Katastrophe sorgte vielleicht für eine Veränderung des Flusslaufes, wodurch das Orakel, das aus der Wasserlautstärke gedeutet wurde, nicht mehr rekonstruierbar, zumindest aber nicht mehr funktionsfähig war. Erst ein weiteres Desaster viele Jahrzehnte später korrigierte dies dann wieder, sodass Restauration und Wiederinbetriebnahme des Orakels möglich waren.[61]

52 TAM II 905; vgl. Sonnabend: Zeugnisse (Anm. 3), 1223. – Ausführlich zum Opramoas-Dossier Kokkinia, Christina: Die Opramoas-Inschrift von Rhodiapolis. Euergetismus und soziale Elite in Lykien. Bonn 2000, die in Kap. IIIb den Text und in Kap. IIIc eine deutsche Übersetzung der umfangreichen Inschrift bietet.

53 *Or. Sib.* 3,439–441; vgl. 4,109–113, bes. 112.

54 Şahin: Kragos (Anm. 33), 340–345 sowie Şahin: Bauinschrift (Anm. 1), 105 – Wegen der Angaben der sibyllinischen Orakel kommt das Beben von 69 eher nicht in Frage, kann aber nicht ausgeschlossen werden.

55 Vgl. allgemein Jones: Earthquakes (Anm. 6), 60.

56 Şahin: Leuchtturm (Anm. 1), 15.

57 Suet. *Vesp.* 16,3; vgl. Andreu: Bauwesen (Anm. 20), 384, 388.

58 Eck: Struktur (Anm. 43), 198; Eck u. a.: Leuchtturm (Anm. 2), 114.

59 Vgl. Şahin: Kragos (Anm. 33), 337 m. Anm. 2.

60 Şahin: Kragos (Anm. 33); Or. Sib. 3,439–441 und 4,112.

61 Diese Idee entwickelte Jörn Kobes am 20.02.2014 in persönlicher Kommunikation mit dem Verfasser. Freilich ist er sich darüber im Klaren, dass die Belastbarkeit seiner inhaltlich plausiblen These durch eine naturwissenschaftlich-geologische Untersuchung zu überprüfen wäre. – Einige skeptische Überlegungen zur Annahme eines Bebens ums Jahr 140, in dessen Folge Opramoas aktiv geworden sei, äußert Ganzert, Joachim: Doch kein Erdbeben in Patara? In: AW 39/4 (2008), 45–51. Allerdings bezieht er sich ausschließlich auf das Theatergebäude. Dass es keinen konkreten Hinweis gibt, ob das Theater von einem Erdbeben beschädigt wurde, und es sich daher wohl vielmehr um einen Neubau handelte, lässt nicht den Schluss zu, in Patara habe die Erde nicht gebebt. Doch auch wenn Ganzert Recht hat, heißt dies nicht, dass nicht lange nach 68/69, aber eben vor 140 ein Beben stattgefunden haben könnte, das dem reichen

ZUSAMMENFASSUNG

Der vorliegende Beitrag entstand aus der Beschäftigung mit der Doppelinschrift vom patarensischen Druckrohraquädukt. Galt das Interesse zunächst lediglich der Frage, wann das auf der Inschrift genannte Erdbeben stattgefunden hat, knüpften sich schnell weitere Fragen, insbesondere zur kaiserlichen Intervention nach der Katastrophe daran. Die Angaben der Inschrift schließen einen Zusammenhang zwischen dem dort genannten Beben und dem literarisch überlieferten von 68 n. Chr. aus. Vielmehr ist von einem seismischen Ereignis um die Mitte des Jahres 69 auszugehen, wenngleich weitere Belege hierfür einstweilen fehlen. Die Reparatur des Aquäduktes finanzierte der propraetorische Legat Priscus aus Mitteln des lykischen Bundes sowie der bereits eingezogenen Kopfsteuer. Als Stifter ließ er den Usurpator Vespasian auf dem Stein verewigen. Ein Statthalter konnte demnach selbst in der außergewöhnlich angespannten Konstellation eines Zusammentreffens von Naturkatastrophe und Bürgerkrieg nicht auf eine höhere Autorität verzichten, um seine Maßnahmen zu legitimieren. Zugleich zeigt sich, wie sehr der Kaiser, und zumal ein Usurpator, auf die Unterstützung selbst militärisch marginaler Provinzen wie Lykien angewiesen war. Um deren Rückhalt zu gewinnen, genügte indes offensichtlich der Verzicht auf der Reichszentrale zustehende Mittel als Ausweis der großzügigen Fürsorge. Damit ließe sich auch plausibel erklären, weshalb auf der Thermeninschrift in Patara der Name des ursprünglichen Stifters vollständig getilgt und durch den des Prätendenten ersetzt wurde: Wegen des technischen Zusammenhanges zwischen Aquädukt und Therme sollte Vespasian auch als freigiebiger Bauherr der letzteren erscheinen. Weshalb dann aber auf die Rekonstruktion des Apollon-Heiligtums mit seinem bekannten Orakel verzichtet wurde, lässt sich derzeit nicht befriedigend beantworten.

ANHANG: DER TEXT DER INSCHRIFT[62]

Αὐτοκράτωρ Καῖσαρ Φλάουιος Οὐεσπασιανὸς Σεβαστὸς vac
2 τὸ τοῦ ὑδραγωγίου ἀνάλημμα συμπεσὸν σεισμοῖς ἐκ θεμελίων ἀποκατέστησε
τοῖς ἐπ᾽ αὐτῷ λιθίνοις ἐκ τετραπέδου λίθου σωλῆσι προστεθένος καὶ ἑτέρου παρὰ τὸ
4 ἀνάλημμα θλειμματικοῦ ὑδραγωγίου διὰ τριστίχων σωλήνων ὀστρκίνων παλαιστ-
αίων ὥστε δυεῖν ὄντων, εἰ θάτερον ἐπισκευῆς δεηθείη, μὴ ἐνποδιζέσθαι τὸν δρόμον
6 ἀδιαλείπτου μενούσης τῆς χρήσεως. ἐπεσκεύασε δὲ καὶ τὰ λοιπὰ τοῦ ὑδραγωγίου

Lykier Wohltätigkeit und dabei insbesondere die Wiederherstellung des Orakels ermöglichte. Auch hier sollte die Quellenlage zu antiken Naturkatastrophen bedacht werden.

62 Eck u. a.: Leuchtturm (Anm. 2), 116; Şahin: Bauinschrift (Anm. 1), 103; Şahin: Leuchtturm (Anm. 1), 18; İplikçioğlu: Provinz (Anm. 2), 7, Nr. 05; SEG 57 (2007), 1673. – Die Textform gibt die Inschrift auf der Westseite wieder. Der Text entspricht wesentlich dem von Eck u. a. Die Majuskeln in Zeile 9 jedoch wurden – wie bei Şahin und anders als bei Eck u. a. – bewusst nicht aufgelöst, weil im vorliegenden Aufsatz die Frage, welche Lesart zutrifft, weder gelöst werden kann noch soll.

καὶ τὸ ὕδωρ μετὰ μῆνας δ´ παραπεσεῖν εἰσήγαγεν διὰ Σέξτου Μαρκίου Πρείσκου πρεσβευ

8 τοῦ αὐτοῦ ἀντιστρατήγου ἐκ τῶν συντηρηθέντων τῇ πόλει χρημάτ[ω]ν ἀπὸ κεφαλαίων καὶ

τὸ ἔθνος συνήνενκε Χ^{vac} , μηδεμιᾶς Κ̣Α̣Τ̣Α̣Ν̣Λ̣ΡΑ ἐπιγραφῆς γενομένης. τοῦ ἔργου καταρχ

10 θέντος μὲν ὑπὸ Οὐιλίου Φλάκκου πρεσβευτοῦ Κλαυδίου Καίσαρος Σεβαστοῦ ἀντιστρατήγου

συντελειωθέντος δὲ καὶ εἰσαχθέντος τοῦ ὕδατος ἐπὶ Ἐπρίου Μαρκέλλου πρεσβευτοῦ Κλαυδίου

12 Καίσαρος Σεβαστοῦ ἀντιστρατήγου.^{vac}

OSWALD SPENGLER, ATLANTIS UND DER PLAN EINES *ATLAS ANTIQUUS* [1]

Sebastian Fink

Oswald Spengler ist weithin für sein geschichtsphilosophisches Werk „Der Untergang des Abendlandes“ bekannt. Dieses Buch, das in den Jahren von 1911 bis 1917 verfasst wurde, machte Spengler zu dem Star der deutschen Philosophie nach dem ersten Weltkrieg.[2] Doch Spengler betätigte sich nicht nur als Geschichtsphilosoph, sondern ebenso als politischer Publizist, als Wissenschaftsorganisator – so war er etwa an der Herausgabe der Nietzsche-Gesamtausgabe des C.H. Beck-Verlages maßgeblich beteiligt – und als Historiker. Beim Plan des *Atlas antiquus* sah sich Spengler sowohl als Wissenschaftsorganisator, als auch als konzeptionell arbeitenden Historiker, der seinen Kollegen neue Wege der Forschung aufzeigt. Möglich gemacht wurde diese Tätigkeit Spenglers durch seine Berühmtheit, die ihm laufend Einladungen zu jeder Art von Kongressen bescherte, deren Organisatoren sich durch einen Auftritt „des Propheten des Untergangs“ ein wenig Aufmerksamkeit erhofften. Zudem war Spengler in den Jahren nach der Veröffentlichung seines Hauptwerkes zu einem einflussreichen Mann geworden, der von ihm geschätzte Wissenschaftler durch seine zahlreichen Kontakte zu politisch und finanziell einflussreichen Kreisen massiv unterstützen konnte.

Spengler nahm Kongress- und Vortragseinladungen relativ häufig an, da er diese Gelegenheiten nutzte, so geht es aus dem veröffentlichen Briefwechsel hervor, um sich ein Netzwerk an Informanten und Experten für die ihn interessierenden Epochen aufzubauen. So findet sich im Index zu den Briefen Spenglers ein regelrechtes „Who is Who“ der deutschsprachigen Altertumskunde, wobei an dieser Stelle angemerkt werden muss, dass die Briefausgabe keineswegs vollständig ist. So zählen etwa der Ausgräber Assurs Walter Andrae,[3] der Universalhistoriker Eduard Meyer,[4] die Leipziger Altorientalisten Heinrich Zimmern[5] und Alfred Jeremias, der Wiener Altorientalist Viktor Christian,[6] der Meyer-Schüler Hans-Erich Stier,[7] der Altphilologe und Althistoriker Franz Altheim und der Berliner Professor für

1 Ich danke Reinhold Bichler für eine kritische Lektüre dieses Textes.

2 Zum Erfolg Spenglers und zu seiner Zusammenarbeit mit dem C.H. Beck-Verlag siehe Rebenich 2013, 287–310 mit weiterer Literatur.

3 Zu Andrae siehe Bär 2003.

4 Die Beziehung Spenglers zu Eduard Meyer schildert Demandt 1990.

5 Zu Zimmern siehe den Nachruf von Landsberger 1931.

6 Eine wissenschaftliche und politische Verortung Christians unternimmt Leitner 2010. Einen Nachruf bietet Weidner 1966.

7 Die wissenschaftliche und politische Karriere Stiers stellt Gauger 2007 dar.

historische Geographie Albert Herrmann zu den Korrespondenzpartnern Spenglers. Doch zunächst zu Spenglers Plan eines *Atlas antiquus*.

Spengler hielt am 02.10.1924 einen Vortrag mit dem Titel „Der Plan eines neuen Atlas antiquus“ auf dem deutschen Orientalistentag in München.[8] Der Orientalistentag wurde von der Deutschen Morgenländischen Gesellschaft veranstaltet, der Spengler im gleichen Jahr beigetreten war.[9] Er blieb bis zu seinem Tod Mitglied der Gesellschaft und korrespondierte mit vielen Gelehrten, die er wohl zum größten Teil auf dem Orientalistentag kennengelernt hatte. In den Mitgliedernachrichten vom Jahre 1937 ist in der Rubrik „Verstorbene Mitglieder“ ein „Dr. Oswald Spengler, München, Widenmeyerstr. 26“ vermerkt.[10] Ein Nachruf von Seiten der Gesellschaft ließ sich nicht finden. Vielleicht blieb Spengler dieser verwehrt, da er spätestens mit seinem Buch „Jahre der Entscheidung“ und seiner wiederholten Weigerung für die Nationalsozialisten publizistisch tätig zu werden bei den neuen Machthabern in Ungnade gefallen war und sein Name – auf Weisung des Reichspropagandaministers – in der Presse nicht mehr genannt werden durfte.[11]

Interessant erscheint hier, dass Spenglers Teilnahme am Orientalistentag ein Vortrag von Carl Heinrich Becker, gehalten am 07.02.1923, im Rahmen der Ortsgruppe Leipzig der Deutschen Morgenländischen Gesellschaft über „Die magische Kultur Oswald Spenglers“ vorausgegangen ist, der auch im selben Jahr publiziert wurde.[12] Möglicherweise führte dieser Kontakt zu Becker zu dem Vortrag auf dem Orientalistentag. Ob Becker auch an dem bei Koktanek erwähnten Vortrag Spenglers im Mai 1921[13] in Leipzig teilnahm, ist nicht klar. Im publizierten Briefwechsel findet sich kein Schreiben an oder von Becker, allerdings kommt Becker in einigen Schreiben Albert Augusts von Lecoq und Eduard Meyers vor, wobei vor allem Lecoq Spengler bittet, seinen Einfluss auf den „Staatssecretär Becker“ geltend zu machen, um den Erhalt der asiatischen Sammlung im Berliner Völkerkundemuseum zu sichern.[14] Spengler entsprach der Bitte und so kam es am 5. Juli 1924 zu einem Treffen mit Becker (Spengler wurde von Eduard Meyer, Heinrich Lüders, Otto Franke und Walter Schultze begleitet), in dessen Verlauf Becker zustimmte, sich für den Erhalt der asiatischen Abteilung einzusetzen. Zudem wurde eine Schriftenreihe geplant, die der Öffentlichkeit die Bedeutung der asiatischen Funde bekannt machen und an der Spengler mitwirken sollte.[15] Koktanek ist der Ansicht, dass der Vortrag am Orientalistentag nicht zuletzt dazu diente, diese Schriftenreihe zu bewerben, zu deren Verwirklichung es allerdings nie kam.[16]

8 Zeitschrift der Deutschen Morgenländischen Gesellschaft 78 (1924), LXXX.

9 Zeitschrift der Deutschen Morgenländischen Gesellschaft 78 (1924), XXV. Hier findet sich unter der Rubrik „Neue Mitglieder“ folgender Eintrag: „2433 Herr Dr. Oswald Spengler, München, Agnesstr. 54.“

10 Zeitschrift der Deutschen Morgenländischen Gesellschaft 91 (1937), *3*.

11 Vgl. Rebenich 2013, 303.

12 Becker 1923.

13 Koktanek 1968, 354–355.

14 Spengler 1963, 320. Die anderen Briefe, die Becker erwähnen, finden sich im Register, siehe Spengler 1963, 794.

15 Koktanek 1968, 356.

16 Koktanek 1968, 357–361.

In seinem halbstündigen Vortrag stellt Spengler zunächst das Problem dar, dass das vorliegende Kartenmaterial der aktuellen Forschung um „mindestens 30 Jahre" hinterherhinke und es daher dringend notwendig sei, die aktuellsten Forschungsergebnisse in einem Geschichtsatlas darzustellen. Zudem sollten auch Informationen über „Siedlungs- und Rasseverhältnisse an Hand von Gräber- und Höhlenfunden, über Fauna und Flora, über die großen Kultur- und Handelswege" nicht fehlen. Die Fachwelt zeigte sich interessiert, wies aber gleichzeitig auf die zahlreichen Probleme hin, die so einem interdisziplinären Unternehmen im Wege stehen:

> Nachdem er den ganzen Vortrag hindurch immer wieder die Notwendigkeit der Zusammenarbeit von Historikern, Geographen, Biologen und Prähistorikern betont hatte, wandte er sich an die deutsche Gelehrtenwelt, die allein genügend wissenschaftliche Vielseitigkeit und Tiefe der Forschung besitzt, um dieser Aufgabe gewachsen zu sein, und bat die anwesenden und nicht anwesenden Gelehrten zu erwägen, ob seine Ideen [...] in der Wirklichkeit umgesetzt werden könnten. – In der Diskussion, an der sich die Herren Herrmann, Lehmann-Haupt, Eisler, v. Duhn und der Sektionsleiter [E. Kornemann-Breslau – S. F.] beteiligten, wurde unter Anerkennung der großen Bedeutung des vorgelegten Grundplanes auf die enormen Schwierigkeiten hingewiesen, die der Verwirklichung entgegenstehen. Kornemann wies auf das Fehlen eines zuverlässigen Corpus topographicum hin, dessen Schaffung Camille Jullian schon vor 20 Jahren mit Recht gefordert hat (Klio II, 1902, S. 1–13).[17]

Der Spenglersche Vortrag wurde erst nach dessen Tod im Jahr 1937 in einem von seiner Nichte Hildegard Kornhardt herausgegebenen Band „Reden und Schriften" publiziert und somit allgemein zugänglich gemacht. Spengler beklagt im publizierten Text, dass die „historische Geographie in Deutschland und anderswo so gut wie ausgestorben ist" und sich deshalb in der „Geschichtsforschung ein Notstand" entwickelt habe. Die Wichtigkeit von gutem Kartenmaterial kann nach Spengler kaum überschätzt werden:

> Das Ergebnis jeder geschichtlichen Forschung ist ein geschichtliches Bild, welches sich nach gewissen Ideen in kartographischer Form fürs Auge darstellen läßt. Die Karte sieht vom Persönlichen und Einmaligen in der Regel ab, aber sie arbeitet dafür die allgemeine Form um so schärfer heraus und ermöglicht damit ein Zusammensehen und Zusammenfühlen mit einem einzigen Blick, wie es kein noch so deutlich geschriebener Bericht vermag.[18]

Er betont, dass gerade in den neu entwickelten Fächern wie der Altorientalistik und der Ägyptologie gute Karten weitgehend fehlen und viele Bücher damit unverständlich blieben, da der Leser geographisch nicht im Bilde sei und ihm somit „entscheidende Zusammenhänge" entgehen würden.[19] Um die angesprochenen Zusammenhänge richtig zu verstehen, bedürfe es jedoch neuer „Ideen der Kartographie". Vieles könne erst erkannt und verstanden werden, wenn man die Geländeformen, das Vorkommen von Rohstoffen, die Meeresströmungen und die Windrichtungen vor Augen habe. Man müsse sich die Behinderung des Verkehrs durch den einstmals weit verbreiteten Urwald immer vor Augen halten, denn die „Waldmassen sind in der Frühzeit das einzige und entscheidende Hindernis für Verkehr und Wanderung",

17 Zeitschrift der Deutschen Morgenländischen Gesellschaft 78 (1924), LXXX.
18 Spenlger 1951, 96.
19 Spengler 1951, 97.

während das Meer das verbindende Element der Völker sei.[20] Ebenso wichtig sei es auch, so Spengler, Klimaveränderungen in Betracht zu ziehen, um geschichtliche Veränderungen zu verstehen. Wenn Karten mit solch einer Vielfalt von Informationen erzeugt werden sollen, sind Vertreter verschiedenster Wissenschaftsdisziplinen gefordert:

> Es ist klar, daß Karten dieses Stils nur durch eine Zusammenarbeit von Vertretern der verschiedensten Wissensgebiete hergestellt werden können. Ein einzelner ist der Aufgabe nicht mehr gewachsen. Aber wenn der Geologe, der Tier- und Pflanzengeograph zunächst eine Unterlage schaffen, in welche der Anthropologe und der Prähistoriker seine Eintragungen vornimmt, um endlich dem Vertreter der politischen und Wirtschaftsgeschichte Platz zu machen, ist ein Anschauungsmaterial denkbar, das durch sein bloßes Vorhandensein fürs Auge zu Entdeckungen führt, welche der reinen Bücherarbeit verschlossen geblieben sind.[21]

Ganz im Sinne Eduard Meyers betont Spengler, dass der Umfang eines solchen Kartenwerkes heute von keiner „Ostgrenze" mehr beschränkt werden darf, es müsste also die ganze alte Welt miteinbezogen werden. Die zeitliche Obergrenze für den Atlas sieht Spengler in den Kreuzzügen und dem Mongolensturm.[22] An der vorgeschlagenen Gliederung des Kartenwerks ist bemerkenswert, dass Spengler eine Karte der „primitiven Welt" fordert, die Auskünfte über „Handels- und Wanderungssrouten" geben soll, die seiner Ansicht nach über lange Zeit konstant geblieben sind und daher auch für die Lösung von Problemen der späteren Geschichte hilfreich sind, sowie die geforderte gemeinsame Bearbeitung der Schicksale „Ägypten und Babyloniens und der sie umgebenden Welt."[23]

Wie bereits eingangs geschildert, zeigte sich die Fachwelt interessiert, wies aber auf die fehlenden Vorarbeiten, vor allem auf das Fehlen eines Verzeichnisses aller bekannter Toponyme und der Informationen zu deren Lokalisation und die zu erwartenden Schwierigkeiten eines solchen Werkes hin. Im Jahr 1930 ging Albert Herrmann anlässlich eines Vortrages beim sechsten Deutschen Orientalistentag in Wien nochmal auf Spenglers Plan ein:

> Am 2. Okt. 1924 legte Oswald Spengler auf dem Münchner Orientalistentage den Plan eines neuen Atlas Antiquus vor; er forderte dort stärkere Berücksichtigung der physisch-geographischen und Rasseverhältnisse, vor allem Einbeziehung der prähistorischen Ergebnisse und Ausdehnung der Darstellung über die ganze Erde. Das Ziel war zu hoch gesteckt, so daß der an sich begrüßenswerte Plan leider nicht zur Ausführung kam.[24]

Herrmann betonte, dass es nun, da der große Plan nicht umgesetzt werde, umso wichtiger sei, wenigstens Teilgebiete zu bearbeiten und stellt sein Projekt eines „Historical and Commercial Atlas of China" vor, den er mit Hilfe eines nordamerikanischen Instituts herauszubringen gedenke. Herrmann konnte diesen Plan auch in die Tat umsetzten, der Atlas erschien im Jahr 1935.[25] Damit hat Spenglers Vortrag

20 Spengler 1951, 98.
21 Spengler 1951, 99.
22 Spengler 1951, 100.
23 Spengler 1951, 101.
24 Herrmann 1930, *113*–*114*.
25 Herrmann 1935.

zumindest Herrmann – mit dem Spengler ja auch in Briefkontakt stand – dazu angeregt ein Atlasprojekt in Angriff zu nehmen.

Anhand eines wissenschaftlichen Großprojektes, des „Tübinger Atlas des Vorderen Orients", soll versucht werden aufzuzeigen, inwieweit Ideen Spenglers in der modernen Kartographie umgesetzt wurden, wobei ich hier nur ganz am Rande darauf eingehen werde, ob Spenglers „Plan eines neuen Atlas Antiquus" die Arbeiten direkt oder indirekt beeinflusst hat.

Im Jahr 1993 fand das Projekt „Tübinger Atlas des Vorderen Orients" seinen Abschluss, nachdem Spezialisten aus den Fächern „Geographie und Spezielle Botanik, Urgeschichte, Vor- und Frühgeschichte, Altorientalistik, Ägyptologie, Biblische Archäologie, Judaistik, Alte, mittlere und neuere Geschichte, Islamkunde und Christlicher Orient"[26] ab 1969 daran gearbeitet hatten. Allein aus dieser Fächerliste erkennt man, dass auch der Tübinger Atlas den von Spengler geforderten breiten Ansatz vertritt. In seiner endgültigen Fassung enthält der Tübinger Atlas zwei Teile. Der Teil A – Geographie – bietet Informationen über Relief, Gewässer, Siedlungen, Geologie, Geomorphologie, Klima, Hydrogeographie, Vegetation, Zoologie, naturräumliche Gliederung, Bevölkerung, Siedlung, Wirtschaft und Verkehr. Der Teil B – Geschichte – bietet Karten von der Steinzeit bis in die Zeit nach dem ersten Weltkrieg. Beide Kartenwerke erschienen in Lieferungen von 1977 bis 1993 und sind von zahlreichen Beiheften begleitet worden, die Erläuterungen zu den Karten sowie Indizes bieten, die das Kartenwerk erschließen.[27] Wolfgang Röllig beschreibt das Ziel des Projekts folgendermaßen:

> Ein wesentliches Ziel von TAVO ist es, den geschilderten Raum unter unterschiedlichen Gesichtspunkten zu behandeln. [...] Zu selten nämlich wurden bisher die historischen Fakten, so wie sie aus archäologischen Befunden, antiken oder modernen Quellentexten erschließbar sind, korreliert mit natürlichen Gegebenheiten des geographischen Raumes, mit der Landschaft, der Vegetation, ihren klimatischen Bedingungen, ihren wirtschaftlichen Möglichkeiten oder ihren natürlichen Grenzen.[28]

Ideengeber hierfür war nach Rölligs Auskunft nicht etwa Oswald Spengler, sondern Ernst Kirsten (1911–1987),[29] ein Experte für die historische Geographie Griechenlands und Italiens, der „immer wieder programmatisch auf den Zusammenhang zwischen geographischen Voraussetzungen und historischen Ereignissen"[30] hingewiesen habe und der dem Tübinger Projekt auch als Berater zur Verfügung stand. Ob Kirsten von Spengler beeinflusst war, ist mir nicht bekannt, allerdings lässt seine Zusammenarbeit mit Hans Erich Stier,[31] deren sichtbares Ergebnis zwei historische Kartenwerke sind, eine gewisse Bekanntschaft mit den Ideen Spenglers annehmen. Der Eduard Meyer-Schüler Stier, der Spenglers Aufsätze auch in der Zeit des Nationalsozialismus in seiner Zeitschrift „Die Welt als Geschichte" abgedruckt hat und

26 Röllig 1974, 531.
27 Zu Ansatz und Ausführung des Tübinger Atlas siehe auch Röllig 1991.
28 Röllig 1991, 3.
29 Zu Kirsten siehe Olshausen 2012.
30 Röllig 1991, 3.
31 Von Gauger 2007 wird die im Spengler-Briefband enthaltene Korrespondenz zwischen Spengler und Stier nicht berücksichtigt.

der mit Spengler in einem regen Ideenaustausch stand, gab mit Kirsten zusammen den in vielen Auflagen verbreiteten Westermann-Atlas zur Weltgeschichte heraus.

Grundlage jeglicher historischen Karte bilden Texte und archäologische Zeugnisse aus der entsprechenden Epoche. Die Ortsnamen aus Texten müssen zusammengestellt und mit archäologischer Evidenz in Zusammenhang gebracht werden. Der einfachste Fall liegt vor, wenn man Monumente mit Beischriften findet, die den Ortsnamen erwähnen, bei einer Fundlage ohne Schriftzeugnisse bleiben Identifikationen oft unsicher, was dazu führt, dass aus dem alten Orient zwar viele Orts- und Gewässernamen aus Texten bekannt sind, aber nicht lokalisiert werden können.[32] Das 1952 von Leemans, Kupper und Nougayrol begründete „Répertoire geographique des textes cunéiformes" stellte sich dieser Aufgabe des Sammelns und Kommentierens der Orts- und Gewässernamen aus den Keilschrifttexten. Nach zügiger Sammelarbeit in den Anfangsjahren des Projekts erschien es jedoch bald so, als ob das Répertoire eingestellt worden sei. Mit dem Bekanntwerden des großen Tübinger Atlas-Projekts übertrugen die Begründer die Herausgabe desselben an den Tübinger Altorientalisten Wolfgang Röllig, unter dessen Leitung auch zahlreiche Bände erschienen sind, wobei das Projekt heute noch nicht endgültig abgeschlossen ist.[33]

Es kann festgestellt werden, dass der Tübinger Atlas einen Teil der Spenglerschen Forderung nach einem neuen Atlas Antiquus erfüllt hat, wobei er natürlich nur den Bereich des Vorderen Orients umfasst.[34] Ägypten wird vom Tübinger Atlas auch aufgenommen und somit ist auch die von Spengler geforderte gemeinsame Behandlung Ägyptens und Babyloniens gegeben. Auch die geographischen Verhältnisse – man denke an die obige Aufzählung – wurden im Tübinger Atlas solide aufgearbeitet. Man wagte sich auch an eine kartographische Aufarbeitung des nach Spengler dunkelsten Gebiets der „Geschichte wie der Geographie" des Vorderen Orients „von Alexander dem Großen bis zum Mongolensturm"[35], und legte somit eine völlig neue Grundlage für das historisch-geographische Verständnis des Vorderen Orients, wenn auch viele Detailprobleme noch nicht gelöst werden konnten. Der von Spengler geforderte Atlas Antiquus für die gesamte antike Welt – von Spanien bis Japan und von Skandinavien bis Südafrika – ist in der von ihm gewünschten Form wohl noch nicht erschienen.

KORRESPONDENZ ZUR HISTORISCHEN GEOGRAPHIE

Einen Einblick in Spenglers Beschäftigung mit den Fragen der historischen Geographie gibt seine Korrespondenz. Mit Albert Herrmann (1886–1945), Professor für historische Geographie in Berlin, diskutierte Spengler über Fragen der Loka-

32 Röllig 1991, 12 stellt fest, dass nur fünf Prozent der aus altbabylonischer Zeit bekannten Ortsnamen lokalisierbar sind.

33 Zur Geschichte und Ausrichtung des Répertoirs vgl. Röllig 1986.

34 Zur räumlichen Abgrenzung siehe Röllig 1991, 2.

35 Spengler 1951, 103.

lisierung von Atlantis,[36] von Orten aus der Odyssee und über die Entstehung der Ilias. All diesen Fragen ist eine gewisse Aktualität nicht abzusprechen, da neben dem populärwissenschaftlichen Dauerbrenner Atlantis[37] auch die Lokalisation von Troja in den letzten Jahren lebhaft diskutiert wurde.[38]

Ein anderer Korrespondenz- oder Diskussionspartner war Paul Borchardt (1886–1957). Borchardt wird in den erhaltenen Briefen immer wieder erwähnt, Briefe an oder von ihm sind jedoch entweder nicht erhalten oder nicht in die Briefausgabe aufgenommen. Er war vor allem als Atlantisforscher bekannt und machte sich mit der These, dass sich Platons Atlantis in Tunesien lokalisieren ließe, einen – zweifelhaften – Namen.[39]

Das Schreiben von Albert Herrmann an Spengler nimmt Bezug auf ein nicht erhaltenes Schreiben Spenglers, in dem dieser Herrmann seine Thesen zu Odyssee und Ilias darlegte.[40] Herrmann stimmt Spengler zunächst zu, dass die uns überlieferten Fassungen von Odyssee und Ilias „verhältnismäßig junge Gebilde sind“, hält jedoch Spenglers These, dass in den beiden Texten auch „Märchen aus Etrurien“ eingeflossen seien, für gänzlich unwahrscheinlich.[41] Dann geht es um Fragen der historischen Geographie, genauer um die Frage, ob die Odyssee zur Rekonstruktion der „geographischen Anschauungen“ der, leider nur unsicher zu bestimmenden, Zeit ihrer Komposition genutzt werden kann:

> Ich gehöre nicht zu den Homererklärern, die nach Tagen und Kilometern die Örtlichkeiten lokalisieren, und ich lege immer Wert darauf, daß die Dichter keine Geographie schreiben wollten. Aber das muß ich immer wieder betonen, daß sie auch Kinder ihrer Zeit waren und daher unbewußt geographische Anschauungen in die Dichtungen hineintrugen, die ihrer Zeit entsprachen.[42]

Herrmann gehörte zu den Gelehrten, die meinten, verschiedene Schichten der Odyssee feststellen zu können, und er war der Meinung „eine Ur-Odysee festgestellt zu haben, die sich aus den Abenteuern bei den Kyklopen, auf Trinakria (mit dem nachfolgenden Schiffbruch) und bei den Phäaken (in einer kürzeren Fassung, z. B. ohne Nausikaa, den Tempel Poseidons und den Palast Alkinoos) zusammensetzt.“[43] Herrmann versucht Spengler davon zu überzeugen, dass seine Ergebnisse in Übereinstimmung mit denen führender Gelehrter – wie Eduard Schwarz oder Paul Borchardt – stehen würde, aber unabhängig von diesen erlangt worden seien, und geht noch auf einige Detailfragen der Lokalisierung verschiedener Örtlichkeiten aus der Odyssee ein. Seinen Brief schließt er mit einem Hieb auf seine Gegner:

36 Das in der Fragmentsammlung „Frühzeit der Weltgeschichte“ (Spengler 1966) oft vorkommende Atlantis hat nichts mit dem Platonischen Atlantis zu tun, sondern ist, neben Kasch und Turan, eine Bezeichnung für die drei, von Spengler postulierten, Kulturkreise der Frühgeschichte.

37 Einen Überblick über die Rezeptionsgeschichte gibt Bichler 1999.

38 Siehe etwa Ulf / Rollinger 2012.

39 Siehe Borchardt 1927.

40 Spengler 1963, 495–498 (Albert Herrmann an Spengler, Charlottenburg, 29.01.1927).

41 Spengler 1963, 495 (Albert Herrmann an Spengler, Charlottenburg, 29.01.1927).

42 Spengler 1963, 496 (Albert Herrmann an Spengler, Charlottenburg, 29.01.1927).

43 Spengler 1963, 496 (Albert Herrmann an Spengler, Charlottenburg, 29.01.1927).

> Jedenfalls komme ich immer mehr zu der Überzeugung, daß wir die ältesten griechischen Nachrichten wesentlich anders erfassen und mehr mit prähistorischen und anderen Quellen in Verbindung bringen müssen. Diese Wurschtelei, wie sie bisher besonders von den Nurphilologen getrieben wird, führt nicht zum Ziele.[44]

Spengler diskutierte diese Fragen auch mit Eduard Meyer, der sich über Herrmann und Borchardt recht deutlich äußert:

> Heute bekomme ich einen Aufsatz von Paul Borchardt über Atlantis, der sich dabei auf Sie beruft. Es ist mir unbegreiflich, wie irgend jemand, der Platos Schriften kennt und ruhig durchliest, überhaupt auf den Gedanken kommen kann, daß dieses von ihm geschaffene Utopia irgendwelche reale Grundlage habe oder eine alte Überlieferung benütze. Aber solche phantastischen Kombinationen sind ja jetzt in Mode, wie z. B. A. Herrmann sich in derselben Weise über die Phäaken und die Irrfahrten des Odysseus ergeht. Ich muß Ihnen offen bekennen, daß es mir ganz unmöglich ist, mich zum Lesen derartiger Schriften zu zwingen.[45]

Spengler antwortet Herrmann am 07.02.1927 und geht gleich auf die Frage der Entstehung der Odyssee ein:

> Ich glaube nicht, daß es eine Ur-Odyssee überhaupt gegeben hat. Wenn man sich mit derartigen Fragen beschäftigt, verfällt man leicht in den Fehler, die Entstehung solcher altertümlichen Dichtungen sich nach Analogie heutiger gedruckter Dichtwerke vorzustellen. [...] Was die homerische Dichtung betrifft, so muß man die Skaldenzeit und die der literarischen Epik sorgfältig auseinanderhalten. Jene Sagas, von denen der einzelne Sänger einen ganz persönlichen Vorrat besaß, die also niemals von zwei verschiedenen Sängern in derselben Form gelernt wurden, sind ohne Zweifel in ganz unbekümmerter Weise nach Ort und Gelegenheit inhaltlich variiert worden, so daß man wohl von verwandten Motiven sprechen kann, aber nicht von einer Urfassung.[46]

Er führt weiter aus, dass die Verschriftlichung wohl zu einer Vergrößerung des Umfangs der Texte sowie zu einer „gewissen Ordnung der Motive“ geführt habe, und stellt schließlich fest, dass es so viele Ur-Odysseen gegeben habe, „als es Sänger gab“. Für Spengler ist die uns bekannte Odyssee „das Ergebnis einer Auswahl von Texten, über welche wir uns heute kaum noch ein Urteil bilden können.“[47] Dann wendet sich Spengler – ganz im Sinne Meyers, obwohl nicht sicher ist, ob er Meyers Brief bei der Abfassung seines Schreibens an Herrmann bereits vorliegen hatte – gegen eine geographische Auswertung der Irrfahrt des Odysseus. Dazu führt er ins Treffen, dass viele Ortsnamen der Odyssee aus alter Zeit stammen würden und bereits in der Entstehungszeit der Odyssee keine feste Lokalisation mehr gehabt hätten.[48] Diese Ortsnamen hätten ihre Lokalisation in der realen Welt verloren, da sich der geographische Horizont seit der Skaldenzeit eingeschränkt habe:

> Der geographische Horizont der Verfasser der heutigen Odyssee war nicht größer, sondern kleiner als der der Skalden der minoischen Zeit, denn seitdem war die Kenntnis des westlichen Mittelmeeres wesentlich zurückgegangen. Ich glaube deshalb nicht, daß sich mit der Ordnung der Begebenheiten in dem heutigen Texte bestimmte geographische Abmessungen verbunden

44 Spengler 1963, 498 (Albert Herrmann an Spengler, Charlottenburg, 29.01.1927).
45 Spengler 1963, 501 (Eduard Meyer an Spengler, Berlin-Lichterfelde, 06.02.1927).
46 Spengler 1963, 501–502 (Oswald Spengler an Albert Hermann, München, 07.02.1927).
47 Spengler 1963, 502 (Oswald Spengler an Albert Hermann, München, 07.02.1927).
48 Diese Meinung vertritt Spengler auch in späteren Schriften. Siehe dazu Pesditschek 2004, 903.

> haben, wenigstens nicht in dem Grade, daß man daraus eine Irrfahrt ursprünglichster Form wieder herstellen könnte.[49]

In einem Brief an Spengler äußert sich Eduard Meyer abermals in bereits bekannter Weise über einen Aufsatz, den Borchardt ihm zugeschickt habe: „Ganz entsetzt bin ich, wie ich Ihnen offen aussprechen muß, über die Arbeiten von Borchardt."[50] Borchardt würde die „elendesten Machwerke der späthellenischen Literatur" sowie die „tollsten Phantasien über Phäaken, Kyklopen u. a." ohne jeden Anflug einer Quellenkritik verwenden und für bare Münze nehmen. Meyer fällt das vernichtende Urteil:

> Das ist der wüsteste Dilettantismus, der, wenn es so weitergeht, das Ende aller Wissenschaft bedeuten würde.[51]

Spengler erörtert in einem längeren Brief an Meyer hauptsächlich die Frage nach den Zusammenhängen zwischen dem „westlichen Mittelmeer und der Ägäis", geht dann abermals auf die Frage nach der Lokalisierung verschiedener Orte ein, wobei er eine zunehmende Wüstenbildung im Nordafrika des dritten und zweiten Jahrtausends annimmt, die seiner Meinung nach zur Aufgabe einiger städtischen Zentren geführt habe. Auf die Frage, die sich dem Leser an dieser Stelle aufdrängt, ob die von ihm angenommene Wüstenbildung und die dadurch erzwungene Aufgabe von Städten seiner Ansicht nach den historischen Hintergrund von Platos Atlantiserzählung bildet, geht er in seinem Brief an Eduard Meyer vorsichtshalber nicht ein:

> Ich halte es für möglich, daß etwa zur Zeit Homers auch das städtische Handelszentrum in Tunis ebenso zugrunde ging wie die jetzt gefundenen Städte in Ostturkistan. Jedenfalls können darüber nur Ausgrabungen entscheiden. Die Spuren in den griechischen Sagen sind aber deutlich genug, um einen Versuch zu rechtfertigen. Ich würde mich deshalb freuen, wenn Borchardt, statt in griechischen Quellen dilettantische Studien zu machen, endlich auf seine Versuchsgrabung eine systematische Ausgrabung folgen ließe, was letzten Endes eine Geldfrage ist. Die Kindereien von Herrmann haben der Sache natürlich nur geschadet.[52]

Mit diesen Bemerkungen über die „Kindereien" von Herrmann endet auch die dokumentierte Beschäftigung Spenglers mit der historischen Geographie. Wenn auch Spenglers Plan eines *Atlas antiquus* nicht verwirklicht wurde, zeigt sich anhand dieser Detailuntersuchung, dass Oswald Spengler innerhalb der Altertumswissenschaften erstaunlich gut vernetzt war. Die Frage, wer wen beeinflusste, ist dabei wohl nicht immer leicht zu klären. Spenglers quellenkritische Beschäftigung mit Herrmanns und Borchardts Ideen mögen wohl dem Einfluss Eduard Meyers geschuldet sein.

49 Spengler 1963, 503 (Oswald Spengler an Albert Hermann, München, 07.02.1927).
50 Spengler 1963, 562 (Eduard Meyer an Spengler, Berlin-Lichterfelde, 29.05.1928).
51 Spengler 1963, 562 (Eduard Meyer an Spengler, Berlin-Lichterfelde, 29.05.1928).
52 Spengler 1963, 565 (Oswald Spengler an Eduard Meyer, München, 30.06.1928).

LITERATUR

BAUMGARTNER, W., „Alfred Jermias (1864–1935)“, *Zeitschrift für Assyriologie* 43 (= N. F. 9), 1936, 299–301.

BÄR, Jürgen, „Walter Andrae – Ein Wegbereiter der modernen Archäologie“, in Joachim Mahrzahn, Beate Salje (Hrsg.), *Wiedererstehendes Assur*, Mainz am Rhein, 2003, 45–52.

BECKER, Carl Heinrich, „Spenglers Magische Kultur“, *Zeitschrift der Deutschen Morgenländischen Gesellschaft* 77, 1923, 255–271.

BICHLER, Reinhold, „Atlantis“, in: *Der Neue Pauly* Bd. 13, Stuttgart, 1999, Sp. 333–338.

BORCHARDT, Paul, „Platos Insel Atlantis – Versuch einer Erklärung mit 3 Skizzen und 2 Karten“, *Dr. A. Petermann's Mitteilungen* 73 (1927), Heft 7/8, 19–32.

DEMANDT, Alexander, „Eduard Meyer und Oswald Spengler. Lässt sich Geschichte voraussagen?“, in William M. Calder III, Alexander Demandt (Hrsg.), *Eduard Meyer: Leben und Leistung eines Universalhistorikers* (= Mnemosyne Supplement 112), Leiden, 1990, 159–181.

GAUGER, Jörg-Dieter, „Hans-Erich Stier (1902–1979) – Althistoriker, Mitgründer der CDU, Kulturpolitiker“ *Historisch-Politische Mitteilungen – Archiv für Christlich-Demokratische Politik* 14, 2007, 187–212.

HERRMANN, Albert, „Der Plan eines historischen und Wirtschafts-Atlasses von China und Zentralasien“, *Zeitschrift der Deutschen Morgenländischen Gesellschaft* 84, 1930, *113*–*114*.

HERRMANN, Albert, *Historical and Commercial Atlas of China* (= Harvard-Yenching Institute Monograph Series 1), Cambridge, Massachusetts, 1935.

KOKTANEK, Anton Mirko, *Oswald Spengler in seiner Zeit,* München, 1968.

LANDSBERGER, Benno, „Heinrich Zimmern“ (Nachruf), *Zeitschrift für Assyriologie* 40 (= N. F. 6), 1931, 133–143.

LEITNER, Irene Maria, „‚Bis an die Grenzen des Möglichen‘: Der Dekan Viktor Christian und seine Handlungsspielräume an der Philosophischen Fakultät 1938–1943“ in Mitchell G. Ash, Wolfram Nieß, Ramon Pils (Hrsg.), *Geisteswissenschaften im Nationalsozialismus*, Göttingen, 2010, 49–77.

OLSHAUSEN, Eckart, „Kirsten, Ernst“, in: Peter Kuhlmann, Helmuth Schneider (Hrsg.): *Geschichte der Altertumswissenschaften. Biographisches Lexikon* (Der Neue Pauly 12), Stuttgart, 2012, Sp. 656–657.

PESDITSCHEK, Martina, „Oswald Spengler als Prophet – für das Griechenland der minoisch-mykenischen Zeit“, in: H. Heftner, K. Tomaschitz (Hrsg.): *Ad Fontes! Festschrift für Gerhard Dobesch*, Wien, 2004, 901–905.

REBENICH, Stefan, *C. H. Beck 1763–2013: Der kulturwissenschaftliche Verlag und seine Geschichte*, München, 2013.

RÖLLIG, Wolfgang, „Der Tübinger Atlas des Vorderen Orients und seine altorientalischen Karten“, *Acta Antiqua Academiae Scientarum Hungaricae* 22, 1974, 531–537.

RÖLLIG, Wolfgang, „Aims and Organization of the ‘Répertoire geographique des textes cunéiformes’ and historical Geography“, *Sumer* 42, 1986, 40–43.

RÖLLIG, Wolfgang, „Ziele – Methoden – Ergebnisse der Arbeit am ‚Tübinger Atlas des Vorderen Orients“, in: Wolfgang Röllig (Hrsg.), *Von der Quelle zur Karte: Abschlußbuch des Sonderforschungsbereichs: ‚Tübinger Atlas des Vorderen Orients‘*, Weinheim / Basel / Cambridge / New York, 1991, 1–17.

SPENGLER, Oswald, *Briefe 1913–1936*, Herausgegeben von Anton M. Koktanek in Zusammenarbeit mit Manfred Schröter, München, 1963.

SPENGLER, Oswald, *Frühzeit der Weltgeschichte: Fragmente aus dem Nachlass*, Herausgegeben von Anton M. Koktanek, München, 1966.

ULF, Christoph; Rollinger, Robert (Hrsg.), *Lag Troja in Kilikien? Der aktuelle Streit um Homers Ilias*, Darmstadt, 2012.

WEIDNER, Ernst, „Viktor Christian“ (Nachruf), *Archiv für Orientforschung* 21, 1966, 258–259.

REGEN IN DER GRIECHISCH-RÖMISCHEN ANTIKE

Josef Fischer

1. EINFÜHRUNG

Die Kulturgeschichte des Regens in der klassischen Antike ist noch nicht geschrieben. Eine eingehende und alle Aspekte dieses facettenreichen Themas berücksichtigende Behandlung des flüssigen Niederschlags im griechisch-römischen Altertum stellt vielmehr ein Desiderat der klassischen Altertumswissenschaften dar.[1] Eine derartige umfassende Aufarbeitung kann freilich auch im Rahmen dieses lediglich in das Thema einführenden Artikels nicht geleistet werden, der sich auf drei ausgewählte Aspekte beschränken und anhand exemplarischer Quellentexte mögliche Antworten auf folgende Fragen geben will: Wie erklärte man in der Antike das Entstehen von Regen? Aufgrund welcher Indizien glaubte man Regen vorhersagen zu können? Und schließlich: Wie versuchte man die Entstehung von Regen zu beeinflussen?

2. DIE ENTSTEHUNG VON REGEN

Regen entsteht, vereinfacht gesagt, dadurch, dass Wasser unter dem Einfluss von Wind und Sonneneinstrahlung verdampft und aufsteigt. Infolge der Abkühlung beim Aufstieg bilden sich Tröpfchen und Eiskristalle und verbinden sich zu Wolken. Wird das Gewicht der Tropfen und Eiskristalle zu groß, sinken sie aufgrund der Schwerkraft wieder zu Boden, die Eiskristalle schmelzen erneut, und es regnet. Dieser hier nur sehr simplifizierend skizzierte Ablauf wurde bereits in der Antike erkannt. Im 4. Jh. v. Chr. beschrieb ihn der Philosoph und Naturforscher Aristoteles bereits in seiner *Meteorologie*, wo es heißt:

> Das Feuchte auf der Erde verdunstet unter der Einwirkung der Sonnenstrahlen und der sonstigen von oben kommenden Wärme und steigt nach oben. Aber wenn die Wärme, die es emporsteigen ließ, es verlässt, […] weil sie so hoch in die Luft über der Erde hinaufgeführt wurde, dann kühlt der Wasserdampf ab, kondensiert – eine Folge des Wärmeverlusts wie der hohen Region – und wird aus Luft zu Wasser; danach strebt er wieder der Erde zu.[2]

* Dieser Beitrag entspricht weitgehend meinem auf der Internationalen Konferenz „Wasser im Altertum“ (Berlin, 14.–18. 07. 2014) gehaltenen Referat. Ich danke Herrn Prof. Dr. Klaus Geus für das freundliche Angebot, meinen Beitrag an dieser Stelle zu publizieren.

1 Durchaus wird der Regen natürlich in Studien zur antiken Meteorologie berücksichtigt. Grundlegend und einführend dazu Taub (2003). Vgl. außerdem die Beiträge in Cusset (2003).

2 Aristot. *meteor.* 1, 9 (Übersetzung: H. Strohm).

Und im gleichen Kapitel dieses Werkes hält er noch einmal fest:

> Es steigt also fortwährend das Feuchte auf durch die Kraft des Warmen und fällt wieder zur Erde nieder infolge der Abkühlung.

Auf diesen Kreislauf kommt Aristoteles auch im zweiten Buch seiner *Zweiten Analytik* noch einmal zurück, wo er betont, dass all diese Vorgänge sich notwendigerweise vollziehen und einander bedingen:

> Wenn die Erde feucht geworden ist, entsteht notwendig Dunst, und wenn dieser entstanden ist, Bewölkung, und wenn diese entstanden ist, Regen; und wenn dieser entstanden ist, wird die Erde notwendig feucht. Dieses aber war das am Anfang, so dass man im Kreis herumgegangen ist. Wenn nämlich eines von ihnen, welches auch immer, besteht, so besteht ein anderes, und wenn jenes, ein noch anderes, und wenn dieses, das erste.[3]

Entsprechende Schilderungen des Wasserkreislaufs finden sich auch in anderen antiken Schriften; angeführt sei an dieser Stelle etwa jene Erklärung, die der römische Dichter und Philosoph Titus Lucretius Carus, ein Anhänger der Lehre Epikurs, in der ersten Hälfte des 1. Jh. v. Chr. in seinem Lehrgedicht *Von der Natur der Dinge* gibt:

> Lass mich jetzt auch erklären, wie droben in Wolken des Regens
> Nässe sich sammelt und dann sich als Guss auf die Erde hernieder
> senkt. Und zuerst will ich zeigen, dass Wasseratome in Mengen
> mit dem Gewölke zugleich aus sämtlichen Dingen sich heben,
> und dass beide sich dort gleichmäßig vermehren, die Wolken
> ebensowohl wie das Wasser, das in dem Gewölk ist enthalten:
> Geradeso wie auch bei uns mit dem Blute zugleich sich der Körper
> mehrt und der Schweiß und alle verschiedenen Säfte der Glieder.
> Auch aus dem Meer entnehmen sie oft viel Feuchtigkeit, wie es
> auch bei den wollenen Vliesen geschieht, die am Strande man aufhängt,
> wenn die Winde die Wolken hin über den Ozean jagen.
> Ähnlich erhebt sich die Feuchtigkeit auch aus sämtlichen Flüssen
> bis zum Gewölke empor. Sobald sich die Wasseratome
> viel und vielfach vereinen, von überallher sich vermehrend,
> dann entladen die Wolken, die voll sind von Wasser, den Regen
> um die Wette aus doppeltem Grund: teils presst sie der Winddruck,
> teils drückt eigene Fülle der Wolken, sobald sich die Massen
> stärker geballt, auch von oben und bringt so den Regen zum Ausfluss.
> Übrigens, selbst wenn der Wind die Wolken zerflattert und auflöst
> und auch die Sonne von oben die glühenden Strahlen versendet,
> rieselt aus ihnen noch Regen herab und tröpfelt, wie wenn man
> Wachs auf glühendem Ofen zerschmelzt und die Masse herabtropft.
> Aber ein heftiger Regen entsteht, wenn mit Heftigkeit beide
> Kräfte vereint Druck üben, der Wind und die Pressung der Wolken.
> Langandauernder Regen entsteht nur dann für gewöhnlich,
> wenn sich die Wasseratome in reichlicher Menge versammeln,
> wenn sich Gewölk auf Gewölk noch mit triefendem Nebel vereinigt
> und von jeglicher Seite dies allenthalben herannaht,
> wenn zudem auch die Erde die Feuchtigkeit ringsum empordampft.[4]

3 Aristot. *an. post.* 2, 96a (Übersetzung: W. Detel).
4 Lucr. 495–523 (Übersetzung: H. Diels).

Ebenso beschäftigt sich der römische Architekt Vitruv, der etwa eine Generation jünger als Lukrez war, in seinem Werk *Über die Architektur* mit der Entstehung des Regens, wo es im achten, der Wasserversorgung gewidmeten Buch folgendermaßen heißt:

> Wohin immer aber die aus den Quellen, Flüssen, Sümpfen und aus dem Meer sich zusammenballenden Dünste von den Winden getrieben werden, von der Sonnenwärme werden sie gesammelt, aufgeschlürft und so als Wolken in die Höhe gehoben; wenn sie dann, von der Luftströmung getragen, an die Berge gelangen, lösen sie sich durch den heftigen Widerstand dieser, in Folge ihrer eigenen Fülle und Schwere zu Wasser werdend, auf und ergießen sich so auf die Erde.
> Dass aber Dampf, Nebel und Dünste aus der Erde entstehen, scheint darin seinen Grund zu haben, weil die Erde sowohl glühende Hitze, als heftigen Lufthauch, Kälte und eine große Wassermasse in sich birgt. Sobald daher der in der Nacht abgekühlte Erdkreis plötzlich von der aufgehenden Sonne berührt wird, erhebt sich auch das Wehen der Winde aus den Finsternissen, und aus den feuchten Gegenden steigen die Wolken in die Höhe: die Luft aber, dann von der Sonne erwärmt, hebt mit dem Tau die aus der Erde kommenden Dünste empor.
> Man kann dafür ein Beispiel aus den Bädern nehmen. Es hat doch kein gewölbtes Zimmer in den warmen Bädern Brunnen oberhalb, sondern die Luft in denselben, aus den Öfen durch den Hauch des Feuers erwärmt, zieht vom Fußboden Wasser an sich, trägt es mit sich an die gewölbte Decke empor und hält es dort; weil sich daher immer warmer Dampf in die Höhe drängt und anfangs wegen der geringen Menge nicht wieder herabfällt, so tropft er endlich, sobald sich einmal mehr Dunst angesammelt hat, auf die Köpfe der Badenden herab. Auf diese Art hebt auch die freie Luft, sobald sie von der Sonne Wärme empfangen, die Feuchtigkeit, die sie allenthalben aufschlürft, empor und ballt sie zu Wolken zusammen. Die Erde entsendet auch ebenso in Folge der Hitze die Dünste, wie der menschliche Körper in Folge von Wärme Schweiß absondert.[5]

Neben rein naturwissenschaftlichen Erklärungen finden sich in der antiken Literatur aber auch immer wieder religiöse Begründungen von Niederschlag, wie sie etwa wiederum bei Aristoteles anklingen, wenn er in seiner *Physik* erneut auf den Wasserkreislauf und die Notwendigkeit der damit verbundenen Abläufe zu sprechen kommt, die ohne – wie auch immer geartetes positives oder negatives – Ziel geschehen:

> Warum soll es denn undenkbar sein, dass die Natur ohne alle Finalität und Rücksicht auf das Bessere arbeite, (dass vielmehr alles in der Natur nur der Regel der Notwendigkeit folge,) wie beispielsweise Zeus es regnen lasse, nicht um das Getreide wachsen zu lassen, sondern aus reiner Notwendigkeit – denn die aufgestiegene Luft müsse abkühlen und die abgekühlte Luft müsse, zu Wasser geworden, herunterfallen; sei es aber dazu gekommen, so sei das Wachstum des Getreides bloß eine (beiläufige) Folge der Umstände.[6]

Regen ist für viele ein Werk der Götter, und vor allem der Göttervater Zeus, der Gott des Himmels und des Wetters, bzw. sein römisches Pendant Iuppiter werden dafür verantwortlich gemacht.[7] Die Verehrung als *Zeus Hyetios* oder als *Zeus Om-*

5 Vitr. 8, 2, 2–4 (Übersetzung: F. Reber).
6 Aristot. *phys*. 2, 8 (Übersetzung: H. Wagner).
7 Interessant ist in diesem Zusammenhang, dass Zeus im Mythos auch die Gestalt eines goldenen Regens annimmt, um zu seiner Geliebten Danaë zu gelangen.

brios bzw. als *Iuppiter Pluvialis* ist vielerorts belegt.[8] Wie verbreitet diese Ansicht war, zeigt am besten das Beispiel der Komödie *Die Wolken* des Aristophanes aus dem 5. Jh. v. Chr.: Als Sokrates dort postuliert, dass es keine Götter, und eben auch keinen Zeus gebe, antwortet sein Gesprächspartner, der Bauer Strepsiades: „*Und wer regnet dann?*"[9]

Selbstverständlich wurden nicht nur Zeus bzw. Iuppiter mit Regen in Verbindung gebracht, sondern auch andere Götter und übernatürliche Wesen.[10] Bekannt ist etwa jene Geschichte, die Herodot, ebenfalls im 5. Jh. v. Chr., im ersten Buch seiner *Historien* überliefert: Der lydische König Kroisos sollte, nachdem er dem persischen König Kyros in die Hände gefallen war, auf dem Scheiterhaufen verbrannt werden. Bereits auf diesem stehend, soll Kroisos dann dreimal den Namen des Atheners Solon gerufen haben, der ihn davor gewarnt hatte, sein Leben ohne Kenntnis von dessen Ende glücklich zu nennen. Kyros fragte nach, was dies zu bedeuten hätte, und befahl, nachdem er die Erklärung des Lyders vernommen hatte, diesen zu verschonen. Allerdings ließ sich, wie Herodot kolportiert, das Feuer nicht mehr löschen:

> Als Kroisos, so erzählen die Lyder, die Sinnesänderung des Kyros erfuhr und sah, wie jedermann das Feuer zu löschen versuchte, es aber nicht mehr bewältigen konnte, da schrie er laut und rief zu Apollon, er möge ihm doch beistehen und ihn aus der gegenwärtigen Not erretten, wenn er ihm je etwas Willkommenes geschenkt habe. So rief er in Tränen zum Gott. Da ballten sich am heiteren Himmel und bei Windstille plötzlich Wolken zusammen, ein Unwetter prasselte nieder, und es regnete in Strömen. So erlosch der Scheiterhaufen.[11]

Hier ist es also Apollon, der mit einem plötzlich gesandten Regenguss Kroisos zu Hilfe kommt. In der ältesten Überlieferung dieser Geschichte, die sich im 3. *Epinikion* des Bakchylides findet, ist es freilich abermals Zeus, der den Scheiterhaufen, den in dieser Version Kroisos selbst entzündet hatte, um sich und seine Familie zu verbrennen, löschte, ehe dann Apollon den Lyderkönig und seine Angehörigen zu den Hyperboreern entrückte.

Noch bekannter als die Geschichte um Kroisos ist vermutlich jenes Regenwunder, das sich im 2. Jh. n. Chr. während des Markomannenkrieges des römischen Kaisers Marcus Aurelius im Land der Quaden ereignete.[12] Aufgrund glühender

8 Pausanias berichtet etwa von der Verehrung des *Zeus Hyetios* in Lebadeia (Paus. 9, 39, 4) oder der des *Zeus Ombrios* auf dem Berg Hymettos (Paus. 1, 32, 2). Auf dem Berg Arachnaion waren, wie wiederum Pausanias (Paus. 2, 25, 10) kolportiert, „Altäre des Zeus und der Hera errichtet, und wenn es ihnen an Regen fehlt, bringen sie den Gottheiten hier Opfer dar" (Übersetzung: J. Laager).

9 Aristoph. *nub.* 366 ff. Pausanias (Paus. 1, 24, 3) berichtet, dass sich auf der Akropolis von Athen eine Statue der Göttin Ge (Erde) befand, die „Zeus um Regen bittet, sei es dass die Athener selbst des Regens bedurften, sei es, dass alle Griechen von einer Dürre betroffen waren" (Übersetzung: J. Laager).

10 In den in hellenistischer Zeit entstandenen orphischen Hymnen werden der Südwind (*Notos*) und die Wolken als Regenspender angesprochen (9; 82).

11 Hdt. 1, 87 (Übersetzung: J. Feix).

12 Zu diesem Regenwunder, das auf der Mark-Aurels-Säule in Rom auch bildlich dargestellt ist, vgl. etwa Jobst (1978); Barta (1979); Rubin (1979); Fowden (1987); Sage (1987); Kovács (2008a); Kovács (2008b); Israelowich (2008).

Hitze und Trockenheit war ein römisches Heer in größte Bedrängnis geraten, ehe es durch plötzlich einsetzenden Regen gerettet wurde. Welcher Gottheit für dieses Regenwunder zu danken war, wurde schon in der Antike heftig diskutiert: der frühchristliche Autor Tertullian berichtet in seiner bereits am Ende des 2. Jh. n. Chr. verfassten *Verteidigung des Christentums* davon, dass die Gebete der Christen im Heer diesen Regen bewirkt haben sollen.[13] Und auch der Kirchenvater Eusebius schreibt in seiner im 4. Jh. n. Chr. entstandenen *Kirchengeschichte* das Wunder den christlichen Soldaten zu:

> Von Mark Aurel wird erzählt, dass er im Kampfe mit den Germanen und Sarmaten in große Not geriet, weil sein Heer von Durst gequält wurde. Da knieten sich die Soldaten der sog. melitenischen Legion, als sie schon dem Feinde gegenüber Stellung genommen hatten, auf den Boden, wie es bei uns während des Betens Brauch ist, und flehten zu Gott. Dieser Anblick schon erschien den Feinden wunderbar. Aber es sollte sogleich, wie die Erzählung weiß, noch etwas viel Wunderbareres folgen: ein Unwetter, das die Feinde in Flucht und Verderben trieb, und ein Regen, der über die Truppe mit den Betern sich ergoss und der gesamten Mannschaft, nahe daran, an Durst zu sterben, Erquickung brachte. Diese Geschichte wird sowohl von nichtchristlichen Schriftstellern, welche über die damalige Zeit geschrieben haben, berichtet, als auch von unseren eigenen Geschichtsschreibern mitgeteilt. Aber die heidnischen Schriftsteller erwähnen zwar das Wunder, geben indes, weil dem Glauben fremd, nicht zu, dass es auf unsere Bitten hin erfolgt ist.[14]

Eine der erwähnten heidnischen Versionen überliefert etwa der römische Politiker und Geschichtsschreiber Cassius Dio:[15] nach seinem im frühen 3. Jh. n. Chr. verfassten Bericht soll *Hermes Aerios* aufgrund der Praktiken eines ägyptischen Magiers namens Arnuphis eingegriffen haben. Das byzantinische Lexikon der *Suda* wiederholt diese Angabe, bringt aber auch die Variante, dass der chaldäische Magier Julianos verantwortlich gewesen sei.[16] Nach den im 3. Jh. n. Chr. verfassten *Oracula Sibyllina* soll dagegen das persönliche Gebet des Kaisers an eine namentlich nicht genannte Gottheit, bei der es sich nach allgemeiner Ansicht aber wohl um Iuppiter gehandelt haben dürfte, den Regen bewirkt haben;[17] auch Claudius Claudianus (4./5. Jh. v. Chr.) bringt das Wunder in seinem *Panegyricus auf den sechsten Konsulat des Honorius* mit der Person des Marcus Aurelius und seiner Tugend in Verbindung.[18] Die Fähigkeit, in Kontakt mit den Göttern zu treten und derartige Wunder zu erbitten, wurde in der Antike in vielen Fällen mit Herrschergestalten verbunden; sie gehörte zu ihrem Charisma.

3. DIE VORHERSAGE VON REGEN

Die Beobachtung der Natur und ihrer Abläufe, der Jahreszeiten, des sich regelmäßig verändernden Sternenhimmels und von immer wieder kehrenden Wetterphänome-

13 Tert. *apol.* 5.
14 Eus. *HE* 5, 5 (Übersetzung: P. Häuser).
15 Cass. Dio 71, 8–10.
16 Suda ι 334 s. v. Iulianos.
17 *Or. Sib.* 12, 194–200.
18 Claud. *VI. cons. Hon.* 339–350.

nen hat schon früh zu Versuchen geführt, Gesetzmäßigkeiten im Verlauf des Wetters zu erkennen und allgemeingültige Regeln zu formulieren. Solche Wetterregeln finden sich bereits um 700 v. Chr. in den *Werken und Tagen* des in Boiotien ansässigen Dichters Hesiod.[19] Freilich regte sich auch Widerstand gegen die Gültigkeit dieser aus der Beobachtung des Jahreslaufs und des Sternenhimmels abgeleiteten „Bauernregeln“. So warnt etwa Plinius der Ältere in seiner im 1. Jh. n. Chr. verfassten *Naturgeschichte* davor, Wetteränderungen an bestimmten Tagen zu erwarten,[20] während sein Zeitgenosse Lucius Iunius Moderatus Columella, der Verfasser eines bedeutenden Werkes *Über die Landwirtschaft*, einwandte, dass Wettererscheinungen nicht fest mit bestimmten Daten oder Konstellationen am Sternenhimmel verbunden werden könnten.[21] Dennoch bietet Columella genau solche Angaben, wenn er etwa den Frühaufgang der Plejaden im Mai mit West- bzw. Nordwestwinden und Regenfall verbindet (Colum. 11, 2, 39).

Neben solchen „Bauernkalendern“, die immer wiederkehrende Trends verzeichnen, gibt es auch viele „kurzfristige“, aus der unmittelbaren Beobachtung der Umwelt resultierende Wetterzeichen, die über nahe Wetterereignisse Auskunft geben sollten. Ein wichtiges Werk, das sich mit jenen Erscheinungen und Vorkommnissen beschäftigt, die man in der Antike mit Regen in Verbindung brachte, ist die Schrift *Über Wetterzeichen*, die Theophrast von Eresos (4./3. Jh. v. Chr.), dem Nachfolger des Aristoteles als Vorstand des Lykeion, zugeschrieben wird. Sie listet eine Reihe von Anzeichen auf, die auf kommenden Niederschlag hindeuten, so etwa eine Rötung des Himmels bei Sonnenaufgang oder Sonnenuntergang, dunkle Flecken auf der Sonne, Wolken, aus denen sich die Sonne erhebt oder in die sie untergeht, einen wolkenverhangenen Mond, eine große Anzahl von Sternschnuppen, Blasen, die in großer Anzahl an die Wasseroberfläche von Flüssen steigen, oder das Flackern von Lampen.[22]

Eine weitere wichtige Quelle stellt das im 3. Jh. v. Chr. verfasste astronomische Lehrgedicht *Über die Himmelserscheinungen* des aus Kilikien stammenden, dann aber lange am makedonischen Hof tätigen Dichters Aratos dar, das auf dem verlorenen Prosawerk des bedeutenden Astronomen Eudoxos von Knidos aus dem 4. Jh. v. Chr. basiert. Auch Aratos empfiehlt die Beobachtung von Sonne und Mond: so sei etwa ein Lichtring um den Mond ein Anzeichen für kommenden Regen. Weiter heißt es dort:

> Wenn aber bei ihrem Aufgang oder auch wieder bei ihrem Untergang die Strahlen (der Sonne) zusammengehen und sich um die Mitte verdichten, oder auch wenn die Sonne von Wolken bedrückt, sei es zum Tage kommt von der Nacht her, sei es aus dem Tage zur Nacht geht, so

19 Z.B. Hes. *erg.* 414: *Lindert aber die mächtig sengende Sonne schon ihre schweißtreibende Hitze, schickt der machtvolle Zeus Herbstregen.* 447–450: *Aufgemerkt, sobald du den Schrei des Kranichs hörst, der vom hohen Gewölk wie jedes Jahr seinen Ruf herabschickt, das Zeichen zum Pflügen gibt und die Zeit des Winterregens ankündigt.* 485–488: *Ruft der Kuckuck erstmals im Jahr „Kuckuck“ im Laub der Eiche und erfreut die Menschen auf der unendlichen Erde, dann lässt es Zeus vielleicht drei Tage lang regnen, nicht mehr, als dass es die Spur eines Maultiers füllt, doch auch nicht weniger* (Übersetzung: O. Schönberger).

20 Plin. *nat.* 18, 231.

21 Colum. 11, 1, 32

22 Theophr. *de sign.* 10–25.

dürfte sie wohl bei niedergehendem Regen diese Tage durchlaufen. Auch wenn eine kleine Wolke ihr zuvor aufgeht, hinter dieser dann ohne Strahlen sie selber sich hebt, sei nicht unbekümmert um Regen. [...] Doch wenn es tagsüber Regen gibt, schau darauf nach den Wolken, zur untergehenden Sonne gewandt: Wenn eine Wolke, die wie geschwärzt aussieht, die Sonne verdüstert und die Strahlen sich dahin und dorthin um die Wolke herum, die sich in der Mitte wälzt, teilen, dann wahrlich brauchst du wohl noch bis in den Morgen hinein ein Obdach. [...] Auch wenn sie sich zu erscheinen anschickt und Wolken vor dem Morgenrot, unten gerötet, von hier und von dort erscheinen, werden an diesem Tag die Fluren nicht ungenetzt bleiben.[23]

Die Schrift des Aratos hat in der Antike weitreichenden Einfluss erlangt. Dies bezeugen nicht nur zahlreiche antike Kommentare, sondern auch Übersetzungen ins Lateinische, wie sie etwa für Cicero, Germanicus, den Großneffen des Augustus, oder den Dichter Avienus im 4. Jh. n. Chr. belegt sind. Deutlich sind die Bezüge zu Aratos auch in den *Georgica* des römischen Nationaldichters Vergil (1. Jh. v. Chr.), der ebenfalls dazu auffordert, Sonne und Mond zu beobachten. Auch er nennt Flecken und Wolken vor der aufgehenden Sonne, die Art und Weise wie die Strahlen der aufgehenden Sonne durch die Wolken brechen, oder bestimmte Färbungen der untergehenden Sonne genauso als Hinweise auf kommenden Regen wie trübes Mondlicht. Vergil versucht aber auch aus dem Verhalten von Tieren Hinweise über die Wetterentwicklung zu gewinnen:

Nie schadet prasselnder Regen,
ohne zu warnen. Denn wenn er sich naht, so flieht in die tiefsten
Täler der Kranich, der hoch sonst fliegt, die Färse blickt ängstlich
auf zum Himmel und schnaubt mit geblähten Nüstern die Luft ein,
oder den See umschnellt mit schrillem Gezwitscher die Schwalbe,
quakend plärren die Frösche im Sumpf die uralte Klage.
Oft auch trägt aus den Zellen des Haufens heraus ihre Eier
über den engen Steig die Ameise. Wasser aufsaugend
wölbt der Regenbogen sich weit; aufflatternd vom Fraße
lärmt der Raben krächzendes Heer, dicht Flügel an Flügel.
Allerlei Vögel der See, was sonst noch Kleinasiens Wiesen
rings in süßem Gewässer durchsucht am Lauf des Cayster,
sieht man die Schultern wetteifernd netzen, tautropfenperlend,
bald den Kopf darbieten dem Gischt, bald rennen ins Sturzbad
schäumender Wellen, ruhlos bemüht und vergeblich, zu baden.
Dann schreit krächzend die Krähe, der Unheilsvogel, nach Regen,
schreitet einsam stelzend einher im trockenen Sande.[24]

Eine lange Liste von Wetterzeichen bietet auch Plinius im 18. Buch seiner *Naturgeschichte*. Auch er empfiehlt zunächst die Beobachtung der Sonne:

23 Arat. 840–868 (Übersetzung: M. Erren).

24 Verg. *georg*. 1, 373–389 (Übersetzung: K. Bayer). Gerade die Krähe wird oftmals als Botin kommenden Regens angesehen, so etwa auch bei Hor. *carm*. 3, 17. Vom Raben als Verkünder künftigen Niederschlags spricht Plin. *nat*. 18, 362 (siehe unten). Ein Tier, das vielfach mit Regen assoziiert wird, ist das Schwein, so etwa bei Ail. *nat*. 7, 8. Es beruht dies vermutlich auf einer fälschlichen Verbindung von ὗς und ὕειν. Auch Bienen wurden als gute Wetterpropheten betrachtet, denn man konnte beobachten, dass sie ihren Stock nicht verließen, wenn Regen drohte.

> Geht sie von Wolken verhüllt auf, so kündigt sie Regen an; ebenso Winde, wenn sich vor ihrem Aufgang die Wolken röten; ferner Regen, wenn zwischen roten Wolken auch dunkle erscheinen; wenn ihre Strahlen beim Auf- und Untergang zusammenzulaufen scheinen, so zeigt das Regen an. [...] Wenn die Wolken beim Sonnenaufgang teils nach Süden, teils nach Nordosten verstreut sind, deuten sie, mag es auch noch so klar sein, dennoch Regen und Wind an; wenn beim Sonnenauf- oder -untergang die Strahlen verkürzt erscheinen, bedeutet dies Regen. Wenn es bei Sonnenuntergang regnet oder die Sonnenstrahlen eine Wolke an sich ziehen, kündigt dies für den folgenden Tag raues Wetter an. Wenn beim Aufgang der Sonne die Strahlen nicht leuchtend hervorbrechen, bedeuten sie Regen, auch wenn sie von keiner Wolke umgeben sind.[25]

Als nächstes Wetterzeichen gilt Plinius der Mond:

> Wenn der Mond hell strahlend mit reinem Glanz aufgeht, kündigt er heiteres Wetter an; wenn rötlich, Wind; wenn dunkel, für 15 Tage Regen. Sind seine Hörner abgestumpft, bedeutet dies Regen; aufgerichtet und spitz sind sie stets ein Vorzeichen für Winde, besonders aber am vierten Tage. [...] Wenn ihn am vierten Tag ein rötlicher Kreis umgibt, zeigt dies Winde und Regengüsse an.[26]

Des Weiteren empfiehlt Plinius die Beobachtung der Sterne, aber auch Phänomene wie Regenbogen oder Blitz und Donner sind Indikatoren für zukünftigen Regen:

> Doppelte Regenbögen deuten auf Regen; nach dem Regen auf heiteres, jedoch unzuverlässiges Wetter; ein Wolkenring um einige Sterne zeigt ebenfalls Regen an. Wenn es im Sommer heftiger donnert als blitzt, kündigt dies Winde aus derselben Richtung an; wenn es dagegen weniger gedonnert hat, Regen. [...] Blitzt es nur aus Nordost, so bedeutet dies für den folgenden Tag Nässe; wenn aus dem Norden, bedeutet dies Wind aus dieser Richtung. Blitzt es in heiterer Nacht aus Süden, Nordwesten oder Westen, so zeigt dies Wind und Regen aus denselben Richtungen an. Donner am Morgen bedeutet Wind, Donner am Mittag Regen.[27]

Von besonderer Aussagekraft ist auch die Beobachtung der Wolken:

> Ziehen Wolken am heiteren Himmel dahin, so ist, aus welcher Richtung auch immer dies geschieht, Wind zu erwarten. Wenn sie sich an dieser Stelle zusammenballen, von der herannahenden Sonne zerstreut werden und dies von Norden her geschieht, bedeutet es Winde, wenn von Süden, kündigen sich Regen an. Wenn sie bei Sonnenuntergang von beiden Seiten des Himmels emporsteigen, zeigt dies schlechtes Wetter an. Ziehen schwarze Wolken heftiger von Osten her, so drohen sie für die Nacht Regen an, wenn von Westen, dann erst für den folgenden Tag. Wenn sich die Wolken wie Schafwollflocken zahlreich von oben her zerstreuen, kündigen sie Nässe für drei Tage an.[28]

Schließlich gibt Plinius der Ältere an, wie man aus dem Verhalten von Tieren auf das Wetter schließen könne. Auf Regen lässt ihm zufolge etwa das Krächzen von Raben schließen:

> Wenn die Raben andauernd mit einem eigenartigen Schluchzen krächzen und sich schützen, deuten sie auf einen heiteren Tag; wenn sie aber die Töne teilweise verschlucken, gibt es windreiches Regenwetter.[29]

25 Plin. *nat.* 18, 342–344 (Übersetzung: R. König).
26 Plin. *nat.* 18, 347 (Übersetzung: R. König).
27 Plin. *nat.* 18, 353 (Übersetzung: R. König).
28 Plin. *nat.* 18, 355 (Übersetzung: R. König).
29 Plin. *nat.* 18, 362 (Übersetzung: R. König). Aber nicht nur im 18. Buch, auch an anderen Stellen kommt Plinius der Ältere auf Beobachtungen zu sprechen, die auf die kommende Witte-

Der Sophist Claudius Aelianus (2./3. Jh. n. Chr.) befasst sich in seinen *Tiergeschichten* ebenfalls mit Wetterzeichen, und er betont, dass Hunde, Ochsen, Schweine, Ziegen, Schlangen und andere Tiere das Wetter genauso vorhersagen könnten wie den Ertrag von Ernten, Hungersnöte, Erdbeben und Seuchen, obwohl ihnen selbst der Verstand fehle.[30] Im siebten Buch dieses Werkes kolportiert er, dass bei den Libyern die Ziegen kommenden Regen ankündigen:

> Diese zeigen auch den bevorstehenden Regen an. Denn wenn sie aus den Ställen kommen, eilen sie im schnellsten Laufe nach dem Futter; und wenn sie sich gesättigt haben, kehren sie sich der Heimat zu, und immer den Blick dorthin gerichtet, weiden sie nicht mehr, in der Erwartung, dass der Hirt sie schnell zusammentreiben soll.[31]

Und im selben Abschnitt dieses Werkes schreibt er die gleiche Fähigkeit den Ochsen zu:

> Wenn der Ochs brüllt und die Erde beriecht, so muss es regnen; wenn er viel und mehr als gewöhnlich frisst, zeigt er Unwetter an.

Einige der bekanntesten antiken Wetterzeichen werden von Vögeln gegeben.[32] Dies liegt sicher daran, dass die Menschen der Antike vielfach beobachten konnten, wie sensibel Vögel auf wechselnden Luftdruck und aufkommenden Wind reagieren. Dies zeigt sich etwa daran, dass Möwen vor einem drohenden Sturm ins Landesinnere fliegen, oder dass Krähen und andere Vögel vor Regen lauter sind.[33] Eben diese Phänomene schildert auch Aelianus in seinen *Tiergeschichten*:

> Aristoteles sagt, wie ich höre, dass die Kraniche, wenn sie vom Meere her nach dem Lande fliegen, den Kundigen eine nachdrückliche Warnung des kommenden Sturmes sind; fliegen sie aber ruhig dahin, so verheißen sie günstige Witterung und ruhigen Luftstand, und geben durch ihr Stillschweigen denen, die der Sache nicht unkundig sind, zu erkennen, dass heiterer Himmel ist. Wenn sie aber niedrig fliegen und schreien und Unruhe machen, und selbst Unruhe zeigen, so drohen sie auch hierdurch heftigen Sturm. Wenn der Reiher in der Dämmerung schreit, scheint er dasselbe anzuzeigen, wie Aristoteles aus eigener Beobachtung sagt; fliegt er aber gerade nach dem Meere hin, so zeigt er heftige Regengüsse an. Bei stürmischem Wetter verkündigt die Stimme der Eule Heiterkeit und einen hellen Tag; wenn aber das Wetter heiter ist und die Eule schreit, so muss man Sturm erwarten. Wenn der Rabe schnell und hastig schwatzt und mit den Flügeln schlägt und klatscht, erkennt er zuerst den nahenden Sturm. Ferner lehrt der Rabe, die Krähe und die Dohle, wenn sie am späten Abend schreien, dass ein Sturm im Anzug ist. Wenn die Dohlen nach der Weise des Habichts schreien und bald höher, bald niedriger fliegen, zeigen sie, wie jener sagt, Kälte und Regen an. Wenn die Krähe zur Zeit des Abendessens leise ruft, so verkündigt sie für den folgenden Tag heiteres Wetter. Die Erscheinung zahlreicher und weißer Vögel belehrt über heftigen Sturm. Wenn das Rotkehlchen in den Ställen und Wohnungen umhergeht, so gibt es zu erkennen, dass Sturm im Anzug ist. Hähne und Haushühner zeigen auch Sturm an, wenn sie mit den Flügeln schlagen, sich unruhig gebärden und dumpfe Töne von sich geben. Auch die Hühner drohen mit Wind, wenn sie sich

rung schließen lassen. So leitet er Wettervorhersagen etwa auch von Spinnennetzen ab, deren gehäuftes Auftreten dem römischen Forscher zufolge auf drohenden Regen schließen lassen: Plin. *nat.* 11, 84.

30 Ail. *nat.* 6, 16.

31 Ail. *nat.* 7, 8 (Übersetzung: F. Jacobs).

32 Vögel sind in der Antike auch ein übliches Medium, um den Willen der Götter kundzutun.

33 Vgl. Plin. *nat.* 18, 362.

> baden und kündigen Windstöße an. Im Winter zeigen die Hühner, wenn sie gegeneinander fliegen und unter einander herlaufen, heitere Witterung an. Wenn sich die Vögel um die Seen und an den Ufern der Flüsse versammeln, so haben sie von dem bevorstehenden Winter Kenntnis. Wiederum ist den Meer- und Seevögeln, wenn sie ans Land gehen, nicht unbekannt, dass heftiger Sturm bevorsteht; wenn aber die Landvögel nach den feuchten Gegenden eilen, verkündigen sie heitere Witterung, vorausgesetzt, dass sie schweigen.[34]

Selbstverständlich lässt sich die Wirkung der frühen Schriften über Wetterzeichen bzw. das Fortleben der Traditionen zur Vorhersage von kommendem Regen noch viel länger verfolgen. So finden sich etwa in den sog. *Geoponika*, einer Sammlung von landwirtschaftlichen Schriften, die im 10. Jh. n. Chr. in Byzanz für Kaiser Konstantin VII. Porphyrogennetos angefertigt wurde, ganz analoge Hinweise: auch hier ist davon die Rede, dass man durch die Beobachtung von Sonne und Mond, aufgrund von bestimmten Erscheinungen bei Regenbögen oder anhand des Verhaltens von Vögeln, Insekten, Haus- und Wildtieren auf baldigen Regen schließen könne.[35] Dass viele der überlieferten Wetterzeichen einander entsprechen, hat nicht nur damit zu tun, dass die jeweiligen Autoren ihre Vorgänger gut kannten und rezipierten. Neben selbstverständlich erfolgten literarischen Übernahmen ist auch damit zu rechnen, dass die Kenntnis vieler der angeführten Regeln weitverbreitet war und ein Allgemeingut darstellte.

Die Wetterzeichen wurden nach populärer Auffassung von den Göttern geschickt, um den Menschen zu helfen.[36] Die Fähigkeit, sie zu deuten und damit das Wetter vorherzusagen, wurde in der antiken Überlieferung immer wieder Männern zugeschrieben, die als besonders weise galten. So berichtet etwa Aelianus in seinen *Tiergeschichten* vom Philosophen Anaxagoras, dem wegen seiner Vorhersage von Regen besonderer Ruhm zugekommen sein soll:

> Und als in Olympia Anaxagoras ebenfalls im Regenmantel den Spielen zusah, und ein Regenguss erfolgte, pries ihn ganz Hellas, dass er höhere Einsicht besäße als man menschlichen Naturen zutrauen sollte.[37]

Selbiges berichtet Plinius der Ältere vom vorsokratischen Philosophen Demokrit:

> Man erzählt, dass der bereits erwähnte Demokrit seinen Bruder Damasos, der gerade erntete, bei glühendster Hitze bat, die übrige Saat stehen zu lassen und das bereits Abgeschnittene schnell unter Dach und Fach zu bringen; und nur wenige Stunden später habe ein heftiger Regen seine Vorhersage als richtig erwiesen.[38]

34 Ail. *nat.* 7, 7 (Übersetzung: F. Jacobs).

35 *Geopon.* 1, 3.

36 Dies hält etwa Aratos gleich zu Beginn seines oben bereits besprochenen Werkes fest (Arat. 1–14). Im griechischen Kult wurde dieses Geben von Vorzeichen durchaus gewürdigt; so berichtet Pausanias in seiner *Beschreibung Griechenlands* etwa von einem Altar des *Zeus Semaleos* („Zeus, der Zeichengeber"), der auf dem Parnes in Verbindung mit *Zeus Ombrios*, also Zeus in seiner Funktion als Regengott, verehrt wurde (Paus. 1, 32, 1). Der römische Philosoph Seneca macht sich über diese Ansicht, dass die Götter den Menschen Zeichen zur Deutung des zukünftigen Wetters schicken würden, freilich lustig (Sen. *nat. quaest.* 2, 32, 3).

37 Ail. *nat.* 7, 8 (Übersetzung: F. Jacobs).

38 Plin. *nat.* 18, 341 (Übersetzung: R. König). Dieselbe Geschichte findet sich auch in den *Teppichen* des Theologen und Kirchenschriftstellers Klemens von Alexandrien, eines Zeitgenossen des Aelianus (Clem. Alex. strom. 6, 3, 32, 2): *Demokrit aber wurde weise genannt, weil er auf*

Die meisten antiken Wetterzeichen betrafen Pflanzen, Tiere oder Himmelserscheinungen, aber auch Haushaltsgeräte wurden genau beobachtet, um daraus Erkenntnisse zu gewinnen, so etwa Lampen, bei denen etwa aus der Helligkeit der Flamme oder ihrem Flackern Rückschlüsse gezogen wurden.[39] Lampen spielten auch in der Magie eine wichtige Rolle; der *Lychnomantie*, der Wahrsagerei aus Lampen, kommt etwa in den magischen Papyri eine wichtige Rolle zu. In seinem Roman *Metamorphosen*, der auch unter dem Titel *Der goldene Esel* bekannt ist, lässt Apuleius (2. Jh. n. Chr.) die Hexe Pamphile auftreten, die eine Lampe verwendet, um Regen vorherzusagen:

> Wie es etwas später war, sah Pamphile die Lampe an und rief „Ei, was werden wir morgen für Regen kriegen!" – „Woher weißt du denn das?" fragte der Mann. – „Das sagt mir die Lampe", antwortete sie. Milo fing darüber laut zu lachen an und sprach: „Potz Stern! Hab' ich doch nimmermehr gedacht, dass wir an unsrer Lampe eine so kluge Sibylle hätten, die auf ihrer Leuchterwarte alle Verrichtungen des Himmels und selbst die Sonne beobachtet!" Darauf nahm ich das Wort. „Oh, von dergleichen Weissagungen", sprach ich, „hat man Beispiele die Menge. Sie lassen sich auch leicht erklären. Denn ist gleich dieses Feuer klein und nur durch Menschenhände angezündet, so waltet dennoch zwischen ihm und dem großen himmlischen Feuer (wovon es ursprünglich abstammt) eine starke Sympathie. Es kann daher nicht nur selbst durch geheime Ahnungen die Veränderungen des hohen Äthers vorauswissen, sondern sie uns auch recht wohl vorher verkündigen.[40]

4. DIE MANIPULATION DES REGENS

Die starke Abhängigkeit der antiken Menschen vom Wetter führte auch dazu, dass man nichts unversucht ließ, die Naturgewalten zu beeinflussen.[41] Man flehte zu den Göttern,[42] wenn der Regen ausblieb und – was in der Antike vielerorts und häufig vorkam – Dürre das Land bedrohte. Natürlich war es besonders Zeus, an den man sich in solchen Fällen wandte. Eine bekannte Geschichte findet sich in der griechischen Mythologie. Wegen der Ermordung des Stymphalos durch Pelops hatte Zeus Dürre und Unfruchtbarkeit über Griechenland geschickt, und erst als sein eigener

Grund seiner Beobachtung der Vorgänge am Himmel vieles voraussagte. So sagte er z. B., als ihn sein Bruder Damasos gastfreundlich bei sich aufgenommen hatte, vorher, dass ein starker Regenguss kommen werde, was er aus Beobachtung gewisser Gestirne erschlossen hatte. Wer ihm nun glaubte, der brachte das Getreide in Sicherheit (dies war nämlich, wie zur Sommerszeit üblich, noch auf den Dreschplätzen im Freien), den anderen aber ging alles zugrunde, da unerwartet ein gewaltiger Gussregen herabströmte (Übersetzung: O. Stählin).

39 Vgl. Theophr. *de sign.* 14, 34, 42, 54; Aristot. *meteor.* 347a 20–32; Verg. *georg.* 1, 390; Plin. *nat.* 18, 357

40 Apul. *met.* 2, 11 (Übersetzung: A. Rode).

41 Zum antiken Wetterzauber vgl. etwa; McCartney (1925); Fiedler (1931). Speziell zum Regenzauber siehe Gruppe (1906) 818–834; Samter (1922). Zum antiken und modernen Regenzauber vgl. auch Håland (2001). Dass der Wetterzauber in der antiken Magie eine besondere Stellung einnahm, zeigt sich schon darin, dass, als in konstantinischer Zeit jede Magie verboten wurde, der Wetterzauber und der Heilzauber erlaubt blieben (*Cod. Iust.* 9, 18, 4).

42 Von Gelübden, die Regen herbeirufen, spricht etwa Vergil in seinem *Landleben* (Verg. *georg.* 1, 157).

Sohn Aiakos, der als der frommste der Griechen galt, zu ihm betete und ihm opferte, sandte er Regen, wie etwa Klemens von Alexandria in seinem Werk *Teppiche* berichtet:

> Als einmal eine lange andauernde Dürre Griechenland dem Verderben nahe brachte und Misswuchs herrschte, da kamen, wie erzählt wird, die Griechen, die noch am Leben waren, wegen der Hungersnot als Bittflehende nach Delphi und fragten die Pythia, auf welche Weise sie aus der Not erlöst werden könnten. Sie antwortete ihnen aber, es gebe nur ein Mittel der Hilfe gegen das Unglück, wenn sie nämlich das Gebet des Aiakos verwendeten. Aiakos ließ sich also von ihnen gewinnen, stieg auf den griechischen Berg hinauf, hob seine reinen Hände zum Himmel empor, rief Gott als gemeinsamen Vater an und flehte zu ihm, er möge sich des bedrängten Griechenlands erbarmen. Und wie er noch betete, ertönte ein ungeheurer Donnerschlag, und der Himmel bedeckte sich ringsum mit Wolken, heftige und anhaltende Regengüsse strömten herab und erfüllten das ganze Land. Infolge davon reifte eine ergiebige und reiche Ernte heran, die durch das Gebet des Aiakos herbeigeführt war.[43]

Dieselbe Geschichte überliefert auch Pausanias, der eine Generation älter als Klemens war, in seiner *Beschreibung Griechenlands*. Während beim christlichen Theologen der Name des Gottes, an den sich Aiakos wandte, nicht direkt genannt wird, und man nur vermuten kann, dass Zeus – immerhin gleichzeitig der verbreitetste Regengott und der Vater des Aiakos – gemeint ist, gibt Pausanias genauere Auskunft und nennt *Zeus Panhellenios* als Empfänger der Opfer:

> Eine Dürre lastete einst während langer Zeit über Griechenland, der Gott ließ es weder außerhalb des Isthmos noch innerhalb der Peloponnes regnen. So schickte man nach Delphi, um nach der Ursache zu fragen und zugleich Befreiung von dem Übel zu erbitten. Die Pythia gab ihnen zur Antwort, sie müssten Zeus gnädig stimmen: wenn er ihnen aber Gehör schenken solle, müsse ihn Aiakos anflehen. So schickten sie aus jeder Stadt Abgesandte, um Aiakos darum zu bitten. Dieser opferte dem Panhellenischen Zeus und erwirkte durch sein Gebet, dass über Griechenland Regen fiel.[44]

Der römische Kaiser Marcus Aurelius überliefert im 2. Jh. n. Chr. in seinen *Selbstbetrachtungen* folgendes Gebet der Athener um Regen:

> Gib bald Regen, lieber Zeus, gib Regen den Fluren und Auen der Athener![45]

Neben der Anrufung von Zeus und anderen Göttern suchte man Zuflucht in verschiedenen magischen Ritualen.[46] Ebenfalls im 2. Jh. n. Chr. berichtet der Perieget Pausanias in seiner *Beschreibung Griechenlands* von einem eigentümlichen Ritus zur Beschwörung von Regen, der an einer nach der Nymphe Hagno benannten Quelle im Lykaiongebirge in Arkadien durchgeführt wurde:

43 Clem. Alex. *strom.* 6, 3, 28 (Übersetzung: O. Stählin).

44 Paus. 2, 29, 7 (Übersetzung: J. Laager).

45 Der Kaiser hebt außerdem hervor, dass man entweder gar nicht beten soll oder auf diese Art, so einfach und edelgesinnt (M. Aur. 5, 7).

46 Dies wurde freilich bereits von Seneca im 1. Jh. n. Chr. kritisiert. In seinen *Naturkundlichen Fragestellungen* (Sen. *nat. quaest.* 4, 7, 2) spricht er von Beschwörungen, mit denen man in alter Zeit versucht habe, Regen herbeizuholen oder zu vertreiben, was aber selbstverständlich nicht funktionieren könne. Von einem gescheiterten Versuch, durch Opfer an Zeus, den Regengott, Niederschlag hervorzurufen, berichtet im 2. Jh. n. Chr. auch der Redner und Sophist Alkiphron (*ep.* 3, 35).

> Wenn nun während langer Zeit eine Dürre herrscht und die Samen in der Erde und die Bäume zu verdorren drohen, betet der Priester des Zeus Lykaios zu diesem Wasser, opfert, was Brauch ist, und hält einen Eichenzweig hinunter gerade bis zur Oberfläche der Quelle, aber nicht in ihre Tiefe. Wenn dann das Wasser in Bewegung gerät, steigt ein leichter Dunst auf, der einem Nebel gleicht. Nach kurzer Zeit wird der Dunst zu einer Wolke, und wenn diese ihrerseits andere Wolken an sich gezogen hat, lässt sie über das Land der Arkader Regen fallen.[47]

Das bekannteste Ritual zur Herbeirufung von Regen in Trockenzeiten ist allerdings für die Stadt Rom belegt. Der römische Lexikograph Festus berichtet im 2. Jh. n. Chr. von der alten Praxis, einen bestimmten Stein von außerhalb in die Stadt zu bringen:

> Als *lapis manalis* bezeichnen sie einen bestimmten Stein, der sich außerhalb der *porta Capena* neben dem Tempel des Mars befand, und wenn sie diesen wegen allzu großer Trockenheit in die Stadt zogen, folgten sofort Regengüsse.[48]

Einer kurzen Notiz im Aeneis-Kommentar des Servius ist zu entnehmen, dass die *pontifices* für die Durchführung dieses Rituals verantwortlich waren.[49] Ein Fragment des Gelehrten Cornelius Labeo, wohl aus dem 3. Jh. n. Chr., bringt es mit der *disciplina Etrusca*, der auf die Etrusker zurückgehenden Lehre von der Deutung göttlicher Signale, in Verbindung und berichtet, dass es durchgeführt wurde, wenn die Leber eines Opfertieres eine bestimmte Farbe aufwies.[50] Wiederum Festus ist es, der verdeutlicht, dass solche Rituale als *aquaelicia* bezeichnet wurden:

> Es wird *aquaelicium* genannt, wenn Regen durch bestimmte Mittel herbeigelockt wird, wie einst, wenn man es glaubt, als der *lapis manalis* in die Stadt gezogen wurde.[51]

Die Bezeichnung *aquaelicium* bezieht sich freilich nicht nur auf diese Prozession, sondern auch auf andere, vom Bewegen des *lapis manalis* deutlich zu trennende Rituale, die dem Herbeilocken von Regen dienten. So kennen wir etwa Bittprozessionen zum Iuppitertempel auf dem Kapitol, die aufgrund der Barfüßigkeit der daran teilnehmenden verheirateten Bürgerfrauen, der *matronae*, die entgegen ihrer sonstigen Gewohnheiten auch ihre Haare gelöst trugen, als *nudipedalia* bezeichnet werden.[52] Auch die römischen Staatsbeamten, die bei diesen Prozessionen mitmarschierten, verzichteten auf ihre üblichen Statusmerkmale: sie verzichteten auf ihre mit dem Purpurstreifen eingefasste Toga, die *toga praetextata*, und die Liktoren hielten ihre mit einem Beil versehenen Rutenbündel, die *fasces*, gesenkt. Diese *nudipedalia* werden etwa im 1. Jh. n. Chr. von Petronius in seinem satirischen Roman

47 Paus. 8, 38, 4 (Übersetzung: J. Laager). Von einem anderen bekannten Ritual zum Herbeilocken von Regen berichtet Antigonos in seiner *Geschichte wunderbarer Begebenheiten* (Antigon. Hist. mirabil. 15): Wenn die Leute von Krannon in Thessalien Regen benötigten, schüttelten sie einen großen bronzenen Wagen, den sie in einem Tempel aufbewahrten. Eine Abbildung dieses Wagens mit Gefäßen und zwei Raben auf den Rädern findet sich auch auf den Münzen Krannons.

48 Fest. *epit*. p. 128 (Übersetzung vom Verfasser).

49 Serv. *Aen*. 3, 175.

50 Fulg. p. 559 ed. Helm p. 112, 11.

51 Fest. ep. p. 2 (Übersetzung vom Verfasser).

52 Vgl. dazu RE XVII, 2 (1936) 1239–1241, s. v. Nudipedalia (E. Marbach).

Satyricon erwähnt, wo sich der Freigelassene Ganymedes über den Niedergang der Sitten beklagt:

> Denn was ist zu erwarten, wenn weder Götter noch Menschen sich dieser Stadt erbarmen? Ich glaube ganz gewiss, dass alles vom Himmel herab kommt. Nicht einer glaubt mehr, dass der Himmel sei! Kein Mensch hält die Fasttage! Kein Mensch macht sich aus dem Jupiter so viel! Sondern alle drücken die Augen zu und zählen ihr Geld. Sonst gingen fromme Matronen noch barfüßig auf den heiligen Hügel, mit fliegenden Haaren und reinen Seelen und baten den Jupiter um Wasser – den Augenblick regnete es, als wie mit Krügen [...] Alles war glückselig! Aber jetzt achtet man die Götter wie die Mäuse: die Füße sind ihnen gebunden: und weil wir keine Religion mehr haben, so liegen die Äcker darnieder![53]

Und auch Tertullian spricht in seiner schon erwähnten Schrift zur *Verteidigung des Christentums* von diesen Prozessionen, er freilich beklagt den Vollzug dieser heidnischen Praktiken und die Ungerechtigkeit, mit der Iuppiter unverdientermaßen für das Senden von Regenfällen verantwortlich gemacht wird:

> Wenn nun der Sommer den Winter und seinen Regen aufhält und man um die Ernte in Sorge ist, dann pflegt ihr, vollgegessen und gleich wieder zum Essen bereit, während Badestuben, Kneipen und Bordelle in Betrieb sind, dem Iuppiter Regenopfer darzubringen, sagt dem Volk barfüßige Bittgänge an, sucht den Himmel im Kapitol und erwartet Regenwolken vom Dachgebälk, dem wahren Gott und wahren Himmel aber kehrt ihr den Rücken. Wir dagegen, vom Fasten ausgedörrt und durch jedwede Art von Enthaltsamkeit entkräftet, von jedwedem Lebensgenuss eine Zeitlang ausgeschlossen, wälzen uns in Sack und Asche und bedrängen voller Empörung den Himmel, klammern uns an Gott, und wenn wir ihm Erbarmen abgetrotzt haben – huldigt man dem Iuppiter![54]

5. RESÜMEE

Im Rahmen des vorliegenden Beitrags wurden einige Perspektiven aufgezeigt, welche die Beschäftigung mit dem Thema „Regen“ für die klassischen Altertumswissenschaften bietet. Es wurden sowohl naturwissenschaftliche als auch mythische Erklärungen für die Entstehung von Regen in den Quellen vorgestellt, es wurde angedeutet, an welchen Himmels- und Naturphänomenen man kommenden Regen erkennen wollte, und abschließend wurden einige wenige Beispiele für antiken Wetterzauber präsentiert. Viele weitere wichtige Aspekte des vielschichtigen Themas konnten in diesem Rahmen freilich überhaupt nicht angesprochen werden, so etwa – um nur einige wenige Beispiele anzuführen – die Nutzung des Regenwassers,[55] Vorrichtungen zum Auffangen und Speichern des Regenwassers,[56]

53 Petron. *sat.* 44 (Übersetzung: W. Heinse).

54 Tert. *apol.* 40, 14 (Übersetzung: H. Kellner); vgl. auch Tert. *de ieiun.* 16, 4–6.

55 Man denke etwa an die Nutzung des Regenwassers in der Landwirtschaft: siehe für die griechische Welt Krasilnikoff (2002).

56 Vgl. zu antiken Zisternen Ohlig (2007); Schäfer – Schön – Gerdes – Heinrichs (2014).

der Regenschutz,[57] die Folgen von ausbleibendem oder übermäßigen[58] Regen oder die Auswirkungen von Regen auf bestimmte historische Ereignisse.[59] Auch zu den behandelten Punkten könnte man noch weitere Quellen anführen und verschiedene Gesichtspunkte näher beleuchten. Die Beschäftigung mit dem Regen in der klassischen Antike erweist sich so als ein Thema, das noch viele offene Fragen bietet und damit ein ergiebiges Feld für weiterführende Forschungen darstellt.

6. AUSWAHLBIBLIOGRAPHIE

AGER (2010) = Britta K. AGER: *Roman Agricultural Magic*. Ph.D. dissertation. University of Michigan 2010.

ALDRETE (2007) = Gregory ALDRETE: *Floods of the Tiber in Ancient Rome*. Baltimore 2007.

BARTA (1979) = Gabor BARTA: *Legende und Wirklichkeit das Regenwunder des Marcus Aurelius*. In: R. Klein (Hrsg.): Marc Aurel. Darmstadt 1979, 347–358.

BARTON (1994) = Tamsyn BARTON: *Ancient Astrology*. London 1994.

CUSSET (2003) = Christophe CUSSET (Hrsg.): *La météorologie dans l'Antiquité, entre science et croyance, Actes du Colloque international interdisciplinaire de Toulouse, 2–4 mai 2002*. Saint-Étienne 2003.

DURSCHMIED (2002) = Erik DURSCHMIED: *Als die Römer im Regen standen. Der Einfluss des Wetters auf den Lauf der Weltgeschichte*. Bergisch Gladbach 2002.

FIEDLER (1931) = Fritz FIEDLER: *Antiker Wetterzauber* (Würzburger Studien zur Altertumswissenschaft 1). Stuttgart 1931.

FOWDEN (1987) = Garth FOWDEN: *Pagan Versions of the Rain Miracle of A. D. 172*. In: Historia 36 (1987) 83–95.

GILBERT (1907) = Otto GILBERT: *Die meteorologischen Theorien des Griechischen Altertums*. Leipzig 1907.

GRUPPE (1906) = Otto GRUPPE: *Griechische Mythologie und Religionsgeschichte*. Zweiter Band. München 1906.

HÅLAND (2001) = Evy Johanne Håland: *Rituals of Magical Rain-Making in Modern and Ancient Greece: A Comparativ Approach*. In: Cosmos 17 (2001) 197–251.

ISRAELOWICH (2008) = Ido ISRAELOWICH: *The Rain Miracle of Marcus Aurelius: (Re-)construction of Consensus*. In: Greece & Rome 55 (2008) 83–102.

JOBST (1978) = Werner JOBST: *11. Juni 172 n. Chr. Der Tag des Blitz- und Regenwunders im Quadenlande*. Wien 1978.

KRASILNIKOFF (2002) = Jens KRASILNIKOFF: *Water and Farming in Classical Greece: Evidence, Method and Perspective*. In: Karen Ascani – Vincent Gabrielsen – Kirsten Kvist – Anders Holm

57 Nur am Rande sei in diesem Zusammenhang darauf hingewiesen, dass im Gegensatz zum Sonnenschirm der Regenschirm für die Antike offenbar nicht ausdrücklich belegt ist. Der älteste urkundlich belegte Regenschirm wird im Jahr 800 in einem Schreiben (Brief Nr. 207) des Abtes Alkuin von Tours an den Salzburger Erzbischof Arn erwähnt; und zwar schickte der bedeutende Gelehrte und Berater Karls des Großen seinem Freund, dem ersten Erzbischof von Salzburg, dieses als *tentorium* bezeichnete Gerät, damit es „sein verehrungswürdiges Haupt vor dem Regen beschütze".

58 Man denke etwa an die regelmäßigen Tiberüberschwemmungen, die die Stadt Rom in der Antike heimsuchten. Vgl. dazu SONNABEND (1999) 60–63; SCHNEIDER (2001); ALDRETE (2007).

59 Vgl. dazu die populärwissenschaftlichen Bücher von DURSCHMIED (2002) und KLAGE (2002), die als antikes Beispiel jeweils die berühmte Schlacht zwischen Römern und Germanen im Teutoburger Wald anführen.

Rasmussen (Hrsg.): Ancient History Matters: Studies Presented to Jens Erik Skydsgaard on His Seventieth Birthday. Rom 2002, 47–62.

KLAGE (2002) = Jan KLAGE: *Wetter macht Geschichte. Der Einfluss des Wetters auf den Lauf der Geschichte*. Frankfurt 2002.

KOVÁCS (2008a) = Péter E. KOVÁCS: *Das Regenwunder Marc Aurels*. In: Acta antiqua Academiae scientiarum Hungaricae 48 (2008) 387–404.

KOVÁCS (2008b) = Péter KOVÁCS: *Marcus Aurelius' rain miracle and the Marcomannic Wars*. Leiden 2008.

MCCARTNEY (1925a) = EUGENE S. MCCARTNEY: *Magic and the Weather in Classical Antiquity 1*. In: The Classical Weekly 18 (1925) 154–157.

MCCARTNEY (1925b) = EUGENE S. MCCARTNEY: *Magic and the Weather in Classical Antiquity 2*. In: The Classical Weekly 18 (1925) 163–166.

MCCARTNEY (1928a) = EUGENE S. MCCARTNEY: *Greek and Roman Weather Lore of the Sun and the Moon 1*. In: Classical Weekly 22 (1928) 25–31.

MCCARTNEY (1928b) = EUGENE S. MCCARTNEY: *Greek and Roman Weather Lore of the Sun and the Moon 2*. In: Classical Weekly 22 (1928) 33–37.

MORGAN (1901) = Morris H. MORGAN: *Greek and Roman Rain-Gods and Rain-Charms*. In: Transactions and Proceedings of the American Philological Association 32 (1901) 83–109.

OHLIG (2007) = Christoph OHLIG (Hrsg.): *Antike Zisternen*. Siegburg 2007.

ROBERTSON (1987) = Noel ROBERTSON: *The Nones of July and Roman Weather Magic*. In: Museum Helveticum 44 (1987) 8–41.

RUBIN (1979) = Ze'ev RUBIN: *Weather Miracles under Marcus Aurelius*. In: Athenaeum 57 (1979) 357–380.

SAGE (1987) = Michael M. SAGE: *Eusebius and the Rain Miracle: Some Observations*. In: Historia 36 (1987) 96–113.

SALERNO (2009) = Francesco SALERNO: *Opus manu factum, natura agri et l'utilisation de l'eau de pluie dans la jurisprudence romain*. In: Ella Hermon (Hrsg.): Vers une Gestion Intégrée de l' Eau dans l' Empire Romain. Actes du Colloque International Université Laval, octobre 2006. Roma 2009, 271–275.

SAMTER (1922) = ERNST SAMTER: *Altrömischer Regenzauber*. In: Archiv für Religionswissenschaft 21 (1922) 317–339.

SCHÄFER – SCHÖN – GERDES – HEINRICHS (2014) = Thomas SCHÄFER – Frerich SCHÖN – Andreas GERDES – Jens HEINRICHS (Hrsg.): *Antike und moderne Wasserspeicherung. Internationaler Workshop vom 11.–14. 5. 2011 in Pantelleria (Italien)*. Rahden / Westfalen 2014.

SCHNEIDER 2001 = Helmuth SCHNEIDER: *Überschwemmungen und Hochwasserschutz im antiken Rom*. In: Albrecht Hoffmann (Hrsg.): Wasserwirtschaft im Wnadel. Festschrift zum 60. Geburtstag von Professor Frank Tönsmann. Kassel 2001, 203–207.

SONNABEND (1999) = Holger SONNABEND: *Naturkatastrophen in der Antike. Wahrnehmung – Deutung – Management*. Stuttgart 1999.

TAUB (2003) = Liba TAUB: *Ancient Meteorology*. London – New York 2003.

TAVENNER (1918) = Eugene TAVENNER: *The Roman Farmer and the Moon*. In: Transactions and Proceedings of the American Philological Association 49 (1918) 67–82.

THOMMEN (2009) = Lukas THOMMEN: *Umweltgeschichte der Antike*. München 2009.

FLUSS – STADT – FLUSS/STADT-GOTTHEIT

Herbert Graßl

Die Beschäftigung mit homonymen oder wie die Griechen sagten, eponymen Städten und Flüssen, war bislang allein eine Domäne der Sprachwissenschaft, die neben den Etymologien auch die wechselseitige Abhängigkeit im Blick hatte (der sogenannte hydronymische Namenswechsel)[1]. Nicht nur im Altertum wurden Städte als jüngere und sich auch verändernde geographische Phänomene häufig nach Flüssen benannt, doch ist auch die gegenteilige Praxis nicht unbekannt[2]. Welche historischen, geopolitischen und geographischen Umstände dafür jeweils verantwortlich waren, blieb dabei weitgehend unberücksichtigt. Da in der Antike Flüsse auch (männliche) Gottheiten waren, stellt sich natürlich die Frage der Beziehung dieser göttlichen Wesen zu den gleichnamigen Städten, die in der Antike meist als weiblich verstanden und auch so ins Bild gesetzt wurden. Diese Problemstellung wurde durch einen epigraphischen Neufund im Zentrum des heutigen Salzburg, dem *municipium Claudium Iuvavum*, neu akzentuiert.[3] Ein Weihealtar für das Wohlergehen eines Großhändlers namens Marius Anicetus wurde dem Iupiter und dem (Gott) Iuvavus gestiftet, der hier zum ersten Mal genannt ist:

I(ovi) o(ptimo) m(aximo)
et Iuvavo
[P]ro salute
Mari Aniceti
et negotiationi
ei{i}us

Dieses Denkmal wurde von der Forschung kontrovers diskutiert: ein Teil sieht im Gott Iuvavus den Flussgott der heutigen Salzach. Fluss und Stadt sind auf der Tabula Peutingeriana als *fl. Ivaro* bzw. *Iuavo* verzeichnet. Dem Gott Iuvavus wurde die Weihung zuteil, da er den auf dem Wasserweg abgewickelten Warenverkehr des Großhändlers beschützte[4]. Die meisten Interpreten stellen einen Zusammenhang

1 W. Brandenstein, Fluß und Stadt, in: Corolla Linguistica. Festschrift Ferdinand Sommer, Wiesbaden 1955, 5–11; G. Rasch, Antike geographische Namen nördlich der Alpen (RGA Erg. 47, 2005), 113; 198; H. Krahe, Würzburger Jahrbücher 1 (1946) 212–214.; A. Greule, Der hydronymische Namenswechsel, in: Ortsnamenwechsel, Heidelberg 1986, 312–322; V. Georgiev, Die altgriechischen Flussnamen, Sofia 1958; RGA s. v. Flussnamen; W. Schulze, Zur Geschichte lateinischer Eigennamen, Berlin 1904, 537–539; H. Rix, Medma, Ort und Fluss in Bruttium, Beiträge zur Namensforschung 3, 1951/52, 243–255.

2 Vgl. dazu die Materialzusammenstellung bei J. Tischler, Kleinasiatische Hydronomie, Wiesbaden 1977, 162–163.

3 Erstpublikation von Chr. Farka, Iupiter und Iuvavus. Ein neuer Weihealtar vom Residenzplatz in Salzburg, in: Der Residenzplatz, 2009, 24–27.

4 Farka (wie Anm. 3).

zwischen Fluss und Stadt in Abrede und sehen in Iuvavus einen römischen Stadtgott[5]. Da diese Debatte aber nur aus einer sehr eingeschränkten Perspektive geführt wurde, soll dieses Problem vor dem Hintergrund der dokumentierten antiken Praxis aufgerollt werden. Die literarischen Quellen zur antiken Geographie und Religion stellen genügend Hinweise für eine breitere Debatte zur Verfügung.

Zur Homonymität von Städten und Flüssen sind, neben gelegentlichen Hinweisen bei Historikern[6] und Dichtern[7], vor allem die einschlägigen Zeugnisse bei Strabon, Pomponius Mela, dem Älteren Plinius, Ptolemaios, Vibius Sequester und Stephanos von Byzanz in unterschiedlichem Maß ergiebig. Natürlich finden sich bei diesen Autoren gelegentlich auch fehlerhafte Angaben, so wurden bei Vibius Sequester den Flüssen gleichnamige Städte auch erfunden[8]. Bei mehrfacher einhelliger Bezeugung, Kontinuität der Benennung und archäologischer Siedlungsevidenz kann man aber von einer abgesicherten Quellenlage ausgehen. Städte nach Flüssen zu benennen war weitverbreitete Praxis der phönikischen und später griechischen Kolonisation (von Spanien und Nordafrika bis zum Schwarzen Meer, z. B. Lixus, Tanais oder Phasis[9]). Diese Neugründungen lagen an den Flussmündungen, hatten den Charakter von Meer- und Flusshäfen oder waren in der Form eines *anáplous* mit Schiffen vom Meer aus erreichbar. Wenngleich größere Flüsse dabei im Fokus der Betrachtung stehen (z. B. der Baetis, Istros, Borysthenes oder Tanais), sind in der Mehrzahl kleinere Gewässermündungen zu Stadtgründungen genutzt worden. Solche Benennungen drücken einen engen räumlichen Konnex zwischen Fluss und Stadt aus, aber auch die spezielle Beziehung zum sowie die Kontrolle des Gewässers. Die Stadt erhob damit ihren Anspruch auf den Fluss, auch in seiner göttlichen Qualität. Diese Benennungspraxis findet sich, neben der phönikischen und griechischen Welt, auch bei anderen Kulturen, etwa Illyrern und Thrakern, den keltisch geprägten Donauländern[10], ja Ptolemaios bezeugt dies für viele Kulturen am Rande der Oikumene. Eine nach einem Fluss benannte Stadt konnte dabei auch einen zweiten Namen führen, z. B. Borysthenes = Olbia. Doch nicht allein bei Neugründungen an den Meeresküsten kam diese Praxis zum Tragen. Im griechischen Mutterland, im zentralen Kleinasien, Italien, Gallien, also Altsiedlungsgebieten, sind Namensgleichheiten bekannt. Hier liegen die Städte nicht an Flussmündungen;

5 M. Hainzmann, Iuvavus. Römischer Stadtgott oder norische Flussgottheit? RÖ 33, 2010, 23–34; M. Hainzmann – P. de Bernardo Stempel, Iuvavus und Verwandte, RÖ 34/35, 2011–2012, 51–62; G. Thüry, Neues zur Epigraphik des römischen Salzburg, Diomedes N. F. 5, 2010, 83–85; ders., Das römische Salzburg 2013, 76–77; ders., Die Stadtgeschichte des römischen Salzburg, Oxford 2014, 52–54; E. Weber, Salzburg in der Tabula Peutingeriana, in: Ein kräftiges Halali aus der Römerzeit. Norbert Heger zum 75. Geburtstag, Salzburg 2014, 309–316.

6 Thuk. 6,4,4 zu Akragas; Amm. Marc. 22,8,27 zu Phasis: *Phasis est nomine fluvii dictitata*; Curt. 7,4,31.

7 Verg. *Aen.* 3,702: *Gela fluvii cognomine dicta;* Sil. 14,218; Claud. *rapt. Pros.* 2,58.

8 Z. B. Crustumium, Isamnus, Oaxes.

9 Dazu O. Lordkipanidze, Phasis. The River and City in Colchis, Stuttgart 2000.

10 S. Nemeti, River names in roman *Dacia,* in: Antiquitas Istro-Pontica. Mélanges d'archéologie et d'histoire ancienne offerts à Alexander Sucereanu, Cluj-Napoca 2010, 103–113.

die meergeleitete Routenführung stand hier nicht im Blick.[11] Im Binnenland ist auch die inverse Benennungspraxis verbreitet, dass Flüsse nach Orten benannt wurden. Dieses, vor allem in Kleinasien, aber auch sonst gut dokumentierte Phänomen kommt bei regionalen Gewässern, die nur für eine einzelne Stadt Bedeutung hatten, vor, besonders dann, wenn kein altererbter Gewässername zur Verfügung stand oder tradiert wurde. Interessant ist auch das Faktum, dass ein und derselbe Fluss an verschiedenen Stellen seines Laufes mit verschiedenen am Fluss liegenden Städten in Beziehung stand und deshalb unterschiedlich benannt wurde, z. B. in Kampanien der Volturnus=Casilinus. Dies ist Ausfluss der Regionalität geographischer Horizonte und unterschiedlicher Ansprüche. Nicht unwesentlich ist auch der Blick auf die geschlechtliche Zuordnung dieser Namen. Dabei sind mehrere Möglichkeiten bekannt: Fluss und Stadt tragen einen identischen Namen, aber unterschiedliches Geschlecht, z. B. Sybaris, Xanthos, Phasis oder Tanais. Der Fluss führt einen männlichen, die Stadt einen weiblichen Namen, was im Griechischen leicht mit dem Artikel ausgedrückt werden kann, auch wenn manche antiken Autoren diese Norm nicht beachteten und auch den Städten ein männliches Geschlecht zuordneten. Die zweite Möglichkeit kennt unterschiedliche Endungen, die männliche für den Fluss, die weibliche für die Stadt, z. B. Istros – Istria, Naron – Narona, Limyrus – Limyra. Ein Spezifikum, das vor allem in Italien, doch auch sonst in der römischen Welt verbreitet ist, kennt eine männliche Namensform für den Fluss und ein Neutrum für die Stadt, z. B. Ticinus – Ticinum, Volturnus – Volturnum, Ariminus – Ariminum.

Die Flüsse des Altertums wurden bekanntlich auch in ihrer göttlichen Funktion gesehen, auch wenn die literarischen, epigraphischen und numismatischen Zeugnisse dies nur für eine bestimmte Anzahl dokumentieren.[12] In homonymen Städten spielt diese Gottheit eine wichtige, aber nie die dominierende Rolle im kultischen Geschehen. Es wäre also völlig überzogen, in diesem Zusammenhang von einer „Stadtgottheit" zu sprechen. Die religionshistorische Forschung hat die Rolle sogenannter „Stadtgottheiten" insbesondere für die griechische Welt klargestellt[13]: jede Stadt ist hier gesondert zu betrachten, jede Stadt kennt ihre Spezifika, das Beispiel Athens mit dem herausgehobenen Kult der Athena Polias ist keinesfalls auf ganz Hellas zu übertragen. Eponyme Flussgötter wurden von den Städten kultisch verehrt, übten eine besondere Schutzfunktion aus, wurden als Statuen an promi-

11 Nicht an Flussmündungen lagen die Städte Adranon, Harpasa, Tisua, Longane, Baktra, Kleitor, Aesis, Karesos, Kephisos, Orneai.

12 Zu Flussgöttern: RE s. v.; W. Roscher, Mythologisches Lexikon I 1487; C. Weiss, Griechische Flussgottheiten in vorhellenistischer Zeit, Würzburg 1984, 244–246; dies., LIMC 4,1, 139–148; S. Klementa, Gelagerte Flußgötter des Späthellenismus und der römischen Kaiserzeit, Köln-Weimar-Wien 1993, 253–256. Die kultische Verehrung von Flüssen war schon vor Griechen und Römern bekannt; vgl. dazu etwa Max. Tyr. 2,8; M. Sasel-Kos, Pre-Roman Divinities of the Eastern Alps and Adriatic, Ljubljana 1999, 13–61; 93–119; 134–136.

13 Vgl. dazu S. G. Cole, Civic Cult and Civic Identity, in: Sources for the Ancient Greek City-State (Acts of the Copenhagen Polis Centre Vol. 2, ed. M. H. Hansen, Copenhagen 1995), 292–325; U. Brackertz, Zum Problem der Schutzgottheiten griechischer Städte, Diss. Berlin 1976.

nenten Stätten platziert und symbolhaft im Münzbild[14] präsentiert. Sie konnten aber das kultische Leben einer Polis nie dominieren, wie das auch am Beispiel von Phasis deutlich wird[15]. Ein auch in der antiken Literatur angesprochenes Beispiel dafür liefert Akragas auf Sizilien. Hören wir dazu Älian (*v. h.* 2,33): „Die Akragantiner stellen den Fluss, der ihrer Stadt den Namen gibt, als schönen Jüngling dar und bringen ihm so ihre Opfer. Sie weihten auch in Delphi eine aus Elfenbein geschnitzte Statue, auf die sie den Namen des Flusses schrieben; und diese Statue zeigt einen Jüngling." Nach diesem Zeugnis war eine Statue des Flussgottes Akragas als Jünglingsfigur sowohl im städtischen Raum aufgestellt, wo auch ein entsprechender Kult praktiziert wurde, als auch im panhellenischen Heiligtum von Delphi, hier mit erklärender Inschrift. Münzbilder von Akragas zeigen den Kopf dieser Figur mit göttlichen Attributen und Umschrift[16]. Ab dem Hellenismus wird nach dem Vorbild von Antiocheia am Orontes die vergöttlichte Tyche der Stadt mit dem eponymen Flussgott, der einem städtischen Heros gleichkommt, dargestellt. Der Fluss nimmt dabei die Rolle des Spenders von Fruchtbarkeit und Leben für die Polis ein und wurde als schützender Gott verehrt. Ein schönes Beispiel für diesen Konnex überliefert Cicero, der zugleich die Klammer zur römischen Religion herstellt. Die Stadt Himera auf Sizilien, eine Gründung Zankles, besaß ein hochverehrtes bronzenes Kultbild ihrer Polis und des gleichnamigen Flusses, dargestellt als wunderschöne Frau (Cic. *Verr.* 2,2,87): *Erant signa ex aere complura; in his eximia pulchritudine ipsa Himera in muliebrem figuram habitumque formata ex oppidi nomine et fluminis*. Auf Münzen Himeras ist die Göttin Himera am Altar opfernd dargestellt, dahinter ein Silen in einem Nymphäum; ein Flussfisch im unteren Bildabschnitt weist auf das gleichnamige Gewässer hin[17]. Nach der Zerstörung Himeras durch die Karthager 409 v. Chr. konnten die Überlebenden das Kultbild retten, bis es ins Visier des Statthalters Verres geriet. Doch als der seine lebendige Gastgeberin zu Gesicht bekam, ließ sein Interesse an der Statue schlagartig nach. In diesem Fall waren Stadt und Fluss in einer Frauenstatue personifiziert, die (weibliche) Stadt dominierte den (männlichen) Fluss. Nach hellenistischem Vorbild lagerte auch in römischen Denkmälern der Fluss zu Füßen der personifizierten städtischen Tyche, etwa in Aquileia[18]. In der mythischen Überlieferung wurde die 181 v. Chr. gegründete latinische Kolonie mit einem Fluss Aquilis, eine Erfindung von kaiserzeitlichen Poeten und Mythographen[19], in Verbindung gebracht.

14 Zu Flussgöttern auf Münzen: G. K. Jenkins, The Coinage of Gela I, Berlin 1970, 165–175; W. Leschhorn, Lexikon der Aufschriften auf griechischen Münzen, Wien 2002, 409.

15 Lordkipanidze (wie Anm. 9) 62 zu Kulten in Phasis.

16 BMC Sicily 19; zu einer ähnlichen Bildstatue des eponymen Flussgottes in Selinus: Ph. L. Lehmann, A numismatic approach to the sculpture, Diss. New York 1943, 30.

17 BMC Sicily 79; vgl. dazu F. W. Hamdorf, Griechische Kultpersonifikationen der vorhellenistischen Zeit, 1964, 29.

18 LIMC s. v. Aquileia; M. M. Tinauer, Die Bildtypen der Flussgötter in der griechischen und römischen Kunst, Dipl. Univ. Graz 1985, 54.

19 Vgl. Zos. 5,29,4; Sozom. 1,6,5; dazu A. Grilli, Il mito del fiume ΑΚΥΛΙΣ, RIL 113, 1979, 21–24.

In der römischen Religion nehmen Schutzgottheiten für Städte einen auch architektonisch prominenten Platz ein[20]. Ihre Rolle kommt dabei der eines *genius loci* sehr nahe. In einigen spanischen und gallischen Städten verehrte man die Schutzgöttin *Tutela* (auch zusammen mit dem *Genius* oder der *Fortuna*). Weiblich gedachte Städte kennen ebensolche Kultpersonifikationen, in Noricum z.B. *Teurnia* oder *Celeia*. Einen speziellen Fall bilden Städte mit neutralem Namen. Hier sind weibliche Schutzgottheiten zwar durchaus bekannt, treten aber zugunsten männlicher Gottheiten in Gestalt von Flüssen, Bergen oder Gründungsheroen deutlich zurück[21]. Von der Stadt Volturnum in Kampanien ist der Flussgott Volturnus nicht zu trennen[22]. In diese Gruppe wird auch der Gott Iuvavus zu stellen sein: er war mit der Stadt Iuvavum in besonderer Weise verbunden, schützte die Bewohner und ihr Fortkommen, auch das geschäftliche. Die Herkunft des Stadtnamens Iuvavum wird in der sprachwissenschaftlichen Forschung sehr unterschiedlich erklärt[23]. Der enge geographische Zusammenhang mit dem Fluss steht aber außer Zweifel. Eine Ableitung des Ortsnamens aus dem Flussnamen ist daher angesichts der antiken Regularien nicht von der Hand zu weisen. In der Tabula Peutingeriana scheint der Fluss als Ivaro auf; es könnte aber sein, dass dies eine spätantike Namensform darstellt. Aus kaiserzeitlichen Namen mit v(b) in der Endung wird in der Tabula Peutingeriana gerne ein Name mit r: Ad Bivium wird zu Ad Birium, Alabon zu Alarante, Abusina zu Arusena[24]. Diese spätantike Namensform für den Fluss findet sich dann auch in frühmittelalterlichen Quellen. Es ist also gut denkbar, dass Fluss und auch Flussgott in der frühen Kaiserzeit, dem Zeitpunkt des Weihesteines, als Iuvavus bekannt waren. Oder sollte man sich ernstlich die gleichzeitige Verehrung eines Gottes Iuvavus und eines Flussgottes Ivarus in der Stadt Iuvavum vorstellen? Es hat wohl nur einen lokalen Schutzgott gegeben, den Flussgott Iuvavus. Wie in vielen anderen Fällen auch verlieh der Fluss der Stadt seinen Namen.

ANHANG: HOMONYME FLÜSSE UND STÄDTE BEI

Strabon: Aigospotamoi, Aisis – Aision, Ariminos – Ariminon, Aternos – Aternon, Baitis, Baktros – Baktra (auch Zariaspa genannt), Belon, Borysthenes, Druentios-Druentia, Himera, Ilibirris, Istros, Karesos, Kephisos – Kephisia, Killaios – Killa (Fluss nach Stadt benannt), Kyparisseis, Laos, Limyros – Limyra, Liternos – Li-

20 M. Boos, *In excelsissimo loco* – An Approach to Poliadic Deities in Roman Colonies, in: TRAC 2010. Proceedings of the Twentieth Annual Theoretical Roman Archaeology Conference, Oxford 2011, 18–31.

21 In Noricum stehen die Städte Cetium mit dem *mons Cetius* und Virunum mit dem Gründungsheros Virunus in Beziehung. Die Provinz Noricum, die als einzige einen nicht weiblichen Namen trägt, wurde auf Münzen Kaiser Hadrians systemgerecht als Frau abgebildet; dazu C. Houghtalin, The Personifications of the Roman Provinces, 1996, 432–434.

22 Stat. *silv*. 4,3,67–94.

23 Zusammenstellung der Forschungsliteratur bei bei G. Thüry, Stadtgeschichte (wie Anm. 5) 52–54.

24 Die sprachlichen Varianten der Tabula Peutingeriana zu Flussnamen sind bei E. Weber, Tabula Peutingeriana. Kommentar, Graz 1976, 66–67 zusammengestellt.

ternon, Lixos, Matrinos – Matrinon, Medoakos, Naron – Narona, Okalee, Orneai, Paisos, Phasis, Ruskinon, Selinus, Siris, Sukron, Sybaris, Tanais, Tartessos, Tikinos – Tikinon, Timavos – Timavon, Truentinos, Volturnos – Volturnon, Xanthos.

Pomponius Mela: Baetulo, Gna, Himera, Hypatos, Limyra, Lixus, Phasis, Volturnus – Volturnum, Xanthus.

Plinius d. Ältere: Absarrum, Acheron, Aeminium, Arachosia (die Stadt hieß auch Kufis), Arbium, Bactrus – Bactrum (früher Zariastes genannt), Barbesula, Borysthenes (Stadt hieß mit alten Namen Olbiopolis und Miletopolis), Caere (Fluss nach Stadt benannt), Chadisia, Cios, Harpasos – Harpasa, Hierum, Himera, Laus, Limyra, Liquentia, Lixus, Maenuba, Malaca, Narona, Penius, Phasis, Philocalia, Pisaurus – Pisaurum, Sala, Salduba, Scamander – Scamandria, Sucro, Tamuda, Titanus, Tripolis, Truentus – Truentum, Tyra (Ort hieß früher Ophiusa), Volturnus – Volturnum, Xanthus.

Ptolemaios: Albanos – Albana, Apsorros, Baelo, Besyngas – Besynga, Borysthenes, Chaberis , Karkina, Kartenna, Phasis, Sados – Sada, Sala, Salathos, Siga, Tanais, Temalos – Tamala.

Vibius Sequester: Agragas, Aternus, Camicos, Casilinum (Fluss hieß sonst Volturnus), Crustumium, Gela, Helorus, Himera, Ilerda, Isamnus (nach Siedlung benannt), Lycastus, Oaxes – Oaxia, Pisaurus – Pisaurum, Selinus, Tauromenius – Tauromenium (Fluss nach Stadt benannt), Volturnus – Volturnum, Xanthus.

Stephanos von Byzanz: Adria, Arachotos – Arachotoi, Ariminon, Baitis, Belon, Bolbaiotes – Bolbai (Fluss nach Stadt benannt), Borysthenes, Bouprasion, Elea, Eloros, Eunaios – Eunai, Gela, Gerros, Harpasos – Harpasa, Himeras – Himera, Hypanis, Hypios, Idymos – Idyma, Idyros, Kephisos – Kephisia, Killaios – Killa (Fluss nach Stadt benannt), Lametos, Lamyros – Lamyra, Laos, Limyros – Limyra, Linx, Lixos – Lixa, Mieza, Myros – Myra, Narnos – Narnia, Nestos, Orneai, Pydes, Rhizon, Salgas – Salga, Sebennytos, Siris, Sybaris, Tamese, Tanais, Taras, Tartessos, Terina, Tikinos, Thymbrios – Thymbra (Fluss nach Ort benannt), Tyras, Xanthos.

VON AQUA MAGNA BIS DILUVIUM

Jasmin Hettinger

EINE SYSTEMATISCHE ANNÄHERUNG AN DEN HOCHWASSERBEGRIFF IN DEN ANTIKEN LATEINISCHEN SCHRIFTQUELLEN

Es ist hinlänglich bekannt, dass die Stadt Rom schon in der Antike in regelmäßigen Abständen von Tiberüberschwemmungen heimgesucht wurde. Denn da sie als Schaltzentrale eines riesigen Reiches ohnehin im Brennpunkt der zeitgenössischen Literatur stand, liegen zur Tiberflut für einen langen Zeitraum genügend Zeugnisse vor, die umfangreiche Studien der Hochwasser und der daran gekoppelten Bewältigungsmechanismen ermöglichen.[1] Für den Rest der Apenninhalbinsel ebenso wie für die Provinzen steht jedoch eine derartige Studie noch aus. Zwar ist dies zum Teil sicherlich der weniger vollständigen und inkohärent anmutenden Quellenlage geschuldet, doch lassen sich aus den vielen, in verschiedensten Texten und archäologischen Befunden verstreuten Hinweisen trotzdem genügend Informationen über den Umgang mit Hochwassern in Italien und den Provinzen ziehen, denen nachzugehen Aufgabe künftiger Forschungen sein muss.[2] Ein erster Schritt zu dessen Erforschung ist die Sichtung und Zusammenstellung von Quellenbelegen, insbesondere aus dem schriftlichen Material. Dabei fällt recht schnell auf, dass es unterschiedliche Begrifflichkeiten gab, mittels derer die Konzepte „Hochwasser" oder „Überschwemmung" in der lateinischen Sprache ausgedrückt wurden. Nicht verwunderlich ist die zunächst sehr allgemeine Feststellung, dass sich Begriffswahl und Ausdrucksweise je nach Sprecher, Kontext und Textgattung unterscheiden. Bei eingehender Betrachtung lassen sich jedoch eine gewisse Systematik und verschiedene Begriffsgenesen ausmachen, die im Folgenden vorgestellt werden sollen. Insgesamt konnten vier mehr oder weniger in sich geschlossene Begriffsfelder identifiziert werden. Diese sind: 1. das Substantiv *inundatio*, 2. Verben und Partizipial-

1 Die aktuellste Monographie, die sich den Tiberüberschwemmungen in und um Rom widmet, ist die Studie von G. S. Aldrete, Floods of the Tiber in Ancient Rome, Baltimore 2007. Sie enthält neben einer Zusammenstellung der literarischen Quellenbelege einen Überblick über die Hydrologie des Flusssystems des Tibers, kartographische Rekonstruktionsversuche der Reichweite verschieden starker Hochwasser für das augusteische Rom sowie eine Zusammenfassung der getroffenen Schutzmaßnahmen und der mentalen Bewältigungsformen.

2 Der vorliegende Beitrag ist Bestandteil einer solchen Arbeit. In meinem Dissertationsprojekt gehe ich aus althistorischer Perspektive den verschiedenen uns bekannten Maßnahmen der römischen Hochwasservorsorge und -bewältigung nach und versuche, diese zueinander in Beziehung zu setzen, sofern sich Pfadabhängigkeiten feststellen lassen.

konstruktionen, 3. *aqua magna* und begrifflich ähnliche Ausdrucksweisen, und schließlich 4. der Begriff *diluvium* und Abwandlungen davon.

Trotz der nun folgenden systematischen Erörterung der vier Begriffsfelder erhebt der vorliegende Beitrag nicht den Anspruch, eine vollständige Begriffsgeschichte darzustellen. Vielmehr soll lediglich auf Tendenzen im Wortgebrauch der gängigsten lateinischen Vokabeln für „Hochwasser“ und „Überschwemmung“ aufmerksam gemacht werden. Daran anknüpfend ist zu betonen, dass das Lateinische als „Mutter“ der romanischen Sprachen ein sehr anders geartetes Sprachgefühl und -verständnis als das Deutsche voraussetzt. Diese grundsätzliche Andersartigkeit des Lateinischen im Gegensatz zum Deutschen muss bei den im Folgenden vorgestellten Beobachtungen und Erklärungsansätzen immer mitbedacht werden.

I. *INUNDATIO*

In den heutigen romanischen Sprachen lässt sich die Vokabel für Überschwemmung direkt aus dem lateinischen *inundatio* herleiten, so etwa *inondation* (frz.), *inondazione* (it.), *inundación* (sp.), *inundação* (port.) oder *inundaţie* (rum.). Ebenso wird im Englischen – neben *flood* – häufig auf das Wort *inundation* zurückgegriffen. Davon ausgehend wäre zu erwarten, auch in den antiken lateinischen Flutbeschreibungen in der Regel auf diese Vokabel zu treffen. In der Tat ergibt die Wortsuche nach *inundatio* eine recht hohe Trefferquote, doch sticht sofort das fast gänzliche Fehlen historiographischer Quellenstellen ins Auge, die etwa von den berüchtigten Tiberüberschwemmungen in Rom oder anderen Flutereignissen berichten. Stattdessen finden sich unter den am meisten vertretenen Textgattungen Rechtstexte, Schriften der römischen Feldmesser, einige (natur-)philosophische Abhandlungen sowie spätantike Werke religiösen Inhalts. Einige wenige Beispiele sollen genügen, um den hauptsächlichen Gebrauch des *inundatio*-Begriffs in den antiken Quellen herauszustellen und zu charakterisieren.

Innerhalb der Rechtstexte findet sich der Begriff vor allem in den justinianischen Korpora, darunter besonders häufig in den Digesten. Die folgende Auswahl soll vor allem die Bandbreite der juristischen Themenfelder aufzeigen, in denen Hochwassersituationen offenbar expliziter Berücksichtigung bedurften:

Dig. 7,4,24 (Auf welche Weise der Nießbrauch oder das Gebrauchsrecht verlorengeht):
… si flumen inundatione hortum occupavit…
… wenn der Fluss wegen einer Überschwemmung den Garten überflutete…

Dig. 18,6,11 (Über das Risiko und den Nutzen der verkauften Sache):
Fundi nomine emptor agere non potest, cum, priusquam mensura fieret, inundatione aquarum aut chasmate aliove quo casu pars fundi interierit.
Der Käufer kann wegen des Grundstücks nicht klagen, wenn, bevor die Vermessung durchgeführt wurde, durch eine Überschwemmung mit Wasser oder einen Erdrutsch oder durch einen anderen Vorfall ein Teil des Grundstücks zugrunde gegangen ist.

Dig. 41,1,6 (Von der Erwerbung des Eigentums der Sachen):
… namque inundatio speciem fundi non mutat…
… denn eine Überschwemmung verändert nicht das Aussehen eines Grundstücks…

Dig. 43,19,1,9 (Von Privatwegen):
… Si quis propter inundationem usus non sit itinere actuque hoc anno…
… Wenn jemand wegen einer Überschwemmung in jenem Jahr keinen Gebrauch von einem Feldweg oder Durchgangspfad [zum Viehtreiben] gemacht hat…

Aus diesen Beispielen wird deutlich, dass die Erwähnung von Überschwemmungen hier eher als Ausgangslage für die darauf folgenden Erörterungen von Rechtsfragen dient, die völlig unterschiedlicher Natur sein können (Wegerecht, Grenzstreitigkeiten, Besitzrechte, etc.). Im modernen Sinne gesprochen, handelt es sich bei diesen Gesetzen also am ehesten um verschiedentliche Nachsorgeregelungen für die Zeit nach einem Flutereignis. Was allen diesen Textstellen jedoch gemeinsam ist, ist deren exemplarischer Charakter, womit auch erklärt werden kann, warum gerade in der vorliegenden Textgattung die sehr allgemeine Bezeichnung *inundatio* besonders häufig anzutreffen ist.[3] Die Erklärung liegt auf der Hand. Gerade weil schon im Vorfeld einer Flut Nachsorgearbeiten festgelegt und normiert werden sollten, mussten die im Text beschriebenen Überschwemmungen rein fiktive, noch nicht eingetretene sein, oder anders formuliert, dann musste die verwendete Begrifflichkeit schlicht zur Bezeichnung von Hochwassern als Naturerscheinung im Allgemeinen dienen. Davon ausgehend wird auch schnell klar, in welchen anderen Textgattungen der *inundatio*-Begriff noch auftaucht. Angesprochen wurden schon die Landvermesserschriften, die sich bekanntlich ebenfalls, wenn auch unter anderen Gesichtspunkten, mit Grenzstreitigkeiten und Besitzrechten auseinandersetzen. Zusätzlich zu erwähnen ist in diesem Zusammenhang die oft thematisierte Bestimmung der rechtlichen Qualität eines Grundstücks, die unter anderem vom Gutachten der *agrimensores*, der Landvermesser, abhing und die für die Entscheidungsfindung in anfallenden Rechtsstreitigkeiten von enormer Relevanz war. Denn nur ausgehend vom Rechtsstatus des von Flutschäden betroffenen Landes konnten von den Behörden die institutionellen Zuständigkeiten geklärt sowie die Verantwortlichen benannt und zur Rechenschaft gezogen werden. Vor diesem Hintergrund nimmt es nicht Wunder, dass in den Vermessungsschriften der hypothetisch gemeinte *inundatio*-Begriff auftritt.[4]

3 Es bleibt hier darauf hinzuweisen, dass an vielen Stellen innerhalb der Digesten und anderer Rechtskorpora, bei denen Überschwemmungen eine Rolle spielen, auch auf alternative Ausdrucksweisen zurückgegriffen wird. So etwa in *Dig.* 2,11,2,8: *Cum tamen (…) flumen sic abundasset, ut implesset omnem locum, in quo sisti oportuit…* – „Wenn aber (…) der Fluss so sehr über seine Ufer getreten sein sollte, dass er die gesamte Stätte [mit Wasser] auffüllt, an welchem [dem Vorgeladenen] vor Gericht zu erscheinen angeordnet worden war …"

4 C 34,20; 40,6; 66,12. Insbesondere im Kontext der *controversiae de alluvione*, also Rechtsstreitigkeiten, die wegen Anschwemmung oder Erosion von Erde an flussnahen Grundstücken aufkommen, werden in den Feldmesserschriften Überschwemmungen thematisiert. Für neueste Studien s. P. Castillo, Die Controversia de alluvione bei Gromatikern und Juristen, in: eTopoi 2 (2012/2013), 1–23. Viel häufiger als *inundatio* wird in diesem Zusammenhang allerdings *alluvio* oder *adluvio* verwendet, was jedoch nicht die Überschwemmung selbst bezeichnet, sondern auf den Vorgang des Wegschwemmens von Erde durch eine Flut Bezug nimmt, also auf den Erosionsprozess. Vgl. Isid. *De differentiis verborum* 40: *Alluvium consumptio est riparum ex aquis, colluvium autem congregatio sordium, quae fit a fluxione multa.* – „*Alluvium* ist die Aufzehrung der Ufer durch Gewässer, *colluvium* aber ist die Ansammlung von Schlamm, die aus der starken Strömung erwächst."

Kein rein hypothetisches, dafür aber regelmäßig auftretendes Hochwasser war die sogenannte Nilschwelle in Ägypten. Dieses sich jährlich im Sommer wiederholende Naturereignis, das den Nil über mehrere Wochen hinweg anschwellen ließ und das flussnahe Ackerland bewässerte, hatte die antiken Menschen schon lange vor der römischen Provinzialisierung Ägyptens fasziniert. Dementsprechend reich ist der schriftliche antike Fundus an mehr oder weniger phantastischen Erklärungsversuchen für dieses Phänomen.[5] Auch unter römischer Herrschaft brach die literarische Auseinandersetzung mit der Nilschwelle nicht ab, sicherlich nicht zuletzt deshalb, weil vom Ausmaß der Nilflut die Getreideernte abhing, mit deren Hilfe in der Kaiserzeit wiederum die Getreideversorgung der stadtrömischen Bevölkerung sichergesellt wurde. Es war also kein rein naturwissenschaftliches Interesse, welches die römische Oberschicht dazu veranlasste, sich über Entstehung, Ausmaß und Wirkung der Nilschwelle, die *Nili inundatio* oder auch *inundatio Nili*, auszutauschen. Insofern ist die Benutzung des Nomens *inundatio* konsequent, denn meist wurde auf kein konkretes Nilhochwasser in einem bestimmten Jahr Bezug genommen. Vielmehr diente es zur Bezeichnung des Phänomens als solches.[6]

Neben der Bezugnahme auf die Nilschwelle oder auf Überschwemmungen als Naturphänomen im Allgemeinen konfrontiert der Politiker und Philosoph Seneca seine Leser mittels des Begriffs *inundatio* mit der stoischen Denkfigur des Kataklysmos, also der weltumfassenden, alles zerstörenden Sintflut, die, gemäß der Lehre der Stoa, in regelmäßigen Abständen über die Erde hereinbrechen und die Menschheit auslöschen würde.[7] Ähnlich gingen die christlichen Autoren mit dem Begriff um. Sie nutzten ihn ebenfalls, um eine weltumfassende Überschwemmung, namentlich die biblische Sintflut oder die noch zu erwartende endzeitliche Flut, zu benennen.[8] Des Öfteren findet sich in der christlichen Literatur Roms der *inundatio*-Begriff auch zur Beschreibung eines wohltuenden Wasserüberflusses, wie man ihn sich etwa im Paradiesgarten Eden vorstellte,[9] oder aber als Metapher, um damit der Vorstellung einer energetischen Erfüllung der Welt durch den heiligen Geist (*inundatio spiritus sancti*[10]) in angemessener Wortgewalt Ausdruck

5 Die umfassendste Quellenkompilation und Abhandlung zum Nilkult und den Nilfluttheorien von der Zeit des 4. Jh. v. Chr. bis ins 7. Jh. n. Chr. stellt wohl noch immer D. Bonneau, La crue du Nil, divinité égyptienne, à travers mille ans d'histoire (332 av. –641 ap. J.-C.) d'après les auteurs grecs et latins, et les documents des époques ptolémaïque, romaine et byzantine (= Études et commentaires 52), Paris 1964, dar.

6 Zur Nilschwelle z. B. Plin. *nat.* 6,166; 9,179; 24,169; Sen. *nat. quaest.* 6,8,3; *Cod. Iust.* 7,41,2.

7 Sen. *nat. quaest.* 3,5,34; 3,29,1; 6,30,3; Sen. *Lucil.* 71,15. Häufiger verwendete Seneca allerdings einen anderen Begriff zur Bezeichnung der Sintflut, *diluvium*, der in Abschnitt IV dieses Artikels erörtert wird. Für eine kurze Darstellung des zyklischen Sintflutgedankens bei Seneca, auch im Vergleich mit anderen Stoikern, s. J. Wildberger, Seneca und die Stoa: Der Platz des Menschen in der Welt Bd. 1 (= Untersuchungen zur antiken Literatur und Geschichte 84/1), Berlin 2006, 56–58.

8 Lact. *epit. div. inst.* 66,4; Hieron. *tract. psalm.* 103,78; August. *civ.* 20,16; August. *gen. contra Manich.* 2,3.

9 Z. B. Cypr. *ep.* 73,10: *Has arbores rigat quattuor fluminibus id est evangeliis quattuor, quibus baptismi gratiam salutari et caelesti inundatione largitur.* Vgl. auch Lact. *div. inst.* 7,3,25.

10 August., *serm.* 32,41,115. Vgl. Aug. *enarr. in psalm.* 38,45,8,3; Verec. Iunc. *Comm. super cant. eccl., In cant. Habacuc* 10,17.

zu verleihen. Aber letztere nicht „wässrige" Überschwemmungen sind hier nicht von Belang.

Mit dem oben Festgestellten soll nun keinesfalls argumentiert werden, im Lateinischen habe man generell auf die Verwendung des Substantivs *inundatio* verzichtet, um über ein real vorgefallenes Flutereignis zu reden – mehrere Zitate aus der kaiserzeitlichen und spätantiken Historiographie beweisen das Gegenteil.[11] Trotz allem scheint die übergeordnete Bezeichnung in der gehobenen Literatur, worunter neben der Dichtung eben die Historiographie zählt, eher wenig Verwendung gefunden zu haben. Die alternativen Ausdrucksweisen sowie die Beweggründe der Autoren werden im nächsten Abschnitt separat behandelt. Damit endet der kurze Überblick über den Gebrauch der Vokabel *inundatio* in den literarischen Quellen, doch findet sich der Begriff auch im epigraphischen Material wieder, wovon hier exemplarisch drei kaiserliche Inschriften aus der Umgebung von Ostia, dem Hafen der Stadt Rom, aufgrund ihres kontextuellen Zusammenhangs vorgestellt werden sollen. Alle drei Inschriften nehmen auf Flusshochwasser des Tibers in der Nähe seiner Mündung Bezug und stellen zudem exzellente Beispiele für kaiserlichen Euergetismus im unmittelbaren Umfeld Roms dar. Außerdem dokumentieren sie das Patronageverständnis sowie die Herrschaftsrhetorik im Dienste des römischen Kaisertums.

Das erste Zeugnis stammt aus der Regierungszeit des Kaisers Claudius und datiert ins Jahr 46 n. Chr. Erwähnt werden die Kanäle, die Claudius im Zuge des Hafenausbaus als zusätzlichen Verkehrsweg neben dem natürlichen Tiberarm hatte anlegen lassen.[12] Ausdrücklich wird darin darauf hingewiesen, dass der Kaiser durch diese Arbeiten die Stadt von „Überschwemmungen befreit" habe:

> Ti(berius) Claudius Drusi f(ilius) Caesar / Aug(ustus) Germanicus pontif(ex) max(imus) / trib(unicia) potest(ate) VI co(n)s(ul) design(atus) IIII imp(erator) XII p(ater) p(atriae) / fossis ductis a Tiberi operis portu[s] / caussa emissisque in mare urbem / inundationis pericylo liberavit[13]

Neben der eindeutig militarisierten Rhetorik fällt auf, dass erstens nicht etwa die wirtschaftlichen Interessen, die bei der Anlage des Hafens und der Kanäle eine gewichtige Rolle gespielt hatten, erwähnt werden, sondern ausschließlich der fürsorgliche Charakter der kaiserlichen Baumaßnahmen gegenüber der Stadtbevölkerung Roms.[14] Die Flutgefahr (*periculum inundationis*) sei durch sie nun gebannt. In

11 Vgl. z. B. Tac. *hist.* 1,86,2; Suet. *Aug.* 28,3 und *Otho* 8,3; SHA *Hadr.* 21,6 sowie *Ant.* 9,3 und *M. Aurel.* 8,4.

12 Suet. *Claud.* 20,1,3; Tac. *ann.* 12,43,1; 15,18,2; Cass. Dio 60,11,2. Die Neuanlage des Hafens war im Jahr 42 n. Chr. begonnen worden und wurde erst unter Claudius' Nachfolger Nero abgeschlossen. Vgl. M. Horster, Bauinschriften römischer Kaiser. Untersuchungen zu Inschriftenpraxis und Bautätigkeit in Städten des westlichen Imperium Romanum in der Zeit des Prinzipats (= Historia Einzelschriften 157), Stuttgart 2001, 296, mit weiterführender Literatur.

13 CIL XIV 85 = ILS 207. „Tiberius Claudius, Sohn des Drusus, Caesar Augustus Germanicus, usw. befreite die Stadt von der Gefahr der Überschwemmungen, indem er im Zuge der Arbeiten am Hafen Kanäle vom Tiber ins Meer leiten ließ."

14 Dass mit *urbs* ohne weiteren Namenszusatz nur die Stadt Rom und nicht Ostia gemeint sein kann, versteht auch A. Wilson, The Mediterranean Environment in Ancient History: Perspectives and Prospects, in: W. V. Harris (Hrsg.), The Ancient Meditarranean Environment between

diesem Fall wirkt die Verwendung des Nomens gleichsam als Bekräftigung für die Endgültigkeit der kaiserlichen Bauvorsorge. Von nun an sei die Stadt also frei von Überschwemmungen. Dass diese Aussage, wie zu erwarten war (es gibt keinen vollständigen Hochwasserschutz), nicht gänzlich zutraf, zeigen die beiden jüngeren Inschriften. Die eine lässt sich auf die Jahre 102–109 n. Chr. in die Regierungszeit Trajans datieren, die andere stammt aus dem Jahr 190 n. Chr. und dokumentiert Brückenbauarbeiten unter Commodus. Obgleich recht fragmentiert, lässt die trajanische Inschrift noch die Bürde der regelmäßigen Tiberüberschwemmungen ebenso spüren wie die immensen Anstrengungen seitens des Kaisers, dieser Herr zu werden:

> [Imp(erator) Caes(ar) divi] / Ne[rvae fil(ius) Nerva] / Tra[ianus Aug(ustus) Ger(manicus)] / Dac[icus trib(unicia) pot(estate) ---] / im[p(erator) --- co(n)s(ul) --- p(ater) p(atriae)] / fossam [fecit] / [q]ua inun[dationes Tiberis] / [a]dsidue u[rbem vexantes] / [rivo] peren[ne instituto arcerentur][15]

Die häufig wiederkehrenden Überschwemmungen waren laut Aussage der Inschrift der Auslöser für die Bauarbeiten Trajans und standen bei deren Ausführung angeblich im Mittelpunkt. Leider ist das Ende des Inschriftentextes weggebrochen, doch das letzte Wortfragment *peren[---]* gibt immerhin einen Hinweis auf eine auf Langfristigkeit angelegte Schutzmaßnahme: Die Maßnahme wurde im Text wohl als *perennis*, also als „immerwährend“ bezeichnet. Jedenfalls wird aus dieser Inschrift gleich in doppelter Hinsicht ersichtlich, dass der römische Kaiser, in diesem Fall Trajan, auf einen möglichst lange währenden Flutschutz aus war, was zum einen im eben genannten *perennis* zum Ausdruck gebracht wird, zum andern in der Verwendung der Bezeichnung *inundatio*. Das Phänomen der Tiberüberschwemmung im Allgemeinen sollte also durch die Baumaßnahme im wahrsten Sinne des Wortes eingedämmt werden, auch für kommende Zeiten.

Die Inschrift von Kaiser Commodus schließlich berichtet vom Bau einer Brücke, die zur Eindämmung von Überschwemmungen errichtet worden sein soll:

> [Imp(erator) Caes(ar) M(arcus) Aur(elius) Commodus] / [Antoninus Pius Fe]lix Aug(ustus) / Sarm(aticus) Germ(anicus) max(imus) Brit(annicus) p(ontifex) m(aximus) trib(unicia) / [p]otest(ate) XV imp(erator) VIII co(n)s(ul) VI p(ater) p(atriae) / pontem arcendae / [i]nundationis gratia / fecit dedicavitque[16]

Science and History, Leiden / Boston 2013, 271. Er sieht in der Inschrift deshalb einen Beweis für das profunde hydrologische Verständnis römischer Ingenieure, die am Bau der claudischen Kanäle beteiligt waren. Denn die Überzeugung, dass Baumaßnahmen am Flussunterlauf zur Reduktion der Flutanfälligkeit der weiter stromaufwärts gelegenen Stadt Rom habe beitragen können, setze eine gute Kenntnis der vorherrschenden Eigenschaften des Wasserhaushalts im Flusssystem des Tibers voraus.

15 CIL VI 964 = CIL XIV 88 = ILS 5797a = AE 2002, 279. „Imperator Caesar Nerva Trajanus Augustus Germanicus usw. ließ einen Graben anlegen, durch den die Überschwemmungen des Tibers, die wiederholt die Stadt erschütterten, mittels eines für die Ewigkeit angelegten Kanals in Schranken gehalten werden sollten.“

16 AE 1909, 67. „Imperator Caesar Marcus Aurelius Commodus usw. ließ (diese) Brücke zur Überschwemmungsabwehr errichten und weihen.“

Im vorliegenden Fall scheint es fraglich, ob mit *pons* wirklich von einer Brücke im Sinne einer Flussüberführung die Rede ist, oder nicht vielmehr eine Art Wehr oder ein wie auch immer geartetes Schutzverdeck gemeint sein soll. Auch fehlt jeder Hinweis auf die Art der Überschwemmung. Wir erfahren nicht, ob es sich um Flusshochwasser oder Überschwemmungen durch Meerwasser handelt. Ausgehend vom Fundort der Inschrift und in Ermangelung genauerer Informationen über den Fundkontext wären beide Hochwasserarten denkbar. Fest steht immerhin, dass auch hier wieder explizit als Auslöser das Problem der Überschwemmung (*inundatio*) und als Baumotivation deren Abwehr (in Form des Gerundivums *arcenda*) genannt werden. Es versteht sich von selbst, dass der fertige Bau vor noch nicht eingetretenen, zukünftigen Überschwemmungen schützen sollte und deshalb der allgemeine Begriff *inundatio* verwendet wird.

II. VERBEN UND PARTIZIPIALKONSTRUKTIONEN

Im vorigen Abschnitt wurde gezeigt, dass – trotz der etymologischen Verwandtschaft der Vokabel für „Überschwemmung“ in den modernen romanischen Sprachen mit dem lateinischen *inundatio* – in der gehobenen römischen Literatur nur äußerst selten und in bestimmten Zusammenhängen von diesem Begriff Gebrauch gemacht wurde. Innerhalb der historiographischen Werke, ebenso wie im Großteil der Dichtung der lateinisch sprachigen Antike, wurde sehr viel öfter auf Verben und Partizipialkonstruktionen zurückgegriffen, um eine Überschwemmung zu beschreiben. Während es für die Dichtkunst im Grunde genommen selbstverständlich ist, lieber poetische Umschreibungen zu benutzen als ein klar definiertes, feststehendes Wort, ist dasselbe Vorgehen in der historiographischen und anderer Prosaliteratur nicht unbedingt sofort nachvollziehbar. Freilich hatten alle großen Geschichtsschreiber der Römer eine erstklassige rhetorische Schulbildung genossen und so nimmt es nicht Wunder, dass auch deren literarischen Produkte von hohem sprachlichem Niveau zeugen. Dennoch macht die soeben genannte Beobachtung es wünschenswert, sich wenigstens ein paar Gedanken über die Eigenart des sprachlichen Stils dieser Texte zu machen.

Schon einige wenige Quellenstellen zu Überschwemmungen des Tibers in und um Rom genügen, um einen Einblick in Stil und Ausdrucksweise der Flutbeschreibungen zu geben:

Livius, *Ab urbe condita* 4,49,2:
… quorum agros Tiberis super ripas effusus maxime ruinis villarum vastavit.
… deren [Etrusker] Felder und insbesondere deren Villen der sich über die Ufer ergießende Tiber in Trümmer legte und verwüstete.

Livius, *Ab urbe condita* 30,38,10:
… nam ita abundavit Tiberis, ut ludi Apollinares circo inundato extra portam Collinam ad aedem Erycinae Veneris parati sunt.
… denn der Tiber war nämlich dermaßen angestiegen, dass die Apollinischen Spiele, da der Circus unter Wasser stand, außerhalb des Collinischen Tors beim Tempel der Venus Erycina abgehalten wurden.

Tacitus, *Annales* 1,76:
Eodem anno continuis imbribus auctus Tiberis plana urbis stagnaverat...
Im selben Jahr setzte der durch anhaltende Regenfälle angeschwollene Tiber die ebenen Flächen der Stadt unter Wasser...

Aurelius Victor, *Epitome de Caesaribus* 13:
... Tiberis inundavit magna clade aedium proximarum...
... der Tiber schwoll zum großen Schaden der nächstliegenden Tempel an...

Ammianus Marcellinus 29,6,17:
Tiberis (...) cloacis et fluviis abundantibus multis (...) effusione imbrium exuberans nimia...
Der Tiber (...) während viele Abwasserkanäle und Bäche überliefen (...) war durch den Zufluss der starken Regenfälle so übermäßig aufgefüllt...

Augustinus, *De civitate Dei* 3,18:
Nam exundante nimis ultra morem fluvio Tiberino paene omnia urbis plana subversa sunt...
Denn dadurch, dass der Fluss Tiber über das übliche Maß hinaus bis zum Überlaufen angeschwollen war, wurde fast die Gesamtheit der ebenen Flächen der Stadt zerstört...

Alle zitierten Textstellen bieten eine sehr anschauliche und griffige Schilderung des Fluthergangs und teils auch der unmittelbaren Flutursache. Starke Regenschauer (*imbres*), brachten den Tiber immer wieder zum Ausufern durch Überfülle (ausgedrückt z.B. mit den Verben *super ripas exundare*, *abundare*, selbstverständlich auch *inundare*) und zum Ergießen (*effundere*) auf große ebene Flächen (*plana*, *agres*), wo oft tagelang alles unter Wasser stand (*stagnare*). Außerdem kam es oft zu Rückstau, wohl durch verstopfte Abwasserkanäle (*cloacae*), was auch die Kanalreinigungsarbeiten unter Augustus und seinen Nachfolgern oder die Eingliederung der Kloakenaufsicht in die Tiberbettkommission unter Trajan nahelegen.[17] Allerdings ist zu bedenken, dass die meisten der geschilderten Flutereignisse sich lange vor dem Verfassen des jeweiligen historiographischen Werkes ereignet haben und die Autoren dementsprechend keine Augenzeugenberichte liefern, sondern höchstens aus eigenen Erfahrungen das Zustandekommen und den Verlauf der Flut extrapolieren konnten. In manchen Fällen mag das sehr wohl auf der Grundlage älterer Quellen, vielleicht sogar zeitgenössischer Berichte geschehen sein, doch bleibt dennoch immer der Verdacht einer übermäßigen literarischen Ausschmückung zur Steigerung der Dramaturgie bestehen. Immerhin kann für die Beschreibungen der Tiberüberschwemmungen davon ausgegangen werden, dass die Einzelheiten nicht völlig fiktiv sein können, da die in Rom lebenden und wirkenden Autoren den Tiber und seine Charakteristika kennen mussten. Der erzählende Charakter der Texte liegt selbstverständlich in den Erfordernissen der jeweiligen literarischen Gattung begründet und nicht in einem Mangel an eindeutigen Begriffen zur Bezeichnung

17 Zur Kanalreinigung unter Augustus: Suet. *Aug*. 37,1. Die Eingliederung der Kloakenaufsicht in die *cura riparum et alvei Tiberis* in trajanischer Zeit legt die Titulatur auf verschiedenen *cippi*, Steine zur städtischen Tiberbegrenzung, nahe, die sich während des Principats Trajans in *cura riparum et alvei Tiberis et cloacarum* u.ä. erweitert. Hierzu C. Masi Doria / C. Cascione, *Cura riparum*, in: E. Hermon (Hrsg.), Riparia dans l'Empire Romain. Pour la définition du concept; proceedings of the Quebec seminar, 29–31 October 2009, Oxford 2010, 288. Für eine aktuelle Gesamtstudie zur *cura riparum* s.A. Lonardi, La *cura riparum et alvei Tiberis*. Storiografia, prosopografia e fonti epigrafice (= BAR International Series 2464), Oxford 2013. Vgl. auch weiter unten Fn. 38.

des Konzepts „Überschwemmung". Außerdem waren die antiken Autoren, neben dem erzählenden Stil ihrer Texte, auf maximale sprachliche Gewandtheit und Wortvarianz bedacht.

Epigraphische Quellen, die Wiederaufbauarbeiten nach einem Flutereignis dokumentieren, weisen ebenfalls sehr häufig eher umschreibende Flutberichte auf, jedoch aus völlig anderen Gründen. Hier ist es allein der euergetische Kontext, der die Art und Weise der Flutschilderung festlegt. Im städtischen Umfeld der klassischen Antike war es üblich, dass die Honoratioren und andere wohlhabende Bürger der Stadt ebenso wie etwaige auswärtige Patrone für die Finanzierung öffentlicher innerstädtischer Anlagen aufkamen. Nicht zuletzt war es dabei von Wichtigkeit, den eigenen Namen bekannt zu machen und mit möglichst positiven Errungenschaften für die Gemeinde in Zusammenhang zu bringen, um sich etwa bei Wahlen eine günstigere Ausgangsposition zu verschaffen. Neben dem Namen des Wohltäters musste gerade bei Wiederaufbaumaßnahmen im Gegensatz zu einem kompletten Neubau genau auf die von ihm gestifteten Gebäudeteile verwiesen werden, um eine eindeutige Zuschreibung der verschiedenen Bauelemente überhaupt möglich zu machen. Da sich die am häufigsten von Flutschäden betroffenen Gebäude jedoch eher außerhalb der Stadt oder an den Stadtgrenzen befanden, z. B. Überlandstraßen und Brücken sowie Stadtmauern und Aquädukte, lag die Zuständigkeit für die Reparaturarbeiten oft beim Kaiser bzw. bei dessen Vertretern in der Provinz.[18] Wahrscheinlich nennen deshalb fast alle inschriftlichen Nachrichten, die uns über die Behebung von Flutschäden informieren, den Kaiser, Angehörige des Kaiserhauses oder zumindest kaiserliche Kuratoren als Stifter. Dennoch stehen auch ihre Inschriften in klassischer euergetischer Tradition, weisen ähnliche oder identische Formulierungen auf und folgen den gängigen epigraphischen Konventionen. Schließlich musste es als besonderes Verdienst gegenüber der Gemeinde gelten, wenn gerade die Teile, die beim letzten Hochwasser besonders schwerwiegende Schäden davon getragen hatten, durch Intervention des Kaisers oder seiner Vertreter verstärkt wiederaufgebaut und mit zusätzlichen Schutzvorkehrungen versehen wurden. Wohl aus diesem Grund wurde selbst bei den Inschriftentexten an kaiserlich finanzierten Bauten auf solche Detailfragen großer Wert gelegt.

So informiert etwa ein Inschriftenstein aus Axima[19] in der Alpenregion über ein Flutereignis, bei dem die über ihre Ufer getretenen Sturzbäche die Teilstrecke

18 Jedenfalls gilt dies für die Reichsstraßen und die damit verbundenen Brücken. Die Oberaufsicht über das Straßennetz lag innerhalb Italiens bei den Kaisern als *curatores viarum*. Außerhalb waren zwar die Straßenanrainer für die Finanzierung der Instandhaltungsarbeiten zuständig, doch blieb der Kaiser auch dort aufgrund seines *imperium maius* oberster Straßenbauer. Dazu M. Rathmann, Untersuchungen zu den Reichsstraßen in den westlichen Provinzen des Imperium Romanum (= Beihefte der Bonner Jahrbücher 55), Mainz 2003, 58.

19 Gefunden wurde der Stein im heutigen Bourg-Saint-Maurice. Die genauen Fundumstände sind unbekannt. Dazu R. Bernard, Les dangers de la vie au bord de l'eau. A propos de deux inscriptions évergétiques de Savoie, ILN Vienne, 536, à Gilly; ILN Alpes Graiae, 54, à Bourg-Saint-Maurice, in: R. Bedon (Hrsg.), Vicinitas aqvae. La vie au bord de l'eau en Gaule romaine et dans les régions voisines. Actes d'un colloque tenu à la Faculté des lettres et des sciences humaines de Limoges, le 1er et le 2 juin 2007, 263–276.

einer Straße beschädigten, nämlich die Strecke, die durch das Gebiet der *Ceutrones* führte:

> [Imp(erator) Caes(ar) Marcus] / [Aurelius Antoninus] / [Aug(ustus) trib(unicia) potest(ate) XVII] / [co(n)s(ul) III pont(ifex) max(imus) p(ater) p(atriae) et] / Imp(erator) Caes(ar) Lu[cius] / Aurelius Verus Au[g(ustus)] / [tr]ib(unicia) potest(ate) III co(n)s(ul) II / [vi]as per fines Ceutro/[n]um vi torrentium / [ev]ersas exclusis / [flu]minibus et in na[tu]/[ra]lem alveum redu[ctis] / [m]olibus plurib[us locis] / [opp]ositis item po[ntes] / [tem]pla et balin[ea] / [pec(unia)] sua resti[tuer(unt)][20]

Recht anschaulich wird hier erst der Grund für die Zerstörung angegeben, nämlich die „Wucht der Sturzbäche" und die „über die Ufer getretenen Flüsse", um gleich darauf nicht minder anschaulich die Wiederaufbaumaßnahmen und die Anlage zusätzlicher Dammkonstruktionen zu schildern. Der Beweggrund für eine derart ausführliche Beschreibung ist schnell identifiziert: je eindrucksvoller die Überschwemmungskatastrophe geschildert wird, desto verdienstvoller erscheint im Nachhinein die Baumaßnahme der beiden Kaiser und umso mehr Dank und Respekt wird ihnen entgegengebracht. Ein etwaiges Streben nach einer sprachlich ausgefeilten Ausdrucksweise sucht man hier jedoch vergebens. Dass die Kaiser schließlich auch für die Rückführung der Flüsse in ihr „natürliches Flussbett" sorgten, entspricht klar der Herrschaftskonzeption der römischen Kaiser, die sich als oberste Aufseher und Herrscher nicht nur über Menschen und Länder, sondern ebenso über die Naturgewalten verstanden.

Ein ähnliches Zeugnis liegt uns in einer Bauinschrift aus Samnium vor. Zwar ist der Anfang des Textes und somit die gesamte Kaisertitulatur, die vermutlich dort zu lesen war und zur Datierung hätte dienen können, weggebrochen, doch ist immerhin der Teil komplett erhalten geblieben, in dem ein Flutschaden und dessen Behebung thematisiert werden:

> ------] / pontem portus Curensis violen/tia torrentis ablatum dextra / laevaque a solo exstructis / pilarum molibus usui / commeantium amplificato / opere restituit / cura agente L(ucio) Baebio Celso c(larissimo) v(iro)[21]

Verursacht wurde der Schaden auch hier durch die „Gewalt eines Sturzbaches" (*violentia torrentis*). In beiden Fällen kann also ein möglicherweise durch Schmelzwasser in den Bergen verursachtes Hochwasser konstatiert werden oder alternativ ein durch exzessiven Regen im Gebirge hervorgerufenes. Jedenfalls weist die Vokabel *torrens* (Sturzbach), die jeweils als unmittelbare Schadensursache genannt wird, auf eine relativ plötzlich entstandene, gewaltige Sturzflut hin, wie sie gerade

20 CIL XII 107 = ILS 5868 = AE 1996, 981 = AE 2009, 803 = ILN Alpes Graiae 54. „Die Kaiser Marcus Aurelius und Lucius Verus etc. stellten die Straße, die durch Ceutronengebiet führt und von der Wucht der Sturzbäche aufgewühlt worden war, wieder her, indem sie aus eigenen Geldmitteln die übergetretenen Flüsse wieder in ihr natürliches Flussbett zurückführten und ihnen mehrere Dämme entgegenstellten, ebenso stellten sie die Tempel und Thermen aus eigenen Geldmitteln wieder her." Der Stein datiert ins Jahr 162/3 n. Chr.

21 AE 1958, 269: „… ließ die Brücke von portus Curensis, die von der Gewalt eines Sturzbaches weggespült worden war, wieder herstellen, indem er rechts und links der Pfeiler die Molen, die zum Durchgangsverkehr verbreitert worden sind, von Grund auf errichten ließ unter der Aufsicht des Lucius Baebius Celsus, einem ausgezeichneten Mann."

in bergigen Regionen häufig auftritt. Dieser Befund passt sehr gut mit den Fundorten der beiden Inschriften zusammen. Das letztere Beispiel zeigt außerdem, dass Neu- und Wiederaufbauarbeiten gelegentlich dazu genutzt wurden, ohnehin wünschenswerte Veränderungen am Bau vorzunehmen und somit sogar einen gewissen Nutzen aus dem Schadensereignis zu ziehen.

Des Weiteren liegt uns aus der hispanischen Provinz *Baetica* eine ins 3. Jh. datierte Ehreninschrift für den Statthalter vor, die neben anderen Dingen von der Erneuerung des durch anhaltende Regengüsse (*adsidui imbres*) zerstörten Straßenbelags durch den Statthalter informiert:

> Q(uinto) Torio Q(uinti) f(ilio) Culleoni / proc(uratori) Aug(usti) provinc(iae) Baet(icae) / quod muros vetustate / collapsos d(e) s(ua) p(ecunia) refecit solum / ad balineum aedificandum / dedit viam quae per Castul(onensem) / saltum Sisaponem ducit / adsiduis imbribus corrup/tam munivit signa Vene/ris Genitricis et Cupidi/nis ad theatrum posuit / HS centies quae illi summa / publice debebatur addito / etiam epulo populo remisit / municipes Castulonenses / editis per biduum circens(ibus) / d(ecreto) d(ecurionum)[22]

Imber ist, im Gegensatz zu *pluvia*, ein starker Platz- oder Gewitterregen.[23] Solche Starkregen sind in der Fundregion bis heute typisch und sorgen dort noch immer für plötzlich eintretende extreme Flusshochwasser. Es ist also anzunehmen, dass im vorliegenden Fall die Straße durch den vom Dauerregen stark angeschwollenen Fluss, möglicherweise den heutigen Guadalimar, so stark beschädigt worden war, dass eine Erneuerung der Fahrbahn notwendig wurde.[24] Aus dem Text ist also nicht nur zu erfahren, dass ein Flutschaden eingetreten ist, sondern auch wie er zustande gekommen ist, wo er sich ereignet hat und wie er durch wen behoben wurde. In diesem Punkt überschneidet sich der Informationsgehalt aus der Literatur mit dem aus den epigraphischen Zeugnissen, wobei letztere insbesondere über strukturelle Maßnahmen zur Fluthilfe und zum Flutschutz exaktere Daten liefern, wenn auch nur sehr punktuell.

Aus der Sicht des heutigen Historikers kann die sprachliche Eigenart, sowohl der antiken Literatur als auch der epigraphischen Praxis, als Glücksfall gelten. Denn die Benennung der unterschiedlichen Auslöser für die beschriebenen Über-

22 CIL II 3270 = ILS 5513 = AE 1975, 526: Dem Quintus Torius Cuellonis, Sohn des Quintus, kaiserlicher Prokurator der Provinz Baetica, (…), weil er die Straße, die durch den *saltus Castulonensis* nach *Sisapo* führt und von anhaltenden Regenfällen beschädigt war, befestigen ließ, usw."

23 L. Doederlein, Lateinische Synonyme und Etymologieen Bd. 2, Leipzig 1827, 87–89 glaubt sogar, im Gegensatz zum deutlich harmloseren *pluvia*, in *imber* einen zerstörerischen, wolkenbruchartigen Winterregen identifizieren zu können, der von kaltem, sehr stürmischem Wetter begleitet zu werden pflegt.

24 Der Fundort *Castulo*-Cazlona südlich der modernen Siedlung Linares ist am Fluss Guadalimar gelegen. Die Finanzierungslage ist bei derart kleinen Straßen und Teilstrecken unklar. Vgl. Rathmann, Op. cit., 141 Fn. 807. Offenbar finanzierte der Statthalter der baetischen Provinz die Straßenarbeiten als Privatperson, da die anderen genannten *munera* eher euergetische Charakterzüge gegenüber der Stadtgemeinde von *Castulo* tragen und der Zusatz *de sua pecunia* (die Finanzierung aus eigenen Geldmitteln), sofern er sich auf alle Maßnahmen und nicht nur auf die Reparatur der Stadtmauern bezieht, auch eher für ein privates Geschenk und nicht für eine statthalterliche Pflichterfüllung spricht.

schwemmungen und deren Verlauf liefern uns wertvolle Hinweise auf die damaligen hydrologischen Charakteristika einer Region und können bis zu einem gewissen Grad Aufschluss geben über die verschiedenen Hochwasserarten, mit denen der antike Mensch konfrontiert wurde. Der Gebrauch einer universalen Vokabel zur Bezeichnung von letzten Endes doch recht unterschiedlichen Flutereignissen würde da mehr verschleiern als aufdecken.

III. *AQUA MAGNA, AQUA INGENS, MAGNITUDO AQUARUM*

Was dieses Begriffsfeld charakterisiert, ist die explizite Benennung des Stoffes „Wasser", *aqua*, dem ein (scheinbar) rein quantifizierender Zusatz beigegeben ist, wie dies etwa bei *aqua magna* oder *magnitudo aquarum* der Fall ist. In den literarischen Werken aus der Zeit der ausgehenden Republik und den Anfängen des Principats ist diese Art der sprachlichen Bezugnahme auf Hochwasser besonders häufig anzutreffen. So finden sich beispielsweise bei T. Livius, dessen historiographisches Werk *Ab urbe condita* immerhin mehrere Jahrhunderte der römischen Geschichte umfasst, verschiedene derartige Bezeichnungen für die häufigen Tiberhochwasser in Rom: *aquae magnae*,[25] *aquarum magnitudo*[26] *und aquae ingentes*.[27]

Eine ähnliche Wortwahl ist uns aus den *commentarii* zum Bürgerkrieg von C. Iulius Caesar bekannt. Im Umfeld der Entscheidungsschlacht beim hispanischen *Ilerda* zwischen Caesar und den Legaten des Pompeius ereignete sich ein außergewöhnlich starkes Flusshochwasser. Dieses lähmte die Mobilität der caesarischen Truppen ebenso wie deren Versorgung. Zwar sahen sich auch gegnerische Truppenteile durch die Flut eingeschränkt handlungsfähig, doch wohl in weitaus geringerem Maße als es bei Caesar selbst der Fall war, sofern wir seinen Ausführungen Glauben schenken können. Seine Truppen waren für mehrere Tage von den Wassermassen regelrecht eingeschlossen, zumal beide kurz zuvor gebaute Brücken stark beschädigt und daher unpassierbar waren:

> Zwei Tage nach dieser Schlacht ereignete sich ein plötzliches Unglück. Es brach nämlich ein solches Unwetter herein, daß man sicherlich in dieser Gegend noch nie größere Wassermassen (*maiores aquas*) gesehen hatte. Infolge dieses Unwetters stürzte von allen Bergen das Schmelzwasser herab, überflutete die Flußufer und riß an einem einzigen Tag die beiden Brücken weg, die C. Fabius hatte bauen lassen. Dieser Umstand brachte Caesars Heer in große Schwierigkeiten. Da sich das Lager, wie oben erklärt wurde, zwischen den zwei Flüssen Sicoris und Cinga befand, die 30 Meilen voneinander entfernt und zur Zeit beide nicht zu überqueren waren, wurden notgedrungen alle auf dem engen Zwischenraum festgehalten. Die Gemeinden, die sich Caesar angeschlossen hatten, konnten kein Getreide herbeischaffen, und die Soldaten, die zum Futterholen weiter vorgedrungen waren, konnten nicht zurückkehren, da sie durch die Flüsse vom Heer abgeschnitten waren. Auch war es unmöglich, den umfangreichen Nachschub, der aus Italien und Gallien kam, ins Lager zu bringen.[28]

25 Liv. 24,9,6.
26 Liv. 30,26,5.
27 Liv. 35,9,2 und 38,28,4.
28 Caes. *BC* 1,48,1–4 (Übersetzung nach M. Deißmann): *Accidit etiam repentinum incommodum biduo, quo haec gesta sunt. tanta enim tempestas cooritur, ut numquam illis locis maiores*

Die Stelle des caesarischen Truppenlagers am Zusammenfluss zweier Flüsse erwies sich unter diesen Umständen als besonders ungünstig. Zusätzlich machten Caesar die Angriffe der feindlichen Truppenteile zu schaffen, die vom gegenüberliegenden Ufer aus jegliche Wiederaufbaumaßnahmen der Versorgungswege erfolgreich zu stören versuchten:

> Diese Überschwemmung (*aquae*) dauerte einige Tage. Caesar versuchte, die Brücke wieder herstellen zu lassen. Der Hochwasser führende Fluß (*magnitudo fluminis*) verhinderte dies jedoch. Auch die feindlichen Cohorten, die am anderen Ufer aufgestellt waren, ließen es nicht zu. Denn es war leicht für sie, das zu verhindern, einmal wegen der Gestalt des Flußbetts und der Höhe des Wasserstandes (*ipsius fluminis natura atque aquae magnitudine)*, dann aber vor allem, weil sie von der gesamten Uferstrecke aus mit ihren Wurfgeschossen auf einen einzigen und engbegrenzten Platz zielen konnten; und es war schwer, gleichzeitig in dem reißenden Fluß den Bau zu vollenden und den Geschossen auszuweichen.[29]

Bezieht sich Caesar allgemein auf die allgegenwärtigen Wassermassen, verzichtet er weitgehend auf ein quantifizierendes Adjektiv und begnügt sich damit, die Pluralform *aquae* zu verwenden. Bezieht er sich hingegen auf ein bestimmtes Gewässer, den Fluss, zieht er durchaus das Adjektiv *magnus* hinzu, um auf dessen außergewöhnlich hohen Pegelstand hinzuweisen.

Nach diesem kurzen Einblick in das livianische und das caesarische Werk könnte nun argumentiert werden, dass es bei einem so herausragenden Schriftsteller wie Livius nicht sehr verwunderlich sei, eine derart poetisch anmutende Wortwahl zur Bezeichnung eines Hochwassers anzutreffen. Bei den *commentarii* Caesars liegt die Sache jedoch etwas anders. Obgleich auch sein Werk von Beginn an zur Publikation verfasst worden war, ist die Textgattung eine andere, die wesentlich weniger sprachliche Ausschmückungen verlangt.[30] Außerdem ist die relative Schlichtheit

aquas fuisse constaret. tum autem ex omnibus montibus nives proluit ac summas ripas fluminis superavit pontisque ambos, quos C. Fabius fecerat, uno die interrupit. quae res magnas difficultates exercitui Caesaris attulit. castra enim, ut supra demonstratum est, cum essent inter flumina duo, Sicorim et Cingam, spatio milium XXX, neutrum horum transiri poterat, necessarioque omnes his angustiis continebantur. neque civitates, quae ad Caesaris amicitiam accesserant, frumentum supportare neque ii, qui pabulatum longius progressi erant, interclusi fluminibus reverti neque maximi commeatus, qui ex Italia Galliaque veniebant, in castra pervenire poterant.

29 Caes. *BC* 1,50,1–3. (Übersetzung nach M. Deißmann) *Hae permanserunt aquae dies complures. conatus est Caesar reficere pontes, sed nec magnitudo fluminis permittebat, neque ad ripam dispositae cohortes adversariorum perfici patiebantur. quod illis prohibere erat facile cum ipsius fluminis natura atque aquae magnitudine, tum quod ex totis ripis in unum atque angustum locum tela iaciebantur; atque erat difficile eodem tempore rapidissimo flumine opera perficere et tela vitare.*

30 W. Richter, Caesar als Darsteller seiner Taten, Heidelberg 1977, 46–48. Zwar waren die *commentarii* keine eindeutig definierte Textgattung, doch zeichneten sich so benannte Texte durch ihre strenge Sachlichkeit und „das völlige Fehlen jeder künstlerischen Absicht“ aus (ibid., 47). Speziell Caesars *commentarii* sind allerdings in erster Linie als politische Schrift zu verstehen, mit deren Hilfe gezielt auf aktuelle machtpolitische Fragen zu seinen eigenen Gunsten Einfluss genommen werden sollte. Er hatte sie also unter erheblichem Zeitdruck verfassen müssen, womit auch erklärt werden kann, warum er sich gerade für die Textgattung der *commentarii* entschied. Mehr zu den caearischen *commentarii*, den Schwierigkeiten bei deren literarischen

von Caesars Ausdrucksweise charakteristisch für sein Werk.[31] Dennoch verzichtet auch er weder auf sprachliche Varianz noch auf die Konstruktion *Wasser / Gewässer + Adjektiv, das auf Größe bzw. Masse anspielt*. Auffällig ist zudem, dass beide Autoren auf die Benutzung eines allgemeineren aber griffigeren Nomens, wie etwa *inundatio*, von vorn herein verzichteten. Was bewegte die beiden Autoren dazu, sich genau dieses Musters zu bedienen? In der späteren, kaiserzeitlichen Historiographie findet sich die Ausdrucksweise *aqua magna* o.ä. zur Bezeichnung eines Hochwassers nicht mehr, wohingegen eine weitere Textstelle aus dem Briefkorpus des Staatsmanns Cicero die Verbreitung derartiger Ausdrucksweisen für die (späte) Republik bestätigt. In einem Brief an seinen Bruder Quintus berichtet er über eine verheerende Überschwemmung in der Stadt Rom im Jahre 54 v. Chr.:

> In Rom, und zwar besonders auf der Via Appia um den Marstempel, gibt es eine erstaunliche Überschwemmung (*diluvium*; dazu mehr in Abschnitt IV). Die Wandelhalle des Crassipes ist weggespült, ebenso die Gärten und viele Werkstätten und Wirtshäuser. Die große Wassermasse (*magna vis aquae*) reicht bis zum öffentlichen Fischteich.[32]

Freilich ist darauf hinzuweisen, dass in allen angeführten Textstellen weniger die Flutung bestimmter Plätze oder Güter im Vordergrund steht als vielmehr die schiere Menge an Wasser. *Aqua magna*, „großes Wasser", entspricht also am ehesten unserem Begriff „Hochwasser" und weniger dem Konzept der Überschwemmung. Auch scheint es kein feststehendes Begriffspaar dafür gegeben zu haben. Die Erwähnung von Wasser, zusammen mit einem quantifizierenden Adjektiv, reichte zum Verständnis wohl aus, was aus den Beispielen ersichtlich wird.

Ein ebensolches Bild ergibt sich aus den zwei nicht literarischen Belegstellen, die erhalten geblieben sind: einer Inschrift und einem Rechtstext. Sie enthalten die Wortlaute *aqua magna* bzw. *aquarum magnitudines* und beziehen sich damit jeweils auf ein Hochwasserereignis. Bei der Inschrift handelt es sich um die berühmten *fasti Ostienses*, einem Marmorkalender aus der an der Tibermündung gelegenen Hafenstadt Ostia. Dieser vermutlich im Jahre 2 n. Chr. begonnene monumentale Kalender, von dem heute noch fragmentarisch die Ereignisse aus den Jahren 49 v. Chr. bis 175 n. Chr. enthalten sind, hält als besonderes Ereignis für das Jahr 147 n. Chr., neben anderen Ereignissen, eine Tiberüberschwemmung fest: *[---]X K(alendas) April(es) aqua magna fuit.*[33] Dieser nüchternen Feststellung folgt kein weiterer Kommentar zu Ausmaß oder Folgen des Hochwassers. Die Wortkargheit

Bewertung sowie zu *commentarii* als Gattungsbegriff s. J. Rüpke, Religiöse Erinnerungskulturen. Formen der Geschichtsschreibung in der römischen Antike, Darmstadt 2012, 63–67.

31 So urteilt etwa schon sein Zeitgenosse Cicero über die *commentarii* Caesars, sie seien „nackt, gerade und anmutig, ohne Redeschmuck wie ein Körper ohne Gewand" (*valde quidem probandos; nudi enim sunt, recti et venusti, omni ornatu orationis tamquam veste detracta*). Zitiert nach W. Richter, op. cit., 11.

32 Cic. *Ad Q. fr.* 3,5,8 (3,7,1). *Romae, et maxime Appia ad Martis, mira proluvies. Crassipedis ambulation ablata, horti, tabernae plurimae; magna vis aquae usque ad piscinam publicam.*

33 CIL XIV 244 (p. 481, 773) = CIL XIV 245 = CIL XIV 4531 = CIL XIV 4532 = CIL XIV 4532a = CIL XIV 4533 = CIL XIV 4534 = CIL XIV 4535 = CIL XIV 4536 = CIL XIV 4537 = CIL XIV 4538 = CIL XIV 4539 = CIL XIV 4540 = CIL XIV 4542 = CIL XIV 4543 = CIL XIV 4544 = CIL XIV 4546 = CIL XIV 5354 = CIL XIV 5355 = ILS 6126.

liegt in der Natur der Quellengattung begründet, die generell nur knapp die das Jahr kennzeichnenden Ereignisse festhalten soll, nicht aber darüber hinausgehende Maßnahmen.[34] Was den Rechtstext angeht, handelt es sich um eine Regelung aus den justinianischen Digesten. Im 50. Buch ist dort Folgendes zu lesen:

> (…) Unglücks- und Todesfälle von Tieren aber, die sich ohne eigenes Verschulden ereignen, für die Flucht von Sklaven, die nicht unter Aufsicht zu stehen pflegen, Diebstahl, Aufstände, Brände, Hochwasser (*aquarum magnitudines*) und Überfälle von Räubern werden von niemandem verbürgt.[35]

In beiden Fällen mag es zunächst verwundern, dass, trotz des späten Entstehungszeitraums der Texte, überhaupt auf die antiquierte Ausdrucksweise zurückgegriffen wird, statt einer zeitgemäßeren Vokabel wie etwa *inundatio*, die sich gehäuft allerdings erst in der Kaiserzeit findet, den Vorzug zu geben. Doch Rechtstexte, ebenso wie Inschriften, bedienen sich häufig einer antiquierten Sprache, sodass dies auch hier als teilweise Erklärung geltend gemacht werden kann. Zudem ist zu bedenken, dass der Gesetzestext aus einem Kapitel des Buches 50 stammt, das mit *De diversis regulis iuris antiqui* überschrieben ist. Das Kapitel enthält also eine Zusammenstellung älterer Rechtsregeln. So ist zu vermuten, dass eine wörtliche Übernahme stattgefunden hat und der Text im originalen Wortlaut vorliegt. Dies scheint umso wahrscheinlicher, als der Paragraph einem Buch des spätklassischen Juristen Ulpian entnommen ist, der sich, neben dem Verfassen relativ ausführlicher Rechtskommentare, vor allem auf das Sammeln und Kompilieren von spätrepublikanischen, früh- und hochklassischen Rechtsregelungen konzentrierte.[36]

Abschließend kann festgehalten werden, dass *aqua magna* als Begrifflichkeit für Hochwasser in der Literatur nur in republikanischer Zeit (Caesar, Cicero) und zu Beginn der Principatszeit (Livius) öfter auftritt, wohingegen die Quellen aus der Kaiserzeit und der Spätantike sich auf je eine Textstelle beschränken. Anscheinend hat ein Wandel in der Sprechweise und möglicherweise auch im Empfinden der antiken Menschen stattgefunden. Denn das Adjektiv *magnus* meint nicht nur quantitativ „groß“ oder „gewaltig“, sondern kann auch eine gewisse Ver-, wenn nicht gar Bewunderung gegenüber der als *magna* gekennzeichneten Sache ausdrücken. Dieses Erstaunen vor den gewaltigen Wasserfluten musste in den kaiserzeitlichen Texten einem Ausdruck weichen, der viel mehr den Aspekt des Unterwasserstehens

34 Die *fasti Ostienses* gehören zwar zu den ausführlichsten inschriftlichen Jahresaufzeichnungen, die aus der römischen Antike auf uns gekommen sind, und erinnern in ihrem Stil schon beinahe an die livianischen Epitome, doch orientieren sie sich formal noch immer an früheren *fasti*. J. Rüpke, op. cit., 99.

35 *Dig*. 50,17,23. *Animalium vero casus mortesque, quae sine culpa accidunt, fugae servorum que custodiri non solent, rapinae, tumultus, incendia, aquarum magnitudines, impetus praedonum a nullo praestantur.*

36 Denkbar wäre auch, dass Ulpian durch die ausgiebige Lektüre und Kompilationsarbeit früherer Rechtstexte deren Stil und Ausdrucksweise übernommen hat. Vgl. T. Honoré, Ulpian, Oxford 1982, 247: „His [Ulpian] importance lies in the part he played in the transmission of the Roman legal heritage. (…) Indeed his work amounts to a precodification. But he was not just a compiler. His Latin prose is cool and lucid. If readers of the *Digest* have the impression that the texts are composed in the language of the late republic or early empire, this is in large measure because so much of it is excerpted from Ulpian.“

beinhaltet, nämlich *inundatio*. Ist dieser Wandel am Ende auf eine Veränderung der Vulnerabilitätsmuster der stadtrömischen Gesellschaft in der Kaiserzeit zurückzuführen, deren Nutz- und Wohnbauten sich immer weiter an die Ufer des Tibers heranwagten,[37] was schließlich zur Gründung einer Tiberbettkommission im Jahre 15 n. Chr. führen sollte, deren Aufgabe es war, die Ufergrenzen auszumessen, zu kennzeichnen und zu überwachen?[38]

IV. *DILUVIUM / DILUVIES, PROLUVIES, INLUVIES* ETC.

> Cataclysmus, quod nos diluvium vel irrigationem dicimus, cum factum est, omne genus humanum interiit praeter Deucalionem et Pyrrham …[39]

Der Kataklysmos der stoischen Lehre sowie die biblische Sintflutgeschichte wurden oben bereits angesprochen, da beide hin und wieder als *inundatio* bezeichnet wurden. Die gängigere Bezeichnung für das Konzept einer alles umfassenden Flut war jedoch *diluvium* oder *diluvies*.[40] Auch außerbiblische mythische Sintfluten wurden in der lateinischen Literatur, allen voran die Sintflut mit Deukalion und Phyrra, als *diluvium* ausgewiesen, was auch im obigen Text des Fabelschriftstellers Hygin vom Beginn der Principatszeit deutlich wird. *Irrigatio* wird von ihm als zweite alternative Bezeichnung genannt. Tatsächlich aber kommt sie in der klassischen antiken Literatur im Zusammenhang mit der Sintflut nicht vor und wird hier daher beiseitegelassen.[41] Den Ursprung hat dieses Nomen im Verb *diluere*, was so viel wie „auflösen“, „aufweichen“, „entfernen“ bedeutet. Somit wohnt der Bezeichnung von

37 A. Wilson, op. cit., 269, stellt bereits für die Zeit der mittleren und späten Republik fest, dass die fortschreitende Urbanisierung Roms und insbesondere die Umgestaltungsarbeiten auf dem Marsfeld durch Pompeius und Caesar die Größe der von Flut betroffenen, intensiv genutzten Flächen der Stadt beständig erweiterte. Eine zunehmende Anfälligkeit Roms für Überschwemmungen durch den Tiber aufgrund der laufenden Vergrößerung der bebauten und versiegelten Flächen sieht auch Aldrete, op. cit., 85–89.

38 Cass. Dio 57,14,7–8. Zum Entstehungskontext der *cura riparum* s. H. Schneider, Überschwemmungen und Hochwasserschutz im antiken Rom, in: A. Hoffmann (Hrsg.), Festschrift zum sechzigsten Geburtstag von Universitätsprofessor Frank Tönsmann und dem zehnjährigen Bestehen des Fachgebiets Wasserwirtschaft (= Kasseler Wasserbau-Mitteilungen 12), Kassel 2001, 206; P. Leveau, Les inondations du Tibre à Rome: politiques publiques et variations climatiques à l’époque romaine, in: Vers une gestion intégrée de l’eau dans l’Empire Romain: actes du colloque international Université Laval, octobre 2006, Rom 2008, 139. Für Fragen zur Gründung und Funktionen des Kuratorenkollegiums s. C. Masi Doria / C. Cascione, op. cit., insbesondere 286–290. Für Quellenmaterial und umfassende Studien s. die Monographie von A. Lonardi, op. cit. Eine Auswahl an Grenzsteinen der Tiberbettkommission findet sich auch bei Dessau: ILS 5925–5934.

39 Hygin. *fab. Deukalion et Pyrrha*: „Als der Kataklysmos (*cataclysmus*), den wir *diluvium* oder *irrigatio* nennen, sich ereignete, kam die gesamte Menschheit um, außer Deukalion und Pyrrha …“

40 Zur biblischen Sintflut etwa Boeth. *de fide cath.* 60,130; Tert. *adv. Iud.* 2,81; Ambr. *de Noe*, passim. Zum Kataklysmos bei Seneca s. *Naturales quaestiones*, Buch 3, passim.

41 Erst in der spätantiken christlichen Tradition findet *irrigatio* häufiger als Bezeichnung der biblischen Sintflut Verwendung, so z. B. bei Aug. *enarr. in psalm.* 40,3,8.

Anfang an ein totalitärer, fataler Unterton inne, der auf die Hörer bzw. Leser unter Umständen geradezu beängstigend wirken konnte. So kann es denn auch kaum verwundern, dass in den wenigen aus der Antike überlieferten Briefen, in denen ein vom Verfasser unmittelbar selbst erlebtes Flutereignis geschildert wird (es sind gerade einmal zwei!), auf dieses Wortfeld zurückgegriffen wurde. Selbstverständlich sind die Briefe, ähnlich den *commentarii* Caesars, keine tatsächlichen Egodokumente, sondern waren von vorn herein Teil eines zur Veröffentlichung vorgesehenen Textkorpus, also literarische Produkte. Dementsprechend geben sie keine authentischen Erfahrungsberichte wieder, sondern enthalten geschönte, sprachlich wie inhaltlich sehr künstlerisch komponierte Narrationen. Für die Begriffsanalyse sollte diese Tatsache allerdings nicht weiter stören, geht es hier doch auf sehr allgemeiner Ebene um die Bandbreite an Möglichkeiten, Überschwemmungsereignisse in der lateinischen Sprache in Worte zu fassen. Und schließlich ist bei zwei rhetorisch so exzellent ausgebildeten Autoren wie Cicero und Plinius dem Jüngeren gerade in besonderem Maße zu erwarten, dass sie mit einer klugen Wortwahl gezielt Emotionen bei ihren Lesern hervorzurufen im Stande waren.

Die Textstelle aus dem Cicerobrief wurde in Abschnitt III schon abgedruckt, sodass hier nur noch einmal der entsprechende Satz zitiert werden soll: *Romae, et maxime Appia ad Martis, mira proluvies*.[42] Auch aus dem Pliniusbrief soll, obwohl der gesamte Brief ausschließlich eine schwere Tiberüberschwemmung zum Thema hat, nur der das Stichwort enthaltende Satz zitiert werden: *Num istic quoque immite et turbidum caelum? Hic assiduae tempestates et crebra diluvia*.[43] Während durch die Vorsilbe *pro-* bei *proluvies* wahrscheinlich vor allem auf die Anschwemmungen von Treibgut, Schlamm und Abfall verwiesen wird (*proluvies* kann auch Unrat bedeuten),[44] verstärkt die Benutzung der Vokabel *diluvium* im Plural noch einmal den zerstörerischen Wasserüberschuss, auf den hier möglichst anschaulich hingewiesen werden soll. In Anbetracht der Tatsache, dass *diluvium* hauptsächlich für allumfassende Sintfluten verwendet wurde, mögen die beiden Briefzitate etwas übertrieben wirken. Ihre Wirkung beim Publikum dürften sie aber nicht verfehlt haben. Zudem scheint auf der Grundlage der Briefe die Annahme bezeugt, dass die Benutzung des *diluvium*-Begriffs an keinen bestimmten zeitlichen Kontext gebunden war, sondern ausschließlich durch die Situation und Aussageabsicht der Sprecher bestimmt wurde. Immerhin schreibt Cicero im 1. Jh. v. Chr. (die von ihm beschriebene Überschwemmung ereignete sich im Jahre 54 v. Chr.), während der Brief Plinius des Jüngeren auf einen unbestimmten Zeitpunkt nach 107 n. Chr. datiert wird.[45]

Schließlich könnte der Text einer Ehreninschrift aus der *Africa proconsularis*, die ins späte 3. oder ins 4. Jh, datiert, auch für die Annahme sprechen, dass vor al-

42 Cic. *ad Q. fr.* III 5,8 (III 7,1): „In Rom, und zwar besonders auf der Via Appia um den Marstempel, gibt es eine erstaunliche Überschwemmung." Vgl. S. 122.

43 Plin. *epist.* 8,17,1: „Ist denn dort bei euch das Wetter auch so rau und stürmisch? Hier kommt es zu anhaltenden Unwettern und wiederholten Überschwemmungen."

44 In ähnlichem Sinne verwendet bei Lucr, *rer. nat.* 6,1200.

45 Zur Datierung des Plinius-Briefs s. A. N. Sherwin-White, The letters of Pliny. A historical and social commentary, Oxford 1985 (Erstauflage 1966), 467.

lem Augenzeugen und unmittelbar von der Flut Betroffene die Verwendung des stark emotional aufgeladenen Nomens *diluvium* vorzogen. Die Inschrift ist dem Mitglied einer bekannten städtischen Honoratiorenfamilie gewidmet, einem gewissen Lucius Octavius Aur[elianus?] Didasius, der sich nach einer schlimmen Überschwemmung mit besonderem Engagement um die Wiederaufbauarbeiten kümmerte. Der Inschriftentext lautet:

> Didasi // L(ucio) Octavio Aur[eliano?] / Didasio c(larissimo) v(iro) m[---]o / civi genitali ob [sin]gu[la]/rem in protegendis [civib(us)] / fidem et paratum [er]ga [o]mn[e]s / amorem thermas [et aquam? corrup]/tam post diluviem [---]/to servato [---] / propria liberalitate [ex] o[rnavit] / excoluit perfecit dedi[c]avit / bene merito civi et pa[tr]ono [splen]/didissimus ord[o et] populus [mun(icipii)] / [U]ruensium statuam [posueru]nt d(ecreto) d(ecurionum) [p(ecunia) p(ublica)][46]

Die an mehreren Stellen stark beschädigte Inschrift gibt immerhin die Information preis, dass es sich bei dem geehrten Wohltäter um einen gebürtigen Bürger der Stadt (*civis genitalis*) handelt. Entsprechend herzlich fiel laut Inschrift denn auch sein Bemühen um die wohl arg durch die Flut in Mitleidenschaft gezogene Gemeinde aus. Da seine Familie schon seit längerem im Stadtrat engagiert war und daher nicht nur fest in der Gegend etabliert, sondern auch ansässig gewesen sein muss,[47] ist es nicht unwahrscheinlich, dass im Zuge der Überschwemmung auch eigener Familienbesitz beschädigt worden war. Eine Augenzeugenschaft kann nicht ausgeschlossen werden, ist aber in diesem Fall nicht zwingend. Da die Dedikanten die Stadtbewohner von *Ureu* waren, lag die Textgestaltung letzten Endes bei ihnen bzw. beim Rat der Stadt, dessen Mitglieder sicher mehrheitlich die Flut selbst miterlebt und mit ihren Folgen im Privaten oder als Räte zu kämpfen hatten. Durch den Gebrauch des Wortes *diluvium* jedenfalls wird der Anschein erweckt, die Flut sei tatsächlich ein außergewöhnlich traumatisches Ereignis gewesen, wodurch das Verdienst des Geehrten wiederum erhöht wird.

Eine andere Erklärung ist allerdings auch denkbar. Der Schlüssel liegt im Aufstellungsort und -zeitpunkt des Inschriftensteins. War *diluvium* in früherer Zeit, wie bereits festgestellt, eher für Sintflutsagen verwendet worden, finden sich in der Spätantike endlich ein paar Texte, die das Nomen im Zusammenhang mit einer historischen Flut oder auch der Nilschwelle benutzen:

> Sextus Aurelius Victor, *de Caesaribus* 32:
> Tiberis adulta aestate diluvii facie inundavit.
> Der Tiber stieg im Spätsommer einer Sintflut gleich an.
>
> Cassiodor, *Variae* 3,52 (Brief Theoderichs an Cassiodor):

46 AE 1975, 880: „(Statue des) Didasius. Dem Lucius Octavius Aur[elianus?], einem (ausgezeichneten?) Mann und gebürtigen Bürger unserer Stadt, wegen seiner einzigartigen Treue, die Bürger zu beschützen, und seiner bereitwilligen Zuwendung zu allen. Er hat aus der ihm eigenen Großzügigkeit heraus die Thermen (und den) nach der schlimmen Überschwemmung (beschädigten Aquädukt?) erhalten(?) und (ausgeschmückt?), verbessert, vollendet und geweiht. Auf Beschluss des Stadtrats stiften der ehrwürdige *ordo* und das Stadtvolk der Uruenser diese Statue dem wohlverdienten Bürger und Patron (aus öffentlichen Mitteln)."

47 M. Vannesse, Les usages de l'eau courante dans les villes romaines. Le témoignage de l'épigraphie, Latomus 71 (2012), 476.

… ubi Nili fluminis superveniente diluvio indicia finium vastissimus gurges abradit…
… wo das unermesslich tiefe Wasser des plötzlich zur Flut ansteigenden Nils die Grenzmarkierungen herausreißt…

Aurelius Victor wurde um 320 n. Chr. in *Africa* geboren, das seinerzeit bereits stark christianisiert war. Und Cassiodor, der vermutlich der Spross einer syrischstämmigen Familie war, stand zeit seines Lebens verschiedenen Kirchenvertretern in einem ohnehin schon dem christlichen Glauben verschriebenen Reich sehr nahe. Daher ist zu vermuten, dass der Eingang des ursprünglich biblisch-mythischen Ausdrucks *diluvium* in die Alltagssprache durch die zunehmende Christianisierung im Römischen Reich vorangetrieben worden sein könnte. Schließlich erfreute sich das Noahmotiv in den ersten nachchristlichen Jahrhunderten wachsender Beliebtheit, insbesondere als Schmuckmotiv auf Kasten und Deckeln, aber auch auf Sarkophagen. Die Präsenz des Mythos war also auch visuell erfassbar, was die Sintflutsage den Menschen mehr ins Bewusstsein rückte, wobei umgekehrt die Abbildung der Flutgeschichte auf Reliefs bereits andeutet, dass die Kenntnis des Mythos weit verbreitet und die Erzählung beliebt war.[48] Da nun der Inschriftenstein aus *Ureu* aus einer zum Entstehungszeitpunkt schon teils christlich geprägten Provinz stammt, könnte es ebenso gut möglich sein, die Benutzung des *diluvium*-Begriffs eben diesen Umständen zuzuschreiben.

ZUSAMMENFASSUNG

Die Möglichkeiten der lateinischen Sprache, das Phänomen „Hochwasser" oder auch ein einzelnes Flutereignis in Worten auszudrücken, sind, wie gezeigt, sehr vielfältig und lassen sich zum einen in verschiedene Phasen der römischen Geschichte einordnen, zum anderen eher den unterschiedlichen Textgattungen und Sprechsituationen zuordnen. So ist beispielsweise deutlich geworden, dass der neutrale, wenn nicht gar eher bewundernde Ausdruck *aqua magna* nur in republikanischer Zeit verbreitet war und möglicherweise in Zusammenhang mit den summarischen Jahresberichten, den *annales*, steht. Diesen Schluss legt zumindest dessen Gebrauch in den *fasti Ostienses* nahe. Dessen Verschwinden könnte indes mit der zunehmenden Urbanisierung der ebenen Flächen der Stadt Rom und der damit einhergehenden Steigerung der Vulnerabilität der stadtrömischen Gesellschaft zusammenhängen, doch ist bei dermaßen gewagten umwelthistorischen Interpretationen immer große Vorsicht geboten. Schließlich stehen die großflächige Inanspruchnahme und Bebauung des Marsfeldes schon für republikanische Zeiten fest.[49]

48 G. Koch, Frühchristliche Sarkophage, München 2000, 138. In der vorkonstantinischen Epoche ist der Gebrauch des Motivs noch hauptsächlich auf die Stadt Rom begrenzt, wo es aber recht beliebt gewesen sein muss, da es dort sehr häufig auftaucht. Ab Konstantin findet es sich dann auch auf Sarkophagen, jedoch meist nur an untergeordneter Stelle, etwa auf Nebenseiten oder als Beimotiv.

49 Eine detaillierte Darstellung der baulichen Entwicklung des Marsfeldes als zusammenhängendes Areal liefert neuerdings die archäologische Untersuchung von J. Albers, Campus Martius:

Die Flutberichte aus der Geschichtsschreibung, die sich fast ausschließlich auf Tiberhochwasser beziehen, bedienen sich schon aufgrund ihres erzählenden Charakters einer Vielzahl an Ausdrucksweisen, die sich nicht auf ein einzelnes Wort für „Überschwemmung“ beschränken, sondern eher von Verben und Partizipien Gebrauch machen. Im Gegensatz zu den *annales* oder *fasti* steht hier nicht das bloße Faktum im Vordergrund, dass sich eine Flut ereignet hat, sondern, je nach Absicht des Autors, bestimmte Einzelheiten aus dem Flutgeschehen, etwa die Überflutung bestimmter Gebäude und Plätze oder der Verlauf des die Flut verursachenden Unwetters. Derselben Ausdrucksmöglichkeit bedienen sich auch Inschriftentexte, wenn auch aus anderen Gründen und auf sehr viel knappere Art und Weise, da der Platz auf einem Inschriftenstein zumeist begrenzt und zudem teuer war und sich gerade deshalb Partizipialkonstruktionen anboten. Auf diese Weise konnte viel Information auf relativ engem Raum untergebracht und trotzdem auf Details zur Flut und zum Wiederaufbau eingegangen werden, die insbesondere den Stiftern wichtig waren. Trotz der offenkundigen Verschiedenartigkeit der beiden Textgattungen (Historiographie und Ehren-/Bauinschriften) ist beiden gemeinsam, dass sie sich erstens auf konkrete Ereignisse und Gewässer beziehen und sich zweitens für bestimmte Detailfragen zum Flutverlauf interessieren. Eben darin liegt letztlich auch der Mehrwehrt dieser Texte für geographische Fragen in der altertumswissenschaftlichen Forschung begründet, da sie immerhin Hinweise auf Flutursachen und -charakteristika für bestimmte Regionen und Gewässer liefern können.

In den Schriften der römischen Feldmesser hingegen wird, ebenso wie in Gesetzestexten oder naturphilosophischen Abhandlungen, viel häufiger das übergeordnete Substantiv *inundatio* verwendet. In Anbetracht der Tatsache, dass in den ersten beiden Textarten nur selten schon vorgefallene Flutereignisse beschrieben werden und der Fokus stark auf dem Umgang mit fingierten zukünftigen Überschwemmungen liegt, ist ein solcher Befund nicht weiter verwunderlich. Und schließlich erklärt sich auch die Verwendung des *inundatio*-Begriffes in naturphilosophischen Betrachtungen damit, dass darin, bis auf wenige Ausnahmen, keine konkreten Hochwasserereignisse thematisiert werden, sondern generell Hochwasser als Naturphänomen. Dabei spielt es letzten Endes keine Rolle, ob sich die Anschauungen auf ein bestimmtes Gewässer, wie z. B. den Nil oder den Tiber, beschränken oder nicht, da selbst in jenen Fällen kein einmaliges, sondern ein sich regelmäßig wiederholendes Ereignis beschrieben wird. Daran anknüpfend suggeriert die Verwendung des Nomens auf kaiserlichen Bauinschriften an Flutschutzbauten eine dauerhafte Gefahrenbeseitigung, was wiederum die Selbstdarstellung des Kaisers als Bezwinger von Mensch und Natur gleichermaßen unterstützt.

Ähnliches gilt schließlich für Schriften, die mythische Sintfluten zum Thema haben, da auch diese Art von Überschwemmungen weder historisch sind, noch geographisch begrenzt auftreten. Mehr noch als *inundatio* findet zu deren Bezeichnung allerdings der Begriff *diluvium* Verwendung, der aufgrund seiner Exklusivität (er kommt in den nüchternen Feldmesserschriften oder den Rechtstexten nicht vor)

Die urbane Entwicklung des Marsfeldes von der Republik bis zur mittleren Kaiserzeit (= Studien zur antiken Stadt 11), Wiesbaden 2012.

die Besonderheit und Einmaligkeit des Ereignisses zusätzlich betont. Aufgrund seiner Herleitung vom Verb *diluere* (auflösen) setzt er beim Leser zudem negative Emotionen frei, die den Eindruck des Außergewöhnlichen sowie insbesondere den Eindruck des totalitären, unumkehrbar Katastrophalen erzeugen. In diesem die Emotionen stark ansprechenden Potential des Begriffs wird denn auch der Grund dafür zu finden sein, warum ausgerechnet beide auf uns gekommenen Zeugnisse von Fluterlebnissen aus erster Hand (die Briefe von Plinius dem Jüngeren und von Cicero) gerade wegen ihres hohen literarischen Anspruchs darauf zurückgreifen. In dieselbe Kategorie könnte auch der spätantike Inschriftentext aus Nordafrika eingeordnet werden, wobei hier, wie gezeigt wurde, alternativ die Möglichkeit besteht, die Verwendung des Nomens *diluvium* im Kontext der zunehmenden Christianisierung des Römischen Reiches zu verstehen. Dahingehend sind denn sicherlich auch die anderen spätantiken, nicht christlich-theologischen Texte zu interpretieren, die *diluvium* als bildhafte Beschreibung eines realen Flutereignisses benutzen. Daran wird wieder einmal deutlich, dass ein kultureller Wandel sich auch in der Veränderung der Wortwahl für bestimmte Konzepte manifestieren kann und daher philologische Studien zur Erforschung historischer Fragestellungen nicht vernachlässigt werden dürfen.

ZUR AUSSAGEKRAFT GEOGRAPHISCHER TEXTE UND KARTEN IM 16. UND 17. JAHRHUNDERT – EIN EVALUATIONSVERSUCH ANHAND DREIER TEXTE

Volker Kaminske

EINLEITUNG

Es ist von jeher fraglich gewesen, in welchem Umfang die schriftlich überkommenen Aussagen zur Gestalt der Erde auf Realität beruhen oder auf einer phantasievollen Ausschmückung von Gehörtem. Durch die Seefahrt waren zumindest Ansätze für eine selektive kartographische Lagebestimmung von Festland und Inseln vorhanden, wie auch für die Beschreibung des Küstenverlaufs. So werden in portugiesischen und holländischen Quellen etwa die Umrisse Afrikas schon für die Zeit um 1600 „relativ genau“ beschrieben (z. B. Vallseca 1439, Ortelius 1570). Mit dem Begriff der „relativen Genauigkeit“ werden topologische Phänomene wie „rechts“, „links“ von irgendetwas, „darüber“, „darunter“ bzw. „davor“ oder „dahinter“ erfasst und, im Gegensatz zur genauen Entfernungs-, Höhen- oder Winkelangabe heutiger Karten, eben nur relationsmäßig beschrieben.

Aber gerade die damalige Seefahrt war daran interessiert, zusätzliche und genauere Hinweise und Karten zu erlangen, um Reiserisiken zu erkennen und zu minimieren sowie um die Reisedauer genauer abschätzbar zu machen. Interesse hierfür bestand seinerzeit vor allem in Spanien, Portugal, den Niederlanden, England und Frankreich. Wissenschaftliche Versuche, die Erkenntnisse der Seefahrt durch eigene Reisen oder durch eine systematische Aufarbeitung vorhandener Quellen zu ergänzen, begannen zu Beginn des 17. Jahrhunderts.

In dieser Studie geht es darum, damals publizierte Berichte – unabhängig vom tatsächlich vorhandenen Wissensstand – hinsichtlich ihrer Richtigkeit zu verifizieren oder zu falsifizieren und dies in einen Zusammenhang zum Kenntnisstand des jeweiligen Autors zu bringen.

Es stellt sich dabei auch die Frage, wie bei der Beschreibung des Begriffsfeldes „Raum“ Worte im Sinne von erklärenden, systematisierenden Begriffen zur Verfügung stehen bzw. genutzt werden und inwieweit diese Begriffe im Laufe der Zeit sich änderten oder erhalten blieben.

Ein zu diesem Zwecke aus (autor-)typischen Texten extrahierter Kriterienkatalog sollte in der Lage sein, hierzu belastbare Hinweise zu erbringen, die gegebenenfalls auch auf andere Arbeiten dieser Autoren bzw. auf andere Autoren dieser Zeit übertragbar sein sollten.

Ein Versuch, anhand dieser Kriterien, ihrer Lokalisierung und Interpretationsmöglichkeiten, eine (textstatistische) Textinterpretation historischer Zeitphasen über formale, sprachliche und inhaltliche Kriterien für die Texte in vergleichender und bewertender Form vorzunehmen, dürfte durchaus realisierbar sein. Danach wären im Textvergleich von selbst bereisten und unbekannten Bereichen in der Beschreibung typische Unterschiede zu erwarten, die sich in der sprachlichen Darstellung und im Verhältnis von Informationsgehalt und Textfülle nachweisen lassen müssten.

In der Form von bewerteten 6×6-Felder-Tafeln lassen sich gegenseitige Einflüsse und Abhängigkeiten dieser Kriterien feststellen und quantifizieren. Über ihre Primärinformation hinaus ergeben diese Bewertungstafeln auch noch die Rohdaten für die weiteren Auswertungsschritte in 4.3. In einem ersten Schritt wird auf Ergebnisse zurückgegriffen, deren Werte graphisch veranschaulicht werden. Dabei lassen sich Aussagen zum Verhältnis von Informationsgehalt und dazu benötigter Textfülle der einzelnen Faktoren ableiten, und dies im Vergleich von bekanntem und unbekanntem Gebiet. Diese Aussagen sollten auf andere Arbeiten der jeweiligen Autoren übertragbar sein.

1. DER BEGRIFF DER KOGNITIVEN KARTE

Wenn Menschen ihre Umwelt wahrnehmen, entsteht als geistiges Äquivalent dieses Vorgangs ein Vorstellungsbild, das die Umwelt sehr subjektiv und stark selektiv repräsentiert: eine sog. kognitive Karte über „die Welt in unseren Köpfen“ (vgl. GOULD/WHITE 1974 sowie DOWNS/STEA 1982). Hierbei finden zwar alle Sinnesreize Berücksichtigung, doch werden vor allem die visuellen Reize zu kartenartigen Vorstellungsbildern verarbeitet (vgl. KUIPERS 1982, BURENHOLT 2009, MARK et al. 2011). Dieser Vorgang findet in zwei Schritten statt: Beim ersten Schritt entstehen zunächst kognitive Karten, die auf der direkten, aber selektiven Wahrnehmung der Welt beruhen. In einem zweiten Schritt werden diese objektivierbaren Eindrücke ergänzt um subjektiv basierte, affektiv-emotionale Vorstellungsbilder zur Welt, wie sie in der eigenen Vorstellung existieren. Solche Vorstellungsbilder können demnach als realistische Eindrücke auf *eigenen Anschauungen* beruhen, aber auch auf *gedanklichen Konstruktionen* des eigenen, lückenhaften Weltbildes und auf darauf bezogenen Informationssplittern sowie deren phantasievoller Verknüpfung (vgl. KUIPERS et al. 2003, KÖCK 2013). Diese Feststellung galt im 16. Jahrhundert schon genauso, wie sie auch heute noch im Zeitalter der exakten Erdvermessung gilt.

2. DIE BEDEUTUNG VON KARTEN IM 16. UND 17. JAHRHUNDERT

Die Verfügbarkeit von Karten und ihr Genauigkeitsgrad galten schon von der Antike bis zum Mittelalter und noch lange danach als wichtiges Machtwissen, das für die Erforschung und Nutzung der Erde von größtem Vorteil war. Die heute als „Historische Geographie“ bezeichnete Disziplin war damals als Kosmologie, Kosmographie, Geographie oder Kartographie eine hoch anerkannte Wissenschaft und zeugt über die namentliche Verknüpfung mit so namhaften Wissenschaftlern wie

Behaim, Mercator, Münster, Varenius oder Newton auch heute noch von ihrer damaligen Bedeutung.

Was war das Ziel von Karten und Landschaftsbeschreibungen? Nach heutiger Auffassung ist das Ziel primär in der Lagebeschreibung von Punkten bzw. der Verlaufsbeschreibung von Linien, sowie der Charakteristik der Orte und einer generellen Landschaftsbeschreibung zu suchen, wodurch bei etwaigen Reisewegen die Lage von Orten aufgrund des Verlaufs von Straßen und Flüssen oder der Einpassung in die Landschaft antizipierbar und nachvollziehbar wurde.

Wieviel jedoch war ein derartiges Wissen zwischen 1500 und 1700 n. Chr. wert? Auf der Basis von genauen Vermessungen, Luft- und Satellitenbildern lässt sich heute trefflich über die damaligen Karten und die erkennbaren Differenzen zu heutigen Kartenwerken räsonieren. Für die wissenschaftsgeschichtliche Einordnung und den Erkenntniswert damaliger Werke ist deren sachliche Richtigkeit als Ausdruck des damaligen Wissensstandes von entscheidender Bedeutung. Es stellt sich somit vor allem die Frage, inwieweit die textlich beschriebenen und durch Karten dargestellten Erdbereiche sachlich richtig und dem wissenschaftlichen Erkenntnisstand gemäß wiedergegeben werden. Die Beantwortung hängt ab vom Wissensstand der damaligen Zeit, von der Zuverlässigkeit und der Reputation des jeweiligen Autors, bei der man zwischen Selbsterfahrungen und Mitteilungen anderer unterscheiden muss, von seiner Reisetätigkeit und von seiner Fähigkeit, räumliche und zeitliche Zusammenhänge zu erkennen und darzustellen. Insofern entspricht das hier verfolgte Vorgehen im Prinzip einem hermeneutischen Ansatz, bei dem die sachliche Richtigkeit räumlicher Darstellungen überprüft wird. Durch eine Inhaltsanalyse von Texten damaliger Autoren könnten die Intentionen dieser Zeit bei der Erstellung von Karten und Reiseberichten herausgestellt werden.

Aus verkehrstechnischen und finanziellen Gründen ist grundsätzlich davon auszugehen, dass die Geographen der damaligen Zeit in noch sehr viel geringerem Umfang die Welt bereist haben, als das heute der Fall ist, und dass über diese Welt in noch viel höherem Maße als heute spekulative Vorstellungen im Umlauf waren. Beispiele aus Karten gibt es genügend, aber auch gedruckte Vorstellungsbilder zum Aussehen von Phantasielebewesen wie dem (nie existierenden) Einhorn oder einem (so nie aussehenden) Elefanten.

Will man weiterhin heutige Verhaltensweisen im Umgang mit der Vorstellung von Räumen auf damalige Verhältnisse übertragen und zur Rekonstruktion des damals möglicherweise existierenden Weltbildes nutzen, so kommt man aus grundsätzlichen wahrnehmungspsychologischen Überlegungen um nachfolgende Feststellungen nicht herum:

- Die Darstellung eines Raumes orientiert sich an der Charakteristik der Orte und wichtigen Landmarken.
- Die Spezifität der Sichtweisen führt zu Toponymen ein und desselben Raumes.
- Die räumliche Position des Autors und einwirkende Referenzrahmen beeinflussen die Art einer räumlichen Darstellung in starker Weise mit (z. B. Levinson/Wilkins 2006, Ilyushechkina et al. 2014). So ist bei der Ausrichtung früherer Karten im Vergleich zu heute von einer mentalen Rotation auszugehen, wie sie

auch Piaget bei der Ausbildung der räumlichen Wahrnehmungsfähigkeit schon feststellte und für die Ausbildung einer allgemeinen Entwicklungsstufung verwendete (PIAGET/INHELDER 1971).

- Die Bedeutung eines Objekts oder eines Vorgangs nimmt i. d. R. mit der Distanz exponentiell ab. Das heißt: Ein bekanntes Bauwerk im aktiven Wahrnehmungsraum besitzt einen deutlich höheren Bekanntheitsgrad und auch deutlich höhere Bedeutung als ein gleich großes bzw. gleich bedeutendes, adäquates Bauwerk in Ungarn, Spanien, Nordafrika. Eine leichte Erderschütterung in der Heimatregion erzeugt mehr persönliche Betroffenheit als ein verheerendes Erdbeben in China oder Indonesien.
- Der Umfang der Bedeutungsabnahme hängt ab von der gesellschaftlichen Relevanz des Objekts bzw. des Vorganges. So erreicht der sog. Distanzwiderstand, also der Hemmungsgrad bei der Weitergabe von Informationen in Bereichen hoher Lebensnotwendigkeit (z. B. Orte mit hoher Infektionsgefahr, Orte mit hohem Arbeitsplatzangebot) geringere Werte als in solchen mit geringerer Lebensnotwendigkeit (z. B. Orte mit hohem Erlebniswert, vielen Museen o. ä.).

Bei einer Anwendung auf die Zeit des 16. und 17. Jahrhunderts muss für die Raumwahrnehmung der unterschiedlichen Sichtweise der Zeitalter Rechnung getragen werden. Die Aufgabe bestand damals vor allem darin, gängige Reiserouten, ihre Stationen und Wegabschnitte hinreichend genau zu beschreiben sowie mögliche Risiken und positive Aspekte der Reiseziele aufzuführen.

Das Grundprinzip einer Bedeutungsabnahme mit der Distanz, und dies abhängig vom Anlass einer Beobachtung, dürfte allerdings schon damals in ähnlicher Weise bestanden haben, wie die Kognitionspsychologie als grundlegende Erkenntnis zum räumlichen Wahrnehmungsvorgang herausstellt (KAMINSKE 2012, KUIPERS 1982, 2001).

3. ANSATZ ZUR ABSCHÄTZUNG DES REALITÄTSBEZUGES

Karten und Texte der damaligen Zeit bieten als Interpretationshilfen Informationen durch deren Art, Inhalt und Dichte dieses Inhaltes an. Dabei geht es entweder um den Karteninhalt einer physisch vorhandenen Karte oder das verbale Äquivalent von Karteninhalten einer (nur) kognitiv vorhandenen Karte. Hier die objektivierbaren Informationen von den übernommenen, spekulativen Informationen zu trennen, gibt eine Hilfe zur Feststellung des damaligen Wissensstandes bzw. seiner Diffusion in weitere Kreise.

Zunächst: Über die Wissensgenerierung, die Wissensweitergabe und –auswertung räumlichen Wissens der damaligen Zeit ist nur wenig bekannt. Vor allem fehlen systematische Hinweise, wann welche Erkenntnisse vorlagen und ob diese einen – im heutigen wissenschaftlichen Sinn – validen oder allenfalls spekulativen Hintergrund besaßen. Die Frage nach Differenzierung zwischen authentischem Wissen und übertragenem Wissen ist daher nicht nur logisch, sondern auch sehr wohl berechtigt.

Um die Authentizität von textlichen und kartographischen Angaben abschätzen zu können, waren Kriterien in zeitgenössischen Texten zu finden, mit deren Hilfe ein Ansatz zur bestmöglichen Feststellung der zu prüfenden Aussagekraft formuliert werden konnte. Ziel dieses Ansatzes sollte sein, bei Autoren mit bekannter Biographie und bekannter Ortskenntnis in einem Vergleich, aufgrund der sprachlichen Darstellung (Exaktheit, Vollständigkeit, Umschreibungen, Ausschmückungen), Hinweise für eigene Kenntnis bzw. übernommene Informationen zu finden, um diese so gefundenen Variablen dann an anderen Werken der gleichen Autoren bzw. anderer Autoren auf allgemeine Anwendbarkeit zu überprüfen.

Die verfügbaren Variablen sind also in einem Evaluationsversuch hinsichtlich ihrer Aussagekraft auf Validität zu überprüfen. Hierzu wurde ein textstatistischer Ansatz entwickelt, der auf formal-inhaltlicher Basis die Wertigkeit solcher Variablen, aufgrund der Häufigkeit ihres Auftretens und ihrer Bedeutung, testet und beurteilt. Der über eine Mehrfeldertafel zu testende Ansatz beruht auf einer Kreuztabelle, wie sie Vester 1990 vorgestellt hat, um qualitative Einflussfaktoren quantifizieren zu können. Dabei wird für jede Variable ihr Einfluss auf alle anderen Variablen ermittelt und mit einem Gewichtungsfaktor (hier: 1–3 für nicht vorhanden bis sehr stark auftretend) versehen. Aus der jeweiligen Zeilensumme lässt sich dann der Einfluss auf die anderen Variablen ableiten, aus der Spaltensumme die Beeinflussbarkeit jeder Variablen (siehe genaue Erklärung und Anwendungsbezug in Kap. 3.7).

Aufgrund der Zielsetzung von Texten und Karten finden sich derartige Variablen etwa in der absoluten Angabe der Lage (z. B. „40 km nördlich von Frankfurt“), aber auch in relativen Lagebezeichnungen (z. B. „den Main aufwärts, eine halbe Tagesreise auf der rechten Mainseite“). Ebenso gehen aus Zusatzinformationen die über Städte verfügbaren bzw. über Linienverläufe vorhandenen Informationen hervor. Mit authentischen Kenntnissen werden auch Kurven, Krümmungen oder Schlingen bei Flüssen sowie starke Anstiege bei Straßen zu wichtigen Informationen, während bei übertragenen Kenntnissen die Beschreibung dieser topographischen Eigenschaften in der Regel dem Luftlinienverlauf entspricht.

Die hier kurz skizzierten Kriterieneigenschaften sollten, zusammen mit formalen Kriterien aus der Darstellung des Textes und einem möglichen Abgleich mit Quellenangaben und Biographie, in der Lage sein, auch andere Bücher des Autors bzw. andere Autoren nach Art einer Check-Liste oder zur Kontrolle in der Art eines sprachlichen „Fingerabdruckes“ überprüfbar zu machen. In der Folge wird dieser Kriterienkatalog genauer vorgestellt.

3.1 Kriterienbereich A: Kontrollhinweise auf Reisetätigkeit aus der Biographie

Die Geographen der damaligen Zeit gelten heutzutage aufgrund der von ihnen durchgeführten Reisen als topographische Experten – generell oder zumindest für bestimmte Länder. Da die Reisetätigkeit zu damaliger Zeit zeitaufwendig und kostspielig war, ist aber von nur selektiven Kenntnissen innerhalb Europas, selbst Mitteleuropas auszugehen. Insofern ist der explizite Hinweis auf die Eigenkenntnis bestimmter Bereiche, aufgrund durchgeführter Reisen, ein wertvoller Kontrollhinweis für den Vergleich von verschiedenen Gebietsbeschreibungen.

3.2 Kriterienbereich B: Quellenangaben

Topographische Beschreibungen basieren sehr häufig auf Angaben der antiken Schriftsteller (Pomponius Mela, Strabon u. a.), was sich dann zumeist auch im Titel erkennen lässt („Geographia antiqua"). Zeitgenössische Autoren sind in dieser Zeit selten.

In der Regel werden Quellenangaben oder Verweise auf andere Autoren nicht in der heute üblichen Form verwendet. Üblich ist allenfalls zu Beginn oder am Schluss eines Werkes der Hinweis, dass bestimmte Bücher bzw. die Aussagen eines bestimmten Autors (egal ob antiker oder zeitgenössischer Autor) hilfreich waren.

Damit entfällt aber auch die Möglichkeit, zwischen eigenen Erfahrungen eines Autors und übernommenem Wissen zu unterscheiden, denn die Anführung von Städtenamen, wie das lautmalerisch an die Sprechweise angepasste „Göppinga", „Reutlinga" oder „Menz" (= Mainz), kann in einer Karte sowohl für einen eigenen Aufenthalt am jeweiligen Ort sprechen, bei dem die dort übliche Dialektform als authentisch unverändert übernommen wurde, aber auch für eine unreflektierte Übernahme einer in der damaligen Zeit üblichen, verbal verballhornten Weitergabeform. Tatsache ist zudem, dass die vielen verschiedenen Schreibweisen von Namen aller Art auf einer derartig fehlerbehafteten, mündlichen Weitergabe beruhen.

3.3 Kriterienbereich C: Formale Kriterien

Hilfe gibt die Schreibweise mit einer formal-strukturellen Differenzierung von Orten und ihren Eigenheiten durch unterschiedliche Schrifttypen, z. B.:

Lateinisch:	Normalschrift
	Normalschrift gesperrt
	Kursivschrift
	Kapitälchen
Deutsch:	Normalschrift
	Gotische Schrift

Die Interpretation, welche Schrift für welche Aussage bzw. Hierarchieebene steht, ist zunächst autorspezifisch zu erarbeiten.

3.4 Kriterienbereich D: Sprachliche Darstellung

Entscheidend für die Lösung des oben genannten Problems ist die Beantwortung der Frage: Was ist als Eigenkenntnis belegbar und was ist vom Hörensagen ergänzt? Die Darstellung selbst scheint beide Lösungen zuzulassen. Als Interpretationsmöglichkeiten für die scheinbar authentische Beschreibung stehen hierfür zwei Versionen mit zwei unterschiedlichen Ursachen gleichwertig nebeneinander.

a) Der Autor war dort und übernimmt Begriffe und Lokalitäten gemäß den dort gemachten Erfahrungen, also auch der dort üblichen Aussprache der Leute.

b) Der Autor war nicht dort und muss die Beschreibung und Aussprache unkontrolliert von Berichterstattern übernehmen und ist hinsichtlich der Wahrheit auf deren Genauigkeit angewiesen.

Lassen sich Kriterien finden, die für die Richtigkeit von a) bzw. b) sprechen?

Als Hilfe lassen sich Erkenntnisse aus der heutigen Zeit zur Arbeit mit kognitiven Karten heranziehen, etwa aus Erkenntnissen der Kognitionspsychologie oder der Wahrnehmungsgeographie (vgl. GÖRZ et al. 2013, KAMINSKE 1996, 2006, 2012,). Hiernach nimmt mit zunehmender Kenntnis auch das Bedürfnis zu, darüber zu berichten und die Bedeutung der jeweiligen Institution, Stadt, Region verbal aufzuwerten.

3.5 Kriterienbereich E: Inhaltliche Kriterien

Nr.	Kriterien	Hinweise auf Primärwissen	Hinweise auf Sekundärwissen
1	Dichte der Information	Ergänzung durch zahlreiche weitere Ortsangaben (z. B. zentrale Burg)	Nur wenige Angaben, meist klischeehafter Art
2	Relative Ortslage	Richtige örtliche Einordnung an Fluß, Grenze, Berg	Lage ungenau: z. B. hinter den Bergen…
3	Absolute Ortslage	Richtige Einordnung hinsichtlich N-S- bzw. W-O-Lage; richtige Abstände zwischen den Orten (bzw. Distanz- oder Zeitangaben)	Lage ungenau: deutliche Abweichungen von den N-S- bzw. W-O-Koordinaten
4	Ergänzung durch andere Angaben (Funktion, Herrscher o. a.)	z. B. Marktfunktion, Hauptstadtfunktion	Zusätzliche Angaben willkürlich; Stellenwert daher unterschiedlich; oft ohne geographischen Wert; Verwendung der Floskel: „ist berühmt für…“
5	Verwendung offizieller Namen, nicht der im Volk gebräuchlichen	Angabe der offiziellen Bezeichnung mit Ergänzung: vulgo …	Nur der zugängliche Namenshinweis
6	Rückgriff auf historischen Hintergrund	Hinweise auf frühere Bautätigkeit der Römer	Ergänzung bei gezielten Rückgriffen auf Konkurrenzwerke

Die separate Nutzung einzelner Kriterien ist in der Regel wenig genau und wenig aussagekräftig und könnte zufallsbestimmt sein. Durch eine Kriterienkombination kann die Genauigkeit allerdings schon deutlich gesteigert werden. Auch die Aussagekraft der Kriterien ist ungleichwertig. Manche sind von eher bestätigender Art, andere sind Differentialkriterien. Darunter versteht man Kriterien, die eindeutig auf Primär- bzw. Sekundärkenntnisse weisen, so z. B. die „absolute Ortslage“ (Nr. 3).

Die Anzahl aufzählbarer Funktionen kann zwar durchaus ein Kriterium für Kenntnis sein, doch hängt diese Variable von der Größe der zu beschreibenden Stadt ab. Kleinere Städte weisen hier von vornherein weniger aufzählbare Punkte auf. So kann zwar die Verortung auf eine direkte Kenntnis verweisen. Es muss daraus aber nicht zwangsläufig auf eine nachfolgende Schilderung von Lageeigenschaften und Funktionen geschlossen werden. Kommen diese jedoch vor, ist zu unterscheiden, ob diese Angaben zeitlich und räumlich logisch erfolgen oder willkürlich und additiv.

Je mehr Variablen positiv nutzbar sind (z. B. Informationsdichte/Absolute Ortslage/mehrere Ortsnamensformen), spricht dies für eine Authentizität der Beschreibung.

3.6 Prinzip des Evaluationsansatzes

Da die Kriterienbereiche A und B nicht durchgängig vorkommen, ist ihre prinzipiell wertvolle Aussagekraft kaum nutzbar. Ihr Wert beschränkt sich deshalb – wo vorhanden – in erster Linie auf Kontrollen und Bestätigungen der anderen Kriterien. Auch das formale Kriterium kann in Einzelfällen Interpretationshilfen leisten, lässt sich aber nicht über mehrere Autoren hinweg als Standard transferieren. Insofern bezieht sich die nachfolgende Evaluation in erster Linie auf die generell verfügbaren, immer direkt überprüfbaren Kriterien von Sprache und Inhalten.

In der Logik einer in sich fundierten Landschaftsbeschreibung bzw. Reisebeschreibung erfolgt eine richtige Schreibweise (= Kenntnis) am ehesten in Kombination mit den Faktoren:

1) Charakteristik von Orten und Landmarken durch Nutzung damaliger Toponyme
2) Bei Toponymnutzung: Namensnennung mehrfach (lateinisch, deutsch, Dialektform)
3) Absolute, richtige kartographische Lage, topologische Exaktheit
4) Relative richtige Lage innerhalb des gewählten Referenzrahmens
5) Zusatzinformationen
6) Angaben nicht additiv, sondern in sinnvollem Kontext
7) Angaben wie „Seinerzeitige aktuelle Bedeutung von Stadt bzw. Region“.
   Wie gerade erwähnt, stehen diesen, doch mehr oder weniger autorspezifischen Angaben, als **Kontrolle** gegenüber:
   a) Objektiv nachprüfbare Angaben (etwa durch nachgewiesene Reisetätigkeit in den beschriebenen Raum)
   b) Berufung auf konkrete andere Quellen (Literaturverweis).

3.7 Überprüfungsansatz

Oben wurde schon das Prinzip eines Überprüfungsansatzes vorgestellt, durch den qualitative Daten in gewissem Umfang quantifizierbar gemacht werden können. Mit Hilfe der vorgestellten Variablen kann nun ein, auf Texten von drei Autoren basierender, Test durchgeführt und auf die obige Fragestellung einer Objektivierbarkeit von Aussagen und einer Differenzierbarkeit von authentischen und übertragenen Informationen überprüft werden (siehe 4.1).

Die oben genannten sechs Variablen werden in einer Bewertungsmatrix nach Vester gewichtet, wodurch ihr Aussagewert quantifiziert werden kann (Vester 1990). Dazu werden die Variablen in einer Kreuztabelle überprüft, inwieweit sie die anderen Variablen beeinflussen bzw. selbst durch diese beeinflusst werden. Das Maß der Beeinflussung kann dann durch Einsatz von Gewichtungsfaktoren (0 bis 3: 3 für starke, 0 für fehlende Einwirkung) eingebracht werden, die durch Summierung jeweils als Zeilensumme (= Aktivsumme AS, weil hier das Beeinflussungspotential einer Variablen auf andere festgehalten wird) oder Spaltensumme (= Passivsumme PS, weil hier die Beeinflussbarkeit einer Variablen durch alle anderen überprüft wird) entsteht.

Aus dieser Anlage ergibt sich fast zwangsläufig, dass wegen der Einordnung der Variablen in eine jeweils gegensätzliche Formulierung („wird beeinflusst“ bzw. „beeinflusst selbst“) eine spiegelbildliche Anordnung der Matrixwerte nicht erwartet werden kann.

Wird die Aktivsumme einer Variablen durch ihre Passivsumme dividiert, lässt sich ihr Beeinflussungspotential abschätzen: Ein hoher Wert entspricht einem hohen Potential, ein geringer Wert entsprechend einem geringen.

Wird hingegen die Aktivsumme mit ihrer Passivsumme multipliziert, wird durch einen hohen Wert eine Schwankungsbereitschaft erkennbar, bei der durch Veränderung des Variablenwertes die Gesamtaussage völlig auf den Kopf gestellt werden kann. Entsprechend zeigt ein geringer Wert an, dass diese Variable gegenüber Wertveränderungen kompatibel ist.

Die Variablen lassen sich entweder beschreibenden Texten entnehmen, aus Karten, oder aber einer Kombination von Text und erläuternder Karte. Im vorliegenden Fall soll die Überprüfung anhand von Texten erfolgen, weil mit deren Interpretation bisher die meisten Probleme verbunden waren und mit diesem Ansatz eine standardisierte Hilfe möglich zu sein scheint.

Tab. 1 *Prinzip der Bewertung in einer 6 × 6-Feldertafel*

Variable in Zeile x beeinflusst Variable in Spalte x wird beeinflusst durch	1)	2)	3)	4)	5)	6)	Aktivsumme AS⌴Q
1) richtige Namensnennung							
2) verschiedene Namensformen							
3) absolut richtige Lage							
4) relative richtige Lage							
5) Auftreten von Zusatzinformationen							
6) Angaben in sinnvollem Kontext							
Passivsumme: **PS** P							

Tab. 2 *Gewichtungsfaktoren und ableitbare Indices*

Wobei : 0 = keine Einwirkung 1 = schwache Einwirkung 2 = mittlere Einwirkung 3 = starke Einwirkung	PS = Passivsumme P = Aktivsumme x Passivsumme AS = Aktivsumme Q = Aktivsumme/Passivsumme

Tab. 3 *Indices und ihre Aussagenintervalle*

Aussage der Q-Zahl:	**Aussage der P-Zahl:**
Aussage: im Intervall von 0,1 bis 14,9	**Aussage:** im Intervall von 0,1 bis 225
Hoher Wert: Zeichen für aktives Element	**Hoher Wert:** Kritisches Element
Geringer Wert: Zeichen für passives Element	**Geringer Wert:** Pufferndes Element

Aus der **Q-Zahl** lässt sich ableiten, in welchem Umfang Variablen auf andere einwirken (quantitativer Faktor). Das heißt: Über z. B. eine „absolut richtige Lage" (= unabhängige Variable) lassen sich in der Regel die anderen Variablen in unterschiedlichem Maße mitbestimmen. Das „Auftreten von Zusatzinformationen" (abhängige Variable) ist dagegen in nahezu allen Fällen abhängig vom Vorhandensein und der positiven Aussage anderer Variablen. Selbst aber kann diese Variable allenfalls eingeschränkt Aussagen zu Lage oder Lageeigenschaften oder zur Namensrichtigkeit von Orten, Flüssen oder Regionen geben. Es kommt demnach auf die Zusammensetzung von unabhängigen und abhängigen Variablen an, deren Zusammenspiel letztendlich erst eine sinnvolle Aussage erlaubt.

Die **P-Zahl** hingegen gibt Hinweise auf den inhaltlichen Aussagewert der jeweiligen Variablen insofern, als erkennbar wird, inwieweit diese tatsächlich nutzbar werden (qualitativer Faktor). So ist etwa die Angabe einer „relativen Lage“, wie z. B. „Rothenburg flussabwärts von Lauda“, zwar richtig, aber wegen der kartographischen Ungenauigkeit und des hohen Spekulationscharakters von nur geringem Wert. Eindeutig besser wäre die Angabe: 30 km bzw. 1 Wegstunde flussaufwärts von Lauda (= absolut richtige geographische Lage). Entsprechend sagt „richtige Namensnennung“ (nur), dass außer der puren (richtigen) Information keine weiteren Angaben gemacht werden, also eine Vernetzung mit anderen Variablen unterbleibt. Die formale Richtigkeit lässt sich demnach nur mit anderen Angaben zusammen als authentische Information validieren.

4. AUSSAGEWERT DURCH ÜBERPRÜFUNG AN ORIGINALTEXTEN

4.1 Ausgewählte Autoren in ihrem Umfeld und ihrer Zeit

An Texten der historischen Geographen Varenius, Cellarius und Cluverus sollen diese Kriterien untersucht und auf ihre Aussage- bzw. Differentialkraft überprüft werden. Allen drei Geographen billigt man einen innovativen Zugang zur Geographie zu, sei es, dass erstmals eine neue, allgemeine Geographie vorgestellt wurde und diese durch eine spezielle, länderkundliche Darstellung ergänzt wurde (Varenius), sei es, dass die übliche Länderkunde aus ihrem Zusammenhang mit antiken Darstellungen gerissen wurde (Cluverus), oder sei es, dass durch Vergleiche oder Heraushebungen erstmals didaktische Bezüge erkennbar wurden (Cellarius). Kurz: Die drei Geographen versprechen zunächst durch ihre Herangehensweise eine valide Aussage bei der Schaffung eines „Weltbildes“, wie es auch bei anderen historischen Geographen wiederzufinden ist. Weiterhin interessant ist die Tatsache, dass sie ihre Werke in zeitlichen Abständen veröffentlichen, die zur Frage Aufschlüsse geben, inwieweit sich neue Kenntnisse oder Einsichten verbreiten bzw. im Fachkollegenkreis verwendet werden (vgl. Anhang).

4.2 Ergebnisse

Die Texte: *Introductio in Universam Geographiam* (Cluverus), *Geographia generalis* (Varenius) und *Geographia iuxta et Nova* (Cellarius) werden auf die obigen Variablen untersucht. Dabei wird berücksichtigt, dass der jeweilige Beeinflussungsfaktor in unterschiedlichem Maße wirksam wird, je nachdem er sich auf direkt bekannte Gebiete oder unbekannte Gebiete bezieht. Infolgedessen werden für jeden Autor zwei Texte mit unterschiedlichem Bekanntheitsgrad auf der Basis biographisch nachgewiesener Eigenkenntnis untersucht.

4.2.1 Qualitative Auswertung

In qualitativer Analyse kann etwa festgehalten werden, dass manche Fehler und Ungereimtheiten auftauchen, die einerseits auf ein allgemein unzulängliches Weltbild, aber auch auf subjektiv vorhandene Weltbildlücken zurückzuführen sind.

Am Beispiel der Beschreibung Afrikas, dessen Binnengliederung ja erst im Laufe des 19. Jahrhunderts endgültig geklärt wurde, lässt sich aber feststellen, dass sich die vorhandenen Wissenslücken durchaus auf die formale Darstellung durchpausen. Es sollen hier einige Beispiele aus qualitativer Betrachtung wiedergegeben werden, die zeigen, wie wichtig eine standardisierende, systematisierende Übersicht nach gemeinsamen Kriterien zur Beurteilung dieser Texte ist.

So wird zunächst bei Cluverus, aber auch bei Varenius und Cellarius die Beckenstruktur Afrikas nicht erkannt. Die Randschwelle des zentralen Westafrikas bleibt unerwähnt, im Gegensatz zu der in Südafrika. Niger und Senegal werden zu einem im heutigen Senegal mündenden Fluss zusammengefasst. Der Nil erhält Zuflüsse aus der libyschen Wüste.

Bei der Siedlungsgeographie wird die Existenz der Garamanten im Süden Libyens von antiken Autoren übernommen. Die Bezeichnung „Äthiopien“ wird auf Gesamtafrika bezogen und insgesamt drei Mal genutzt:

1) als Überbegriff für Nordafrika, v. a. für die W-O-verlaufende Guinea-Sudan-Zone
2) für die Region beim Zusammenfluss von Rotem und Weißem Nil
3) als Bezeichnung des atlantischen Ozeans zwischen Guineaküste und Südspitze Afrikas.

Die Bezeichnung „Afrika“ tritt nur an der afrikanischen Gegenküste von Sardinien und Sizilien auf und entspricht damit der Lokalisierung des damaligen Karthago durch die Römer bzw. dem heutigen Tunesien.

Der Tschadsee wird ohne Namen eingezeichnet und durch Gebirge vom Süden abgetrennt. Es liegt ein direkter Zusammenhang mit dem Lauf des Niger vor. Die Benennungen Guinea, Benin und Biafra sind oft etwas nach Osten hin lageverschoben.

Die Autoren geben häufig Interpretationshinweise über die Schreibweise oder den Umfang der Kapitel. Bei Cluverus beginnt seine Länderkunde ab dem zweiten Band, zunächst mit Europa. Im vierten Buch werden Südosteuropa und Osteuropa (auch Preußen) auf 32 Seiten behandelt. Asien findet seinen Platz im fünften Buch auf 34 Seiten und im sechsten Buch werden *Afrika, Barbaria* auf 13 Seiten, Amerika und Südamerika auf sechs Seiten aufgeführt. Den (unbekannten) Inseln *novi orbis*, den *Magellanica*, der *Terra australis* und weiteren Südseeinseln sind weitere zwei Seiten gewidmet.

Wenn bei Cellarius das (bis heute unvollständige) Straßburger Münster wegen seiner Großartigkeit 1687 Erwähnung findet, das 1513 aber schon fertig erbaute Freiburger Münster nicht, zeigt das zumindest seine Unkenntnis des Breisgaues, wenn nicht gar der Gesamtregion des Oberrheingrabens.

Auch die Frage der räumlichen Sachlogik bzw. der Erkennung von Strukturen und den daraus resultierenden Wechselwirkungen kann entsprechend qualitativ bewertet werden.

Wenn in einem Bericht A beispielsweise formuliert wird: „Die Küste von X ist unpassierbar, das Land in Y ist unfruchtbar und in Z sprechen die Einwohner eine ungewöhnliche und lächerliche Sprache“, ist aufgrund dieser additiv aneinandergereihten Aufzählung kein direkter Zusammenhang erkennbar und es drängt sich die Annahme auf, dass hier lediglich erkennbare räumliche und kulturelle Auffälligkeiten wiedergegeben werden.

Wenn es dagegen in einem Bericht B über die gleiche Region heißt: „An der Küste von X kann man wegen einer Steilküste nicht landen, dagegen besitzt das Nachbarland eine Flachküste. Hier aber ist das Y-Hinterland unfruchtbar. Die deswegen nur ab und zu bis zur Küste vordringenden Bewohner von Z sprechen eine ungewöhnliche und unverständliche Sprache mit Schnalzlauten, weshalb man sich mit ihnen kaum verständigen kann“, so wird im Vergleich beider Darstellungen sofort deutlich, dass sich die zweite Version aus einem höheren Kenntnisstand ableitet. Wie lässt sich dieser Unterschied formalisiert ausdrücken?

4.2.2 Quantitative Auswertung

Das Vorkommen der einzelnen Variablen bzw. ihr Zutreffen innerhalb der geprüften Texte sowie das Auftreten bestimmter Reizworte in sinnvollem bzw. nicht sinnvollem Zusammenhang war die Basis für eine Bewertung im Rahmen der 6×6-Feldertafeln. Diese ergab sich aus dem Maß des logischen Zusammenhanges zwischen beeinflussender und beeinflusster Variablen.

Wiederholend ist darauf hinzuweisen, dass sich aus der Q-Zahl ableiten lässt, in welchem Umfang Variablen auf andere einwirken (quantitativer Faktor). Das heißt: Über eine „absolut richtige Lage“ lassen sich in der Regel die anderen Variablen in unterschiedlichem Maße mitbestimmen. Die P-Zahl hingegen gibt Hinweise auf den inhaltlichen Aussagewert der jeweiligen Variablen insofern, als erkennbar wird, inwieweit diese tatsächlich nutzbar werden (qualitativer Faktor).

Tab. 4 Cluverus: A *(Gebiete mit eigener Kenntnis)*

Variable in Zeile x beeinflusst Variable in Spalte x wird beeinflusst durch	1)	2)	3)	4)	5)	6)	Aktiv-summe AS	Q
1. richtige Namensnennung		2	1	2	1	2	**8**	**1,14**
2. verschiedene Namensformen	1		2	2	2	2	**9**	**0,82**
3. absolut richtige Lage	2	2		3	2	2	**11**	**1,38**
4. relative richtige Lage	2	3	2		2	2	**11**	**0,92**
5. Auftreten von Zusatzinformationen	1	2	0	2		1	**6**	**0,60**
6. Angaben in sinnvollem Kontext	1	2	3	3	3		**12**	**1,33**
Passivsumme: **PS**	**7**	**11**	**8**	**12**	**10**	**9**		
P	56	99	88	132	60	108		

Tab. 5 Cluverus: B *(Gebiete ohne eigene Kenntnis)*

Variable in Zeile x beeinflusst Variable in Spalte x wird beeinflusst durch	1)	2)	3)	4)	5)	6)	Aktiv- summe AS⎵Q	
1. richtige Namensnennung		1	0	2	2	2	**7**	**0,88**
2. verschiedene Namensformen	2		2	2	2	2	**10**	**1,43**
3. absolut richtige Lage	2	1		3	2	2	**10**	**2,50**
4. relative richtige Lage	1	1	1		2	2	**7**	**0,64**
5. Auftreten von Zusatzinformationen	1	2	0	1		1	**6**	**0,55**
6. Angaben in sinnvollem Kontext	2	2	1	2	3		**10**	**1,11**
Passivsumme: **PS** P	**8** 56	**7** 70	**4** 40	**11** 77	**11** 66	**9** 90		

Tab. 6 Varenius: A *(Gebiete mit eigener Kenntnis)*

Variable in Zeile x beeinflusst Variable in Spalte x wird beeinflusst durch	1)	2)	3)	4)	5)	6)	Aktiv- summe AS⎵Q	
1. richtige Namensnennung		1	2	3	2	2	**10**	**0,91**
2. verschiedene Namensformen	2		2	2	2	2	**10**	**1,11**
3. absolut richtige Lage	3	2		3	2	2	**12**	**2,00**
4. relative richtige Lage	2	1	1		2	1	**7**	**0,58**
5. Auftreten von Zusatzinformationen	2	2	0	2		1	**7**	**0,64**
6. Angaben in sinnvollem Kontext	2	2	1	2	3		**11**	**1,38**
Passivsumme: **PS** P	**11** 110	**9** 90	**6** 72	**12** 144	**11** 77	**8** 56		

Tab. 7 Varenius: B *(Gebiete ohne eigene Kenntnis)*

Variable in Zeile x beeinflusst Variable in Spalte x wird beeinflusst durch	1)	2)	3)	4)	5)	6)	Aktiv- summe AS⎵Q	
1. richtige Namensnennung		1	0	2	2	0	**5**	**0,62**
2. verschiedene Namensformen	1		0	2	2	0	**5**	**0,56**
3. absolut richtige Lage	2	2		2	1	1	**8**	**2,67**
4. relative richtige Lage	2	2	1		2	1	**8**	**0,80**

	1)	2)	3)	4)	5)	6)	AS	Q
5. Auftreten von Zusatzinformationen	1	2	1	2		1	**7**	**0,70**
6. Angaben in sinnvollem Kontext	2	2	1	2	3		**10**	**3,33**
Passivsumme: **PS**	**8**	**9**	**3**	**10**	**10**	**3**		
P	40	45	24	80	70	30		

Tab. 8 Cellarius: A *(Gebiete mit eigener Kenntnis)*

Variable in Zeile x beeinflusst Variable in Spalte x wird beeinflusst durch	1)	2)	3)	4)	5)	6)	Aktiv- summe AS	 Q
1. richtige Namensnennung		2	2	1	2	2	**9**	**0,69**
2. verschiedene Namensformen	2		2	1	2	2	**9**	**0,75**
3. absolut richtige Lage	2	2		3	1	2	**11**	**1,38**
4. relative richtige Lage	3	2	1		2	3	**11**	**1,22**
5. Auftreten von Zusatzinformationen	2	3	1	2		2	**10**	**1,25**
6. Angaben in sinnvollem Kontext	3	3	2	2	1		**11**	**1,00**
Passivsumme: **PS**	**13**	**12**	**8**	**9**	**8**	**11**		
P	11	108	88	99	80	121		

Tab. 9 Cellarius: B *(Gebiete ohne eigene Kenntnis)*

Variable in Zeile x beeinflusst Variable in Spalte x wird beeinflusst durch	1)	2)	3)	4)	5)	6)	Aktiv- summe AS	 Q
1. richtige Namensnennung		2	0	1	2	1	**6**	**0,67**
2. verschiedene Namensformen	2		1	2	2	1	**8**	**1,11**
3. absolut richtige Lage	3	1		3	2	2	**11**	**3,33**
4. relative richtige Lage	2	1	1		2	1	**7**	**0,10**
5. Auftreten von Zusatzinformationen	0	2	0	2		1	**5**	**0,10**
6. Angaben in sinnvollem Kontext	2	3	1	2	2		**10**	**0,17**
Passivsumme: **PS**	**9**	**9**	**3**	**10**	**10**	**6**		
P	54	72	33	70	50	60		

Die Bewertungstafeln sagen aus, dass schon durch sechs kartographisch wichtige, jedoch nicht quantifizierte, Angaben eine Aussage über den Autor und seine Kenntnis über das beschriebene Gebiet möglich ist. So wird bei Text A von Cluverus anhand der hohen Q-Werte bei „richtiger Namensnennung“, „absolut richtiger Lage“

und „Angaben in sinnvollem Kontext“ deutlich, dass die hier aufgrund der Biographie erwartbaren logischen Interdependenzen tatsächlich auf einer guten Kenntnis des Gebietes beruhen dürften.

Die Q-Werte bei der Beschreibung von unbekanntem Gebiet deuten daraufhin, dass er – durch Nutzung verschiedener Toponyme – großes Gewicht auf möglichst viele Quellen legt, aus denen er dann die möglichst „genaue Lage“ ableitet.

Entsprechend deuten in Text B von Cluverus die Q-Werte durch „verschiedene Namensformen“, trotz „richtiger Lagebeschreibung“ und „Einbindung in den Kontext“, auf eine gewisse Unsicherheit in der begrifflichen Festlegung hin, die als Ausdruck mehrerer, ungleichwertiger und nicht nachkontrollierbarer Informationsquellen zu deuten ist.

Bei Varenius hingegen, dessen Schwerpunkte eher auf der Feststellung allgemeiner Gesetzmäßigkeiten liegen, die er an regionalen Beispielen zu verdeutlichen sucht, spielt (auch in ihm bekannten Gebiet) die „richtige Namensgebung“ nicht die überragende Rolle. Die „absolut richtige Lage“ und die „Angaben in sinnvollem Kontext“ besitzen mehr Gewicht. Interessant ist, dass hier die Nennung von Toponymen eher als Nachweis eigener Ortskenntnis angesehen wird, statt einer Unsicherheit bei der Namensfestlegung.

Da Varenius auch unbekannte Gebiete seiner Zielsetzung (Nachweis allgemeiner Gesetzmäßigkeiten) unterwirft und hier die gleichen Textstrukturen anwendet, ist die Unterscheidbarkeit zwischen „bekannt“ und „unbekannt“ bei diesem Autor nicht trennscharf genug.

Cellarius hingegen, der sich wieder eher als „Lehrmeister“ versteht, weist im Text A (bekannte Gebiete) hohe Q-Werte bei vier Variablen auf, was daraufhin deutet, dass er mit seinen Informationen aus dem Vollen schöpfen kann.

Bei Text B (unbekannte Gebiete) konzentriert sich die Hauptaussage auf die Richtigkeit der Verortung. Ähnlich wie bei Cluverus tritt die Bedeutung von Toponymen in den Vordergrund. Im Zusammenhang mit den geringen Q-Werten der anderen Variablen muss auch hier dieses Angebot als Zeichen begrifflicher Unsicherheit eingestuft werden.

Die Bewertungstafeln ergeben über ihre Primärinformation hinaus auch noch die Rohdaten für die weiteren Auswertungsschritte in 4.3. In einem ersten Schritt wird auf die AS- und PS-Daten zurückgegriffen, deren Werte zur besseren Beurteilung innerhalb einer 4-Felder-Tafel veranschaulicht werden. Hier lassen sich Aussagen zum Verhältnis von Informationsgehalt und dazu benötigter Textfülle der einzelnen Faktoren ableiten und dies im Vergleich von bekanntem und unbekanntem Gebiet.

In einem zweiten Schritt werden die Q- und P-Werte benutzt, um – wiederum im Vergleich von bekanntem und unbekanntem Gebiet – den Zusammenhang zwischen den beiden Werten zur Charakterisierung eines Autors einzuschätzen. Für eine regelrechte Korrelation reicht die Datenbasis nicht aus. Trotzdem sind die (nicht signifikanten) Regressionsgeraden als Orientierungslinien in die Graphiken eingetragen.

4.3 Autorspezifische Darstellungsformen und ihre Interpretation

Auf die formal-strukturelle Einordnung und Differenzierung von Orten, aufgrund unterschiedlicher Schrifttypen, wird hier nicht detailliert eingegangen. Stattdessen sei auf die Ausführungen zu diesem Punkt in Tab. 10 verwiesen. Den Schwerpunkt dieses Kapitels stellen die beiden Analyseschritte und ihre inhaltliche Interpretation dar.

Da die herangezogenen Variablen (= Erklärungsfaktoren) in sehr unterschiedlichem Maße für Informationsgehalt und textliche Repräsentanz stehen, müssen in der oben angedeuteten 4-Felder-Tafel zwei verschiedene Regionsbeschreibungen der jeweiligen Autoren (mit bzw. ohne Eigenkenntnis) entlang der Achsen „Informationswert"(= AS) und „Textfülle" (= PS) bewertet werden. Die Verteilung der Punkte lässt dann in variablem Maße Hinweise zur generellen Bedeutung und Interpretation der jeweiligen Variablen sowie bei diesem Autor zu.

Benutzt werden die Summenzahlen aus den Bewertungstafeln, die ja das Beeinflussungsmaß (= Zeilensumme: AS) sowie das Maß für den Grad des Beeinflusstwerdens (= Spaltensumme: PS) wiedergeben. Für beide Summen gelten die Intervalle von 0 bis 15.

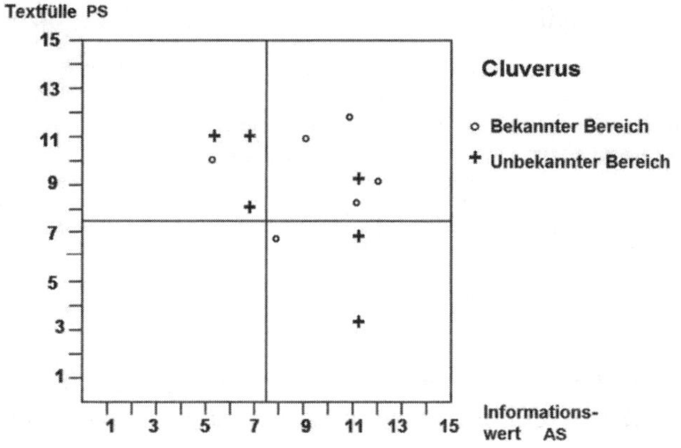

Abb. 1: *Auswertung der AS- und PS-Werte von Cluverus durch eine Vier-Felder-Tafel*

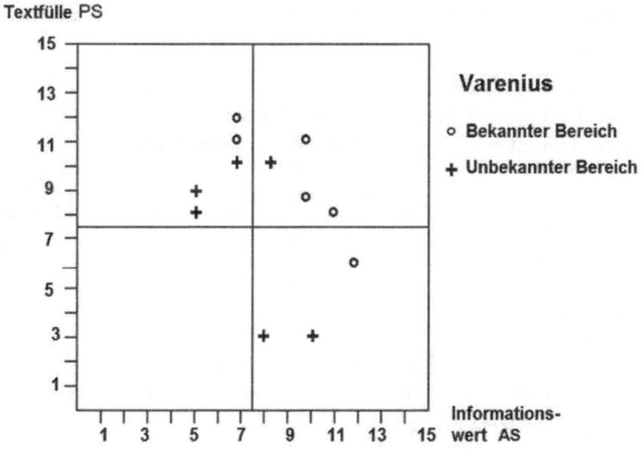

Abb. 2: *Auswertung der AS- und PS-Werte vonVarenius durch eine Vier-Felder-Tafel*

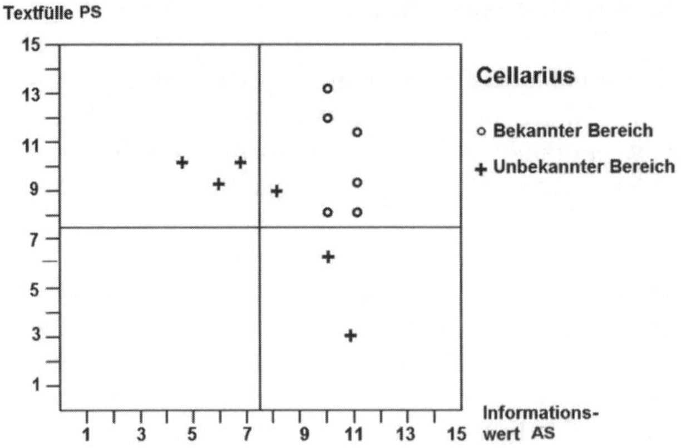

Abb. 3: *Auswertung der AS- und PS-Werte von Cellarius durch eine Vier-Felder-Tafel*

Optimal wäre eine verbale Einbindung sämtlicher Sachinformationen in den Text, weil dadurch eine kausale oder zeitliche Sachlogik erzeugt wird, die sich von einer puren Aneinanderreihung von Ortsnamen eindeutig unterscheiden lässt. Solche Fälle finden sich etwa im rechten oberen Quadranten wieder.

Durch ihre Lage im entsprechenden Quadranten ergibt sich für die Autoren und ihre Gebietskenntnisse folgende Indikatorvariable für „bekanntes Gebiet“ bzw. „Unbekanntes Gebiet“:

Autor	**Indikatorvariable für „Bekanntes Gebiet“**	**Indikatorvariable für „Unbekanntes Gebiet“**
Cluverus	1	2,3
Varenius	3	3,6
Cellarius	2	3,6

Im Gegensatz dazu finden sich im linken oberen Quadranten Hinweise auf eine Verbindung von viel Text ohne die wichtige topographische, geschichtliche bzw. administrative Einbettung, die diese Information erst relevant werden lässt. Die Variablen werden ganz offensichtlich sowohl in „bekanntem“ als auch „unbekanntem“ Gebiet gleichwertig eingesetzt.

Autor	**Unbedeutendes Füllmaterial in „Bekanntem Gebiet“**	**Unbedeutendes Füllmaterial in „Unbekanntem Gebiet“**
Cluverus	5	1, 4, 5
Varenius	4, 5	1, 2, 5
Cellarius	-	1, 4, 5

Im rechten unteren Quadranten wiederum finden sich dann vor allem Items, die auf eine entsprechende Informationsdichte ohne viel zusätzlichen Text zurückgehen, d. h. deren Information für sich allein stehen kann.

Es sollte hieraus erkennbar werden, dass Items mit hohen Werten, nämlich 1, 4 und 5, Hinweise auf Füllmaterial beinhalten. Deren Anzahl – vom weithin bekannten Deutschland ausgehend – nimmt für Beschreibungen eines damals eher unbekannten Europas und Außereuropas zahlenmäßig zu. Dies dürfte die erste Behauptung („Ersatz konkreter Informationen durch Umschreibungen, Füllmaterial usw.") argumentativ unterstützen. Konkret auf die Texte bezogen bedeutet das, dass Varenius selbst in Deutschland keine belastbaren Kenntnisse aufweisen kann, was durch die hohen Werte der Variablen von 4 und 5 nahegelegt wird. Die Variablen 3 und 6 können dagegen als valide Hinweise auf einen hohen Informationswert angesehen werden, da sie nur in bekannten Gebieten Anwendung finden und entsprechende Bedeutung besitzen können. Für Europa und Außereuropa sind sie entsprechend nicht mehr nachweisbar.

In einem *zweiten Analyseschritt* wird versucht, die Wertigkeit und Aussagekraft dieser Faktoren genauer festzulegen, indem ihre aus den Bewertungstafeln stammenden P- sowie Q-Werte als Grundlage genutzt werden. Für die sich anbietende graphische Darstellung ist zu berücksichtigen, dass sich die Intervalle für beide Wertbereiche stark ausdehnen (P: 0,1–225, Q: 0,1–14,9). Da sich die Punktwolken aber meist in engen Grenzen befinden, bietet sich für die Darstellung und Auswertung von P, statt einer linearen, eine logarithmische Skalierung an, wodurch auch die wichtigen „Ausreißer" aus der Punktwolke sinnvoll integriert und bewertet werden können. Es wird erwartet, dass die Position der Punkte einen Hinweis auf den oben formulierten Zusammenhang zwischen Informationsgehalt und Füllmaterial liefern kann.

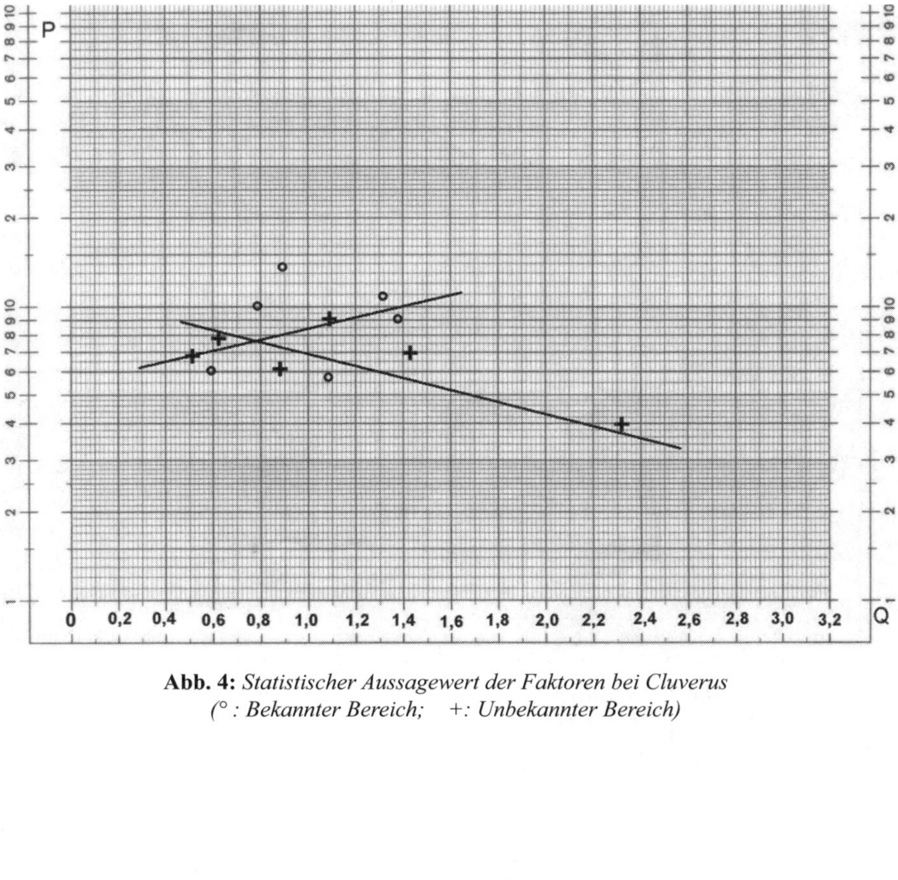

Abb. 4: *Statistischer Aussagewert der Faktoren bei Cluverus*
(° : Bekannter Bereich; +: Unbekannter Bereich)

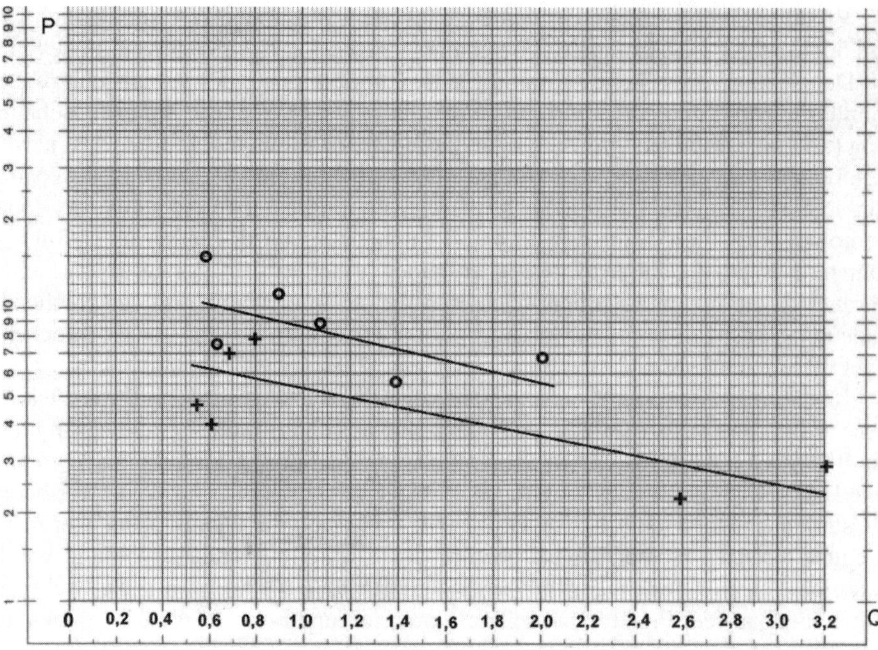

Abb. 5: *Statistischer Aussagewert der Faktoren bei Varenius*
(° : Bekannter Bereich; +: Unbekannter Bereich)

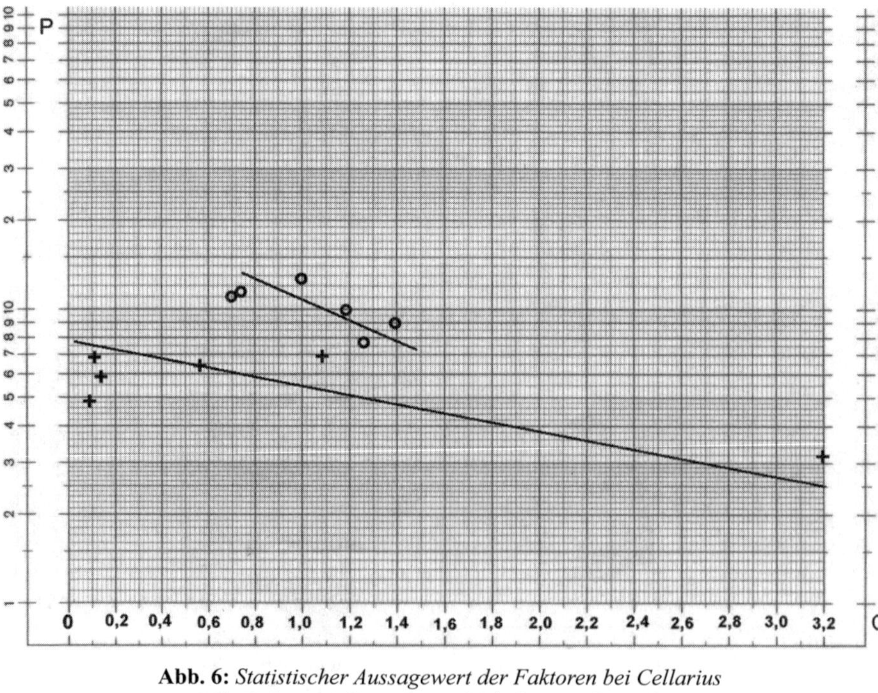

Abb. 6: *Statistischer Aussagewert der Faktoren bei Cellarius*
(° : Bekannter Bereich; + : Unbekannter Bereich)

Erkennbar wird, dass in allen Fällen eine leicht negative Korrelation zwischen den einbezogenen P- und Q-Werten zu bestehen scheint. Der dadurch ausgedrückte Zusammenhang besteht somit darin, dass mit zunehmender Textfülle der Informationsgehalt abnimmt. Die in Abb. 4–6 eingezeichneten Regressionsgeraden können aber aufgrund der geringen Fallzahlen nur zur optischen Orientierung dienen und sind nicht signifikant.

Bei allen drei Autoren unterscheiden sich beide Punktwolken dahingehend, dass die P-Werte eine höhere Wertigkeit aufweisen für „bekannte Gebiete“ im Vergleich zu „unbekannten Gebieten“. Ebenfalls weisen erstere eine sehr viel stärkere Agglomeration der Punktwolke im Vergleich zum „unbekannten Gebiet“ auf. Das bedeutet einmal, dass die Aussagen zu „bekanntem Bereich“ mit höherer Bestimmtheit gemacht werden. Zum Zweiten bedeutet dies, dass die Aussagen der einzelnen Variablen (bei bekanntem Bereich) sehr viel stärker logisch aufeinander bezogen sind. Die Ausreißer sind die Variablen 3, 2 und 5, welche dadurch eine bedeutende Ausnahmestellung hinsichtlich ihrer Erklärungsfunktion andeuten.

Fasst man die bisher besprochenen Kriterienbereiche nach ihrem Aussagewert zusammen und stellt sie wie in Tab. 10 einander gegenüber, lassen sich abschließend folgende Feststellungen treffen: Biographie und Quellenangaben sind oftmals nur ergänzender Natur. Sie lassen sich nicht generell und systematisch auswerten. Das gilt in ähnlicher Form für die Verwendung von Schrifttypen. Hier kann allerdings auf einheitliche Verwendung bei einzelnen Autoren (z.B. Cellarius) gebaut werden. Autorübergreifend sind die Unterschiede jedoch zu groß.

Tab. 10 *Übersicht über die Verwertbarkeit der Kriterien*

Kriterienbereiche	**A) Cluverus**	**B) Varenius**	**C) Cellarius**
Biographie (soweit nutzbar)	* (Seine Reisen sind bekannt)	-	* (Seine Reisen sind bekannt)
Quellenangaben (soweit vorhanden)	+ (nur antike Autoren)	+ (meist antike Autoren)	* (baut auf Cluverus, Varenius und anderen namentlich Genannten auf)
Schrifttypen	Landessprachliche Namen/Begriffe kursiv	-	* werden systematisch zur Differenzierung benutzt: Ortsnamen kursiv; landessprachliche Ortsnamen gotisch; administrative Angaben: Kapitälchen
Inhalt (aus dem Überprüfungsansatz)	+ (in bekannten Gebieten Angaben zu knapp) + (in unbekannten Gebieten größere inhaltliche Breite, aber nicht detailliert); Kartenbeigabe	+ weitgehend; ab Buch II, Kap. 22	* bei bekannten Gebieten: hohe Zuverlässigkeit in den Aussagen; sprachlich gut eingebunden. + bei unbekannten Gebieten

Dabei ist: * verwertbar, + eingeschränkt verwertbar, – zu lückenhaft

Am ehesten kann die textstatistische Struktur der Inhalte genutzt werden, da aufgrund der autortypischen Schreibeigenschaften ähnliche Zusammenhänge zwischen den einzelnen Textstrukturen und –inhalten nicht nur zu erwarten sind, sondern in der Tat auch auftauchen, wie Abb. 4 bis 6 zeigen. Das gilt sowohl für eine Darstellung von bekannten wie unbekannten Gebieten.

Es zeigt sich, dass Kriterien und Variablen aus dem Überprüfungsansatz durchaus sinnvolle Aussagen ermöglichen, die hinsichtlich der ursprünglichen Fragestellung: „Darstellung aus eigener Kenntnis heraus" oder „Wiedergabe fremder Erfahrungen" eine genügende Trennschärfe zulassen. Dass mit nur drei Autoren und jeweils einem Text die Aussagebasis relativ schmal ist, sei eingeräumt. Jedoch ist dieser Ansatz mit den verwendeten Kriterien durch andere oder zusätzliche Variablen und Einbeziehung anderer Autoren jederzeit erweiterbar.

5. ZUSAMMENFASSUNG

Diese Studie hatte zum Ziel – abhängig vom damals vorhandenen Wissensstand – anhand dreier Textbeispiele die dort gemachten Aussagen zu verifizieren bzw. zu falsifizieren. Ein zu diesem Zwecke erstellter Kriterienkatalog sollte hierzu belastbare Hinweise erbringen, die gegebenenfalls auch auf andere Arbeiten dieser Autoren bzw. selbst auf andere Autoren dieser Zeit übertragbar sein sollten.

Der Versuch, eine (textstatistische) Textinterpretation historischer Zeitphasen über formale, sprachliche und inhaltliche Kriterien in vergleichender und bewertender Form vorzunehmen, kann aufgrund seiner sinnvollen, nachvollziehbaren Ergebnisse als realisierbar angesehen werden. Die hier verwendete Auswahl von sechs Kriterien aus dem Text heraus erbrachte interpretierbare und logisch begründbare Hinweise zum damaligen Kenntnisstand des jeweiligen Autors.

Der Versuch kann auch deshalb als erfolgreich angesehen werden, weil über die hier verwendeten Kriterien hinaus noch weitere Angaben zur Evaluation der jeweiligen Texte herangezogen werden können (z. B. aus der Biographie heraus), die diese Ergebnisse bestätigen. Danach sind im Vergleich von selbst bereisten und unbekannten Bereichen in der Beschreibung typische Unterschiede festzustellen, die sich in der sprachlichen Darstellung und im Verhältnis von Informationsgehalt und Textfülle nachweisen lassen.

Die gewonnenen Ergebnisse stehen somit in den vorliegenden Fällen in Einklang mit den (hier nicht verwerteten) biographischen Angaben. Insofern ist zu vermuten, dass dieser Ansatz auch bei weiteren Werken der drei Autoren bzw. auch bei anderen Autoren der damaligen Zeit eine aussagekräftige Interpretationshilfe zum damals vorhandenen, wissenschaftlichen Kenntnisstand liefern kann.

Als Randnotiz sollte zum Schluss erwähnt werden, dass der wissenschaftlichen Arbeit der damaligen Autoren oft keine adäquate Belohnung gegenüberstand. So gibt es in Nachworten und Extraveröffentlichungen deutliche Hinweise auf wirtschaftliche Not der Autoren (z. B. Heinsius 1624, Newton 1672 u. 1693, Kaminske/Staffhorst 1988 u. 2000). Das mag auch erklären, warum eigene Reisen zur Erkundung des Themenfeldes eher die Ausnahme bleiben mussten.

LITERATUR:

a) Originalwerke:

Varenius, B. (1650): Geographia Generalis in qua affectiones generales Telluris explicantur. Amsterdam (Elzevir).

Varenius, B. (1693): Geographia Generalis, bearbeitet von Isaac Newton, 4. Aufl. Jena (Croker).

Cellarius, Ch. (1687): Christophori Cellarii Smalcaldiensis Geographia Antiqua iuxta et Nova. Jena 31698.

Cellarius, Ch. (1676): Nucleus geographicus. Jena.

Cellarius, Ch. (1706): Notitiae Orbis Antiqui sive Geographiae Plenioris. Tomus alter. Leipzig.

Cluverus, Ph. (1624): Philippi Cluveri Introductio in Universam Geographiam, tam Veterem quam Novam. Braunschweig2 (ed. Vorstius J. 1641).

Cochlaeus, J. (1512): Cosmographia Pomponii Melae, authoris nitidissimi, tribus libris digesta, parvo quodam Compendio Joannis Coclei Norici adaucta. Brevis quoque Germaniae descriptio. Nürnberg 1512, hrsg. u. bearb. durch Langosch K.: Johannes Cochlaeus. Brevis Germaniae descriptio. Darmstadt 1960.

Heinsius, D. (1624): Zum Tode des hochberühmten und gefeierten Geographen Philippus Cluverius. Einhefter in: Philippi Cluveri Introductio in Universam Geographiam, tam Veterem quam Novam. Braunschweig2 (ed. Vorstius J. 1641).

Newton, I. (1672): Med. D. B. Varenii Geographia Generalis aucta et illustrata. Cambridge.

Ortelius, A. (1570): Africae tabula nova, Antwerpen.

Vallseca, G. de (1439): Àfrica del norte, Palma de Mallorca.

b) Sekundärliteratur:

Blanke, H. W. (2001): Marco Polo, Bernhard Varenius und Engelbert Kaempfer. Oder: Vom Hörensagen über die gelehrte Recherche zum Autopsie-Bericht. In: Veröff. d. Eutiner Landesbibl., Bd. 5, S. 29–67.

Burenhult, N. (Ed. 2008). Language and landscape: Geographical ontology in cross-linguistic perspective [Special Issue]. Language Sciences, 30(2/3).

Burenhult, N. (2008). Streams of words: Hydrological lexicon in Jahai. Language Sciences, 30(2/3), 182–199. doi:10.1016/j.langsci.2006.12.005.

Burenhult, N. (2008). Spatial coordinate systems in demonstrative meaning. Linguistic Typology, 12(1), 99–142. doi:10.1515/LITY.2008.032.

Burenhult, N., Levinson, S. C. (2008). Language and landscape: A cross-linguistic perspective. Language Sciences, 30(2/3), 135–150. doi:10.1016/j.langsci.2006.12.028.

Downs, R. M., Stea, D. (1982): Kognitive Karten: Die Welt in unseren Köpfen. Harper & Row. New York (UTB 1126).

Gerisch, L. (2001): Varenius und seine Zeit. In: Veröff. d. Eutiner Landesbibl., Bd. 5, S. 19–28.

Görz, G., Ilyushechkina, E., Thiering, M. (2013): Towards a cognitive-linguistic reconstruction of 'mental maps' from ancient texts – The example of Dionysios Periegetes. In: Sharp, B., Zock, M. (Ed.): Proceedings of NLPCS 2013. 10th International Workshop on Natural Language Processing and Cognitive Science. Marseille: CNRS/LIF, 2013, S. 20–34.

Gould, P., White, R. (1974): Mental Maps. Penguin. Harmondsworth.

Haversath, J.-B. (1997): Die Anfänge der Geographie und des Erdkundeunterrichts in der Antike und im Humanismus. In: Frank, F., Kaminske, V., Obermaier, G. (Hrsg.): Die Geographiedidaktik ist tot, es lebe die Geographiedidaktik. (= Münchner Stud. z. Did. d. Geographie Bd. 8), München, S. 1–15.

Ilyushechkina, E., Görz, G., Thiering, M. (2014): Towards a cognitive-linguistic reconstruction of the spatial orientation in ancient texts – The example of Dionysios Periegetes. In: Geus, K., Thiering, M. (Ed.): Common Sense Geography and Mental Modelling. Münster: Mentis Verlag, 2014, 231–249.

KAMINSKE, V. (1981): Zur systematischen Stellung einer Geographie des Freizeitverhaltens. In: Geographische Zeitschrift 69. Wiesbaden, H. 3, S. 217–223.

KAMINSKE, V. (1996): Die verzerrte Welt in den Köpfen. Erklärung und Korrektur – Aufgabe der Geographie? In: Schulheft 82, S. 20–33.

KAMINSKE, V. (2006): Raumwahrnehmung und Raumvorstellung – Rahmenbedingungen und Entwicklung. In: Geographie u. Schule 28, H. 6, S. 12–19.

KAMINSKE, V. (2012): Die räumliche Wahrnehmung. Wiss. Buchgesellschaft, Darmstadt.

KAMINSKE, V., STAFFHORST, U. (1988): Die Geographia generalis des Bernhard Varenius. Ein disziplingeschichtlicher Erklärungsansatz der Geographie auf der Grundlage eines ehemals verwendeten Schulbuches im Gymnasium illustre, Karlsruhe. In: Mat. z. Did. d. Geographie 11. Geogr. Ges. Trier, H. 12, S. 7–28.

KAMINSKE, V., STAFFHORST, U. (1998): Cluverus und Cellarius – zwei Begründer der historischen Geographie im Vergleich. In: Orbis Terrarum, Bd. 4, S. 185–203.

KAMINSKE, V., STAFFHORST, U. (2000): Der Nachruf des Humanisten Heinsius auf den Geographen Clüver. In: Jahresbericht 1999/2000 des Bismarck-Gymnasiums Karlsruhe, Karlsruhe, S. 130–139.

KASTROP, R. (1982): Die Bedeutung des Varenius innerhalb der Entwicklung des geographischen Denkens in Deutschland. In: Büttner, M. (Hrsg.): Zur Entwicklung der Geographie vom Mittelalter bis zu Carl Ritter (= Abh. u. Quellen z. Gesch. d. Geogr. u. Kosmologie Bd. 3), S. 79–95.

KÖCK, H. (2005): Räumliches Denken. In: Praxis Geographie 35. H. 7/8, S. 62–64.

KÖCK, H. (2013): Editorial. In: Geographie u. Schule 35, H. 201, S. 3.

KUIPERS, B. (1982): „The Map in the Head – metaphor“. In: Environment and Behavior 14, S. 202–220.

KUIPERS, B. (2001): The Skeleton in the Cognitive Map. A computational hypothesis. In: PEPONIS, J., WINEMAN, J., BAFNA, S. (Eds.): Space Syntax. Proceedings of the Third International Symposium. Ann Arbor, University of Michigan. S. 10.1–10.7.

KUIPERS, B., TECUCI, D., STANKIEVICZ, B. (2003): The Skeleton in the Cognitive Map. A Computational and Empirical Exploration. In: Environment and Behavior 35. No. 1, S. 81–106.

LEHMANN, K. (2001): Das kurze Leben des Bernhard Varenius (1622–1650). In: Veröff. d. Eutiner Landesbibl., Bd. 5, S. 9–18.

LEVINSON, S. C. (2003). Language and cognition. In: W. FRAWLEY (Ed.), International Encyclopedia of Linguistics, S. 459–463. Oxford: Oxford University Press.

LEVINSON, S. C. (2003). Space in language and cognition: Explorations in cognitive diversity. Cambridge: Cambridge University Press.

LEVINSON, S. C. (2003). Spatial language. In: L. NADEL (Ed.), Encyclopedia of cognitive science, S. 131–137. London: Nature Publishing Group.

LEVINSON, S. C., & WILKINS, D. P. (2006). Patterns in the data: Towards a semantic typology of spatial description. In: S. C. LEVINSON, & D. P. WILKINS (Eds.), Grammars of space: Explorations in cognitive diversity; S. 512–552. Cambridge: Cambridge University Press.

LEVINSON, S. C., & WILKINS, D. P. (2006). The background to the study of the language of space. In: S. C. LEVINSON, & D. P. WILKINS (Eds.), Grammars of space: Explorations in cognitive diversity (pp. 1–23). Cambridge: Cambridge University Press.

MARK, D. M., TURK, A., BURENHULT, N., & STEA, D. (2011). Landscape in language: An introduction. In: D. M. MARK, A. G. TURK, N. BURENHULT, & D. STEA (Eds.), Landscape in language: Transdisciplinary perspectives (pp. 1–24). Amsterdam: John Benjamins.

PIAGET, J., INHELDER B. (1971): Die Entwicklung des räumlichen Denkens beim Kinde. Stuttgart.

SCHWEIZER, K., HORN, M. (2006): Raumwahrnehmung und Raumvorstellung – theoretische Überlegungen und empirische Befunde aus Psychologie und Geographie. In: Geographie u. Schule 28, H. 164, S. 4–11.

STAFFHORST, U. (2007): Miszellen zur Geographia generalis des Bernhard Varenius. In: SCHUCHARDT, M. (Hrsg.): Bernhard Varenius (1622–1650). In: Brill's Studies in Intellectual History Bd. 159, Leiden, S. 215–226.

VESTER, F. (19906): Unsere Welt – ein vernetztes System. München.

Anhang:

Philippus Cluverus (= Philipp Clüver, auch Klüwer), (*1580 in Danzig; †31. Dezember 1622 in Leiden/Niederlande), war Geograph und Historiker, der sich zunächst mit der antiken Geographie beschäftigte und weithin als Begründer der historischen Geographie gilt. Um 1607 begann er – zunächst über kartographische Studien – mit Publikationen zur Geographie, deren Grundlage er auf mehreren Reisen durch England, Skandinavien und Italien erarbeitet hatte.

Als sein wichtigstes Werk sieht man die *Introductio in Universam Geographiam* (*Einführung in die universelle Geographie*) von 1624 an, die auf einer Vorlesung beruht und mehrfach überarbeitet wurde. Der erste von sechs Bänden befasst sich mit einer allgemeinen Einführung über Geographie, Gradnetz und Ozeane. Die weiteren Bände stellen die Erde länderkundlich dar. Alle Bände enthalten kurze Beschreibungen der Länder, mit oft willkürlich gesetzten Schwerpunkten (de campania, de fluviis, de montibus, de variis Italiae incolis). Diese *Introductio* blieb bis zur Mitte des 18. Jahrhunderts ein Standardwerk der Geographie.

Bernhardus Varenius (= Bernhard Varen), (*1622 in Hitzacker, Hannover; †1650/51 in Leiden, Niederlande), ist einer der ersten und bedeutendsten Geographen im Europa des 17. Jahrhunderts. Seine Lehre der allgemeinen Geographie setzte für mehr als hundert Jahre den Maßstab für Lehrinhalte und Anforderungsniveau in Universitäten und Gymnasien.

Varenius studierte zunächst Medizin, bevor er über Kontakte mit Geographen zur Geographie kam. 1649 veröffentlichte er sein Werk Descriptio Regni Japoniae („Beschreibung des Königreiches Japan"). Sein wichtigstes Werk *Geographia generalis* erschien 1650. Hierin werden die grundlegenden Prinzipien der Geographie auf einer breiten wissenschaftlichen Basis unter Nutzung der damals aktuellen Erkenntnisse erläutert. Die allgemeine systematische Geographie wird hier erstmals als Verständnisbasis für global ablaufende Vorgänge und Prozesse dargestellt. Daneben wird in einem zweiten Teil eine Art Regional-Geographie eingeführt. Regionale Darstellungen beruhten bis dahin meist auf Karten und Beschreibungen der Antike. Varenius' Geographie stellt somit die Basis für die spätere thematische Geographie, die Länderkunde und die historische Geographie dar. Das Werk wurde oft überarbeitet und verbessert, die (einzige in Deutschland gedruckte) Ausgabe von 1672 enthält Erweiterungen und Einträge von Isaac Newton.

Christophorus Cellarius Smalcaldensis (= Christoph Martin Keller), (*22. November 1638 in Schmalkalden; †4. Juni 1707 in Halle), wurde schon 1673 zum Rektor des Gymnasiums in Weimar ernannt. Mit dem Jahr 1693 wurde er zum Professor für Geschichte und Beredsamkeit an der neu gegründeten Friedrichs-Universität in Halle berufen.

Während dieser Zeit verfasste er neben Schulbüchern, wie der *Geographia iuxta et Nova* (Zeitz/Jena 1687) oder dem *Nucleus geographicus* (1676), eine Reihe wissenschaftlicher Länderkunden unter dem Titel *Geographia Antiqua*. Cellarius gilt als der erste Geograph, der ein umfassendes systematisches Werk über die alte

Geographie lieferte. Er war dabei, hierzu eine Reihe von Karten zu erstellen. 18 solcher Karten wurden nach seinem Tod 1776 veröffentlicht. Durch seine Werke wurde Cellarius zum „Begründer des universitären Geographieunterrichts“.

DIE GÖTTIN „VOM HEILIGEN WASSER“

Andreas Klingenberg

ÜBERLEGUNGEN ZUR BEDEUTUNG DES WASSERS FÜR DIE AUSBREITUNG UND VEREHRUNG DER PERSISCHEN ANĀHITĀ IM WESTEN KLEINASIENS*

Im Jahr 1894 entdeckte Karl Buresch auf einer Forschungsreise durch Lydien im Ort Kara Silendi, dem antiken Silandos, nahe des Selendi Çay, einen interessanten Grabstein, der sich bereits damals nicht in situ befand und heutzutage verschollen ist.[1] Es handelte sich nach der Beschreibung Bureschs um eine trapezförmige Marmorstele mit leerem Giebel und Kranz über der Inschrift, die nach der sullanischen Ära auf das Jahr 193/194 n. Chr. zu datieren ist und, außer dem Hinweis auf die Bestattung eines Onesiphoros, durch seine Mutter Atalante noch einen Grabfluch enthält, der den potentiellen Grabschänder vor der Bestrafung durch Anaitis warnt:[2]

ἔτους σοη´, μη(νὸς) Γορπιαίου ε´ ἀ(πιόντος). Ἀταλάν-
τη Ὀνησιφόρου τὸν υἱὸν ζήσαντα
ἔτη κε´ ἐτείμησεν. εἴ τις δὲ πα-
ραμάρτῃ τῷ τάφῳ μετὰ τὸν θάνατόν
μου, τὴν Ἀναεῖτιν τὴν ἀπὸ τοῦ ἱεροῦ ὕ-
δατος κεχολωμένην ἕξει.

Weder die äußere Erscheinung noch der grundsätzliche Inhalt der Inschrift sind dabei sonderlich eindrucksvoll. Die Form des Monuments ist in Lydien häufig anzutreffen, und an Grabflüchen besteht im westlichen Kleinasien kein Mangel.[3] Auch Anaitis als strafende Gottheit in Grabflüchen ist nicht singulär.[4] Überhaupt wurde die Göttin in Lydien weithin verehrt, davon künden zahlreiche Belege in Form von Inschriften, Münzen und literarischen Hinweisen.[5]

* Dieser Aufsatz ist hervorgegangen aus einem Vortrag, den ich im Sommer 2014 auf der internationalen Tagung „Wasser in der Antike“ in Berlin gehalten habe. Mein Dank gilt Prof. Dr. Klaus Geus für die Einladung und die anschließende Diskussion. Für kritische Hinweise und hilfreiche Anmerkungen danke ich überdies Prof. Dr. Walter Ameling und Dr. Gregor Staab.

1 Publiziert in Buresch 1898, 117 Nr. 56; s. jetzt TAM V.1, 64; vgl. außerdem Diakonoff 1979, 149 Nr. 23; de Hoz 1999, 134 Nr. 3.15.

2 Text nach TAM (wo außer der Abschrift Bureschs noch eine aus der Hand W.M. Ramsays herangezogen wurde).

3 Das bestätigt bereits ein kurzer Blick in den entsprechenden IK-Band Johan Strubbes (IK 52: Arai epitymbioi); vgl. auch Strubbe 1991.

4 So gibt es etwa ein Beispiel aus dem nahegelegenen Saittai (TAM V.1, 172).

5 Das Material kann hier nicht im Einzelnen ausgebreitet werden; vgl. einstweilen für die Inschriften de Hoz 1999, (Nr. 3.1–3.82) und für die Münzen Moga 2012, 192 Anm. 12.

Wirklich spannend wird es indes bei dem Epitheton, mit dem die invozierte Göttin Anaitis in der Inschrift aus Silandos näher charakterisiert wird. Darin wird sie nämlich Ἀναεῖτις ἀπὸ τοῦ ἱεροῦ ὕδατος genannt, „*Anaitis vom heiligen Wasser*". Dieser Beiname ist in dieser Form einzigartig und einer näheren Betrachtung wert. Der Name Anaitis geht zurück auf die iranische Göttin Anāhitā, von der man seit der Zeit der achaimenidischen Herrschaft im Westen Kleinasiens Kenntnis hatte. Daher liegt es nahe, auch das ungewöhnliche Epitheton mit diesen Ursprüngen in Verbindung zu bringen. Immerhin war Anāhitā im Kern eine Wassergottheit, und die Heiligkeit des Wassers ein zentrales Element in den iranischen Glaubensvorstellungen, wie bereits Herodot wußte.[6] Dafür lassen sich zudem iranische Belege anführen, etwa aus dem Awesta und den bekannten Keilschrifttäfelchen aus Persepolis.[7] Folglich scheint sich in diesem Fall tatsächlich eine Verbindung zwischen der Anaitis und der iranischen Anāhitā zu offenbaren, was in dieser allgemeinen Form in der Forschung weitgehend anerkannt ist.[8]

Der Bezug auf das Wasser tritt bei den fast ausschließlich kaiserzeitlichen Belegen für Anaitis in Lydien sonst kaum derart evident zutage.[9] Vielmehr zeigt sich das Erscheinungsbild der Göttin in den lydischen Zeugnissen als höchst vielfältig – sie wird mal als Anaitis allein, mal als Meter Anaitis, Artemis Anaitis oder persische Artemis angesprochen.[10] In der gelehrten Diskussion sind deswegen seit einigen Jahren mehrfach Stimmen laut geworden, die das persisch-iranische Erbe bei der Verehrung der Anaitis in Lydien insgesamt als oberflächliche Iranisierung bereits vorhandener Kulte betrachten und weitgehend auf die Übernahme des Namens reduzieren.[11]

Dem möchte ich mich an dieser Stelle nicht anschließen. Die folgenden Ausführungen zielen vielmehr auf eine konträre Interpretation ab. Ausgehend von der iranischen Göttin Anāhitā und ihrem aquatischen Charakter (I.) werden im Folgenden die in der eingangs erwähnten Anaitis vom heiligen Wasser fortdauernden iranischen Vorstellungen dargestellt (II.), an die sich einige weitergehende Überlegungen zur Etablierung und Verehrung der Anāhitā in Lydien insgesamt (III.) sowie zur Frage nach deren Verbindung mit dem, wie erwähnt, uneinheitlichen Befund des Anaitiskultes in der Kaiserzeit anschließen (IV.).

6 Hdt. 1,131. Vgl. de Jong 1997, 101 sowie Briant 1994; Wiesehöfer 2004.

7 Vgl. de Jong 1997, 138; Henkelman 2008, 377

8 S. etwa Diakonoff 1979, 161: „*What better proof do we have of her original character than the valuable and unique stele* [...] *testifying the persistence* [sic!] *of this very Iranian aspect several centuries after the disappearance of the Persian rule.*" Vgl. Boyce/Grenet 1991, 245; de Hoz 1999, 74 sowie ferner Ricl 2002, 201 Anm. 12, die auf den seit parthischer Zeit belegten Frauennamen Āb-Nāhīd hinweist, der soviel wie „*(A)nāhīd of water*" bedeute.

9 Eine fragmentarische *lex sacra* im Museum von Manisa (ETAM 19,24; vgl. BE 1995, 497) könnte ebenfalls auf Anaitis zu beziehen sein, wofür die Erwähnung von immer brennendem Feuer und Wasser ein Indiz wäre.

10 S. die Übersicht bei de Hoz 1999, 11

11 In methodischer Hinsicht problematisch sind die Ausführungen von Brosius 1998, zu Recht kritisiert durch Briant 2001, 178 und Moga 2012, 192; verhalten hinsichtlich religiöser Einflüsse durch die Perser ist auch de Hoz 1999, 73–79 ; s. für Hypaipa ebenso Nollé 2012.

I.

Obwohl im Hinblick auf die Ursprünge der iranischen Anāhitā und die Entwicklung ihrer Verehrung im Iran nicht alles abschließend geklärt ist und zum Teil kontrovers diskutiert wird, so lassen sich doch einige Grundlinien skizzieren. Das älteste sicher datierbare Zeugnis für den Namen der Anāhitā in Persien sind die Inschriften Artaxerxes' II. (405/4–359 v. Chr.).[12] Trotzdem kann man doch festhalten, dass ihre Verehrung nicht erst unter diesem Herrscher, sondern ohne Zweifel bereits deutlich vorher einsetzte.[13] Dasselbe gilt übrigens für Mithra, der darin ebenfalls das erste Mal in Persien direkt bezeugt, jedoch in anderen Quellen bis ins 14. Jh. v. Chr. zurückzuverfolgen ist.[14] Ursprünglich handelte es sich bei Anāhitā um eine Wassergottheit, genauer gesagt um eine Flußgöttin. Im Awesta ist ihr der fünfte Götterhymnus gewidmet, der als Ābān Yašt, als Wasserhymnus bezeichnet wird.[15]

Darin wird die Göttin mit drei Attributen benannt, Arədvī Sūrā Anāhitā,[16] die im Sprachgebrauch dann zu Eigennamen wurden, wie wir es in den erwähnten Inschriften des Artaxerxes beobachten können. In sāsānidischer Zeit war dann Ardvīsūr geläufig. Arədvī Sūrā Anāhitā ist die göttliche Repräsentation eines mythischen Flusses, der vom Berg Hukarya im Haragebirge herabströmt, das als Zentrum der Welt verstanden wurde. Sie speist und reinigt den See Vourukaša, dessen Ausflüsse in alle sieben Erdteile strömen und somit letztlich ihren Ursprung in Anāhitā haben. Sie ist aber nicht nur für die Reinigung, sondern insbesondere auch für Fruchtbarkeit zuständig, wie einige kurze Zitate aus dem Ābān Yašt demonstrieren:[17]

1. Es sprach der Weise Herr zu Spitama Zarathustra: verehre sie mir, die heldenhafte reine Arədvī Sūrā Anāhitā, die weit sich ausbreitet, heilkräftig [...], die die Wassergräben fördert, [...], die die Herden fördert, [...], die die Äcker fördert, [...], die das Land fördert. [...]
3. Die große weitberühmte, die von solcher Größe ist wie alle Gewässer, die hier auf Erden fließen, die mit Macht vom Berg Hukarya herabströmt zum See Vourukaša.
5. Und der Abfluss von diesem einen Wasser verteilt sich [...] über alle die sieben Erdteile, und von diesem einen Wasser strömt es [...] gleichmäßig herab im Sommer und im Winter. Sie läutert [...] die Gewässer [...]

Diese Vorstellungen finden erstaunliche Parallelen in der altindischen Kosmologie, die ebenfalls einen Berg als Mittelpunkt der Welt kennt, Meru oder Sumeru[18]. Zu Anāhitā gibt es ein Pendant in Sarasvatī, die im Rigveda ebenfalls eine Flußgöttin ist und in alle Welt fließt.[19] Diese Reihe ließe sich noch fortsetzen. Jedenfalls läßt sich daraus schließen, daß die Verehrung der Anāhitā und der Kern ihrer Darstellung im Ābān Yašt sehr weit zurückreichen müssen und als gemeinsames indo-

12 A²Ha (Schmitt 2009, 186–188); A²Sa (ebd., 191); A²Sd (ebd., 194).
13 Vgl. z. B. Briant 1996, 264; Jacobs 2001, 89
14 CIMRM 16.
15 Vgl. besonders Kellens 2002/2003 (ap- als eigentlicher Name für Anāhitā).
16 Zur Deutung dieser Attribute s. die Diskussion bei Kellens 2002/2003, 321–324.
17 Übersetzung Lommel 1927, 32 (leicht adaptiert).
18 Boyce 1985, 811.
19 S. dazu Lommel 1954.

iranisches Erbe zu betrachten sind.[20] Ihr späterer Name hat sich dann, wie gesehen, aus einem Epitheton entwickelt, ihr eigentlicher Name könnte *ap* – Wasser – gewesen sein. Jedenfalls hat Jean Kellens gezeigt, daß bei der Erwähnung von Wasser im Singular ohne weiteres Attribut an Anāhitā zu denken ist,[21] die den anderen mit Flüssen oder Wasser verbundenen Göttern in dieser Funktion bei weitem den Rang abgelaufen hat. Dieser Umstand ist nicht zuletzt an der Übertragung ihres Kultes in Gebiete außerhalb des Irans deutlich abzulesen, insbesondere nach Pontos, Kappadokien, Armenien und Lydien.

II.

Damit sind wir wieder beim Ausgangspunkt angelangt, der *Anaitis vom heiligen Wasser* aus Silandos. Dabei stellt sich die Frage, in welchem Sinn dieses ἀπὸ τοῦ ἱεροῦ ὕδατος zu verstehen ist. Einen missglückten Genitiv kann man in diesem Fall sicher ausschließen. Daß hier eine geographische Referenz vorliegt, hat Karl Buresch vermutet, der als „*heiliges Wasser der Anaitis*" einen Bach, eine Quelle oder einen See annahm.[22] Bergbäche sowie Gebirgsflüsse sind in Lydien ein typisches Landschaftsphänomen. Die Ruinen der Stadt Silandos liegen am Hang eines Berges, von dem ein Fluß hinab ins Tal fließt und in einen Zufluss des Hermos mündet. In Bureschs Zeit durchfloss überdies ein Bach die antiken Ruinen.[23] Diese landschaftliche Disposition ist keineswegs irrelevant (dazu weiter unten); allerdings hat Louis Robert die simple Identifikation des heiligen Wassers mit einem bestimmten Gewässer sicher zu Recht verworfen.[24] Vielmehr liegt hierin ein Bezug auf den essentiellen aquatischen Charakter der Anāhitā vor, wie er im Awesta geschildert ist.

Dass ein Bezug auf das Awesta auch für Kleinasien nicht gar so weit hergeholt ist, zeigt eine ebenfalls kaiserzeitliche Inschrift aus Ortaköy im westlichen Kappadokien, nach der eine Flavia Prima der „*Großen Göttin Anaitis Barzochara*" drei Hierodulen weihte.[25] Der iranische Beiname Barzochara ist mit „*von der hohen Hara stammend*" zu übersetzen und somit ein Verweis auf die oben skizzierte Darstellung Anāhitās im Ābān Yašt.[26] Diese Inschrift ist überdies noch in anderer Hinsicht aufschlussreich. Wie die Weihung der Hierodulen an die Gottheit verdeutlicht, muß es dort ein Heiligtum gegeben haben. Der awestische Beiname läßt im Grunde nur den Schluss zu, daß die Verehrung an diesem Ort auf die Perser zurückzuführen ist. Die Lage des Heiligtums passt jedenfalls dazu, denn Ortaköy befindet sich am Fuße des Ekecik Dağı, der sich über 2000 m erhebt. Diesem entspringen zudem

20 Ebd.; zur indo-iranischen Religion insgesamt vgl. Gnoli 2006.
21 Kellens 2002/2003, 324
22 Buresch 1898, 118.
23 Ebd.
24 Robert 1976, 46; vgl. Boyce/Grenet 1991, 245.
25 SEG 41, 1417: Ἀγαθῇ vvv τύχῃ | Θεᾷ Μεγίστῃ Ἀναείτιδι | Βαρζοχάρᾳ vv Φωτίδα | καὶ Θέωνα καὶ Πρεῖμαν || τὴν καὶ Γαρσην ἱεροδούλους | ἀνεπηρέαστους ἐκ πάντων | μετ' ἐπιγονῆς διὰ βίου | Φλαουία Πρεῖμα; vgl. auch BE 1968, 538.
26 Schmitt 1970.

mehrere Flüsse. Nicht nur nach der Perzeption der Griechen pflegten die Perser nämlich ihren Göttern bevorzugt auf Bergen zu huldigen.[27] Auch Heiligtümer sind auf Anhöhen entstanden, wie Strabon für das Anaitisheiligtum in Zela berichtet.[28] Nach dem Awesta fand eine Verehrung der Götter idealerweise *„auf der höchsten Höhe, auf der hohen Harati, die Hukarya genannt ist"*[29], statt. Im erwähnten Wasserhymnus sind verschiedene Opfer erwähnt, die Heroen der Anāhitā auf dem Hukarya, aber auch auf anderen Bergen darbrachten.[30] Dagegen verehrten nur böse Menschen Gottheiten wie Anāhitā unterirdisch und würden folgerichtig auch nicht erhört.[31]

Diese Schlußfolgerung läßt sich auch auf Silandos übertragen, das nicht nur hoch gelegen, sondern auch von Bergen umgeben war. Hier wird sich ebenfalls ein Kultort aus persischer Zeit befunden haben, auf dem die rudimentäre Kenntnis des ursprünglich aquatischen Charakters der Anāhitā basierte. Dass sich eine solche Spur ausgerechnet in Silandos etwas deutlicher als anderswo in Lydien erhalten konnte, ist mit der eher periphären Lage abseits der großen Zentren zu erklären.

III.

Die Etablierung und Verbreitung der Anāhitā in Kleinasien lässt sich insofern ganz gut erklären, wenn man die iranischen Vorstellungen von der Göttin einerseits und die besonderen naturräumlichen Strukturen der wichtigsten Kultorte andererseits in die Gleichung einbezieht. Grundsätzlich ist von einer privaten, nicht einer zentral gesteuerten Kultübertragung unter den Persern auszugehen. Bekanntlich bedeutete die persische Eroberung Kleinasiens nicht nur eine stetige Präsenz von Truppen, vielmehr wurden bereits in frühachaimenidischer Zeit Siedlungen gegründet und zu diesem Zweck Iraner angesiedelt.[32] Die herrschaftliche Durchdringung des eroberten Landes war zwar sicherlich ein nicht unwesentlicher Grund, wichtiger wird indes die Bebauung und Kultivierung des Landes gewesen sein.[33] Für die nun als Bauern tätigen Siedler hatte eine Fruchtbarkeit garantierende Göttin wie Anāhitā zweifellos einen besonderen Stellenwert. Bei der Landschaftsstruktur Lydiens, mit seinen vielen Bergen und diesen entspringenden Flüssen, werden diese Siedler sicher an Anāhitā und ihre mythologische Gestalt gedacht haben, deren Wirken sie in den fruchtbaren Tälern erkennen konnten.[34] So boten sich in Lydien viele Orte für die Verehrung der Göttin an, aus denen dann zum Teil die später bekannten Kultorte der Anaitis entstanden sein werden. Das bedeutet freilich nicht, daß jeder verein-

27 Aisch. *Pers.* 610–618; Hdt. 1,131; Xen. *Kyr.* 8,7,3; Strab. 15,3,13 (732C). Die bekannteste Reminiszenz dieses Brauches ist das Opfer Mithradates' VI. nach seinem Sieg über Licinius Murena, App. *Mithr.* 66 (276–279).

28 Strab. 11,8,4 (512C.).

29 Yt. 10,88 (Lommel 1927, 77); vgl. de Jong 1997, 97.

30 Z. B. Yt. 5,25–27 (Yama opfert auf dem Hukarya und wird erhört).

31 Yt. 5,41–43.

32 Dazu s. zuletzt Klingenberg 2014.

33 Ebd. 315

34 Vgl. Diakonoff 1979, 161 zur Kastolosebene und dem Hermostal.

zelte Beleg für Anaitis auf ein ursprüngliches Anāhitāheiligtum zurückgehen muß. Aber für die Orte, an denen sich die Belege häufen und bestenfalls weitere Indizien hinzukommen, darf man diesen Schluss ziehen.

So war es etwa im Falle Hierakomes, des späteren Hierokaisareia. Das dortige Heiligtum der Artemis Persike[35] führten die Bewohner der Stadt in tiberischer Zeit auf Kyros den Großen zurück.[36] Der Ort lag in der nach iranischen Siedlern benannten Hyrkanischen Ebene[37], die frühesten bekannten Einwohner aus dem späten 4. Jh. v. Chr. tragen zudem iranische Namen.[38] Noch im 2. Jh. n. Chr. soll es dort, laut Augenzeugenbericht des Pausanias, Kultrituale nach iranischem Muster gegeben haben, die er in demselben Atemzug auch für Hypaipa konstatiert.[39] Das findet Bestätigung in epigraphischen und numismatischen Quellen. So ist in Hypaipa ‚*magos*' noch in der Kaiserzeit als Priestertitel belegt,[40] und eine erst vor kurzem publizierte Münze wird als Darstellung des Magierrituals gedeutet.[41] Ob dieses in der Kaiserzeit aber noch inhaltlich verstanden wurde, oder ob man mit Pierre Briant nur von „*survivances folkloriques*"[42] auszugehen hat, muß wie für Hierokaisareia offenbleiben. Hypaipa lag aber am Südrand des Tmolosgebirges, auf dem die Perser, nach Strabon, eine Exedra aus weißem Stein errichtet hatten, von der aus man in die umliegenden Ebenen schauen konnte.[43] Man hat darin lange Zeit wenig mehr als einen Wachtposten gesehen,[44] obwohl diese Deutung architektonisch keinen Sinn ergibt. Der Bau konnte jedoch mittlerweile mit den archäologischen Funden auf dem Kel Dağ identifiziert und als „*outdoor, sacred precinct*" erkannt werden.[45] Die Existenz vieler Quellen, der persische Bau und die Bedeutung der Anaitis im nahegelegenen Hypaipa erlauben den Schluss, dass die Perser dort die Anāhitā verehrten.

35 Daß die Bezeichnungen ‚Artemis Persike' und ‚Anaitis' austauschbar sind, zeigt u. a. eine Inschrift aus Hypaipa (IK Ephesos, 3825 = de Hoz 3.58), in der ein Priester der „Anaitis Artemis" erwähnt ist, obwohl dort sonst der Name Artemis Persike gebräuchlich war. Vgl. auch SEG 31, 998 = DE HOZ 1999, Nr. 3.59.

36 Tac. *ann.* 3,62,3: *altius Hierocaesarienses exposuere, Persicam apud se Dianam, delubrum rege Cyro dicatum*. Nach CORSTEN 1991, 171 Anm. 43 hat man dabei an Kyros d. J. zu denken; s. dagegen aber JACOBS 2001, 89 m. Anm. 25, der insgesamt klarmacht, daß dies durchaus keine zwingende Konjektur ist; so auch schon Sekunda 1985, 21.

37 Strab. 13,4,13 (629C.).

38 IK Ephesos 2, Z. 41 (Sisines und Eumanes), dazu ZWANZIGER 1973, 144; BRIANT 1996, 723.

39 Paus. 5,27,5 Der Bezug auf (Artemis) Persike ist eindeutig; allerdings wird in vielen Editionen leider einer sicher falschen Lesart der Vorzug gegeben (ἔστι γὰρ Λυδοῖς ἐπίκλησιν Περσικοῖς ἱερὰ ἔν τε Ἱεροκαισαρείᾳ καλουμένῃ πόλει καὶ ἐν Ὑπαίποις), obwohl die zutreffende und zum übrigen Quellenbefund passende Variante ebenfalls überliefert ist (…ἐπίκλησιν Περσικῆς ἱερὰ…). S. dazu bereits BURESCH 1898, 66 m. Anm. 3; vgl. ROBERT 1976, 29.

40 IK Ephesos 3817a. 3820c.

41 SNG Turkey VII.1, 271; zur Deutung ALTINOLUK 2013, 41.

42 BRIANT 1985 (Zitat S. 180).

43 Strab. 13,4,5 (625C.).

44 Z. B. FOSS 1978, 47–49. Vgl. BENGISU 1996, 2 m. Anm. 4 zu dieser Fehldeutung, die ihren Ursprung in der Beschreibung Strabons hat.

45 BENGISU 1996, hier 13; vgl. DUSINBERRE 2003, 72; ROOSEVELT 2009, 126 und 249, Nr. 17.1.

Mit dem Tmoloszug war jedoch auch der Kult der Kybele verbunden, die im nahen Sardeis ein wichtiges Heiligtum besaß. Ähnliches gilt für die Artemis Sardiane und die Artemis Ephesia. Da sie über ähnliche Wesenszüge wie Anāhitā verfügten, nicht zuletzt an die Funktion als Fruchtbarkeitsgöttin, aber auch an den Wasserbezug ist hier zu denken, führte das zu einer allgemeinen Identifikation der Göttinnen, von der man schon für die achaimenidische Zeit auszugehen hat. Diese Identifikation hat wohl nicht zuletzt Artaxerxes II. befördert. Dieser stellte nämlich nicht nur einen besonderen Bezug zur Göttin in seinen Inschriften heraus,[46] sondern ließ zudem im ganzen Reich Statuen für sie errichten, unter anderem in Sardeis.[47] Da es in der iranischen Tradition keine Vorlage für rundplastische Götterbilder gab,[48] hat man sich wohl jeweils an lokalen Vorbildern orientiert und die Darstellung einer vergleichbaren Göttin übernommen. In Lydien war dies offenbar die Artemis Ephesia, die in dieser Zeit bei den Persern hoch im Kurs stand und einen Perser als Priester hatte.[49] Das würde erklären, warum spätere Darstellungen der Anaitis im westlichen Kleinasien oft vom Typus der Ephesia abgeleitet sind.[50]

IV.

Nach der Perserzeit führten die aus der interkulturellen Begegnung und den historischen Umbrüchen resultierenden Transformationsprozesse zu dem mannigfaltigen, lokal unterschiedlichen Bild der kaiserzeitlichen Anaitis, bei dem zum Teil nur noch der Name als iranisches Substrat zu identifizieren ist. Während in Hierokaisareia und Hypaipa noch einiges mehr vom iranischen Erbe erhalten blieb, andererseits auch deutliche Hellenisierungstendenzen erkennbar sind,[51] ist der Befund in Ostlydien jedoch uneinheitlicher. In überwiegend privaten Inschriften wird die Göttin mal mit Meter, mal mit Artemis verbunden, mal nur als Anaitis adressiert. Darin zeigen sich parallel existierende divergierende Vorstellungen, deren Charakterisierung als ‚Iranisierung‘ lokaler Kulte die zugrundeliegenden Entwicklungen und Prozesse nicht adäquat wiedergibt. Vielmehr handelte es sich in diesem Fall um

46 S. oben Anm. 12.

47 Beross., *FGrHist* 680, F 11 (= Clem. Al. 5,65,3).

48 Die verbreitete Meinung, daß es bei den Persern aber keine anthropomorphen Götterbilder und -vorstellungen gegeben habe, ist allerdings so nicht aufrechtzuerhalten, wie Jacobs 2001 gezeigt hat.

49 Vgl. Xen. *Hell.* 1,2,6: Tissaphernes verteidigt die Artemis Ephesia gegen die Athener. Zum Priester s. Xen. *anab.* 5,3,6: Megabyxos = pers. Bagabuχša („*der Gott dient*“). Nach Benveniste 1966, 108–115 entwickelte sich der Name erst im Hellenismus zu einem bloßen Titel.

50 Z. B. auf den Münzen von Hypaipa (Altinoluk 2013, App. II, Type 1 30. 32. 40 und mehr) oder als Reliefs auf einer Weihung aus Kula (TAM V.1, 331 = de Hoz 1999, Nr. 3.40; vgl. Diakonoff 1979, 141 Nr. 2 mit Fig. 3a–c) bzw. auf einer Ehrung aus Ayazviran (TAM V.1, 449 = de Hoz 1999, Nr. 3.47).

51 Zu denken ist etwa an die Einrichtung von Spielen griechischer Art, vgl. TAM V.2, 1271–1276 = de Hoz 1999, Nr. 3.7–3.12 (Hierokaisareia); IK Ephesos 3809–3813 = de Hoz 1999, Nr. 3.60–3.65 sowie ebd. 3.66 (Hypaipa).

Kulte „*hybrider Natur*“[52], die über Elemente verschiedenen Ursprungs verfügten. Sie sind durch die Identifikation der genannten Gottheiten und einer daraus resultierenden synkretistischen Vermischung einzelner Wesenszüge dieser Gottheiten entstanden. Dieser Prozeß läßt sich nicht im Detail nachverfolgen, da die Zeugnisse überwiegend aus späterer Zeit stammen (2. bzw. 3. Jh. n. Chr.) und – mit Ausnahme der Inschrift aus dem nahegelegenen Silandos – nur das Ergebnis der Entwicklung offenbaren. Dennoch läßt sich die Sachlage gut in das dargestellte Etablierungsmodell einfügen, zumal die landschaftliche Disposition der Katakekaumene den besprochenen Mustern entspricht. Die Region besaß zudem eine offensichtlich große Bedeutung für die Perser, nicht ohne Grund befand sich dort ein Hauptaufmarschgebiet der persischen Armee.[53] Es ist daher von einer dauerhaften Präsenz iranischer Bevölkerungsanteile auszugehen, als stationierte Einheiten, aber wohl auch als Siedler.[54] Daß die iranischen Elemente der späteren Kulte, angefangen beim Namen Anaitis, auf eine Verehrung der Anahita in dieser Gegend in persischer Zeit zurückgehen, darf man somit voraussetzen.

V.

Kommen wir zum Schluss. Der Kult der Anaitis war in Lydien weitverbreitet und ist insbesondere in der römischen Kaiserzeit gut belegt. Das ist insofern bemerkenswert, als es in benachbarten Regionen so gut wie kein vergleichbares Material gibt. Für den Befund in Lydien lässt sich aber eine Erklärung formulieren. Offensichtlich resultiert er aus dem Zusammenspiel verschiedener Faktoren. Eine ganz wesentliche Rolle kommt in dem hier vorgestellten Erklärungsmodell der naturräumlichen Beschaffenheit Lydiens zu. Dessen Landschaftsstruktur besteht aus vielen Bergen und fruchtbaren Flusstälern, deren Hauptströme durch kleinere Flüsse aus den Bergen gespeist werden. Dies passt zum Bild der Anahita nach der iranischen Vorstellung, die nicht zuletzt aus dem Awesta erkennbar ist. Dieser Zusammenhang ist zweifelsohne bereits von den iranischen Siedlern hergestellt worden. Dass insbesondere in den fruchtbaren lydischen Ebenen Kolonien gegründet wurden, ist naheliegend und für einige Gebiete nachzuweisen, für die übrigen Teile zumindest aber zu erschließen. Dies erklärt die Verbreitung der Anaitisverehrung in ganz Lydien. Nach dem Ende der Perserzeit unterlag diese ganz unterschiedlichen Einflüssen und historischen Entwicklungen, aus denen sich das uneinheitliche Bild der Göttin in der Kaiserzeit ergab, so dass keineswegs jedes Zeugnis auf die Fortdauer nennenswerter inhaltlicher Bezüge zur persischen Religion schließen lässt. Auf der anderen Seite verdanken wir aber gerade diesen Entwicklungen, ob es sich nun um eine Hellenisierung oder um eine synkretistische Vermischung mit den bereits vor Ankunft der Perser verehrten Gottheiten Artemis und Meter handelt, dass sich

52 Mitchell 2007, 170.

53 Es handelt sich um die Kastolosebene (Xen. *anab.* 1,1,2; 1,9,7), die nach TAM V.1, 222 dort zu lokalisieren ist.

54 Vgl. Diakonoff 1979, 161 (oben Anm. 34).

überhaupt Elemente der einstigen iranischen Gottheit erhalten haben, und seien sie nur noch im Namen erkennbar.

LITERATUR

S. Altinoluk, Hypaipa. A Lydian City during the Roman Imperial Period, Istanbul 2013.

R. L. Bengisu, Lydian Mount Karios, in: E. N. Lane (Hrsg.), Cybele, Attis and Related Cults. Essays in Memory of M. J. Vermaseren, Leiden 1996, 1–36.

É. Benveniste, Titres et noms propres en iranien ancien, Paris 1966.

M. Boyce, Encyclopaedia Iranica I, 1985, 811–813, s. v. Alborz. ii. Alborz in myth and legend.

M. Boyce/F. Grenet, A History of Zoroastrianism, Bd. III: Zoroastrianism under Macedonian and Roman rule, Leiden 1991.

P. Briant, Les Iraniens d'Asie Mineure après la chute de l'empire achéménide (À propos de l'inscription d'Amyzon), DHA 11, 1985, 167–195.

P. Briant, L'eau du grand roi, in: L. Milano (Hrsg.), Drinking in Ancient Societies. History and Culture of Drinks in the Ancient Near East, Padua 1994, 45–65.

P. Briant, Histoire de l'Empire perse. De Cyrus à Alexandre, Paris 1996.

P. Briant, Bulletin d'Histoire Achéménide II (BHAch II, 1997–2000), Paris 2001.

M. Brosius, Artemis Persike and Artemis Anaitis, in: M. Brosius / A. Kuhrt (Hrsg.), Achaemenid History XI – Studies in Persian History: Essays in Memory of David M. Lewis, Leiden 1998, 227–238.

K. Buresch, Aus Lydien. Epigraphisch-geographische Reisefrüchte, Leipzig 1898.

T. Corsten, Herodot I 131 und die Einführung des Anahita-Kultes in Lydien, IA 26, 1991, 163–180.

M. P. de Hoz, Die lydischen Kulte im Lichte der griechischen Inschriften, Bonn 1999.

A. de Jong, Traditions of the Magi. Zoroastrianism in Greek and Latin Literature, Leiden u. a. 1997.

I. Diakonoff, Artemidi Anaeiti anestesen. The Anaeitis-dedications in the Rijksmuseum van Oudheden at Leyden and related material from eastern Lydia. A Reconsideration, BABesch 54, 1979, 139–188.

E. R. M. Dusinberre, Aspects of Empire in Achaemenid Sardis, Cambridge 2003.

C. Foss, Explorations in Mount Tmolus, CSCA 11, 1978, 21–60.

G. Gnoli, Encyclopaedia Iranica XIII, 2006, 97–100, s. v. Indo-Iranian Religion.

W. Henkelman, The Other Gods Who Are. Studies in Elamite-Iranian Acculturation Based on the Persepolis Fortification Texts (Achaemenid History XIV), Leiden 2008.

B. Jacobs, Kultbilder und Gottesvorstellungen bei den Persern. Zu Herodot, Historiae 1.131 und Clemens Alexandrinus, protrepticus 5.65.3, in: T. Bakır (Hrsg.), Achaemenid Anatolia. Proceedings of the First International Symposium on Anatolia in the Achaemenid Period (Bandirma 15–18 August 1997), Leiden 2001, 83–90.

J. Kellens, Le problème avec Anāhitā, Orientalia Suecana 51/52, 2002/2003, 317–326.

A. Klingenberg, Die ‚Iranische Diaspora‘. Kontext, Charakter und Auswirkung persischer Einwanderung nach Kleinasien, in: E. Olshausen / V. Sauer (Hrsg.), Mobilität in den Kulturen der antiken Mittelmeerwelt. Stuttgarter Kolloquium zur Historischen Geographie des Altertums 11, 2011, Stuttgart 2014, 309–324.

H. Lommel, Die Yäšt‘s des Awesta, Göttingen 1927.

H. Lommel, Anahita–Sarasvati, in: J. Schubert / U. Schneider (Hrsg.), Asiatica. Festschrift Friedrich Weller. Zum 65. Geburtstag gewidmet von seinen Freunden, Kollegen und Schülern, Leipzig 1954, 405–413.

S. Mitchell, Iranian Names and the Presence of Persians in the Religious Sanctuaries of Asia Minor, in: E. Matthews (Hrsg.), Old and New Worlds in Greek Onomastics, Oxford 2007, 151–171.

I. Moga, Strabo on the Persian Artemis and Mên in Pontus and Lydia, in: G. R. Tsetskhladze (Hrsg.), The Black Sea, Paphlagonia, Pontus and Phrygia in Antiquity. Aspects of Archaeology and Ancient History, Oxford 2012, 191–195.

J. Nollé, Zum Kult der Anaïtis Artemis von Hypaipa und zu einigen Patriatraditionen der torrhebischen Kaÿstertal-Stadt, JNG 62, 2012, 127–195.

M. Ricl, The Cult of the Iranian Goddess Anāhitā in Anatolia before and after Alexander, ZAnt 52, 2002, 197–210.

L. Robert, Monnaies grecques de l'époque impériale, RN 18, 1976, 25–56.

C.H. Roosevelt, The Archaeology of Lydia, from Gyges to Alexander, Cambridge 2009.

R. Schmitt, BAPZOXAPA – ein neues Anāhitā-Epitheton aus Kappadokien, ZVS 84, 1970, 207–210.

R. Schmitt, Die altpersischen Inschriften der Achaimeniden. Editio minor mit deutscher Übersetzung, Wiesbaden 2009.

N.V. Sekunda, Achaemenid Colonization in Lydia, REA 87, 1985, 7–30.

J.H.M. Strubbe, „Cursed be he who moves my bones", in: C.A. Faraone/D. Obbink (Hrsg.), Magika Hiera. Ancient Greek Magic and Religion, Oxford/New York 1991, 33–59.

J. Wiesehöfer, „Das Wasser des Königs": Wohltat, paradiesischer Lebensspender und herrscherlicher Genuß, in: U. Hübner / A. Richter (Hrsg.), Wasser – Lebensmittel, Kulturgut, politische Waffe. Historische und zeitgenössische Probleme und Perspektiven in asiatischen und afrikanischen Gesellschaften, Hamburg 2004, 149–164.

R. Zwanziger, Studien zur Nebenüberlieferung iranischer Personennamen in den griechischen Inschriften Kleinasiens. Ein Beitrag zu dem neuen Iranischen Namenbuch, Diss. Wien 1973.

RIVER BOUNDARIES IN THE CLASSICAL TRADITION ON THE GREAT EGYPTIAN KING-CONQUEROR

Ivan Ladynin

Many Greek and Roman authors were strongly convinced that in a remote past the Egyptians waged war with the Scythians; and in the pre- or early Hellenistic time there appeared a belief that they lost it.[1] The first author to have written about it was, certainly, Herodotus: according to his Book Two (Hdt. 2.103) the Egyptian king-warrior Sesostris subdued the Scythians and the Thracians and put in their lands his famous stelae; after that, says Herodotus, Sesostris turned back and reached the river Phasis, where a part of his army settled and gave origin to the people of Colchi.[2] The sequence of events in Herodotus' narration presumes that the Scythians were subdued by Sesostris earlier than the Thracians; thus, the route of the Egyptian king went from Asia through the Caucasus and European Scythia to Thrace.[3] For Herodotus the Phasis was the boundary between Asia and Europe (Hdt. 4.45); at this point he followed the view of the Classical time, which was shared, for instance, by Aeschylos (Frg. 191 Nauck/Radt, *cf.* Pl. Phaed. 109b).[4] Thus, on his incursion into Scythia Sesostris must have also crossed the Phasis, to step outside Asia that he has subdued before (Hdt. 2.102) and to turn thereby of an all-Asiatic ruler into a real world-conqueror.

1 See the special research of this specific motive: Armayor 1980 (an unreasonably sceptical approach towards Herodotus' competence in Egyptian tradition); Ivantchik 1999 (a compendious, nearly monographic treatment of the theme); Hagan 2011.

2 Thus doing he (Sesostris) marched over the country till he had passed over from Asia to Europe and subdued Scythians and Thracians. Thus far and no further, I think, the Egyptian army went; for the pillars can be seen standing in their country but in none beyond it. Thence he turned about and went back homeward; and when he came to the Phasis river, it may be (for I cannot speak with exact knowledge) that King Sesostris divided off some part of his army and left it there to dwell in the country, or it may be that some of his soldiers grew weary of his wanderings, and stayed by the Phasis" (translation after Godley 1920, 391; διεξήιε τὴν ἤπειρον, ἐς ὃ ἐκ τῆς ᾽Ασίης ἐς τὴν Εὐρώπην διαβὰς τούς τε Σκύθας κατεστρέψατο καὶ τοὺς Θρήικας. ᾽Ες τούτους δέ μοι δοκέει καὶ προσώτατα ἀπικέσθαι ὁ αἰγύπτιος στρατός· ἐν μὲν γὰρ τῇ τούτων χώρῃ αίνονται σταθεῖσαι αἱ στῆλαι, τὸ δὲ προσωτέρω τούτων οὐκέτι. ᾽Ενθεῦτεν δὲ ἐπιστρέψας ὀπίσω ἤιε· καὶ ἐπείτε ἐγίνετο ἐπὶ Φάσι ποταμῷ, οὐκ ἔχω τὸ ἐνθεῦτεν ἀτρεκέως εἰπεῖν εἴτε αὐτὸς ὁ βασιλεὺς Σέσωστρις ἀποδασάμενος τῆς ωυτοῦ στρατιῆς μόριον ὅσον δὴ αὐτοῦ κατέλιπε τῆς χώρης οἰκήτορας, εἴτε τῶν τινες στρατιωτέων τῇ πλάνῃ αὐτοὶ ἀχθεσθέντες περὶ Φᾶσιν ποταμὸν κατέμειναν). Cf. Lloyd 1988, 21–22.

3 Armayor 1980, 57–58; West 1992, 120.

4 Ivantchik 1999, 415, 429.

The next author to have written about Sesostris' Scythian war might have been Theopompus of Chios (*FGrHist* 115, F 46 = *Schol. Ap. Rhod.* 4.272–274);[5] but the piece of evidence here is very fragmentary. In the early Hellenistic time this episode was narrated briefly by Megasthenes, whose statement is preserved by Strabo at more (*FGrHist* 715, F 11A = Strab. 15.1.6–8)[6] and by Arrian at less (*FGrHist* 715, F 11b = Arr. *Ind.* 5.4–7)[7] length. Following Strabo's quotation one can say that Megasthenes told about Sesostris campaign from Iberia to Thrace and to Pontus; and the context of this statement, namely the mention of Nabocodrosor's and Tearco/Taharqa's[8] fictitious campaigns to the pillars of Heracles might make an impression that Megasthenes had in mind a roundabout route of Sesostris via the Pillars and Iberia, i. e. Spain, through Europe to Scythia and the Black Sea. This impression would certainly be false: though in Strabo's Book Three Spain was really called Iberia, the country in question here must be Caucasian Iberia which was known to Strabo (Strab. 1.3.21) and, according to him, connected in one place to Colchis through the Phasis valley (id. 11.3.4). Strabo also knew about the alleged Egyptian settlement at Colchis (id. 1.3.21; 11.2.17); thus, in slightly different terms he accounted the same route of Sesostris into Scythia, as Herodotus had done before, and presented Iberia, i. e. the land at the region of the Phasis, as a base for this campaign.

In fact, Megasthenes spoke not only about Sesostris' campaign at Scythia but also about Scythian attack on Egypt; this shows that Megasthenes' evidence belonged to the same "pro-Scythian" tradition ascribing to the Scythians a success over the Egyptian king, as the later evidence by Pompeius Trogus.[9] This evidence

5 "… Sesonchosis, the king of entire Egypt after Oros (Horus), (the son) of Isis and Osiris, having attacked Asia subdued (it) all, as well as the most of Europe. It is more precise about him at Herodotus. And Theopompus calls him in the third (book) Sesostris" (… Σεσόγχωσις Αἰγύπτου πάσης βασιλεὺς μετὰ Ὦρον τὸν Ἴσιδος καὶ Ὀσίριδος παῖδα τὴν μὲν Ἀσίαν ὁρμήσας πᾶσαν κατεστρέψατο, ὁμοίως καὶ τὰ πλεῖστα τῆς Εὐρώπης. ἀκριβέστερον δὲ ἔστι τὰ περὶ αὐτοῦ παρὰ Ἡροδότῳ. Θεόπομπος δὲ ἐν τρίτῳ Σέσωστριν αὐτὸν καλεῖ; scholia according to codex Laurentianus XXXII.9: Keil 1854, 496).

6 "However, Sesostris, the Aegyptian, he adds, and Tearco the Aethiopian advanced as far as Europe; and Nabocodrosor, who enjoyed greater repute among the Chaldaeans than Heracles, led an army even as far as the Pillars. Thus far, he says, also Tearco went ; and Sesostris also led his army from Iberia to Thrace and the Pontus ; and Idanthyrsus the Scythian overran Asia as far as Aegypt" (translation after Jones 1930, 9: καίτοι Σέσωστριν μὲν τὸν Αἰγύπτιον καὶ Τεάρκωνα τὸν Αἰθίοπα ἕως Εὐρώπης προελθεῖν, Ναβοκοδρόσορον δὲ τὸν παρὰ Χαλδαίοις εὐδοκιμήσαντα Ἡρακλέους μᾶλλον καὶ ἕως Στηλῶν ἐλάσαι. μέχρι μὲν δὴ δεῦρο καὶ Τεάρκωνα ἀφικέσθαι, ἐκεῖνον δὲ καὶ ἐκ τῆς Ἰβηρίας εἰς τὴν Θρᾴκην καὶ τὸν Πόντον ἀγαγεῖν τὴν στρατιάν· Ἰδάνθυρσον δὲ τὸν Σκύθην ἐπιδραμεῖν τῆς Ἀσίας μέχρι Αἰγύπτου).

7 "… but on the other hand that Sesostris the Egyptian, after subduing the most part of Asia, and after invading Europe with an army, yet returned back; and Indathyrsis the Scythian who started from Scythia subdued many tribes of Asia, and invaded Egypt victoriously" (translation after Robson 1933, 319; ἀλλὰ Σέσωστριν μὲν τὸν Αἰγύπτιον, τῆς Ἀσίας καταστρεψάμενον τὴν πολλήν, ἔστε ἐπὶ τὴν Εὐρώπην σὺν στρατιῇ ἐλάσαντα ὀπίσω ἀπονοστῆσαι, Ἰδάνθυρσον δὲ τὸν Σκύθεα ἐκ Σκυθίης ὁρμηθέντα πολλὰ μὲν τῆς Ἀσίης ἔθνεα καταστρέψασθαι, ἐπελθεῖν δὲ καὶ τὴν Αἰγυπτίων γῆν κρατέοντα).

8 Radt 2009, 144.

9 Ivantchik 1999, 422, esp. 426.

is preserved in Justin's epitome and in Jordan's *Getica*, which obviously derives its information from Trogus, though the Scythians are replaced there by the Goths. According to Jordan (Jordanus, Getica 6.47), the Gothic king Tanausis defeated the Egyptian king Vesosis in a battle at the river Phasis and pursued him to Egypt (he was prevented from invading it by water barriers and fortifications at the Egyptian border).[10] Justin described the alleged Egypto-Scythian war with much more details, though he did not mention the Phasis as a place of Egyptian defeat (Just. 2.3.8–14); however, at the very start of his *Epitome* (Just. 1.1.6–7) he presented the Egyptian king Vezosis and his Scythian counterpart Tanaus as the rulers of the remote past who waged war in the search of the glory and not territorial benefits; in this context he said that the aim of Vezosis campaign was Pontus.[11] Noteworthy is the analogy with Megasthenes' statement in the mention of Pontus, which proves that it really belonged to the same stem of tradition as the later evidence of Trogus; however, the most important point is that the Phasis was presented by Trogus as the ultimate point of the Egyptian expansion, where the end was put to it by the Scythians (the *Epitome* made it clear that the Scythians attacked the Egyptians before they had time to invade their country). Thus, Megasthenes and Trogus retained, though in a modified way, Herodotus' assumption that the Phasis and the land near it had once served a boundary between the Egyptian domain at Asia and Scythia, which Egyptians invaded or tried to invade.

This boundary is presented otherwise in the *Library of History* by Diodorus, who most probably took his evidence on the point, like on other themes of his Book One, from the treatise on Egypt by Hecataeus of Abdera dating to the late 4th century B.C.[12] Unlike Herodotus, Hecataeus and Diodorus shared the notion that

10 "This was the region where the Goths dwelt when Vesosis, king of the Egyptians, made war upon them. Their king at that time was Tanausis. In a battle at the river Phasis (whence come the birds called pheasants, which are found in abundance at the banquets of the great all over the world) Tanausis, king of the Goths, met Vesosis, king of the Egyptians, and there inflicted a severe defeat upon him, pursuing him even to Egypt. Had he not been restrained by the waters of the impassable Nile and the fortifications which Vesosis had long ago ordered to be made against the raids of the Ethiopians, he would have slain him in his own land" (translation after Mierow 1915, 63; *Hic ergo Gothis morantibus Vesosis, Aegyptiorum rex, in bellum inruit, quibus tunc Tanausis rex erat. Quod proelio ad Phasim fluvium, a quo Fasides aves exortae in totum mundum epulis potentum exuberant, Thanausis Gothorum rex Vesosi Aegyptiorum occurrit, eumque graviter debellans in Aegypto usque persecutus est, et nisi Nili amnis intransmeabilis obstetissent fluenta vel munitiones, quas dudum sibi ob incursiones Aethiopum Vesosis fieri praecepisset, ibi in eius eum patria extinxisset*).

11 "Vezosis, Egyptian king, and Tanaus, king of Scythia, were indeed prior to him in time, the former of whom advanced into Pontus and the latter as far as Egypt; but these princes waged distant, not neighbouring wars, not in struggles with their neighbours and, satisfied with victory, were seeking not dominion for themselves but glory for their peoples" (*Fuere quidem temporibus antiquiores Vezosis Aegyptius et Scythiae rex Tanaus, quorum alter in Pontum, alter usque Aegyptum excessit; sed longinqua, non finitima gerebant bella, contentique uictoria non imperium sibi, sed populis suis gloriam quaerebant*).

12 Schwartz 1885, 1905; Jacoby 1912, 2750–2769; Murray 1970, 144–150; Burstein 1992, 45–49; Dillery 1998, 256 n. 4; Stephens 2003, 32, n. 35; Gozzoli 2006, 193–194; an alternative view: Burton 1972; its fair criticism: Lloyd 1974, 287–289; Murray 1975, 214–215.

the border of Europe and Asia was not the Phasis but the Tanais and Maeotis Lake; this notion coming back to Hecataeus of Miletus (*FGrHist* 1, F 208, 211–215) was common in the Hellenistic time.[13] Accordingly, in Diodorus' narration about the wars and the founding activities of the Egyptian king Sesoosis[14] the river Tanais was presented as a limit of his advance at Scythia (Diod. 1.55.4).[15] Sesoosis did not enter in its course European Scythia, which in fact did not exist at that time at all: according to Diodorus (Diod. 2.43.2), the Scythians resided early between the Tanais and the Caucausus[16] and only in due course, well after the time of Sesoosis, penetrated Europe as far as Thrace (id. 4).[17] Thus, Diodorus "tore away" Sesoosis' campaign at Thrace from the invasion into Scythia: the Egyptian king was said to have returned from Scythia, completed his conquest of Asia and penetrated Europe via the Cyclades (Diod. 1.55.6).[18] Another peculiarity of this evidence is concerned with the place where the Egyptian warriors were thought to have resided after the Scythian campaign: for Diodorus it was not Colchis but the region of Maeotis, in fact, contrary to everything known in the Greek tradition about the location of the Colchi.[19] Thus, accepting the Tanais as the limit of Sesoosis' advance at Scythia and the western border of the Scythian settlement at that time led Hecataeus of Abdera

13 Ivantchik 1999, 415, 429.

14 Burton 1972, 163–159.

15 "For he even passed over the river Ganges and visited all of India as far as the ocean, as well as the tribes of the Scythians as far as the river Tanaïs, which divides Europe from Asia; and it was at this time, they say, that some of the Egyptians, having been left behind near the Lake Maeotis, founded the nation of the Colchi" (translation after Oldfather 1933, 193: καὶ γὰρ τὸν Γάγγην ποταμὸν διέβη καὶ τὴν Ἰνδικὴν ἐπῆλθε πᾶσαν ἕως ὠκεανοῦ καὶ τὰ τῶν Σκυθῶν ἔθνη μέχρι Τανάιδος ποταμοῦ τοῦ διορίζοντος τὴν Εὐρώπην ἀπὸ τῆς Ἀσίας· ὅτε δή φασι τῶν Αἰγυπτίων τινὰς καταλειφθέντας περὶ τὴν Μαιῶτιν λίμνην συστήσασθαι τὸ τῶν Κόλχων ἔθνος).

16 "At first, then, they dwelt on the Araxes river, altogether few in number and despised because of their lack of renown; but since one of their early kings was warlike and of unusual skill as a general they acquired territory, in the mountains as far as the Caucasus, and in the steppes along the ocean and Lake Maeotis and the rest of that country as far as the Tanaïs river" (translation after Oldfather 1935, 27; τὸ μὲν οὖν πρῶτον παρὰ τὸν Ἀράξην ποταμὸν ὀλίγοι κατῴκουν παντελῶς καὶ διὰ τὴν ἀδοξίαν καταφρονούμενοι· ἕνα δὲ τῶν ἀρχαίων ἔχοντες βασιλέα φιλοπόλεμον καὶ διαφέροντα στρατηγίᾳ προσεκτήσαντο χώραν, τῆς μὲν ὀρεινῆς ἕως πρὸς τὸν Καύκασον, τῆς δὲ πεδινῆς τὰ παρὰ τὸν ὠκεανὸν καὶ τὴν Μαιῶτιν λίμνην καὶ τὴν ἄλλην χώραν ἕως Τανάιδος ποταμοῦ).

17 "But some time later the descendants of these kings, because of their unusual valour and skill as generals, subdued much of the territory beyond the Tanaïs river as far as Thrace, and advancing with their armies to the other side they extended their power as far as the Nile in Egypt"(translation after Oldfather 1935, 29; μετὰ δέ τινας χρόνους τοὺς ἀπογόνους τούτων τῶν βασιλέων ἀνδρείᾳ καὶ στρατηγίᾳ διενεγκόντας πολλὴν μὲν πέραν τοῦ Τανάιδος ποταμοῦ χώραν καταστρέψασθαι μέχρι τῆς Θρᾴκης, ἐπὶ δὲ θάτερα μέρη στρατεύσαντας διατεῖναι τῇ δυνάμει μέχρι τοῦ κατ' Αἴγυπτον Νείλου).

18 "In the same way he brought all the rest of Asia into subjection as well as most of the Cyclades islands. And after he had crossed into Europe and was on his way through the whole length of Thrace he nearly lost his army through lack of food and the difficult nature of the land" (translation after Oldfather 1933, 193; ὁμοίως δὲ καὶ τὴν λοιπὴν Ἀσίαν ἅπασαν ὑπήκοον ἐποιήσατο καὶ τῶν Κυκλάδων νήσων τὰς πλείους. διαβὰς δ' εἰς τὴν Εὐρώπην καὶ διεξιὼν ἅπασαν τὴν Θρᾴκην ἐκινδύνευσεν ἀποβαλεῖν τὴν δύναμιν διὰ σπάνιν τροφῆς καὶ τόπων δυσχωρίας).

19 Ivantchik 1999, 416.

and Diodorus to two notable modifications of the initial story known to Herodotus; and these modifications seem to have no parallels in other evidence on this theme.

The classification of sources presented above is in fact not an accomplishment of this paper; it has been largely sketched before by Askold Ivanchik in his compendious study of the fictitious tradition on the war between Egyptians and Scythians (see note 1 *et sq.*). Nevertheless, that study seems to have lacked an explanation of a point, which seems important: why did this tradition necessarily include a notion of some important water limit crossed or approached by the Egyptian king? Incidentally, the nature of this limit is not precisely defined in all parts of the tradition: in Herodotus' narration the Phasis is just a place of alleged Egyptian settlement, and Trogus' evidence allows understanding that it is also a sort of reasonable limit for Egyptian expansion. The evidence that defines it as such quite explicitly is the statement of Hecataeus of Abdera reproduced by Diodorus. Ivanchik did not seek to this detail any explanation other than the fact that the Phasis and the Tanais were both considered by respective authors the borderlines between Asia and Europe. This is undoubtedly true; however, this means that the tradition discussed had some reason to stress that the Egyptian king of the past established his power rather successfully over the entire Asia and then stepped out of its borders and either also had some success or was defeated from the very start. As it has already been said in connection to Herodotus' evidence crossing the boundary of Asia was certainly an effort that would have made the Egyptian king not only master of this continent but the world-conqueror. However, is ascribing to the Egyptian king the dominion over Asia a mere conventionality derived from purely geographical notions?

An important point of Trogus' evidence found at the start of the *Epitome* is the juxtaposition of the Egyptian and Scythian kings to the eventual rulers of Asia, whose line was started with the Assyrian Ninus: while Ninus and his successors were striving for the territorial dominion, the Egyptian Vezosis and the Scythian Tanaus sought only glory. Nevertheless, according to Trogus, both of them did establish their power over Asia for a period of time; and during it the Scythians collected impost (though moderate) from the subdued peoples (Just. 2.3.13). One would not know for sure if the insistence on the moderation and the pursuit of glory of the Egyptian and the Scythian king came from Trogus himself or from his source, but this does not change the pattern much: anyway, Trogus' tradition put the Egyptian and the Scythian king at the very start of the long sequence of kings and peoples that passed from one another what is regularly labeled in the Classical tradition the "kingdom of Asia" (βασιλεία τῆς Ἀσίας).[20] This idea is made even more explicit in

20 The notion that by the gods' will a single power should exist over Asia at a time is neatly formulated by Aeschylos (Aesch. *Pers*. 762–764); according to Herodotus, it initially belonged to the Assyrians and then was taken over successively by the Scythians, the Medians and the Persians (Hdt.. I 95, 103–104, 130; see the main text of the present article below). Ctesias of Cnidus delineated a largely fictitious though consequent scheme that showed the Medians to have acquired their own domination immediately from Assyria and luckily omitted the contemporary prominence of the Neo-Babylonian kingdom (presenting it as a Median vassal; *FGrHist* 688, F 1b = Diod. 2.24–28; cf. *FGrHist* 90, F 3 = Nic. Dam. *Exc. de insidiis*, p. 4.23 de Boor; *cf.* Llevellin-Jones, Robson 2010); symptomatically, Ctesias' scheme cannot be said belonging exclusively to the Greek thought for it certainly translates the ideological message

Hecataeus/Diodorus' account: here the Egyptian king Sesoosis is said to have imposed regular booty on the submitted peoples "according to their ability" and to have brought captives to the forced labour into Egypt (Diod. 1.55.10),[21] thus forming the practices that became current in the functioning of the historical Asiatic "world empires" of the First Millennium B. C.[22]

Thus, the traditions of Hecateaus and Trogus ascribe in a way the very creation of the historical "kingdom of Asia" to the Egyptian king; in this case it would be certainly natural to believe that this king must have possessed Asia in its geographical extent. However, this explanation would not work for the case of Herodotus: in his scheme the "dominion over Asia" was firstly shared by the two lines of kings – those of Phrygia and Lydia, who possessed Lower Asia (Hdt. 1.7), and those of Assyria, who possessed Upper Asia (id. 95)[23]; the entire Asia was seized for the first time by the Scythians (id. 103–106)[24] and then passed to the Medians and Persians. There is no place in this scheme for Sesostris and, though he is said to have surpassed Darius I with his feats at Scythia (id. 2.110), there are no other comparisons of his realm with the historical "kingdom of Asia". Thus, Herodotus' ascribing to Sesostris the dominion over the entire geographical Asia must have other motives than considering him the creator of this structure.

It is legitimate to look for these motives in Herodotus' sources. Askold Ivanchik, whose research has already been mentioned, thought it as vain to see in the Sesostris story a historical source as to address the Egyptian tale of the King Cheops or the Arthur legend for studying the history of Old Kingdom Egypt or early medieval Britain.[25] This comparison is not ill-taken because it goes, in fact, further than its author intended. No one will deny of sober mind that such stories are the works of specific fiction; thus, they can hardly be the sources of the verifiable historical information that is not available otherwise. This does not eliminate the fact that their pegs are real events; and the purpose of these stories was not merely an amusement or a moralizing but also (and, one might say with a deal of reason, in the first place) forming the picture of the past for their audience. If so, then these

of his Persian informers. At the start of Hellenism Alexander the Great arrogated the succession of the "kingship of Asia" to himself (Fredricksmeyer 2000), hence to the Macedonians, which was retained in the eventual tradition of naming the Seleucid kingdom (as well as, wider, the entire system of the Hellenistic states at the Orient dominated by the Macedonians and the Greeks; so in the Book of Daniel) "the Macedonian realm" (Edson 1958).

21 "He dealt gently with all conquered peoples and, after concluding his campaign in nine years, commanded the nations to bring presents each year to Egypt according to their ability, while he himself, assembling a multitude of captives which has never been surpassed and a mass of other booty, returned to his country, having accomplished the greatest deeds of any king of Egypt to his day" (translation after Oldfather 1933, 193; ἐπιεικῶς δὲ προσενεχθεὶς ἅπασι τοῖς ὑποτεταγμένοις καὶ συντελέσας τὴν στρατείαν ἐν ἔτεσιν ἐννέα, τοῖς μὲν ἔθνεσι κατὰ δύναμιν προσέταξε δωροφορεῖν κατ' ἐνιαυτὸν εἰς Αἴγυπτον, αὐτὸς δ' ἀθροίσας αἰχμαλώτων τε καὶ τῶν ἄλλων λαφύρων πλῆθος ἀνυπέρβλητον ἐπανῆλθεν εἰς τὴν πατρίδα, μεγίστας πράξεις τῶν πρὸ αὐτοῦ κατειργασμένος).

22 E. g. Oded 1979; Briant 1996, 76–82 (on the tribute system in the Achaemenian Empire).

23 Ivantchik 1993, 109–110.

24 Asheri, Lloyd, Corcella 2007, 153–156.

25 Ivantchik 1999, 399.

stories, unhistorical, as they are, in our (and largely even Classical Greek) view, must possess a certain internal logic that will be discernible in comparing them to the historical realities of the peoples that created them.

By now the idea that the plots of Herodotus' Book Two were the inventions of its author or anyway originate from the Greek tradition[26] is generally abandoned: Alan Lloyd did a great job to show that Herodotus' statements have authentic Egyptian background,[27] and this point has been emphatically repeated in a number of more recent works.[28] The important question for the present theme is, certainly, if there is Egyptian background behind the story of Sesostris. Perhaps the most illuminating base for the positive answer to it is the famous topos of the Sesostris' stelae put by the king in the places of his triumphs, which beared the descriptions of these triumphs if the enemies were valiant and had such descriptions added with the depictions of feminine genitals if they were cowards (Hdt. 2.110).[29] It has been long ago accepted by the Egyptologists that the name Sesostris is likely to come back to the Egyptian name "Senwosret" (*Z-n-wsrt*) and that the campaigns of the historical Senwosret III is the most probable peg for the entire story of Sesostris.[30] Symptomatically, the boundary stelae of Senwosret III at Nubia from Semne (Berlin 1157) and from the island of Uronarti (Khartoum 3) describing his victories over the Kushites show a very tell-tale passage with a sort of "military moralizing" on the theme of an encounter with a cowardly enemy.[31] The most significant is the sentence *ḥm pw m3ˁ rwd ḥr t3š.f* (decently translated by W.K. Simpson as "One who is driven from his border is a real coward"): here the word *ḥm* is a pejorative[32] (in fact, an obscenity), which is written with the hieroglyphic sign N_{41} having in

26 *E.g.* Heidel 1987; Armayor 1978; 1980; 1985.

27 Lloyd 1975; 1976; 1988.

28 Gozzoli 2006, 155–190; Moyer 2011, 42–83.

29 "When those that he met were valiant men and strove hard for freedom he set up pillars in their land where on the inscription showed his own name and his country's, and how he had overcome them with his own power; but when the cities had made no resistance and been easily taken, then he put an inscription on the pillars even as he had done where the nations were brave; but he drew also on them the privy parts of a woman, wishing to show clearly that the people were cowardly" (translation after Godley 1920, 391; ῾Οτέοισι μέν νυν αὐτῶν ἀλκίμοισι ἐνετύγχανε καὶ δεινῶς μαχομένοισι περὶ τῆς ἐλευθερίης, τούτοισι μὲν στήλας ἐνίστη ἐς τὰς χώρας διὰ γραμμάτων λεγούσας τό τε ἑωυτοῦ οὔνομα καὶ τῆς πάτρης καὶ ὡς δυνάμι τῇ ἑωυτοῦ κατεστρέψατό σφεας·ὅτεων δὲ ἀμαχητὶ καὶ εὐπετέως παρέλαβε τὰς πόλις, τούτοισι δὲ ἐνέγραφε ἐν τῇσι στήλῃσι κατὰ ταὐτὰ καὶ τοῖσι ἀνδρηίοισι τῶν ἐθνέων γενομένοισι καὶ δὴ καὶ αἰδοῖα γυναικὸς προσενέγραφε, δῆλα βουλόμενος ποιέειν ὡς εἴησαν ἀνάλκιδες).

30 See a handy and still not outdated synopsis of bibliography in Ivantchik 1999, 297 (n. 6).

31 "As for desisting after attacking, it is strengthening the heart of the enemy. Aggression is bravery, and to retreat is vile. One who is driven from his border is a real coward. For the Nubian listens only to fall at a word. It is responding to him which makes him retreat; but if one is aggressive against him, he shows his back" (translation after Simpson 2003, 338; stela from Semne, ll. 9–12: Sethe 1928, 83–84; Barta 1974, pl. 1; stela from Uronarti, ll. 6–10: Janssen 1953, 52–54; *…ḏr-ntt ir gr m-ḫt pḥ* sshm-*ib pw n ḫrw qnt pw 3d ẖzt pw ḥm-ḫt ḥm pw m3ˁ rwd ḥr t3š.f ḏr-ntt sḏm n3ḥs r ḫr n r in wšb.f dd ḥm.f 3dt r.f dd.f s3.f*). See also Obsomer 1989, 181–183.

32 Erman, Grapow 1926–50, III, 80 (7).

it the semantic of feminine genitals;[33] together with the determinative D_{52} (the depiction of phallus)[34] the word means something like "effeminate man".[35] Perhaps, the meaning of the word is still reinforced with its homonym *ḥm* "retreating"[36] that appears in the same passage.

Actually the pathos of the passage is directed against not so much the enemy of Egypt, as the alleged hapless Egyptian general who fails to fight Nubians. Nevertheless its coincidence with Herodotus' statement both in meaning and form is so complete, that it can hardly be random. Probably, the propagandist texts of Senwosret III on his Nubian monuments were circulated non-epigraphically inside Egypt and formed the image of this king-warrior that remained familiar with the Egyptians literally for ages; if so, Herodotus' story of Sesostris certainly had a degree of Egyptian inspiration. However, a thing regularly forgotten in connection with this story is that Senwosret III was not the only real Egyptian king who marked his boundaries with stelae: the Egyptian royal texts make it clear that it was an established and well-propagated practice, to say the least, under Thuthmosis I and III and not only at Nubia but also at Asia, where the stelae of these kings were put on the banks of the Euphrates[37]. This practice was not only similar but, probably, imitative of what had once been done by Senwosret III at his Nubian border; and this similarity must have been a factor that led to merging in later tradition the two stages of the Egyptian expansion (those of the Middle and the New Kingdom) into one and, indeed, the only.[38] These contaminated reminiscences must have been associated with an earlier and a better known figure of a king-conqueror, i. e. with Senwosret III; and thus Herodotus and a number of other authors retain for this king the name-forms derived from the name of Senwosret. However, Herodotus placed him in his relative chronology two generations before the Troyan War[39] (*cf.*

33 Gardiner 1957, 492.

34 Gardiner 1957, 456.

35 See on a similar word (*ḥmty*) used in the *Stela of Piankhi*, l. 15, Grimal 1981, 33 (comm. 75); *cf.* the same text, ll. 157–158. The word was noticed by A. Blasius, who, however, was wrong in its exact transliteration (*ḥmty*, like in the *Stela of Piankhi*) and in indicating the relevant vocabulary gloss (Erman, Grapow 1929, III, 80(8) instead of (7); Blasius 2003, 331).

36 Erman, Grapow 1926–50, III, 79.

37 The *Annals* of Thuthmosis III, Year 33: "… to] the east of this water (the Euphrates), after he had set up another also beside the stela of his father, the King of Upper and Lower Egypt, Okheperkare (Thuthmosis I)" (translation after Redford 2003, 74; Sethe 1907, 697 (3–5): *ḏ3t Pẖr-wr n Nhryn3 in ḥm.f r-ḥ3t mšʿ.f r i3bty mw pn śmn.n.f ky r-gs wḏ n it.f nsw-bity ʿ3-ḫpr-k3-Rʿ*); the Stela at Gebel-Barkal of Thuthmosis III: "Thereupon My Majesty set up my stela on that Mountain of Naharin, a block quarried from the mountain, on the west side of the Great Bender (the Euphrates)" (translation after Redford 2003, 107; Helck 1955, 1232 (11): *ʿḥ3.n smn.n ḥm.f wḏ ḥr ḏw pf n Nhrn m šd m ḏw ḥr gs imnty Pẖr-wr*); the Stela of Armant of Thuthmosis III: "he… set up his triumph stela upon its (the Euphrates') bank" (translation after Redford 2003, 155; Helck 1955, 1246 (2): *smn.f wḏ n nḫtw ḥr gs.f*); "he set up his stela there (at Nubia) as he had done at the ends [of Asia]" (translation after Redford 2003, 155; Helck 1955, 1246 (6): *smn.n.f wḏ.f im mi irt.n.f m pḥwy…*). *Cf.* Galàn 1995, 148–150.

38 So the wars of Senwosret III/Sesostris appear in the famous treatise by Manetho of Sebennytos: Manetho, frgg. 34–36 Waddell.

39 *E. g.* Stubbings 1975, 342–350.

Hdt. 2.111–120), i.e. where in real history Ramesses II would belong; and this is a doubtless tribute to the New Kingdom component in this fictitional image.

Coming back to the presentation of Sesostris as a conqueror of entire Asia, one should say that the Ancient Egyptians of the New Kingdom had in fact a notion of a territorial entity vaster than their own country and encompassing their domain in Asia. One of the major cosmological notions of the Egyptians was the term "land" (*tꜣ*):[40] it denoted the territory fit for agriculture and, respectively, for the production of the sacrificial food of gods, *i. e.*, especially, the Nile Valley, and normally it would not be applied to countries outside Egypt. At the start of Egyptian history these countries were juxtaposed to the Nile Valley; and the fact that the Egyptian idea of normal life was incompatible with them was stressed in their standard designation *ḫꜣswt* ("hill-lands, deserts").[41] There are, however, the occurrences when the term *tꜣ* is applied to the stretch of lands outside the Nile Valley forming Egyptian possessions; and in this respect the texts of Dynasty XVIII are illustrative again: Thuthmosis I said in the Tombos Stela that "he brought together the limits of the land (and) united its two ends with his mighty arm" (*ỉn.n(.f) ḏrw tꜣ ḥr ndb.f ḫnd(.n.f) pḥwy.fy m ḫpš.f nḫt*);[42] and Thuthmosis III started his *Annals* with a depiction of the mutiny "starting at Iursa and ending at the ends of land" (*šꜣꜥ-m Yrḏꜣ nfryt-ir pḥw-tꜣ*).[43] The latter phrase calls for a better look for it makes certain things very clear. Initially the words *pḥw-tꜣ* ("marshlands of the land") denoted the watery northern part of the Nile Delta, which served a northern boundary of Egypt properly spoken.[44] However, Iursa is a town at Southern Palestine; and a mutiny at a small territory between it and Egypt is certainly unfit for a rather vast description found in the phrase of the *Annals*. In this case the designation *pḥw-tꜣ* loses its meaning of denoting a watery region and, largely due to its homonymy with the words "the (northern) ends of land", must have been transferred to a much more distant frontier of Egyptian domain at Asia (probably, at Syria).[45] Its use in this meaning in the *Annals* of Thuthmose III is by far not unique; and, significantly, it attests not only the denotation of the Egyptian domain as "land" but also a rather neat notion of its frontier at the North.

What lay behind that border? For the time of the New Kingdom the idea of a very distant land outside the Egyptian domain at Asia and to the North of it was formed by Egypt's relations with Mitanni, its major enemy in the 15th century B. C.[46] Its Egyptian denotation "Naharina" could be applied to the possessions of Egypt in Northern Syria seized from Mitanni but basically it denoted the kernel land of this state on the left bank of the Euphrates.[47] The presentation of the Euphrates as the frontier of the Egyptian domain and its border to Mitanni was not too stressed;

40 Erman, Grapow 1926–50, V, 212.
41 Erman, Grapow 1926–50, III, 234.
42 Sethe 1906, 85 (7–8).
43 Sethe 1907, 648 (6).
44 Erman, Grapow 1926–50, I, 539 (2).
45 Erman, Grapow 1926–50, I, 539 (notice on the top of the page). The author is grateful for the discussion of this and some other points concerning the New Egyptian expansion at Asia to Dr. Alexander Nemirovsky (Institute for General History, Russian Academy of Sciences).
46 *E. g.* Bryan 2000.
47 Gardiner 1947, 1, 136*, 171*–180*; Helck 1971, 117–118, 277–278, 285; Redford 2003, 16.

nevertheless, it came prominent in some texts of Dynasty XVIII[48], including those of Thuthmosis III, which depicted the erection of the boundary stelae namely at the bank of the Euphrates by this king and by his predecessor Thuthmosis III (see note 37). However, an especially significant image of Naharina is given in the New Egyptian *Tale of Doomed Prince*: its plot develops undoubtedly in the kernel land of Mitanni behind the Euphrates, and it is shown not just as a foreign land different from Egypt but as a fairytale world ruled with totally strange laws.[49] Thus, the New Egyptian picture of the world consisted, roughly speaking, of the "Land" (including Egypt itself and its domain, both at Asia and at Nubia); the frontier region of the notorious "ends of the Land", its configuration defined at certain times with the flow of the Euphrates; and the lands behind the frontier, which are thought somehow opposite to the "Land" as far as their basic laws are concerned. Notably, the terms connected with this picture are attested well after the New Kingdom: not only name of Naharina is found in the royal texts of the mid-1st Millennium B. C., when there has been no Mitanni for centuries already;[50] but the pre- or early Hellenistic text of the so-called *Bentresh Stela* still knows of Ramesses II campaigning at Naharina (which the historical Ramesses has not done, in fact) and accepting there a tribute from the rulers of the "ends".[51]

What has been said above allows an explanation for the notion of Herodotus and later authors on the dominion of the great Egyptian king-conqueror over the entire Asia up to the Phasis or the Tanais. "Asia" in the view of these authors, or better to say of their Egyptian informers, must have been more or less the same as the "Land" of the New Egyptian texts, i. e. a huge domain of the lands to the North-East of Egypt gathered under the sway of its ruler. From the start of the Hellenistic time it was stressed that this domain had gradually developed into what was known to the Greeks as "kingdom of Asia"; but Herodotus have not done that stress yet and conveyed the original Egyptian notion more adequately. The basic difference between the lands of the Egyptian domain and outside it is articulated in this fictitious scheme in two ways: these two entities are juxtaposed to one another as different parts of the world; and the population on the latter of them is defined as the Scythians. There is a good reason in saying that such definition was backed by the Egyptian encounter with Scythians under Psammetichus I (Hdt. 1.105.1)[52] as well as with an allusion to Darius' campaign at Scythia, to which the deeds of the Egyp-

48 See the Tombos Stela of Thuthmosis I: "his (Thuthmosis's) southern border to the south of this land (Nubia), his northern (border) to that turned water flowing down to the south" (SETHE 1906, 85 (13–14): *tꜣš.f rsy r ḫntyw tꜣ pn mḥy r mw pf qd ḫdd m ḫnty*).

49 Publication: GARDINER 1932, 1–9; *cf.* HELCK 1987; LOPRIENO 2003, 44.

50 *E. g.* STOCKFISCH 2006, 264, 268.

51 "When His Majesty was in Naharina according to his annual custom, the princes of every foreign land came bowing in peace to the might of His Majesty, from as far as the marshlands/ends" (follows the description of their gifts; KITCHEN 1979, 285 (1–2): *is ḥm.f m Nhrn mi nt-ꜥ.f ṯnw rnpt wrw n ḫꜣst nb iw m ksw m ḥtpw n bꜣw n ḥm.f šꜣꜥ pḥww*).

52 The historicity of the Scythian march in the Eastern Mediterranean up to the Egyptian boundary according to the Classical tradition is a polemic issue; for the time being it is sufficient to indicate two views of an Egyptologist (SPALINGER 1979) and a non-Egyptologist (IVANTCHIK 2005, 229–241) accepting this evidence.

tian conqueror were compared in Herodotus' and Hecataeus/Diodorus' narratives (Hdt. 2.110.2–3; Diod. 1.58.4);[53] still the Scythians were for Greeks a peculiar and unique people, somewhat like the residents of "Naharina" for the Egyptians of the New Kingdom. As for the river boundaries by the Phasis or the Tanais in this scheme, their delineation certainly depended on the idea that the Egyptian king of the past got hold of the entire Asia; however, the very fact that these boundaries were defined by a river might in a long run come back to the reminiscences of the Euphrates as the border of the Egyptian domain at some period of the New Kingdom.

ABSTRACT

The Classical tradition contains a number of statements about the dominion of the great Egyptian king-conqueror over the entire Asia up to the Phasis or the Tanais. The article presents their interpretation in the light of the Egyptian notions of the New Kingdom: at that time the Egyptian possessions at Asia (*tꜣ*, "Land", in the phraseology of some hieroglyphic texts) comprised Eastern Mediterranean, and its utmost limit (marked, symptomatically, with the triumphal stela alluded to in the famous Classical topos of Sesostris' stelae) was the Euphrates; the land on the other side of the Euphrates, Mitanni, or Naharina, was considered a part of a totally alien world. The boundary between the Egyptian possessions and the unconquered Scythian world marked with the waterways, which also divided Asia and Europe in the Classical geography, might be an allusion to the New Egyptian notion, certainly, void of its initial specifics.

BIBLIOGRAPHY

Armayor 1978: O. K. Armayor, Did Herodotus Ever Go to Egypt?, in: JARCE 15, 59–73.

Armayor 1980: O. K. Armayor, Sesostris and Herodotus Autopsy of Thrace, Colchis, Inland Asia Minor, and the Levant, in: HSPh 84, 51–74.

Armayor 1985: O. K. Armayor, Herodotus'Autopsy of the Fayoum: Lake Moeris and the Labyrinth of Egypt, Amsterdam.

Asheri, Lloyd, Corcella 2007: D. Asheri, A. Lloyd, A. Corcella, A Commentary on Herodotus' Books I–IV, Oxford.

Barta 1974: W. Barta, Das Terminus twt auf den Grenzstelen Sesostris' III in Nubien, in: Festschrift zum 150jährigen Bestehen des Berliner Ägyptisches Museums (Staatlichen Museen zu Berlin, Mitteilungen aus der Ägyptischen Sammlung VIII, Berlin, 51–54, pl. 1–2.

Blasius 2003: A. Blasius, Das Königtum der Ramessiden im Spiegel der griechisch-römischen Überlieferung: Geschichte im Spannungsfeld von Tradition und Konstruktion, in: R. Gundlach, U. Rössler-Köhler (eds.), Das Königtum der Ramessidenzeit: Voraussetzungen – Verwircklichungen – Vermächtnis. Akten der 3. Symposiums zur ägyptischen Königsideologie in Bonn 07.–10. Juni 2001, Ägypten und Altes Testament 31.4, Wiesbaden, 306–348.

Briant 1996: P. Briant, Histoire de l'empire perse. De Cyrus à Alexandre, Paris.

53 Ivantchik 1999, 406–407.

BRYAN 2000: B. BRYAN, The Egyptian Perspective on Mittani, in: R. COHEN, R. WESTBROOK (eds.), Amarna Diplomacy – the Beginnings of International Relations, Baltimore, 71–84.

BURSTEIN 1992: S. BURSTEIN, Hecataeus of Abdera's History of Egypt, in: J. H. JOHNSON (ed.), Life in a Multi-Cultural Society: Egypt from Cambyses to Constantine and Beyond, Studies in the Ancient Oriental Civilisations 51, Chicago, 45–49.

BURTON 1972: A. BURTON, Diodorus Siculus: Book I, Études preliminaires des religions orientales dans l'empire romain 29, Leiden.

DILLERY 1998: J. DILLERY, Hecataeus of Abdera: Hyperboraeans, Egypt and interpretatio Graeca, in: Historia 47, 255–275.

EDSON 1958: CH. EDSON, Imperium Macedonicum: The Seleucid Empire and the Literary Evidence, in: CPh 53, 153–170.

ERMAN, GRAPOW 1926–50: A. ERMAN, H. GRAPOW, Wörterbuch der ägyptischen Sprache I–VI, Leipzig; Berlin.

FREDRICKSMEYER 2000: E. FREDRICKSMEYER, Alexander the Great and the Kingship of Asia, in: A. B. BOSWORTH, E. J. BAYNHAM (eds.), Alexander the Great in Fact and Fiction, Oxford 2000, 136–166.

GALÀN 1995: J. M. GALÀN, Victory and Border: Terminology Related to Egyptian Imperialism in the XVIIIth Dynasty, Hildesheimer ägyptologische Beiträge 40, Hildesheim.

GARDINER 1932: A. H. GARDINER, Late Egyptian Stories, Bibliotheca Aegyptiaca 1, Bruxelles.

GARDINER 1947: A. H. GARDINER, Ancient Egyptian Onomastica, I–II, Oxford.

GARDINER 1957: A. H. GARDINER, Egyptian Grammar, Oxford[3].

GODLEY 1920: A. D. GODLEY (transl.), Herodotus. The Persian Wars, Volume I: Books 1.2, Loeb Classical Library 117, Cambridge, MA.

GOZZOLI 2006: R. B. GOZZOLI, The Writing of History in Ancient Egypt during the First Millenium B. C. (ca. 1070–180 BC). Trends and Perspectives, Golden House Publications Egyptology 5, London

HAGAN 2011: R. J. HAGAN, The Old and the Restless: The Egyptians and the Scythians in Herodotus' Histories, in: Senior Projects Spring 2011. Paper 10 (http://digitalcommons.bard. edu/senproj_s2011/10).

HEIDEL 1987: W. A. HEIDEL, Hecataeus and the Egyptian Priests in Herodotus, Book II, New York; London

HELCK 1955: W. HELCK, Urkunden der 18. Dynastie. Abteilung IV, Hft. 17: Biographische Inschriften von Zeitgenossen Thutmosis' III. und Amenophis' II. Berlin.

HELCK 1971: W. HELCK, Die Beziehungen Ägyptens zu Vorderasien im 3. und 2. Jahrtausend v. Chr., Wiesbaden[2].

HELCK 1987: W. HELCK, Die Erzählung vom Verwunschenen Prinzen, in: J. OSING, G. DREYER (eds.), Form und Maß. Beiträge zur Literatur, Sprache und Kunst des Alten Ägypten. Festschrift für Gerhard Fecht zum 65. Geburtstag am 6. Februar 1987, Wiesbaden, 218–222.

IVANTCHIK 1993: A. IVANTCHIK, Les Cimmériens au Proche-Orient, Orbis Biblicus et Orientalis 127, Fribourg Suisse; Göttingen.

IVANTCHIK 1999: A. IVANTCHIK, Eine griechische Pseudo-Historie. Der Pharao Sesostris und der skytho-ägyptische Krieg, in: Historia 48, 395–441.

IVANTCHIK 2005: A. IVANTCHIK, Am Vorabend der Kolonisation. Das nördliche Schwarzmeergebiet und die Steppennomaden des 8.–7. Jhs. v. Chr. in der klassischen Literaturtradition. Mündliche Überlieferung, Literatur und Geschichte, Moskau.

JACOBY 1912: F. JACOBY, Hekataios aus Abdera, in: RE, 7, 2750–2769.

JANSSEN 1953: J. M. A. JANSSEN, The Stela (Khartoum Museum No. 3) from Uronarti, in: JNES 12, 51–55, pl. LXV.

JONES 1930: H. L. JONES (transl.), Strabo. Geography, Volume VII: Books 15–16, Loeb Classical Library 241. Cambridge, MA.

KEIL 1854: H. KEIL, Scholia vetera e codice Laurentiano, Lipsiae.

KITCHEN 1979: K. A. KITCHEN, Ramesside Inscriptions. Historical and Biographical II, Oxford.

LLEVELLIN-JONES, ROBSON 2010: LL. LLEVELLIN-JONES, J. ROBSON, Ctesias' History of Persia, London; New York.
LLOYD 1974: A. B. LLOYD, Rev.: A. Burton, Diodorus Siculus: Book I, in: JEA 60, 287–289.
LLOYD 1975: A. B. LLOYD, Herodotus, Book II. Introduction, Leiden.
LLOYD 1976: A. B. LLOYD, Herodotus, Book II. Commentary 1–98. Leiden.
LLOYD 1988: A. B. LLOYD, Herodotus, Book II. Commentary 99–182. Leiden.
LOPRIENO 2003: A. LOPRIENO, Travel and Fiction in Egyptian Literature, in: D. O'Connor and St. Quirke (eds.), Misterious Lands, London 31–52.
MIEROW 1915: CH. MIEROW, The Gothic History of Jordanes, Princeton
MURRAY 1970: O. MURRAY, Hecataeus of Abdera and Pharaonic Kingship, in: JEA 56, 144–150.
MURRAY 1975: O. MURRAY, Rev.: A. Burton, Diodorus Siculus: Book I, in: JHS 95, 214–215.
OBSOMER 1989: CL. OBSOMER, Les campagnes de Sésostris dans Hérodote: essai d'interprétation du texte grec à la lumière des réalités égyptiennes, Paris.
ODED 1979: B. ODED, Mass Deportations and Deportees in the Neo-Assyrian Empire. Wiesbaden.
OLDFATHER 1933: C. H. OLDFATHER (transl.), Diodorus of Sicily. Library of History, Volume I: Books 1–2.34, Loeb Classical Library 279, Cambridge, MA.
OLDFATHER 1935: C. H. OLDFATHER (transl.), Diodorus of Sicily. Library of History, Volume II. Books 2.35–4.58, Loeb Classical Library 303, Cambridge, MA.
RADT 2009: ST. RADT (ed.), Strabons Geographika. Bd. 8: Buch XIV–XVII: Kommentar, Göttingen.
REDFORD 2003: D. B. REDFORD, The Wars in Syria and Palestine of Thutmose III, Culture and History of the Ancient Near East, 16, Leiden.
ROBSON 1933 J. I. ROBSON (transl.), Arrian. Anabasis of Alexander, Volume II: Books 5–7. Indica, Loeb Classical Library 269, Cambridge, MA.
SCHWARTZ 1885: E. SCHWARTZ, Hekataeos von Teos, in: RhM 40, 223–262.
SCHWARTZ 1905: E. SCHWARTZ, Diodoros (38), in: RE, 5, 669–672.
SETHE 1906: K. SETHE, Urkunden der 18. Dynastie. Abteilung IV, Hft. 1–4: Historisch-biographische Urkunden. Leipzig.
SETHE 1907: K. SETHE, Urkunden der 18. Dynastie. Abteilung IV, Hft. 9–12: Historisch-biographische Urkunden, Leipzig.
SETHE 1928: K. SETHE, Ägyptische Lesestücke zum Gebrauch im akademischen Unterricht. Texte des Mittleren Reiches. Leipzig[2].
SIMPSON 2003: W. K. SIMPSON (ed.), The Literature of Ancient Egypt: An Anthology of Stories, Instructions, Stelae, Autobiographies and Poetry, New Haven[3].
SPALINGER 1978: A. SPALINGER, Psammetichus, King of Egypt: II, JARCE 15, 49–57.
STEPHENS 2003: S. A. STEPHENS, Seeing Double: Intercultural Poetics in Ptolemaic Alexandria, Hellenistic Culture and Society 37, Berkeley; Los Angeles.
STOCKFISCH 2006: D. STOCKFISCH, Mitanni in den Fremdvölkerlisten, in: D. BRÖCKELMANN, A. KLÜG (eds.), In Pharaos Staat. Festschrift für R. Gundlach zum 75. Geburtstag, Wiesbaden, 259–270.
STUBBINGS 1975: F. H. STUBBINGS, The Recession of Mycenaean Civilization, in: CAH[2], 2.2, 338–358.
WEST 1992: S. WEST, Sesostris' Stelae (Herodotus II.102–106), in: Historia 41, 117–120.

WASSER IN DEN ORAKELN BEI HERODOT[1]

Carmen Sánchez-Mañas

In den Orakelpassagen der *Historien* Herodots nimmt der Begriff „Wasser“ eine prominente Stelle ein. Denn „Wasser“ taucht in verschiedenen Formen auf, als natürliche (z. B. Meere, Flüsse, Seen und Quellen) ebenso wie als künstliche (z. B. Kanäle) Wasserkörper.

Konkret lassen sich siebzehn Passagen im Werk finden, in denen Wasserkörper zu Orakelsprüchen in Beziehung stehen: 1.47.3; 1.174.5; 2.111.2; 2.152.3; 2.156.2; 2.158.5; 4.157.3; 4.158.3; 4.178; 4.179.3; 5.92η.2; 6.76.1; 7.117.2; 8.20.2; 8.135.1; 9.43.2 und 9.93.1.

Je nach ihrer Stellung innerhalb der Orakelpassagen lassen sie sich in vier Gruppen einordnen: 1) Wasserkörper als Teil von Orakelsprüchen; 2) Wasserkörper als Erfüllungsort von Orakelsprüchen; 3) Wasserkörper als Lokalisierungsangabe von Orakelinstanzen; 4) Wasserkörper und die mit Orakelsprüchen verbundenen religiösen Aspekte.

Im Folgenden legen wir diese Klassifizierung unserer Analyse zu Grunde. Durch eine Untersuchung dieser Textabschnitte beabsichtigen wir, die Vielfalt an Formen und Funktionalität der Wasserkörper in den herodoteischen Orakeln herauszuarbeiten. Die gezogenen Schlussfolgerungen sollen für das gesamte hier vorgestellte Orakelmaterial gelten.

1. WASSERKÖRPER ALS TEIL VON ORAKELSPRÜCHEN

Unter dieser Überschrift werden die Passagen zusammengefasst, in denen Wasserkörper innerhalb der Orakeltexte (sowohl in Versform als auch in Prosa) erwähnt werden. Das betrifft die folgenden fünf Textabschnitte: 1.47.3; 2.152.3; 2.158.5; 8.20.2 und 9.43.2, die wiederum in zwei Kategorien eingeteilt werden können:

1.1. Unidentifizierbare Wasserkörper

Das so genannte Probe-Orakel (1.47.3) erscheint als erstes in Versform (Hexameter) in den *Historien.* In dieser delphischen Prophezeiung wird ein natürlicher Wasserkörper als Meer am Ende der ersten von fünf Zeilen sichtbar: Οἶδα δ‘ ἐγὼ ψάμμου τ‘ ἀριθμὸν καὶ μέτρα θαλάσσης. Trotzdem muss man in Rechnung stellen,

1 Diese Arbeit stützt sich auf die Aktivitäten der Forschungsgruppe „Byblíon“ (H 52), die von dem Industrie-und Innovationsrat (Regionalregierung von Aragonien) und dem Europäischen Sozialfonds unterstützt wird.

dass die erste Zeile erst durch den nächsten Vers vervollständigt wird: καὶ κωφοῦ συνίημι καὶ οὐ φωνεῦντος ἀκούω. Zusammen bilden die beiden eine „Allwissenheitserklärung" des Gottes Apollons, die durch die drei letzten Zeilen exemplifiziert wird, indem sie das an sieben Orakelinstanzen von Kroisos gestellte Rätsel lösen. Damit wird Delphis Vorrang vor den anderen bewiesen.[2]

Angesichts der Tatsache, dass dieser Orakelspruch die narrative Grundlage für die Beziehung zwischen Kroisos und Delphi (1.53.3; 1.55.2; 1.85.2; 1.91.1–6) darstellt, ist die Allwissenheitserklärung insofern notwendig, als sie einer von Apollon selbst (d.h. in der ersten Person Singular: οἶδα, συνίημι, ἀκούω) verkündeten und daher unfehlbaren Garantie gleichkommt. Auf den ersten Blick ist es aber nicht so klar, warum gerade ein Wasserkörper hier eine Rolle spielt. Hier muss man zum einen berücksichtigen, dass es sich nicht um ein konkretes Meer handeln, wie das Fehlen des Artikels zeigt. Zum anderen ist der Kontext zu beachten. Da in Wirklichkeit die Zahl der Sandkörner und die Größe des Meeres nicht messbar sind und der „Sprachlose" nicht gehört werden kann, veranschaulichen diese drei Konzepte die rhetorische Figur des *Adynatons*, wodurch Apollons Wahrsagemacht unterstrichen wird und die Messung des Sandes und des Meeres als Charakterisierung des Unmöglichen funktioniert. Interesanterweise fügt Herodot in diesen Orakelspruch eine aus dem Volksmund stammenden Formulierung („das Unzählige Messen") ein, die auch bei Pindar als eine Allwissenheitserklärung des delphischen Apollons (Φοῖβε, Z. 40) zu finden ist.[3]

1.2. Identifizierbare Wasserkörper

Die Orakelsprüche an Psammetich (2.152.3), an dessen Sohn Nekos (2.158.5) und an die Euböer (8.20.2), sowie der über die Schlacht bei Plataiai (9.43.2) zeichnen sich dadurch aus, dass sie bestimmte Wasserkörper beinhalten.

In 2.152.3 und 8.20.2 tritt dasselbe Meer aus verschiedenen Perspektiven auf, nämlich das Mittelmeer, auch wenn es nicht direkt angeführt wird. Denn an der ersten Stelle wird die Aufmerksamkeit auf den östlichen Mittelmeerraum mit Ägypten im Mittelpunkt fokussiert. Psammetich, der wegen einer für ihn günstigen Wahrsagung von seinen Kollegen – den anderen elf Königen – exiliert wird (2.147.4; 151.3; 152.1–2), befragt das Orakel in Buto wegen seiner Rache und erfährt, dass Männer aus Bronze als Instrumente seiner Vergeltung aus dem Meer (ἀπὸ θαλάσσης) kom-

2 Dieses Orakel gehört zu einer detailliert beschriebenen Orakelepisode (1.46–49), in der der Lyderkönig Kroisos durch einen Test (das Zusammenkochen einer Schildkröte und eines Lamms) bestimmt, welche unter die von Delphi, Abai, Dodona, Amphiaraos, Trophonios, Branchidai (Didyma) und (Zeus-)Ammon die glaubwürdigsten Orakelheiligtümer sind. Nur das erste und das vierte bestehen die Probe, aber die delphische Antwort wird ausführlicher berichtet als die von Amphiaraos.

3 Siehe Pind. *Pyth.* 9.45–48: οσθα καὶ πάσας κελεύθους·/ ὅσσα τε χθὼν ἠρινὰ φύλλ' ἀναπέμπει, χὡπόσαι / ἐν θαλάσσᾳ καὶ ποταμοῖς ψάμαθοι / κύμασιν ῥιπαῖς τ' ἀνέμων κλονέονται. Darüber hinaus findet man weitere Allwissenheitserklärungen in Hinsicht auf das Ausmaß des Meers, siehe z.B. Hes. *Op.* 648: δείξω δή τοι μέτρα πολυφλοίσβοιο θαλάσσης. Zur Sprichwortverwendung in Hdt. 1.47.3 siehe Asheri 1988, 292, Crahay 1956, 193 und Fontenrose 1978, 113.

men werden (2.152.3). Nach der Landung von bewaffneten Ioniern und Karern in Ägypten entthront Psammetich mit deren Unterstützung seine Gegner (2.152.4–5).

In der aus zwei Hexametern bestehende Prädiktion des Bakis (8.20.2) handelt es sich um eine von Barbaren unternommene Überbrückung eines unbenannten Meers (ζυγὸν εἰς ἅλα βάλλῃ / βύβλινον), die auf diejenige von Xerxes über den Hellespont (7.25.1; 34.1; 36.3) verweist.[4] Aber zugleich deuten die Erwähnung der Insel Euböa in der Prophezeiung sowie die Tatsache, dass ihre Bewohner die Adressaten des Orakels sind, auf die Ägäis hin.

Die unterschiedliche Perspektive erstreckt sich auch auf die Rolle des Meers in beiden Passagen. Während es im ersten Textabschnitt die Herkunft der Helfer des Rezipienten kennzeichnet und daher eine positive Konnotation hat, markiert es im zweiten einen kritischen Augenblick – in dem die Euböier ihre Ziegen von der Insel treiben sollen –, den die Empfänger außer Acht lassen, was ihnen schließlich Unglück einbringt.[5]

Eine Weissagung unbekannter Provenienz informiert in 2.158.5 Nekos darüber, dass er mit dem Kanalbau eigentlich „dem Barbaren" vorarbeitet (μαντηίου ἐμποδίου γενομένου τοιοῦδε, τῷ βαρβάρῳ αὐτὸν προεργάζεσθαι). Deswegen ist sie im Grunde kein Teil eines Orakelspruches. Aber ihre Zugehörigkeit zu dieser Gruppe lässt sich *lato sensu* rechtfertigen. Einerseits ist der Kanal sehr eng mit der Prophetie verbunden, weil Nekos ihretwegen die Arbeiten unterbricht (Νεκῶς μέν νυν μεταξὺ ὀρύσσων ἐπαύσατο). Zum anderen identifiziert Herodot selbst den betreffenden Wasserkörper als den später von Dareios vollendeten Kanal ins Rote Meer (2.158.1).[6]

Im Gegensatz zu den vorigen Stellen werden die in 9.43.2 figurierenden Wasserkörper – zwei böotische Flüsse –innerhalb der Wahrsagung, bzw. in der ersten von vier Versen, explizit benannt: Thermodon und Asopos, an deren Ufern (τὴν δ‘ ἐπὶ Θερμώδοντι καὶ Ἀσωπῷ λεχεποίῃ) der griechische Sieg in der Schlacht bei Plataiai stattfinden wird.

2. WASSERKÖRPER ALS ERFÜLLUNGSORT VON ORAKELSPRÜCHEN

Zu dieser Gruppe gehören die Passagen, die Information über Wasserkörper geben, an denen die Orakelsprüche verwirklicht werden. Das betrifft 4.157.3; 4.158.3; 4.178 und 4.179.3, die sich alle in Libyen lokalisieren lassen, konkret in zwei Gebieten:

4 Siehe How und Wells 1928, 240–41.

5 Den Euböern wird im Krieg sowohl von den vereinigten Griechen (8.19; 8.20.2) als auch von den Persern (8.23.2) Schaden zugefügt.

6 Ferner erwähnt Herodot, dass 120 000 Ägypter beim Bau ums Leben gekommen sind. Kirchberg 1965, 49–50 und Harrison 2000, 156, Anm. 122 assoziieren die Todesopfer mit einer Orakelbefragung. Es ist aber beachtenswert, dass Herodot keine Orakelbefragung aufführt und dass er die Verstorbenen mit dem Orakelspruch nicht explizit verbindet.

2.1. Kyrenaika

Herodot stellt die theraische Kolonisierung von Kyrene (4.149–159) als eine lange und schwierige Unternehmung dar: die Kolonisten gründen die Stadt Kyrene erst nach zwei vorherigen Landnahmen. Damit ist es ihnen endlich gelungen, eine Reihe von delphischen Weissagungen (4.150.3; 4.151.1; 4.155.3; 4.156.2 und 4.157.2) zu erfüllen.

Zunächst lassen sie sich auf Platea, einer Insel gegenüber der libyschen Küste, nieder (4.156.3). Dann nehmen sie sich gerade in diesem Küstengebiet ihren Wohnsitz, nämlich in einer Enklave namens Aziris, die als ein *locus amoenus* geschildert wird: τὸν νάπαι τε κάλλισται ἐπ‘ ἀμφότερα συγκληίουσι καὶ ποταμὸς τὰ ἐπὶ θάτερα παραρρέει (4.157.3). In der Beschreibung taucht ein unbenannter Fluss auf, was nicht nur dem literarischen Topos angemessen ist, sondern auch zur Situation passt. Denn eine optimale Wasserversorgung gehört zu den Schlüsselfaktoren bei der Beurteilung eines potenziellen Siedlungsortes.[7] Dieselbe Problematik ist in 4.158.3 wiederzufinden. Als die Theraier nach sechs Jahren Aziris verlassen (4.158.1), werden sie von den Libyern an eine für eine Gründung geeignete Gegend geleitet. Die Wasserversorgung ist hier doppelt garantiert, zum einen durch häufige Niederschläge (ὁ οὐρανὸς τέτρηται), zum anderen durch eine Quelle. Dass dieser natürliche Wasserkörper dem Gott Apollon geweiht ist (ἐπὶ κρήνην λεγομένην εἶναι Ἀπόλλωνος), bringt darüber hinaus in unsere Untersuchung ein eindeutiges und bisher abwesendes religiöses Motiv ein, das die (künftige) Wichtigkeit des Apollonkults in Kyrene aufscheinen lässt.[8]

2.2. Tritonsee und dessen Umgebung

In seiner Beschreibung der libyschen Völker (4.168–199) schließt Herodot die Machlyer ein (4.178), die als Nachbarn der Lotophagen auch Lotos essen und am großen Fluss Triton (ἐπὶ ποταμὸν μέγαν τῷ οὔνομα Τρίτων ἐστί) wohnen. Der Fluss mündet in den weiten Tritonsee (ἐς λίμνην μεγάλην Τριτωνίδα), in dem wiederum die Insel Phla liegt, welche die Lakedaimonier einem Orakelspruch unbekannten Ursprungs zufolge kolonisieren sollten. Dennoch darf man nicht vergessen, dass Phla streng genommen deswegen kein Erfüllungsort ist, weil die Prophezeiung sich nicht realisiert.[9] Das wirkt auch auf 4.179.3, wo von einer Prädiktion der Wasser-

7 Siehe Frisone 2012, 87.

8 Es ist trotzdem bemerkenswert, dass Herodot dies nur unter Vorbehalt äußert, indem er das Partizip λεγομένην in den Satz schiebt, was eine gewisse Distanzierung dem Erzählten gegenüber darstellt und der aktiven Rolle der Libyer als manipulative Figuren entspricht; vgl. 4.158.2. Von Interesse ist auch die Tatsache, dass die abundante Niederschlagsmenge von Kyrene und die Existenz von der Quelle des Apollons und von Aziris in der griechischen Literatur belegt sind. Siehe Pind. *Pyth.* 4.52: φτα κελαινεφέων πεδίων; 294: οἶκον ἰδεῖν, ἐπ‘ Ἀπόλλωνός τε κράνᾳ συμποσίας ἐφέπων, sowie Call. *Ap.* 88–89, wo die Quelle „Kyre“ und der zweite Siedlungsort „Azilis“ genannt werden: οδ‘ οὔπω πηγῇσι Κύρης ἐδύναντο πελάσσαι/Δωριέες, πυκινὴν δὲ νάπησ‘ Ἄζιλιν ἔναιον.

9 Man hat diesen unerfüllten Orakelspruch traditionell für problematisch gehalten. Crahay 1956, 134 fragt sich, ob er einen satirischen Charakter hat. Malkin 1987, 79 bringt ihn mit der

gottheit Triton an Jason die Rede ist. Als sein Schiff in die Untiefen des Sees gerät und die Argonauten keinen Ausweg finden, taucht der Gott auf (4.179.2) und zeigt ihnen im Tausch gegen einen Dreifuß die Lösung. Aber vorher weissagt Triton auf dem in seinem Tempel gestellten Dreifuß die Einrichtung von hundert griechischen Städten um den Tritonsee (περὶ τὴν Τριτωνίδα λίμνην), sofern ein Nachkomme der Argonauten den Dreifuß wegnimmt. Daraufhin verstecken ihn die Libyer (4.179.3). Wenn auch deren Absicht nicht enthüllt wird, wollen sie offensichtlich die Verwirklichung des Orakels verhindern. Es wäre natürlich verführerisch, die Nichterfüllung dieser Wahrsagungen als von dem bereits angelegten mythischen Charakter der beiden Wasserkörper[10] abhängige Unrealisierbarkeit zu erklären. Aber man darf nicht davon ausgehen, dass diese auch für Herodot gilt.

Aus mehreren Gründen ist auch Tritons Präsenz beachtenswert. Denn Triton wird nur zweimal in den *Historien* erwähnt (4.179; 4.188) und an der hier untersuchten Stelle bildet er auch das einzige Beispiel für einen dem Frager leibhaftig erscheinenden Orakelgott. Man findet hier also wieder ein religiöses Element, das durch die Verwendung von *verba dicendi* nuanciert ist. Überdies verkündet er allein, spontan, ohne vorherige Befragung und direkt auf einem Dreifuß. Das bringt ihn gewissermaßen der Pythia nahe, wenngleich er wegen seiner Tätigkeit vor allem einem ἅλιος γέρων („Meergreis") wie z. B. Proteus analog ist[11].

3. WASSERKÖRPER ALS LOKALISIERUNGSANGABE VON ORAKELINSTANZEN

Unter dieser Überschrift lassen sich die Stellen einordnen, in denen die Orakelinstanzen sich in der Nähe von Wasserkörpern befinden. Dies ist der Fall bei drei Passagen: 2.156.2; 5.92η.2 und 8.135.1. Nach der Natur des Wasserkörpers kann man zwei Gruppen unterscheiden:

vergeblichen, von Dorieus von Sparta geleiteten Kolonisierungsunternehmung in Libyen (Hdt. 5.42.2–3) in Zusammenhang.

10 Siehe Corcella 1993, 366 und Bichler 2000, 65, die sich mit den verschiedenen Lokalisierungsmöglichkeiten des Tritonsees auseinandersetzen. Die zwei Gelehrten weisen außerdem darauf hin, dass Herodot von der von Pindar und anderen Dichtern überlieferten Tradition abweicht, siehe Pind. *Pyth.* 4. 13–56 und Apol. Rhod. 4.1537–1622.

11 Im Werk scheint Triton speziell die Gottheit des gleichnamigen Sees zu sein und nicht allgemein „*un Triton*", wie Crahay denkt (1956, 134, 136). Darauf deuten die Tatsachen hin, sowohl dass er sein eigenes Heiligtum (ἐν τῷ ἑωυτοῦ ἰρῷ) – offenbar im See – (4.179.3) besitzt, als auch dass er an einer anderen Stelle (4.188) in Kombination mit Athena, Poseidon und diesem See genannt wird. Durch die oben genannte Verwendung von *verba dicendi*, die nicht nur das Kapitel als Ganzes (στι δὲ καὶ ὅδε λόγος λεγόμενος (4.179.1)), sondern auch die Erscheinung Tritons und seine Prophezeiung (λόγος ἐστὶ (4.179.2)) betrifft, distanziert sich Herodot dem Erzählten gegenüber. Für die Verbindung zwischen Triton und der Pythia siehe 4.179.1: τρίποδα χάλκεον … βουλόμενον ἐς Δελφοὺς ἀπικέσθαι. Was den Meergreis angeht, teilt der herodoteische Triton mit dieser Figur ausschließlich die Weissagekunst und zeigt keine Verwandlungsfähigkeit. Siehe Hom. *Od.* 4.365

3.1. Orakelinstanzen in der Nähe von Seen

Zwei Orakelstätten lassen sich neben Seen lokalisieren. Die erste ist die der Göttin Leto geweihte Instanz in Buto (Ägypten). Ein namenloser tiefer großer See unweit des oben genannten Heiligtums (ἐν λίμνῃ βαθέῃ καὶ πλατέῃ … παρὰ τὸ ἐν Βουτοῖ ἱρόν) wird in 2.156.2 erwähnt. Der zweite ist der Kopaissee in Böotien (8.135.1). Da er sich in Griechenland befindet und daher allgemein bekannt ist, wäre seine Beschreibung überflüssig. In der Tat dient seine Nennung nur dazu, die genaue Lage des dem Apollon Ptoos geweihten nahen Orakelsitzes zu schildern (κεῖται δὲ ὑπὲρ τῆς Κωπαΐδος λίμνης πρὸς ὄρεϊ ἀγχοτάτω Ἀκραιφίης πόλιος).[12]

Obwohl die Seen verschiedenen Gebieten angehören und von Herodot ungleich beschrieben werden, teilen sie sich denselben Kontext, da beide in Verbindung zu Wundern erwähnt werden. Einerseits handelt es sich um die schwimmende Natur der in dem See liegenden Insel Chemmis (2.156.1–2) (Χέμμις…λέγεται δὲ ὑπ‘ Αἰγυπτίων εἶναι αὕτη ἡ νῆσος πλωτή); andererseits hat es einen im griechischsprachigen Raum auf karisch verkündeten Orakelspruch zum Gegenstand (φάναι δὲ Καρίῃ μιν γλώσσῃ χρᾶν) (8.135.3).[13]

3.2. Orakelinstanzen in der Nähe von Flüssen

Der korinthische Tyrann Periander befragt durch Boten das Gespenst seiner verstorbenen Ehefrau an der einzigen im Werk auftauchenden Totenorakelstätte (5.92η.2). Herodot weist kurz auf deren Standort hin, nämlich in Thesprotien an den Ufern des Flusses Acheron (ἐς Θεσπρωτοὺς ἐπ‘ Ἀχέροντα ποταμὸν ἀγγέλους ἐπὶ τὸ νεκυομαντήιον). Die Gleichnamigkeit zwischen dem thesprotischen Fluss und dem in der Unterwelt (Hom. *Od.* 10.513 u. a.) ist nicht zufällig. Vielmehr stellt sie den gemeinsamen Bezug zum Totenbereich her.[14]

4. WASSERKÖRPER UND DIE MIT ORAKELSPRÜCHEN VERBUNDENEN RELIGIÖSEN ASPEKTE

Diese Abteilung umfasst die Orakelpassagen, in denen die Wasserkörper in Bezug zu religiösen Angelegenheiten stehen. Das manifestiert sich in 1.174.5; 2.111.2; 6.76.1; 7.117.2 und 9.93.1. Alle fünf Stellen lassen sich in zwei Gruppen teilen:

12 Letos Tempel wird in 2.156.4 explizit als Orakelheiligtum bezeichnet: ἐν Βουτοῖ πόλι ἵνα δή οἱ τὸ χρηστήριον τοῦτο ἔστι. Apollons Kultplatz assimiliert sich in 8.135.1 an andere Orakelstätten: ἐλθεῖν ἄρα τὸν Εὐρωμέα Μῦν, περιστρωφώμενον πάντα τὰ χρηστήρια, καὶ ἐς τοῦ Πτῴου Ἀπόλλωνος τὸ τέμενος.

13 An beiden Stelle benutzt Herodot Termini aus dem semantischen Feld „Wunder“: τὸ δέ μοι τῶν φανερῶν ἦν θῶμα μέγιστον (2.155.3); θωμαστότατον (2.156.1); θῶμά μοι μέγιστον (8.135.1). Zu 2.156 siehe Munson 2001, 238–39. Zu 8.135 siehe Harrison 2000, 70.

14 Obgleich Herodot Ephyra nicht nennt, wird dieser Totenorakelsitz normalerweise mit dem in dieser Stadt Thesprotiens gleichgesetzt, siehe Rosenberger 2001, 110 sowie Johnston 2008, 60, die sowohl die unzivilisierte Qualität der ganzen Region hervorhebt als auch den Erwerb eines tödlichen Gifts von Odysseus in Ephyra (Hom. *Od.* 1.259–62) in Erinnerung bringt.

4.1. Wasserkörper in Bezug zu übernatürlich erklärten Verletzungen

Die Orakelsprüche an die Knidier (1.174.5) und an den Pharao Pheros (2.111.2) haben einen gemeinsamen Hintergrund, indem eine Handlung wider ein Gewässer Augenprobleme verursacht, deren Heilung durch Wahrsagungen angeboten wird.

Zu einem versuchen die auf der Halbinsel von Bubassos (Karien) niedergelassenen Knidier, sich durch einen Kanalbau der persischen Eroberung ihres Heimatlandes zu entziehen (1.174.1–3). Beim Graben werden sie überaus häufig – meistens an den Augen (μάλιστα τὰ περὶ τοὺς ὀφθαλμούς) – von Steinsplittern verletzt und befragen das Orakelheiligtum von Delphi (1.174.4). Von ihm erhalten sie im Namen des Gottes Zeus den Auftrag, mit dem Graben aufzuhören (1.174.5), was diese Stelle mit 2.158.5 verknüpft. Infolge dessen kapitulieren sie (1.174.6).

Zu anderem wirft Pheros während einer außergewöhnlichen Überschwemmung einen Speer in die Mitte des Nilstrudels und erleidet dann eine Augenkrankheit die zur Erblindung führt (καμόντα αὐτὸν τοὺς ὀφθαλμοὺς τυφλωθῆναι) (2.111.1–2). Erst nach zehn Jahren erfährt er durch eine Prophezeiung aus Buto von einem Heilmittel (sich die Augen mit Urin einer keuschen Frau zu waschen) (2.111.2), wodurch er wieder gesund wird (2.111.3–4). Die übernatürliche Ursache der Verletzungen ist an beiden Stellen zweimal zu erkennen, nämlich vor und innerhalb der Orakeltexte. Vor den Orakelsprüchen erscheint diese übernatürliche Ursache entweder direkt als göttliches Eingreifen (θειότερον) (1.174.4) oder als eine den Göttern entgegengesetzte Einstellung (ἀτασθαλίῃ χρησάμενον) (2.111.2).[15] In den Prophezeiungen selbst wird entweder eine bestimmte Gottheit als der Unternehmung feindlich dargestellt (Ἰσθμὸν δὲ μὴ πυργοῦτε μηδ‘ ὀρύσσετε·/ Ζεὺς γάρ κ‘ ἔθηκε νῆσον, εἴ γ‘ ἐβούλετο) (1.174.5) oder das Ende einer Bestrafung signalisiert (ἐξήκει τέ οἱ ὁ χρόνος τῆς ζημίης) (2.111.2). Die übermenschliche Veranlassung wird also nur in 1.174 ausgedrückt, während sie in 2.111 unausgesprochen bleibt. Nichtsdestoweniger deutet die Tatsache, dass die Krankheit des Pharaos als Strafe von einer göttlichen Instanz (bzw. einer Orakelstätte) beschrieben wird, auf einen kausalen Zusammenhang zwischen Blindheit und Aktion hin, der *de facto* auch vom göttlichen Eingreifen ausgelöst wird.[16]

15 Der Komparativ θειότερον erklärt sich aus syntaktischen Gründen, indem er zur Konstruktion μλλον γάρ τι καὶ θειότερον … τοοικότος gehört. Siehe auch dazu Harrison 2000, 177, Anm. 71. Zur Verknüpfung zwischen 1.174. und 2.158.5 siehe S. 183. Zu ἀτασθαλίη siehe Powell 1938, 50: „impiety“. Das Benehmen des Pharaos ähnelt dem von Xerxes in 7.35, wo das verwandte Wort ἀτάσθαλα vorkommt (7.35.2).

16 Siehe Harrison 2000, 64 und Kirchberg 1965, 45. Aber man sollte berücksichtigen, dass Herodot in 1.174 und 2.111 sich vom göttlichen Eingreifen distanziert, indem er die Episoden jeweils den Knidiern (ὡς αὐτοὶ Κνίδιοι λέγουσι) und in 2.107.1 erwähnten ägyptischen Priestern (ἔλεγον (2.111.1), λέγουσι (2.111.1)) zuschreibt. Darüber hinaus wird 1.174 aus einer besonders subjektiven Perspektive deswegen präsentiert, weil die Knidier als Erzähler, Agenten und Interpreten auftreten, wie das Verb ἐφαίνοντο (φαίνω) in 1.174.4 – anstatt von ἦσαν (εἰμί) – demonstriert. So wird die göttliche Intervention nicht als Tatsache, sondern nur als Schein inszeniert. Siehe *LSJ* 19969, 488, s. v. εμί: „of the *real* world, *be*“; 1912, s. v. φαίνω: „*Pass.* […] *appear to* be so and so, c. inf.“. Zu φαίνω an dieser Stelle siehe auch Powell 1938, 370, s. v. φαίνω: „pass. […] w. inf. *seem to* […] 1, […] 174.4“).

4.2. Wasserkörper in Verhältnis zu Kult

Im Zusammenhang mit dem delphischen Orakel an Kleomenes (6.76.1) werden zwei Opfergaben des spartanischen Königs geschildert, die eng mit dem Inhalt der Wahrsagung (seiner Eroberung von Argos) verbunden sind. Sobald er und seine Soldaten am Fluss Erasinos – der aus dem Stymphalischen See fließen soll und die Grenze zwischen Lakonien und der Argolis markiert – ankommen, bringt Kleomenes dem Fluss Opfer dar (ἀπικόμενος ὦν ὁ Κλεομένης ἐπὶ τὸν ποταμὸν τοῦτον ἐσφαγιάζετο αὐτῷ). Da ihm die Überquerung nicht gestattet wird, gratuliert er dem Fluss dazu, seine argeischen Mitbürger nicht verraten zu haben, und schlägt einen anderen Weg ein, nämlich zum Meer nach Thyrea. Dort opfert er dem Meer einen Stier (σφαγιασάμενος δὲ τῇ θαλάσσῃ ταῦρον) und fährt mit Schiffen nach Nauplia und ins Gebiet von Tiryns (6.76.2). An dieser Stelle führt Kleomenes also sehr sorgfältig rituelle Handlungen vor seinem Feldzug durch. Nicht nur verlässt er sich auf den Orakelspruch, sondern respektiert auch das negative Ergebnis der Opferschau zur Flussüberquerung (διαβατήρια). Nachdem er sich für eine alternative Seeroute entschieden hat, führt er beim Besteigen des Schiffes (ἐμβατήρια) Opferungen durch. Seine ehrfurchtsvolle Haltung steht an dieser Stelle im Gegensatz zu der seines Halbbruders Dorieus in 5.42.2, aber auch zu derjenigen, die Kleomenes selber in anderen Orakelepisoden (z. B. 6.66.2–3; 6.77.2–3 und 6.79–80) einnimmt.[17]

Die nicht näher spezifizierten Orakelsprüche an die Akanthier (7.117.2) und an die Bürger von Apollonia (9.93.1) leiten zwei verschiedene Kulttypen ein. Der erste bezieht sich konkret auf den Heldenkult des Achaimeniden Artachaies in Akanthos (Τούτῳ δὲ τῷ Ἀρταχαίῃ θύουσι Ἀκάνθιοι ἐκ θεοπροπίου ὡς ἥρωϊ, ἐπονομάζοντες τὸ οὔνομα), nachdem jener dort an einer Krankheit gestorben war (7.117.1). Die Rolle des Persers als Leiters des (vollendeten) Kanalbaus am Fusse des Athosgebirges (7.22.2) bringt diese Passage zu 1.174.5 und 2.158.5 in Beziehung. In der zweiten Prophezeiung handelt es sich um den heiligen Status der dem Gott Helios geweihten Herde (ἱρὰ Ἡλίου πρόβατα), die unter den Bewohnern von Apollonia in hohem Ansehen steht (περὶ πολλοῦ γὰρ δὴ ποιεῦνται Ἀπολλωνιῆται τὰ πρόβατα ταῦτα ἐκ θεοπροπίου τινός). Das Vieh weidet tagsüber in einem von einem Wasserkörper geprägten Bereich, nämlich am Fluss Aoos, der vom Lakmon-Berg herabkommt, durch das ganze Gebiet Apollonias fließt und ins Meer mündet. Sowohl die Beziehung zu Helios als auch der Wasserkontext erinnern an die auf Thrinakia von den Heliaden gehüteten Tiere (Hom. *Od.* 12.127–33).[18]

17 Zur Lage des Flusses siehe Nenci 2007³, 241–42; zu den von Kleomenes dargebrachten Opfergaben siehe Hom. *Od.* 3.5–6 und Burkert 1977, 119 und 217; zu seinem unterschiedlichen Verhalten siehe Munson 1993, 46.

18 Die physischen Charakteristika des Artachaies (außerordentliche Größe und sehr laute Stimme) entsprechen seiner Qualität als Heros, siehe Hdt. 1.60.4 und Hom. *Il.* 5.785; 6.5; zu seiner und ähnlichen Heroisierungen in den *Historien* siehe Harrison 2000, 158, Anm. 2; zur Verbindung zwischen den homerischen Stellen und Hdt. 9.93 siehe Munson 2001, 71. Die heilige Viehherde wird in beiden Fällen angegriffen, was wiederum negative Folgen für die Gefährten des Odysseus und für Euenios hat, der angeblich als Nachtwächter der Herde versagte, siehe Hom. *Od.* 12.340–419 und Hdt. 9.93.2–3; zur weiteren Entwicklung der Geschichte siehe Hdt. 9.93.4 und 9.94.1–3.

5. SCHLUSSFOLGERUNGEN

Auf den vorigen Seiten haben wir versucht, die Thematisierung des Wassers in den herodoteischen Orakelsprüchen zu systematisieren.

Unter den siebzehn analysierten Passagen haben die meisten (bzw. zwölf) Erwähnungen von Wasserkörpern einen vorwiegend unreligiösen Wert. Das gilt für die Fälle, in denen sie *stricto et lato sensu* in die Orakelsprüche eingeschlossen werden (1.47.3; 2.152.3; 2.158.5; 8.20.2 und 9.43.2) oder sie zum Orakelkontext gehören (2.156.2; 4.157.3; 4.158.3; 4.178; 4.179.3; 5.92η.2 und 8.135.1). An allen diesen Stellen mögen die Wasserkörper proverbialisiert werden (1.47.3) oder als Leitmotiv von Orakelsprüchen funktionieren (2.158.5). Darüber hinaus können sie Informationen über die Herkunft von in Orakelsprüchen benannten Personen (2.152.3) sowie über das in Orakelsprüchen behandelte Agieren (8.20.2) geben. In gleicher Weise setzen sie den Standort von in Orakelsprüchen genannten Schlachten (9.43.2) oder Siedlungen (4.157.3; 4.158.3; 4.178 und 4.179.3), aber auch von Orakelstätten (5.92η.2 und 8.135.1) ins Bild.

Demgegenüber werden Wasserkörper nur fünfmal aus einem rein religiösen Gesichtspunkt betrachtet (1.174.5; 2.111.2; 6.76.1; 7.117.2 und 9.93.1). Sogar in diesen Fällen dreht es sich in 6.76.1 und 7.117.2 ausschließlich um die übliche Kultpraxis. Beachtenswert ist außerdem, dass die Wasserkörper als Kultempfänger (von Opfergaben) allein an der ersten von diesen beiden Stellen beschrieben werden, während sie im zweiten Text zur Charakterisierung des Kultempfängers beitragen. Dieselbe nebensächliche Rolle spielen die Wasserkörper an einer dritten Stelle, wo vom Kult die Rede ist, nämlich 9.93.1. Hier findet man das einzige Beispiel in den *Historien* für heiliges Vieh – das mit dem Helioskult verbunden ist – und wo die Wasserkörper nur dazu dienen, die Weidefläche der Herde zu kennzeichnen. An den restlichen Stellen (174.5 und 2.111.2) handelt es um keine Kultangelegenheit, sondern nur um übernatürlich (bzw. göttlich) erklärte Augenprobleme, die in Zusammenhang mit Wasserkörpern (dem Kanal von Knidos und dem Nil) verursacht worden sind. Also stehen beide Wasserkörper nicht im Mittelpunkt der Erzählung.

Diese Nebenrolle ist aber nicht mit Bedeutungslosigkeit zu verwechseln, weil vielfältige religiöse Aspekte die Präsenz von Wasserkörpern in Orakelsprüchen auf ganz subtile Weise durchdringen, wie die Benennung der Quellen Apollons als Sitz des zukünftigen Kyrene (4.158.3) oder die Tätigkeit der Seegottheit Triton als Orakelgott (4.179.2–3) zeigen.[19]

Das wird aber insbesondere bei den Stellen deutlich, in denen ein Kanal gezogen wird (1.174.5; 2.158.5; 7.117.2). Alle drei Texte sind essentiell, um die Beziehungen zwischen Menschen, Göttern und Natur in den *Historien* zu untersuchen. Wie schon festgestellt, wird das Abbrechen des Kanalbaus explizit aus religiösen Gründen nur in 1.174.5 erläutert, während sie in 2.158.5 durch eine bloße Prognose begründet wird und in 7.117.2 gar nicht erst erscheint. Ferner werden Verletzungen und Todesfälle in 1.174.5 und 2.158.5 mit dem Bau selbst zusammengebracht, aber der Tod des Bauleiters Artachaies wird ohne Zusammenhang damit in 7.117.2 be-

19 Siehe S. 184–5.

richtet. Wegen ihrer gemeinsamen Eigenschaften dürfen sie in Verbindung gesetzt werden. Deswegen sind sie in der Forschung oft durch gegenseitig ausschließende Erklärungen interpretiert worden. Denn sie werden als Beweis dafür angeführt, dass Herodot entweder nicht einverstanden oder einverstanden damit sei, dass Menschen die natürliche Welt ändern.[20] Aber diese nach kompletten Analogien strebenden Hypothesen lassen die oben genannten Unterschiede zwischen den Passagen außer Acht und vereinfachen das von Herodot hergestellte facettenreiche Bild in unzulässiger Weise.[21]

Gerade die Vielschichtigkeit ist das herausragende Merkmal der Präsenz des Wassers in den Orakeln bei Herodot.

6. LITERATURVERZEICHNIS:

Asheri, D. 1988. Erodoto. Le Storie. Vol. 3, Libro III, La Persia. Introduzione e commento di David Asheri; testo critico di Silvio M. Medaglia; traduzione di Augusto Fraschetti. Roma, Milano.

Bichler, R. 2000. Herodots Welt. Der Aufbau der Historie am Bild der fremden Länder und Völker, ihrer Zivilasation und ihrer Geschichte. Berlin.

Burkert, W. 1977. Griechische Religion der archaischen und klassischen Epoche. Stuttgart.

Corcella, A. 1993. Erodoto. Le Storie. Vol.4, Libro IV, La Scizia e la Libia. Introduzione e commento di Aldo Corcella; testo critico di Silvio M. Medaglia; traduzione di Augusto Fraschetti. Roma.

Crahay, R. 1956. La Littérature Oraculaire chez Hérodote. Paris.

Fontenrose, J. 1978. The Delphic Oracle: Its Responses and Operation with a Catalogue of Responses. Berkeley, Los Angeles.

Frisone, F. 2012. "Rivers, land organization, and identity in Greek western Apoikíai." Mediterranean Historical Review 27, 87–115.

Harrison, T. 2000. Divinity and History: The Religion of Herodotus. Oxford.

How, W. and Wells, J. 1928. A Commentary on Herodotus with introduction and appendixes. Volume II (Books V–IX). Oxford.

Johnston, S. I. 2008. Ancient Greek Divination. Oxford.

Kirchberg, J. 1965. Die Funktion der Orakel im Werke Herodots. Göttingen.

Lateiner, D. 1989. The Historical Method of Herodotus. Toronto.

Malkin, I. 1987. Religion and colonization in ancient Greece. Leiden.

Munson, R. V. 1993. „Three Aspects of Spartan Kingship". In: R. M. Rosen und J. Farrell (Hrsg.), Nomodeiktes. Greek Studies in Honor of Martin Ostwald, Ann Arbor. 1993, 39–54.

Munson, R. V. 2001. Telling Wonders. Ethnographic and Political Discourse in the Work of Herodotus. Ann Harbor.

Nenci, G. 2007^3. Erodoto. Le Storie. Vol. 6, Libro VI. La battaglia di Maratona. A cura de Giuseppe Nenci. Roma.

Powell, J. E. 1938. A Lexicon to Herodotus. Hildesheim.

20 Zur ersten Auslegung siehe Lateiner 1989, 129 und auch Crahay 1956, 328, der 7.117.2 ignoriert und 2.158.5 mit 1.174.5 gleichsetzt, wofür er ferner als erste von zwei Interpretationen „précepte de conservatisme religieux" anbietet. Zur zweiten Erklärung siehe Scullion 2006, 193–94, der 2.158.5 und 7.117.2 auf dieselbe Ebene stellt und 1.174.5 als Ausnahme versteht.

21 Siehe Romm 2006, 189: „Competing in Herodotus' mind with a sense of the inviolability of nature, in other words, is an esteem for human technological progress, especially when it achieves monumental changes in the landscape or in the quality of civilised life."

Romm, J. 2006. „Herodotus and the natural world“. In: C. Dewald und J. Marincola (eds.), The Cambridge Companion. Oxford, 178–191.

Rosenberger, V. 2001. Griechische Orakel. Eine Kulturgeschichte. Darmstadt.

Scullion, S. 2006. „Religion in Herodotus“. In: C. Dewald und J. Marincola (Hrsg.), The Cambridge Companion. Oxford, 192–208.

FAUCES RUBRI MARIS: THE GRECO-ROMAN BAB AL-MANDAB (5TH CENTURY B.C. – 2ND CENTURY A.D.)

Pierre Schneider

Je songe que Bab el-Mandeb veut dire 'Porte des larmes', et j'ai bien failli l'apprendre par l'expérience (H. de Monfreid, *Lettres de la mer Rouge* – 12 janvier 1914)*

"It is the sea more than anything else that defines the contours of the land and gives it its shape, by forming gulfs, deep seas, straits, and likewise isthmuses, peninsulas, and promontories."[1] Whereas straits delineate the contours of the *oikoumenê* like capes, gulfs etc., they differ from these coastal elements which shape the inhabited world, as they divide seas without closing them; while separating land masses, they link them by allowing the passage from one side to the other. Straits both separate and connect.

In the Greco-Roman representation of the world, four straits were regarded as being different from all others. The inhabited world, thought to be a enormous island, was said to be surrounded by an external ocean entering the different land-masses through four "passages" and forming four gulfs: the Mediterranean Sea, the Red Sea, the Arab-Persian Gulf and the Caspian Sea.[2] The strait today named Bab al-Mandab was one of these four "effluents" of the peripheral ocean.

This remarkable locality of the Greco-Roman *oikoumênê* remained for a long time beyond the reach of those dwelling in the countries of the Mediterranean, until Alexander the Great swept away the limits of the Greek horizon. From then on, not only did the "strait of the Red Sea" contribute to "define the contours" of the southern *oikoumenê*, but it steadily opened the Indian Ocean to Greco-Roman merchants. For this reason, although we wish we had more evidence, the *fauces Rubri maris* really deserves a proper inquiry.

N.B. 1) There are many discrepancies between the Greco-Roman and modern geographical names, among which the following must be emphasised: the Red Sea was commonly called "Arabian Gulf" or "Erythraean sea"; the western Indian Ocean (*viz.* the Red Sea, the Arab-Persian Gulf, the Gulf of Aden and the sea of Oman) was generally named "Erythraean sea" (*Erythra thalassa*; *mare Rubrum*). In this paper, the name "Red Sea" is employed only in its modern sense. 2) According to a common Greco-Roman conception, the western shore of the Red Sea belonged

* I would like to thank Klaus Geus for inviting me to the seminar "The Horn of Africa" held in Berlin on 25th–26th February 2011. The present article is an improved version of my oral presentation.

1 Str. 2, 5, 17 (transl. JONES).

2 See e.g. Dion. Per. 40–57. The four "passages" are the Strait of Gibraltar, the Strait of Bab al-Mandab, the Strait of Hormuz. It was long believed that the Caspian Sea was connected to the outer ocean through a strait.

to Asia, not Africa. 3) Bold numbers in brackets (e. g., [**12**]) refer to the texts collected in the appendix 7.1.4) Last but not least, in the present article the name "Bab al-Mandab" generally points to the region of Bab al-Mandab (see fig. 3), instead of the strait *stricto sensu*.

1. THE HISTORICAL BACKGROUND: THE GREEKS AND ROMANS IN THE AREA OF BAB AL-MANDAB (FROM EARLY TIMES TO THE 2ND CENTURY A. D.)

1.1 Before Alexander the Great

Did many Greek sailors ever skirt the shores of Bab al-Mandab in the pre-Hellenistic times? Very little textual evidence enables us to answer this question. A couple of scarce allusions from Herodotus' *History* lead us to the conclusion that only a few of them might have reached the strait's area.

Herodotus claims that between the late 7th century and the early 6th century B. C.[3] the Egyptian Pharaoh *Nekôs* (Nechao II [*regn.* 610–595 B. C.]) sent out an expedition manned by Phoenician sailors; they had been ordered to circumnavigate Libya (= Africa): "When he had finished digging the canal which leads from the Nile to the Arabian Gulf, he sent Phoenicians in ships, instructing them to sail on their return voyage past the Pillars of Heracles until they came into the northern sea and so to Egypt." At some point in this unbelievable voyage, the Phoenicians sailed across the Strait of Bab al-Mandab. Yet Herodotus, or his sources (maybe Egyptian priests), does not comment on this fact and does not even mention it. At any rate, this region was not visited by Greek individuals, since only Phoenician sailors were on board. The same *Nekôs* is reported to have created a war fleet in the Red Sea. But where in the Red Sea did these ships sail?[4] Herodotus does not give any clue. What purpose were they built for? Herodotus alludes to an occasional (ἐν τῷ δέοντι) military use (πρὸς στρατηίας). Some scholars think that these vessels supported the expeditions to the mysterious land of Punt, while others surmise that they patrolled as far as the open sea (*viz.* the Gulf of Aden) chasing pirates in different parts of the Red Sea, including around the strait of Bab al-Mandab.[5] That these triremes were Phoenician and that some Greeks had been enrolled as sailors is a reasonable assumption, which unfortunately cannot be proved.[6]

The only documented case goes back to the time of the Persian rule in Egypt, *viz.* several decades after *Nekôs*' reign. In *ca.* 520 B. C., Skylax of Karyanda, a Greek serving Darius I, performed a notable achievement. He led a naval expedition from the mouths of the Indus – he had previously sailed down this river – to the Gulf of Suez's harbour where the Phoenician mariners of *Nekôs* had begun their

3 Hdt. 4, 42.
4 Hdt. 2, 159.
5 Desanges (b) 228; Lloyd (a) 142–148.
6 A relationship between *Nekôs* and the city of Miletus is alluded to by Herodotus (Hdt. 2, 159). See also Delbrueck 11; de Romanis 78–81.

circumnavigation.[7] Skylax had to pass the strait on his way to Egypt but no piece of evidence refers to this event.[8] Shortly after Skylax's achievement, Darius had the Nile connected to the Red Sea (*viz.* the Gulf of Suez) by a canal – this is also attested by inscriptions – and sent a certain number of ships to the Arab-Persian Gulf from Egypt.[9] Supposing that Greek sailors were on board, no traces of what they did and saw remain in the extant textual evidence.[10]

1.2 The time of Alexander the Great

Back in Susa after his incredible expedition to India Alexander commissioned several exploratory missions throughout the Erythraean Sea. The circumnavigation of the whole Arabian Peninsula seems to have received special interest, as Alexander planned naval explorations in two opposite directions.[11] Thus, nearly two centuries after Skylax's voyage, the Greeks – or to be more precise, the Greco-Macedonians – penetrated the southern Red Sea again. Whether or not these expeditions were the prelude to future imperialistic plans, they finally failed because of Alexander's premature death.

In the autumn of 324 B.C., three separate missions led by Archias, Androsthenes and Hiero respectively explored the western coast of the Arab-Persian Gulf. Hiero, in particular, had been ordered to reach Egypt (i.e. the Gulf of Suez) from the bottom of the Arab-Persian Gulf, but actually he failed.[12] He undoubtedly passed the strait of Ormuz, because he could not have claimed that the Arabian Peninsula was "not smaller than India" without sailing beyond this point. He was, however, forced to turn back because of impossible sailing conditions, perhaps before he entered the strait of Bab al-Mandab.[13]

The same year other mariners had been assigned the task of doing this voyage in the opposite direction (from the northern Red Sea to the Arab-Persian Gulf) by Alexander. Although they failed to achieve their goal, they probably sailed as far as the Gulf of Aden, as shown by three documents.[14] First Theophrastus, in a section of the *Historia plantarum* devoted to the frankincense tree and myrrh tree, mentions explorers travelling from the Gulf "of the Heroes" (Ἡρώων κόλπου = the Gulf of

7 Hdt. 4, 44 (that Skylax reached Bab al-Mandab is doubted by some scholars, such as Salles 80).

8 Hecataeus' fragment *FGrHist* 1, F 271 points to the Arabian islands named *Kamarênoi.* It has been believed that this name points to the Kamaran Island (Schiwek 14; Högemann 106).

9 Schiwek 15–16; de Romanis 94–95; Briant 397: "… jamais il n'exista de communication directe et régulière entre la mer Rouge et le golfe Persique tout au long de la période achéménide." Also see Högemann 107 (who surmises that the toponym šb, which appear in the inscriptions of Darius, points to *Sabbatha*, "die Stadt der Sabäer am Bab al-Mandab").

10 Delbrueck 21 states that the Greek colony of *Ampelome* (below, n. 31) was founded by Darius.

11 Arrian. *anab.* 7, 19, 6.

12 Arrian. *anab.* 7, 20, 7–9.

13 Delbrueck 29; Högemann 88–94; Salles 84–85.

14 Theophr. *h. plant.* 9, 4, 4–5; Str. 16, 4, 4 [= Berger III B 48, p. 290]; Arrian. *Ind.* 43, 7. There is also an obscure statement by Plin. *nat.* 2, 168 (*victoriae Magni Alexandri lustravere usque in Arabicum sinum*).

Suez) who reached a country belonging to the Sabaeans. He goes on to say that they managed to observe these trees during the main harvesting season (September). This event is beyond doubt connected to Alexander's plan. Secondly Strabo reports – from Eratosthenes – a measurement of the east coast of the Red Sea (14,000 *stadia*), which had been calculated by "Alexander's companions and Anaxikratês" (οἱ περὶ Ἀλέξανδρον ἀνέγραψαν καὶ Ἀναξικράτη).[15] Thirdly Arrian informs us of anonymous explorers sent by Alexander from Egypt (a name which actually points to the Gulf of Suez) to navigate around the Arabian Peninsula, who were compelled to turn back due to lack of water.[16] Whether Alexander ordered one or two naval expeditions cannot be established on the basis of such evidence. Despite this unsolvable problem, however, both Strabo and Theophrastus indicate that the strait of Bab al-Mandab was reached: on the one hand measuring the Arabian coast implied travelling as far as its southern edge; on the other hand the Greek explorers would not have observed frankincense trees if they had not penetrated the region of produce, which was probably situated somewhere around *Kanê* [Qana'] in the Hadramawt, east of Aden.[17] Supposing then that Alexander's mariners reported something about the strait, nothing appears in our evidence.[18] Maybe the discovery of the trees producing the aromatics praised by Herodotus and highly rated in the entire Greek world outshone the crossing of the strait.[19] This assumption can be inferred from the following quotation, which perhaps refers to Anaxikratês' voyage: "However, there goes with them (sc. Herodotus' statements) a story that under the reflected rays of the sun at midday an indescribable sort of collective odour is given off from the whole of the peninsula (…) and that the first news of Arabia received by the fleets of Alexander the Great was carried by these odours far out to sea."[20]

1.3 The Hellenistic period: the Ptolemaic rule in Egypt (323–30 B.C.)

In 283 B.C. Ptolemy II inherited from his father a kingdom whose core was composed of Egypt and *Koilê* Syria. The northern end of the Red Sea (i.e. the Gulf of Suez), and the Egyptian coast with its harbours connected by tracks to the Nile Valley (*Myos Hormos*, *Berenikê*), were more or less under his control. As several events such as the reopening of the Nile-Red Sea canal and the grand Dionysiac procession presented to Alexandrian citizens in the 3rd *Ptolemaieia* celebration show, the countries lying south of Egypt, and particularly the Arabian Gulf, which

15 Ἀναξικράτη[ς]: Bernhardy's emendation has been accepted by Jacoby (*FGrHist,* 307, n. 2) but rejected by Radt 345 who doubts that the Red Sea was explored in Alexander's times. Berger (a) 290–291 believes that this Alexandros was an explorer / writer, and not the Macedonian king. Bernhardy' correction is defended by Gorman 212–213.

16 About this (or these) expedition(s), see Högemann 80–87; Desanges (b) 245; Mac Donald 246; Pfister 32–34; Tarn (b) 13; Delbrueck 28; Bianchetti (b) 158–159.

17 On this question, see Delbrueck 28; Högemann 87; Amigues 674–675; Potts 6.

18 Significantly Arrian. *Ind.* 43, 1–3, does not refer to the strait in his short description of the "Great Sea".

19 Hdt. 3, 107.

20 Plin. *nat.* 12, 86 (transl. Rackham). See Burstein 160, n. 2.

was almost unknown to Greeks at that time, played an important part in Ptolemy II's imperialistic projects.[21]

It is well known that this dynamic king launched expeditions along the western shores of the Red Sea for military and political purposes, namely to capture African war elephants and improve his own prestige – the quest of commercial profit is unlikely.[22] Several settlements (*Arsinoe*, *Philotera*, *Ptolemais Therôn* [= of the Hunts]) established along the western Red Sea coast formed the backbone of Ptolemy's project, which was so successful that, as Huss states, "der nördliche Teil des Roten Meers war eine *mare Ptolemaeum* geworden."[23] Whether other bases of operation south of *Ptolemais* were founded under Ptolemy II's rule remains unclear. It has been claimed, however, that his explorers reached the Bab al-Mandab strait. Two pieces of evidence are supposed to prove it.[24] First, a certain Timosthenes, an admiral of Ptolemy II who wrote a treatise entitled *On harbours*, gave a figure of the width of the strait, which implies that, unless he borrowed this data from a previous source – for instance Anaxikratês –, he reached the area of Bab al-Mandab. Secondly, we are aware of a hunting base called *Demetriou skopiai* (the "look-out place of Demetrios") situated not far from the strait (around Assab?)".[25] Coincidentally, a certain Demetrios responsible for looking after the elephants is mentioned in a papyrus dated 255 B. C.[26], but, as has rightly been observed, this name is too common for any sound conclusion to be drawn from it. In reality, since the identity of the first explorers is not disclosed by our sources [**2**; **3**] it is impossible to determine whether they were sent by Ptolemy II[27] or, as Wilcken believes, by Ptolemy III.[28]

Ptolemy II also had the eastern Red Sea in mind, that is to say the shoreline skirted by Alexander's explorers on their way to the Arab-Persian Gulf. However, compared with the African side, the situation was different. Not only were the resources of the arid Arabian coast particularly scarce and "in general not suited for maritime activity"[29], but the south-west edge of the Peninsula was the domain of several ancient and well-organized states, namely Ma'in, Saba', Qatabân, which the Greeks believed to be extraordinarily wealthy.[30] Under such circumstances no

21 Desanges (b) 246–252; Huss 288; Fraser (a) 177; Yémen (Cuvigny) 68.

22 Huss 288–290 (*contra*: Fraser (a) 141, 177). On elephant hunting, see e. g. Kortenbeutel 27–28, 35–38; Scullard 126–133.

23 Huss 290. *Arsinoe* was located in the bottom of the Gulf of Suez; *Philotera*'s location remains unknown (Marsa Gawasis?); most scholars agree that *Ptolemais* was founded at or near Aqiq: Huss 289; Sidebotham (b) 186–187.

24 Desanges (b) 275–276. See also Fraser (a) 177–178.

25 Str. 16, 4, 9 (Artemidorus).

26 *Pap. Hibeh* 110 (WChr. 435). See Fraser (b) 305–306, n. 365.

27 Kortenbeutel 27–28; Desanges (b) 275. Fraser (b) 303, n. 353 (and [**12**]) considers it possible that the ductus points to Arsinoe Philadelphus. Therefore, the straits would have been crossed under Ptolemy II's rule.

28 Wilcken 87; Burstein 9.

29 Raschke 657.

30 Diod. 3, 47, 4–8 (= Agatharchides, *GGM* 1, 100–101).

Ptolemaic settlement is likely to have been founded in this area.[31] We are merely aware of a certain Aristôn sent by a king – probably Ptolemy II – to explore the eastern coast of the Arabian Gulf as far as the Ocean, *viz.* the Gulf of Aden (πρὸς κατασκοπὴν τῆς ἕως ὠκεανοῦ παρηκούσης Ἀραβίας)[32], implying that he probably saw the strait of Bab al-Mandab. At least Tarn thinks so, claiming that he provided Eratosthenes with a measurement of the Arabian coast which was more accurate than the previous one established by Anaxikratês (14,000 *stadia*).[33] Much more obscure is Agatharchides' statement in a passage laden with *paradoxa* (marvels) that some travellers, while sailing along the coast of the Sabaeans (τοῖς προσπλέουσι τὴν παράλιον), could smell fragrances coming from trees (*viz.* incense trees, myrrh trees etc.).[34] Who were these travellers: merchants and /or official explorers? When did they skirt the Sabaean coast? Did they sail across the strait and travel on the open sea? As these questions remain unanswered, there is no point in inferring anything from this document.[35]

As regards the African coast in Ptolemy III's time, the Greeks, who had reached the western side of Bab al-Mandab, actively explored this area and went even further, as proven by several reliable documents.[36] First, Agatharchides refers to an ethnographical experiment carried out by Simmias, a *philos* (friend) of *Evergetês*, in order to observe the behaviour of a peculiar tribe of African *Ikhthyophagoi* ("Fish Eaters"). This event took place in the neighbourhood of Tajura.[37] Secondly, proof is offered by two ports lying inside the western side of the strait that were named *Berenikê*, after the name of Ptolemy III's wife.[38] This exploratory activity was mainly motivated by the quest for war elephants, which at this time were hunted from bases set up inside and beyond the strait along the northern Somalian coast; the latter region

31 On *Ampelome oppidum* (Plin. *nat.* 6, 159) allegedly founded by Ptolemy II, see, Tarn (b) 21–22; Kortenbeutel 21; Fraser (a) 177; Raschke 964, n. 1249; Desanges (b) 245); Cohen 44–45, 307, 329.

32 Diod. 3, 42, 1 (= Agatharchides, *GGM* 1, 85). About Aristôn, see Kortenbeutel 20; Tarn (b) 14; Delbrueck 33; Woelk 259; Fraser (b) 300, n. 349. About the exploration of the eastern side, see Bianchetti (b) 158–159. Delbrueck's view that Pythagoras (Plin. *nat.* 39, 34) led a "zweite Erkundung" derives from a misinterpretation: the islands referred to by Pl. are unlikely to lie "innerhalb und außerhalb der Mündung des Golfs".

33 Str. 16, 4, 4. See Tarn *l. c.*, n. 32; Kortenbeutel *l. c.*, n. 32; Woelk 259–260; von Wissmann (d) 391.

34 Diod. 3, 46, 5 (= Agatharchides, *GGM* 1, 97).

35 According to Agatharchides, there were, close to the country of the Sabaeans (πλησίον), some "blessed islands" (νῆσοι δ΄ εὐδαίμονες) where unprotected cities lay. Von Wissmann (a) 304–305 believes that these islands should be identified with Aden (also Högemann 85). However, most scholars (Tkač (b) 1402–14; Kortenbeutel 55, n. 169; Müller (a) 191; Delbrueck 35; Raschke 657; Burstein 169; de Romanis 143) think that Suqutra and the adjacent islands should be favoured. I personally tend to prefer von Wissmann's opinion, since: 1) Aden was also called *Eudaimôn*; 2) the two peninsulas of Aden and Little Aden (Desanges (b) 157, n. 39) might look like islands (see [**23**]). In addition, according to the short description from *P. m. r.* 30–31, Suqutra does not have the characteristics of a "blessed" place. In constrast, *Eudaimôn* / Aden is described as an "Arab Alexandria" [**29**] before it was sacked by *Kaisar*.

36 Kortenbeutel 33–34, 37; Rostovtzeff (a) 302; Huss 366.

37 Diod. 3, 18, 4–7 (= Agatharchides, *GGM* 1, 41). See Desanges (b) 292 (Tajura); Woelk 134 (Somaliland).

38 Desanges (b) 295–296.

was called *Arômatophoros* – the Spice-bearing country – and divided into three parts respectively named *Smyrnophoros* (the "Myrrh-bearing country"), *Libanôtophoros* (the "Incense-bearing country") and *Kinnamômophoros* (the "*Kinnamômon* [a still not clearly identified spice] bearing country").[39] An interesting phrase appears in a dedication to the Samothracian gods perhaps dating to Ptolemy III' rule ([**13**[40]] and below, n. 395): "as he sailed out of the Erythraean Sea" (ἐκπλεύσας ἐκ τῆς Ἐρυθρᾶς θαλάσσης). This expression seems to be synonymous with "to pass the strait".[41]

Most historians agree that the demand for war elephants ceased in the early 2nd century B.C. during the rule of Ptolemy V (205–180 B.C.).[42] Yet this new situation did not put an end to the Greek presence around Bab al-Mandab. The area which the Ptolemaic expeditions had penetrated was from then onwards open to private maritime traders (*emporoi*).[43] – whether, or how, or to what extent this phenomenon was connected to a "ptolemaeische Handelspolitik"[44] does not directly concern the present inquiry. Shipowners (*naukleroi*) and merchants (*emporoi*) who were not afraid of the hazards of the sea began – maybe as soon as the early-2nd century[45] – to import commodities such as myrrh and frankincense from modern-day Eritrea and northern Somalia, and even from the Arabian side of the Bab al-Mandab area[46], perhaps as far as Aden.[47] This was a significant change in the pattern of exchanges, since up until then aromatics and spices – no doubt including some varieties harvested in the Horn of Africa – were transported to the Mediterranean world by the trans-Arabian caravan route through Nabatean territory.[48] A papyrus, probably dated to the mid-2nd cent. B. C., reporting a trade voyage to the *Arômatophoros* country, is in this respect a valuable piece of evidence. Interestingly, this document refers to "the outside sea" [**13bis**], a (common?) expression which probably means

39 Str. 16, 4, 14 (Artemidorus). For Ptolemy III's hunting stations inside and beyond the strait, see e.g. Rostovtzeff (a) 302–304; Fraser (b) 305–308, n. 365–370.

40 Raschke 949, n. 1193 believes that this document is "the earliest documentary evidence for the voyage to the vicinity of Bab al-Mandeb and beyond."

41 Cf. [**6**]; Plin. *nat.* 2, 169; Arrian. *Ind.* 43, 2 (τὴν ἔξω θάλασσαν); *P. m. r.* 30. Thus von Wissmann (b) 430 is wrong writing: "about 115 B.C. the strait of Bab al-Mandab was opened for the first time for Egyptian ships, so that these could directly sail to India avoiding transshipment into foreign vessels."

42 See e.g. Rostovtzeff (a) 303–304; Fraser (a) 179.

43 Kortenbeutel 44; Raschke 659; Huss 425.

44 Delbrueck 44; Rostovtzeff (b) 741–743. There is a large amount of academic literature dealing with the African/Indian trade. The reader can find useful discussions and many bibliographical references in recent monographs (e.g. Sidebotham; Raschke; de Romanis; Young).

45 The earliest piece of evidence is the papyrus published by Wilcken. See Kortenbeutel 44; Otto & Bengtson 200. Dihle (c) 115 argues the "griechische Seeleute" sailed beyond Bab al-Mandab only after Eudoxus' voyage to India, a view rejected by Raschke 940, n. 1962.

46 See [**13bis**] (and Wilcken 90–91); [**12**]; Bernand (c) n° 2 (a dedication offered to Pan by Zenodotos for a safe return from the country of the Sabaeans); 3, 13, 43 etc. (dedications offered to Pan by private individuals returning from the *Trôgodytes'* country, perhaps from the strait area); Diodorus (below, n. 107); *OGIS* 132 (130 B.C.). Also see Desanges (b) 299–300; Bernand (c) 570–571; Raschke 950, n. 1193–1194; Fraser (b) 309–310, n. 378.

47 See Schiettecatte (b) 239–242.

48 About the coexistence (and/or competition) of the sea route and the trans-Arabian route, see Ryckmans 85–89; Groom (a) 189–213; Rodinson 215–216; *Yémen* (Robin) 181.

"the sea which extends beyond the strait" (see below, n. 396). Finally, there is a passage in Diodorus stating that in Agatharchides' time (νῦν), "many" (πολλοί) merchants "often" (πολλακίς) reached a country lying beyond the strait. This is necessarily an allusion to merchants loading aromatics in *Smyrnophoros*, somewhere in northern Somalia.[49]

To Greeks becoming increasingly familiar with the crossing of Bab al-Mandab the next step was to catch the monsoon winds and reach India. This feat was achieved by a certain Eudoxus of Cyzicus – apparently not a professional *emporos* – in the late-2nd century, during the rule of Ptolemy VIII and Cleopatra III, and, after the former's death, of Cleopatra III and Ptolemy X Lathyrus (between 116 and 113 B.C.).[50] This discovery was to have impact on the trade in the Arabian part of Bab al-Mandab. Henceforth commodities could be shipped directly from Indian ports to Alexandria by Mediterranean *emporoi*, eventually decreasing the incomes of south Arabian middlemen.[51] Unfortunately, the scarce and vague textual evidence does not enable us to clearly determine how fast this maritime trade expanded. Strabo [**21**], probably referring to the time of Ptolemy XII (*regn.* 80–51 B.C.), claims that not even twenty ships (a year?[52]) would dare to sail out of the Arabian Gulf "as far as India and the extremities of *Aethiopia*" (= *akrôn tôn Aithiopikôn*?). If, as I tend to believe, Strabo does not take into account the *emporoi* shipping local and re-exported goods from Arabian ports[53], one may surmise that more than twenty vessels a year reached the area of Bab al-Mandab in the early-1st century B.C. At any rate, this period witnessed the appearance of an official title, "*epistrategos* of the Erythraean (= the Arabian Gulf?) and Indian Sea (the Gulf of Aden?)"[54] that maybe implied the presence – still not clearly proven by documents – of royal vessels up to the area of the strait.[55] This novel situation certainly correlates with the increasing maritime trade.

The preceeding short presentation would scarcely be comprehensive if I had neglected two documents, whose actual historical value is, however, disputed. The first of these is the narrative by a certain Iamboulos – this name suggests a Nabataean origin – which may go back to the 3rd century B.C. His narrative was summarized

49 Diod. 3, 18, 3 (= Agatharchides, *GGM* 1, 41). See below, n. 107.

50 See e.g. Dihle (c) 111–112; Thiel 44–55; Fraser (a) 181–184; Desanges (b) 151–173. According to Desanges, Eudoxus sailed with the south-west monsoon wind from the Cape Gardafui.

51 Consensus has it (see e.g. Rostovtzeff (b) 738–740; Sidebotham (a) 2; Desanges (b) 157) that Arab middlemen supplied Alexandrian merchants with Indian commodities untill the latter began to use the monsoon winds. For instance, the Indian precious stones (λιθοὶ Ἰνδικοί) used in Ptolemy IV's house-boat (Athenaeus, *Deipn.* 5, 39 [= Kallixenos, *FGrHist* 627, F 1]) may have been imported by them.

52 Fraser (a) 184.

53 At the time of the *Periplus* such short round trips (from Egypt to *Adoulis, Mouza*, or *Kanê*) were still undertaken by some merchants (Casson 288, n. 15).

54 *SB* 1 n° 2264 (79 B.C.); 1 n° 4084 (62 B.C.); 5.2 n° 8036 (74/73 B.C.?); *OGIS* 186 (62 B.C.); 190 (51 B.C.?). See Kortenbeutel 48–50; Rostovtzeff (a) 305; Fraser (b) 314–315, n. 398–401; Delbrueck 45; Dihle (b) 548; Raschke 971, n. 1292–1295; Fraser (a) 184.

55 Kortenbeutel 45; Rostovtzeff (a) 305: "... der Strateg der Thebais, welcher also auch das Oberkommando über die Küste des Roten Meeres führte."

by Diodorus. On his way to the *Arômatophoros* (the "Spice bearing region", i.e. probably the south-west part of the Arabian Peninsula) through Arabia (ἀναβαίνων δὲ διὰ τῆς Ἀραβίας ἐπὶ τὴν ἀρωματοφόρον), and apparently via the trans-Arabian caravan route[56], this Iamboulos and his fellow traveller were caught by robbers and taken prisoners. Sometime later they were again captured, this time by Ethiopian bandits, and brought to the coastal part of *Aithiopia* (πρὸς τὴν παραθαλάττιον τῆς Αἰθιοπίας = northern Somalia?[57]). The rest of the story has no relevance for the present investigation. The point is that despite their vagueness, these geographical details seem to refer to the area of Bab al-Mandab. Iamboulos was apparently taken to the African side across the strait, in a region where the two opposing coasts were close to each other. Of course, the main problem is to what extent we can rely on this story – an issue already discussed in antiquity. Even if Iamboulos' narrative is partly fictional, it nonetheless clearly points to some local trade traffic (or piracy) having required light boats in order to move from one side of the strait to the other (to be discussed below). In contrast, the author of the second document, Euhemerus of Messenia, can be more precisely dated.[58] He was admitted into the circle of the Macedonian king Cassandros as a "friend" (*philos*) between 318 and 298 B.C.[59], and his *Hiera anagraphê* may have been composed between 280 and 260 B.C. He was sent to Blessed Arabia (ἐκ τῆς εὐδαίμονος Ἀραβίας), – probably on some official duty (βασιλικάς τινας χρείας καὶ μεγάλας ἀποδημίας) – the Sabean country is probably meant[60]. Then he boarded a ship on a voyage of the "ocean" (τὸν πλοῦν δι' ὠκεανοῦ) to an enigmatic island (Suqutra?).[61] As far as Euhemerus' narrative is based upon reliable facts it relates a journey to southern Arabia by caravan routes as well as a voyage that probably began in a harbour near the strait. This may support the view that a Greek individual with no relations to the Ptolemies reached the Arabian area of the strait as early as the early-3rd century. Even if the bulk of Euhemerus' story is fictional[62] it nonetheless surely incorporates a certain amount of reliable data collected from external sources. It is therefore plausible that this narrative echoes the ongoing exploration of the area of the southern Red Sea carried out during the reign of Ptolemy II.

56 Diod. 2, 55. *N.B.*: In Phot. *Bibl.* 250, 457b (= Agatharchides, *GGM* 1, 87) the verb κατάγω refers to the trip from southern Arabia to Palestine.

57 Kroll 681, following Tomaschek, states that the *Arômatophoros* lies on the north Somali coast and that the Ethiopians took him to "einem Stamme der ostafrikanischen Küste". Altheim 84 thinks that the Ethiopians inhabited today's Eritrea. Also see Winiarczyk 130–132, 134–137, 139–141.

58 Eusebius Caesariensis, *Praep. Ev.* 2, 5 (= Diod. 6, 1–11). See also Diod. 5, 41–46. See Jacoby 959; Fraser (a) 289–298; Braunert, especially 258–259, 265–268; *Yémen* (Cuvigny) 68.

59 *Stricto sensu*, Cassandros ruled Macedonia as king from 306 to 298 B.C., but actually he acted as a king before 306.

60 Delbrueck 36 identifies *Eudaimôn Arabia* as Aden.

61 Tkač (b) 1403.

62 Such was Eratosthenes' opinion (Str. 2, 4, 2).

1.4 The Early Roman Empire (from Augustus to the 2nd century A. D.)

With the defeat of Cleopatra VII and Mark Antony (31 B. C.) and the annexation of Egypt, Rome under the leadership of Augustus was from then onwards in contact with the southern edge of the *oikoumenê*. In fact it did not take long before the *arma Romana* were sent southwards to *Aethiopia* (the common designation of the Nubian kingdom of Meroe) and to southern Arabia.[63] In 26 B. C. – or perhaps in 25 – an overland expedition headed by the *praefectus* of Egypt Aelius Gallus, a friend of Strabo, penetrated southern Arabia. If taking control of the Sabaean kingdom and the east side of Bab al-Mandab was part of Octavius' plan [**20**][64], the Roman ruler was unsuccessful, at least from the military point of view. Aelius Gallus' army reached *Mariba* (Mârib) but could not advance further. Still, this arduous campaign was not entirely unsuccessful, as the local Arab kingdoms got a taste of the power of Rome: "Local rulers in southern Arabia certainly would have wanted to stay on good terms with Rome to prevent any future military adventures in their territory".[65] It is, however, pure speculation that the strait came under Roman contral at the same time.[66]

We do not know exactly which goals concerning Arabia Augustus' grandson, Gaius Caesar, had in mind in the early years of the 1st century A. D. Pliny the Elder credits him with mysterious "Arabian operations"[67] (*Arabicas res*) as well as a shadowy expedition in Arabia (*expeditione Arabica*). According to the same author, G. Caesar is also said to have headed a naval unit in the Red Sea ([…] *in Arabicum sinum, in quo res gerente G. Caesare Augusti filio* […]) and explored Arabia (*G. Caesar Augusti filius prospexit tantum Arabiam*).[68] Bowersock may be right in thinking that the name *Arabia* actually designates the country of the Nabataeans, on the north-east coast of the Red Sea.[69] Beside these two passages Gaius Caesar, "whose imagination was fired by the fame of Arabia" (*ardentem fama Arabiae*) is, however, reported to have sought glory in the *turifera* ("frankincense bearing") Arabia (*Gaius etiam Caesar Augusti filius inde gloriam petiit*).[70] In his excerpt Pliny associates Gaius Caesar's enterprise with another Arabian expedition, no

63 See *R. Gest. div*. 26, 5.

64 "The campaign of Aelius Gallus at the beginning of the principate has always been something of a mystery, but it is quite clear that Augustus had some kind of expansionist interest at that stage in controlling the rich trade in spices and perfumes" (Bowersock (a) 227). Rostovtzeff (a) 306–309 claims that the Roman policy – like the Ptolemaic one – was driven by "commercial" interests. See also Delbrueck 231; Raschke 871, n. 901; Sidebotham (a) 121; Desanges (b) 308; Scheid 72; Roddaz 273, among many scholars.

65 Sidebotham (a) 128.

66 According to Rostovtzeff (a) 310–311, the Romans set a customs checkpoint on the Arabian Bab al-Mandab and hencefore were capable of locking the strait by its two opposite sides and levying taxes (see also Kortenbeutel 55, 59). This view is barely supported by firm evidence (Raschke 647).

67 Plin. *nat*. 6, 141. This may have been connected with Parthian affairs (Villeneuve 182, n. 206; see also Raschke 872, n. 906).

68 Plin. *nat*. 2, 168; 6, 160.

69 Bowersock (a) 227: the *sinus Arabicus* is identified with the Gulf of Aqaba.

70 Plin. *nat*. 12, 55–56 (transl. Rackham).

doubt that led by Aelius Gallus. Accordingly, the Arabian country targeted by Gaius Caesar can reasonably be equated with the Sabaean kingdom, which was the goal of Gallus' incursion. At any rate, one can assume that if Gaius Caesar did not actually reach the area of the Arabian side of Bab al-Mandab, at least his plans may have encompassed it.[71]

The author of the *Periplus of the Erythraean sea* – written between A. D. 40 and 70 – refers in vague terms to a Roman military intervention around Bab al-Mandab, not documented elsewhere [**29**]: a certain *Kaisar* is reported to have sacked ("subdued" is, however, an alternative translation of the verb *katastrepsato*) *Eudaimôn Arabia* (here referring to Aden instead of the whole Blessed Arabia) not long before the author's time.[72] The author goes on to say that following this action the maritime road to India, which was previously under Arabian control, was opened – obviously to Alexandrian merchants and sailors. Contrary to the statement of the *Periplus*, this nebulous expedition could not have affected the whole maritime traffic to India, since Strabo states that Greek ships were capable of making the round trip between Egypt and India before and after the battle of Actium.[73] This puzzling text has been much commented on, and the most convincing explanation is one of the following two: either the author alludes in a confused way to Gallus' expedition, mixing the name of the country (*Eudaimôn Arabia*) up with that of the city (*Eudaimôn*); or the *Kaisar*, who allegedly attacked Aden (in *c.* A. D. 2?), must be identified with Gaius Caesar. Be that as it may, a Roman military presence around Bab al-Mandab in the mid-2nd century A. D. is attested by an inscription recently found in the main Farasan island (Saudi Arabia). It is dated to the reign of Antoninus (A. D. 144) and refers to a Roman detachment (*vexillatio*), headed by a *praefectus*, being sent to an outpost at this site.. This Roman presence may be, to some extent, explained by troubles in south Arabian kingdoms making the area of Bab al-Mandab insecure. According to the editors of the inscription, the duties of the *praefectus* of Farasan may have included overlooking the trade traffic, hunting pirates or levying customs at the entry of the Red Sea, as the Farasan islands lie close to Bab al-Mandab[74] From a certain point of view, one could even perceive the *vexillatio* as being garrisoned at its north entrance.

Closely connected to Roman penetration into the Red Sea area is the significant boom in Roman sea trade with East Africa, Arabia and India, which started in the late-1st century B. C. Many ships mainly operated by Alexandrian *emporoi* would sail across the strait towards India or East Africa [**21**]. Since several studies have been devoted to the various aspects of the commerce in the Indian Ocean (above, n.

71 Desanges (b) 319–320.

72 This passage has been much debated. See e. g. Delbrueck 241; Dihle (a) 22–25; Fraser (b) 313; Wellesley; Casson 37; Seland. Some scholars (e. g. Sidebotham (a) 131–132) reject the reading "*Kaisar*" as a corruption.

73 Strabo [**21**] and above, p. 199–200. Note that the *P. m. r.* 57 points to a monsoon voyage starting at the cape Gardafui (*Arômata*), which means that some ships might avoid calling in at an Arabian port. See also Plin. *nat.* 6, 174 (Juba) and Desanges (c) 91.

74 Villeneuve 167–178. On Roman military presence in the Red Sea, see Kortenbeutel 59, 62, 64; Rostovtzeff (a) 306, 310–311 (whose conclusions are partly questionable); Raschke 648–649, 894, n. 964.

44) there is no point in dealing further with this topic. Two important issues must, however, be addressed (even though they still remain beyond answer): Firstly, Pliny informs us of the existence of several maritime routes between the northern Red Sea and the west coast of India, but he does not find it important to say along which side the strait was crossed.[75] Secondly, the question arises as to how many freight ships crossed the strait. The only reliable figure comes from Strabo.[76] As mentioned above 120 ships (a year?) headed to India at the time when Aelius Gallus was *praefectus* of Egypt (from 28 to 25 B.C.) – see also the expression "large fleets" [**21**]. How many ships went to East Africa is, however, not stated by Strabo.[77] Supposing that the Arabian expedition of Aelius Gallus facilitated an easier voyage to India after 25 B.C., the number of ships passing the strait may have increased, but there is no accurate evidence to support this hypothesis.[78]

To sum up: Over a period of about 350 years following Alexander's expedition, the strait of Bab al-Mandab gradually became a spot familiar to more and more Mediterranean people. It took about a century for Greco-Macedonian explorers and officials to discover and cross the strait. These men opened the way for an increasing number of *emporoi* who eventually reached the main places of trade located in the western Indian Ocean. Such is the simplified historical background to which I will repeatedly refer in this investigation.

2. FROM *STOMA* TO *STENA*: THE MAKING OF THE GRECO-ROMAN BODY OF KNOWLEDGE

2.1 Was there a "strait" at the edge of the Arabian Gulf in the Greek geographical knowledge before Alexander the Great?

As stated above, very few Greeks – and no Greek ships – are likely to have reached the area of Bab al-Mandab at the time when Egyptian and Persian rulers prevailed in the northern Red Sea. Accordingly, very little information was available to the Greeks. In fact, Herodotus' *Histories*, which is the only substantial document left from this period, shows that the author was poorly informed about the Arabian Gulf, as were probably also his contemporaries. In fact he managed to gather more information about Arabia, India or Ethiopia than about the Red Sea. What Herodotus knew of the latter region is concentrated within a couple of sentences, the sources of which cannot be traced[79]: his Ionian predecessors (Skylax of Caryanda, or Hecataeus of Miletus) and local sources (Persian or Egyptian informants) are

75 Plin. *nat.* 6, 101–106. Without doubt the third route (via *Muza* /*Ocelis Arabiae* / *Cane turiferae regionis*) passed along the eastern side.

76 Str. 2, 5, 12. RASCHKE 662 states that the figures quoted by Strabo are at best "an indication in the change of scale of the traffic".

77 In my opinion Strabo does not mix India up with *Aithiopia*: see RASCHKE 973, n. 1302.

78 See RATHBONE 187, n. 29.

79 Hdt. 2, 11.

possible candidates.[80] At any rate, Herodotus states that the Arabian Gulf is an extension of the Erythraean Sea – i.e. the outer ocean – penetrating the landmass (ἐσέχων ἐκ τῆς Ἐρυθρῆς καλεομένης θαλάσσης): such a "dynamic" conception was common in Greco-Roman geographical thought.[81] Herodotus is also aware of a measure given in sailing days rather than *stadia* or any other units: the length (μῆκος μὲν πλόου) from the bottom, *viz.* the Gulf of Suez, as far as the "open sea"[82] (ἀρξαμένῳ ἐκ μυχοῦ διεκπλῶσαι ἐς τὴν εὐρέαν θαλάσσαν) is a forty days' trip. The west-east crossing of the gulf (*viz.* its width) takes half a day, at its widest (εὖρος δὲ, τῇ εὐρύτατός ἐστι ὁ κόλπος). This incredible figure proves beyond a doubt that Herodotus imagines the Red Sea as an extremely narrow body of water, the biggest ratio of width to length being 1 to 80![83] Herodotus' Red Sea more or less has the shape of a long river. It must be noted that despite the improvements of geographical knowledge the extreme narrowness of the Arabian Gulf was to remain a *locus communis* in Greco-Roman conceptions [**4**; **19**; **20**[84]].

Now what about the spot where the external ocean flows into the Arabian Gulf, that is to say the strait itself? In his narrative of *Nekôs'* expedition, Herodotus ignores its existence, for he just writes that "the Phoenicians set out from the Erythraean sea and sailed the southern sea" (ὁρμηθέντες (…) ἐκ τῆς Ἐρυθρῆς θαλάσσης ἔπλεον τὴν νοτίην θάλασσαν). Similarly, in the passage devoted to Skylax's voyage no strait is mentioned. Although Herodotus was aware that royal ships sailed out of the Arabian Gulf to reach the external (southern) ocean, he did not, or could not, provide any details about this passage. The Greek historian does not even employ the words *stena* and *stoma* (which on the contrary appear in the passages dealing with the Hellespontus or the Bosphorus[85]). Under such circumstances one can reasonably infer that the Bab al-Mandab strait was unknown to Herodotus. As a matter of fact, given the extreme narrowness of his Arabian Gulf – of which the largest width was actually that of a strait[86] – he probably never imagined a strait lying at this place. As we shall discuss later on, a strait is created by a noticeable shrinking of the sea.

This may explain why Herodotus does not count the crossing of Bab al-Mandab among the achievements of *Nekôs* and Darius – this was to change in Hellenistic times, as will be seen below. From his point of view, which was probably shared by his contemporaries, the true feats of these kings were respectively the successful circumnavigation of Libya (i.e. Africa) and the voyage from the Indus to the Gulf

80 See Berger (b – part 1) 33–37; Lloyd (a) 77–140; Högemann 107–111.
81 See e.g. Mela, 3, 71; Plin. *nat.* 6, 107; Manilius 4, 650–651.
82 See Lloyd (a) 64.
83 Lloyd (a) 64–65 estimates the length at ~28,000 *stadia* (3240 naut. miles). As regards the width, he agrees that "we have no alternative but to assume that H., or his source, has made a very bad mistake." Schiwek 14–15 ("Auch die verhältnismäßig richtigen Maße, die Herodot vom Arabischen Golf anführen kann, rühren vielleicht – über Hekateios – von Skylax her") is purely speculative.
84 See also Phot. *Bibl.* 250, 441b (= Agatharchides, *GGM* 1, 2).
85 See Hdt. 4, 85.
86 See ps.-Skyl. 111 about the Columns of Heracles: the crossing is a one day sailing.

of Suez, for such unprecedented undertakings were a sign of their power and also contributed to an increased knowledge of the inhabited world.[87]

Apart from Herodotus' account, two documentary fragments going back to this period and relating to this area have been preserved. First, Pliny the Elder reports an obscure story borrowed from the historian Ephorus of Kymê[88] (4th cent. B. C.). According to the former, people sailing from the *Rubrum mare* (unfortunately we cannot determine whether this designation covers the Red Sea or the Arab-Persian Gulf) towards the island of *Cerne* (lying opposite the Persian Gulf [*contra sinum Persicum*]) could not go beyond anonymous small islands (*parvae insulae*) referred to as "columns" (*quasdam columnas*), due to the unbearable heat. Whether these islands are located around Bab al-Mandab, and whether these *columnae,* which could be the southern or eastern counterpart of the columns of Heracles, point to the strait cannot be clearly established.[89] The second fragment is attributed to Damastes of Sigeion, a "logographer" flourishing in the 5th cent. B. C., who "is said to have frequently taken Hecataeus as an authority"[90]. Damastes claimed that "the Arabian Gulf is a lake" (τὸν μὲν Ἀράβιον κόλπον λίμνην ὑπολαμβάνοντος εἶναι)[91], denying any connection between it and the outer ocean. In fact no serious conclusion can be drawn on the basis of such vague statements, except that very little was known about the shape and limits of the southern Red Sea before Alexander's time.[92]

In conclusion, in the pre-Hellenistic times most Greeks may have believed that the Arabian Gulf flowed into an external sea. It was probably conceived of as a river discharging its waters into the sea through a mouth (*stoma*) which was not regarded as an actual strait.

2.2 The Hellenistic "invention"

A conqueror laying claim to an empire reaching the limits of the inhabited world (i. e. Alexander the Great) could not ignore the Asian gulfs of the Ocean. Not surprisingly Alexander stated that he had discovered (ἐξεῦρε) that the Arab-Persian Gulf was formed by the "Great Sea" penetrating the land mass. Alexander is also reported to have planned an exploratory mission to determine whether the Caspian Sea was an effluent (ἀναχεῖται) of the external ocean, but he died before its completion. Furthermore, as mentioned above, he ordered Anaxikratês to skirt the western side of the Arabian Peninsula. This voyage, however, is not listed by Arrian among Alexander's glorious feats.[93] The reason may be that the exploration of the Arabian

87 Briant 495 (about Darius).

88 Plin. *nat.* 6, 199 (= Ephorus, *FGrHist* 70, F 172).

89 See Delbrueck 23–24; Högemann 106 ("Sie liegen in der straße von Bab al-Mandab"); Desanges (c) 217–218.

90 Van Paassen 138. See also Berger (b) 65.

91 Str. 1, 3, 1 (= Damastes, *FGrHist* 5, F 8). See Schmitthenner 46. For the sake of comparison, see Str. 1, 2, 31.

92 I do not share Högemann's 86 view that the *polis eudaimôn* (Aristoph. *av.* 144) should be identified as Aden, thus that Aden "war längst in Griechenland bekannt."

93 See Arrian. *anab.* 7, 16, 1–3; *Ind.* 43, 1–10. Plin. *nat.* 2, 168 is too vague to support any conclusion (*victoriae Magni Alexandri lustravere usque in Arabicum sinum*).

Gulf was actually achieved by Ptolemy II Philadelphus and Ptolemy III Evergetes, who committed huge means to reconnoitring the southern part of the *oikoumenê*.[94] In the wake of the expeditions commissioned by these kings a substantial amount of geographical and ethnographical knowledge based upon direct observation (*autopsia*) was made available to Greek scholars. Only autopsy, lacking in Herodotus' times[95], could lead to such a considerable enhancement of knowledge. It is worth recalling here that the dichotomy between *opsis / autopsia* ("observation"; "seeing with one's own eyes"), and *akoê* (the sense of this word may vary according to context but in this case it means "second-hand information"; "hearsay evidence") was common in Greco-Roman antiquity and prevailed before Hellenistic times (for instance it can be traced in Herodotus' *Histories*[96]). With regards to the description of the *oikoumenê*, sight (*opsis*) ranked higher than hearsay (*akoê*), for observation was considered the best source of information available.[97] Indeed several centuries after Herodotus, Strabo still claims that knowledge based on *opsis* is more reliable (*pistis*) than information collected from *akoê*, for instance from previous Greek authors who did not travel in the countries they dealt with[98] or from local informants: *akoê* cannot be deemed entirely reliable, if it is not supported by *autopsia*.

It was in the 3rd century B.C. that a large amount of evidence based upon autopsy was brought to Alexandria: unlike Alexander's explorers who due to circumstances hastily went around Bab al-Mandab, the Ptolemaic observers repeatedly cruised in this area, particularly along the western side of the strait. Thus, they managed to gradually reconnoitre the African side of Bab al-Mandab [**2**; **3**]. Observing the decreasing distance between the two sides of the Arabian Gulf they claimed that the cape *Deirê* (below, p. 215) was the gate to the open sea. Sometime later, however, sailing further southwards (*plospleusasin*) and observing crafts coming from the Arabian coast and landing there, they stated that "from then onwards" (*nuni*) the strait was better located in this place (below, p. 216–217). At any case, by reaching a remarkable place of the inhabited world these men enlarged the geographical knowledge of the time, for the theoretical opening imagined by Herodotus and many others in previous times had been replaced by a real strait identified by the means of trustworthy observation.

Besides these official explorers, numerous men stayed for months in the hunting bases around the area of the strait, being hunters, soldiers, craftsmen etc. sent by the kings to hunt for elephants. These men were repeatedly in contact with various neighbouring tribes. For this reason it is likely that a certain amount of information collected from local informants (*akoê*) supplemented autopsy. We also

94 Gem. *isag*. 16, 24.

95 The only piece of information given by Herodotus about the Red Sea which could be based upon *autopsia* is the tides of the Gulf of Suez (Hdt. 2, 11). On this topic see Lloyd (b) 66.

96 Hdt. 2, 29. See, among later authors, Diod. 1, 4, 1.

97 "*Histôriê* ist ein Bemühen, durch Autopsie (*histôr* bedeutet ursprünglich der Augenzeuge), Verhör und Forschung die Erkenntnis von *ta eonta* zu gewinnen" (Högemann 103, n. 4, quoting B. Snell, "Wie die Griechen lernten", *JHS* 93 (1973) 181).

98 Str. 15, 1, 2–7.

see, as mentioned above, private individuals (ἰδιῶται[99]) not hesitating to sail into this area as early as the 2nd century. Almost all of them may have been merchants purchasing aromatics, although there were a few people out of the common, e.g. the famous Eudoxus of Cyzicus (a literate person keen on "geography" [*historia tôn topôn*[100]], but also an *emporos* and an "adventurer") and Iamboulos. No doubt, these men were capable of gathering information relating to this country. Such private individuals may, however, not have been highly esteemed, even if they dared to travel as far as the edge of the world. Strabo thought that they were driven by private interests, and accordingly they were "of no use as regards the history of the places they have seen" (οὐδὲν πρὸς ἱστορίαν τῶν τόπων χρήσιμοι).[101] Perhaps they were also regarded as poorly educated men.

They certainly contrasted with the Ptolemaic officials[102]: not only did the expeditions commissioned by the Ptolemies require skilled men, important means and appropriate equipment (*khoregia*[103]) that only kings could afford, but unlike the *emporoi*, the search for immediate commercial benefit was not the primary goal of royal explorations.[104] This was especially true of the kings, who were keen on science, e.g. Ptolemy II and probably also Ptolemy III. Consequently, observations supplied by observers serving as official explorers were not deemed as forgeries (*pseusmata*[105]). On the contrary, they would confidently (*pistis*) be accepted by Alexandrian scientists. That said, information coming from traders venturing into the area of the strait was not systematically rejected, even by those who criticized them.[106] An interesting example – the only one to have survived – is offered by Agatharchides: Ptolemy III's officer Simmias happened to meet a tribe of Fish Eaters somewhere beyond the strait, who fascinated him because of their extraordinary nature and behaviour (above, p. 198; below, p. 219). Had this *paradoxon* (marvel) not been based on "official" *autopsia*, it would probably have been regarded as a mere forgery. According to Agatharchides, the account was confirmed by a number of merchants who, sometime after Simmias' voyage, also witnessed their strange behaviour. In this case, the *emporoi* were fully trusted as they did not contradict official reports.[107]

Generally, explorers and observers would not conceal what they had witnessed. Some, such as the abovementioned Alexandrian merchants, probably transmitted their observations by oral accounts. Apart from this case, we can, unfortunately,

99 Str. 15, 1, 4.
100 See Str. 2, 3, 4.
101 Str. 15, 1, 5 (transl. JONES). See JACOB 119–120.
102 Such as Satyros (Str. 16, 4, 5), or Simmias (above, p. 198): the latter is said to have sought accurate information (ἀκριβῶς ἐξήτασε).
103 Diod. 3, 18, 4 (= Agatharchides, *GGM* 1, 41). See also Diod. 1, 30, 3. The huge costs of such expeditions may explain why Polybius rejected Eudoxus' voyages as a fictitious narrative (Pol. 34, 3, 5 [= Str. 2, 4, 2]).
104 See, however, BURSTEIN 9 (about ivory).
105 Str. 2, 4, 5.
106 See e.g. Str. 2, 5, 11.
107 See Diod. [**7**]; 3, 18, 3–4 (= Agatharchides, *GGM* 1, 41). *N.B.*: I believe that μέχρι τοῦ νῦν ("down to this day") points to Agatharchides' time, not that of Diodorus.

hardly trace their presence in textual evidence. Thus, we are not capable of estimating the amount of information which might have been provided by elephant hunters, mariners, soldiers, merchants etc. on their return to Egypt. Others would record what they had seen in written reports: to put it in Greek concepts, one could say that *opsis* was converted into a form of *akoê*, or *epistemê*.[108] This body of knowledge (*akoê* / *epistemê*) resting on trustworthy *opsis* was highly valued. As Strabo explains, scholars and scientists describing the *oikoumenê* (geographers and / or "chorographers") needed such written material, since they were unable to visit the whole world in order to collect information by themselves. Strabo does, however, add that so large an amount of *akoê* required a careful assessment (= *dianoia*; *gnômê*) before it could be used.[109] Unfortunately, almost all of these written reports dealing with the southern Red Sea area are lost. It has only been possible to establish that a number of officials – particularly those serving Ptolemy II and Ptolemy III – published "periplus" (a sort of narrative of a voyage), or *historia* ("inquiry", "knowledge so obtained", "information" [*LSJ* s. v.]). Simmias and Ariston may have composed such works.[110] We are also aware of a genre called *hypomnêmata* ("Berichte", "official report", "archives"; the meaning of the Greek word is not clear) and *anagraphai* ("register", "record", "description", "treatise").[111] This material was probably stored in the royal archives in Alexandria and/or in the *Mouseion* Library.

Such were the written sources available for scientific purposes (*epistêmê*). It seems that Eratosthenes (*ca* 275 – *ca* 195) was the first to compile this material, no doubt with a critical commentary. He certainly drew on this evidence when writing his note on the strait [**2; 3**]. As director of the Alexandria Library (from the reign of Ptolemy III to *ca* 194 B. C.) he may have had an easy access to information.[112] Interestingly, his presentation is found in a general description of Arabia which was included in the third part of his *Hypomnêmata geographika*[113], the whole Arabian Gulf being regarded as part of Arabia. As far as we can rely on Strabo's paraphrase Eratosthenes did not describe the eastern side of the strait, which allows for the assumption that his documentary material mainly related to the western side, and that a minimal amount of information about the "Arabian Bab al-Mandab" had been collected by Anaxikratês and Aristôn. As just important as Eratosthenes, Agathar-

108 Str. 2, 5, 11.

109 Str. 2, 5, 11: "… generals, too, though they do everything themselves, are not present everywhere, but they carry out successfully most of their measures through others, trusting the reports of messengers and sending their orders around in conformity with the reports they hear. And he who claims that only those have knowledge who have actually seen abolishes the criterion of the sense of hearing, though this sense is much more important than sight for the purposes of science." (transl. Jones). See Jacob 155–179, 606–614.

110 On Aristôn, see Tarn (b) 14; Rostovtzeff (b) 741; Woelk 259–262; Högemann 102. *Contra*: Fraser (b) 300, n. 349; Raschke 944, n. 1182. On Simmias, see, Diod. 3, 18, 4 (= Agatharchides, *GGM* I, 41).

111 The word *basilika hupomnêmata* appears in Diod. 3, 38, 1, no doubt from Agatharchides (Müller (a) 165). On *anagraphai*, see Str. 16, 4, 4 (Artemidorus). See Kortenbeutel 9–10; Woelk 192, 255–266; Högemann 95, n. 4; Peremans 443–446; Burstein 29–33; Jacob 152.

112 Str. 2, 1, 5.

113 Aujac 65–86; van Paassen 46–50; Geus 260–276; Marcotte LX.

chides flourished in the third quarter of the 2nd century B. C. in Alexandria.[114] Most of the sources utilised in his *On the Erythraean Sea*[115] date back to no later than the reign of Ptolemy III,[116] but, as seen above (Simmias and the Fish Eaters), he may have obtained several pieces of information from Alexandrian *emporoi*. He apparently did not have much to say much about the area of the strait [**6–11**][117], even though he offers several details, which are absent from Eratosthenes' book. Particularly striking in this respect is the puzzling passage in which the strait (*stoma*) appears more or less as the symmetric counterpart of the *mukhos* of the Arabian Gulf (= the Gulf of Suez) [**9**]: "Its width, at the mouth and at the innermost recess, is about sixteen stades" (see below, n. 346). Besides these two authorities very few authors who may have dealt with the strait of Bab al-Mandab are known to us. We are aware of a certain Pythagoras, who lived under Ptolemy II or Ptolemy III – he served as *Ptolemaei praefectus*, Pliny vaguely claims[118] –, and wrote a treaty devoted to the Erythraean Sea (Περὶ τῆς Ἐρυθρᾶς θαλάσσης). To surmise that he paid some attention to the strait may not be complete nonsense, but his book is almost entirely lost.[119] Several excerpts from Artemidorus (*floruit c.* 110 B. C.) who adds a little more to Eratosthenes's note have also come down to us [**14–17**].[120] Any further statement about other Hellenistic scientists (for instance Poseidonius) would be speculative.

To conclude, the Hellenistic period, especially the reigns of Ptolemy II and Ptolemy III, witnessed the discovery of the "mouth" of the Red Sea actually being a strait. In other words, the theoretical limit between the Red Sea and the external ocean – *stoma* – had become the *stena* of the Arabian Gulf.[121] It was not by accident that Eratosthenes, who first put the strait on the Hellenistic map (*khorographikos pinax*)[122], used the word *stena* seven times. It must also be stressed that the most important features of Bab al-Mandab were depicted in this period, and even that very little data was added after the third century B. C.: Artemidorus, though he remarkably improves the description of the Horn of Africa, offers few fresh details about this area, as we shall later see. Similarly, as we will deal with in the following

114 Woelk 253.

115 Agatharchides confesses that he could not complete his work in the way he wanted (Phot. *Bibl.* 250, 460b [= Agatharchides, *GGM* 1, 110]). See Woelk 256; Burstein 12–28.

116 Woelk 256; Desanges (a) 83; Burstein 33 (*contra*: Delbrueck 33–34).

117 Woelk 258, wrongly states that "Agatharchides bei der Beschreibung der Westküste nicht bei Bab al-Mandab fortgefahren ist" (see also Kortenbeutel 9). Although *Ptolemais Therôn* is the southernmost place referred to in his periplus – apart from an allusion to the Dahlak archipelago (Desanges (b) 274) –, the mention of Simmias' experiment proves that he was aware of the strait.

118 Plin. *nat.* 37, 24.

119 Kortenbeutel 34; Pfister 60, 64; Desanges (b) 278–279. Dihle (c) 109 denies that he was aware of the strait without explaining his view.

120 Desanges (a) 85.

121 *N. B.* The word *stoma* happens to be used after the discovery of the strait, particularly when it applies to the connection *in abstracto*, regardless of its physical shape. See e. g. Strabo [**19**]; 2, 5, 18.

122 Str. 2, 5, 17.

chapter, this Hellenistic body of knowledge was hardly enhanced in the Roman times, since it was probably considered to be sufficiently accurate.[123]

2.3 The minor improvements of the Roman times (from Augustus to the 2nd century A. D.)

After the annexation of Egypt, those who could reach the strait and provide new pieces of information were the Roman army and, to a greater extent, Alexandrian merchants (*emporoi*; *negotiatores*), who travelled further than the armies.[124] Strabo indeed claims that "since the Romans have recently invaded *Arabia Felix* with an army of which Aelius Gallus (…) was the commander, and since the merchants of Alexandria are already sailing with fleets by way of the Nile and of the Arabian Gulf as far as India, these regions also have become far better known to us of today than to our predecessors".[125] Although the area of Bab al-Mandab had become a frequented place, the area still lacked accurate description in the Roman period.

Let us first consider Strabo: In the time of the Early Principate, this author has almost nothing to say about the eastern side of Bab al-Mandab, a sector that had not been much utilised by the Ptolemies but had become increasingly familiar to the people of the Mediterranean.[126] On the contrary, Strabo's description of the strait is based upon Eratosthenes and Artemidorus' books bearing no signs of having been updated. The remaining passages in which Strabo refers to Bab al-Mandab are – at best – allusions to its narrowness: in the narrative of Eudoxus' voyages [**19**], he merely reports that the strait is so narrow that no one would believe that an Indian castaway could enter this channel by accident.[127] When, however, Strabo states that in his time more ships would sail across the strait than in the previous decades [**21**], he remains silent about this remarkable place. Apparently, the narrowness of the strait[128] (and the Red Sea)[129] was in Strabo's eyes its main and perhaps only significant characteristic. One may be under the impression that this area did not need further description, as if Hellenistic explorers had already given an account of its features (*idiômata*[130]). Rather, Strabo focuses on newly discovered peoples and countries, e. g. Germans, *Geti*, Britons, *Arabia Felix*, and, to a lesser degree India[131], intending to emphasize that Rome's power had improved the knowledge of a number of countries (μάλιστα ἄμεινον). From this perspective, the Bab al-Mandab region which had been explored for more than two centuries could not be regarded as an unknown area. One could say that it belonged to

123 *Contra*: Kortenbeutel 56 ("ein Rückgang der geographischen Kenntnisse").
124 Str. 2, 5, 12; 16, 4, 24; Plin. *nat.* 6, 149 (*nostri negotiatores*); 12, 57 (ambassadors). See Pfister 66.
125 Str. 2, 5, 12 (transl. Jones).
126 Str. 16, 4, 22.
127 See Str. 2, 5, 18.
128 Also see Mela, 3, 74. See Desanges (a) 86.
129 Str. 1, 2, 28.
130 On this word, see below, p. 215.
131 Str. 1, 2, 1 2; 5, 12; 15, 1, 2.

Hellenistic geography, rather than to the geography of the *Imperium Romanum.*[132] This is the reason why Strabo does not contradict himself when he states that merchants heading to India enhanced the geographical knowledge while still refering to the strait as it had been described by Eratosthenes and Artemidorus.[133]

Of course no one would deny that the information relating to this region had been partly updated, but apart from a limited number of topographical and political facts the Hellenistic body of knowledge was not modified much. Pliny the Elder, who mainly draws on on Juba's *Arabika*[134], provides several recent toponyms related to the area of Bab al-Mandab. In particular, he mentions Adoulis, a harbour apparently unknown to Artemidorus and accordingly also to Strabo, and *Aualitu* etc. (below, p. 218). Pliny is also aware of several measurements of the width of the strait (see app. 7.4), but it cannot be established which of them had actually been estimated in Roman times. There is no description of the area of the strait in Claudius Ptolemy's *Geographika*, a book completed in *ca.* A. D. 150 and partly based on Marinus of Tyrus' research (early-2nd cent. A. D.): "Ptolemy's geography is geodesy and cartography, and he preferred to leave out all that had no direct connection with that aim".[135] The reader only has access to the coordinates of the strait as well as a series of new toponyms, the location of which (for instance Perim) is sometimes incorrect (below, p. 244).[136] That several of these appear in the famous *Periplus of the Erythraean Sea* written in the mid-1st century A. D. by an Alexandrian *emporos*[137] supports the view that such data were mostly gathered from merchants who visited these places for trade, provisions, etc. In addition, as merchants came across local kings whom they supplied with prestigious goods, they were capable of collecting fresh information about the political situation (below, p. 221, 225). The author, who was informed about the sea routes to India and East Africa partly by experience, partly by hearsay,[138] more than once mentions the strait [**27–29**]. On the one hand, what he reports indicates that he was aware of Hellenistic information on this subject, since his figure of the width of the strait repeats that known to the Greeks and Romans since Eratosthenes' time (below, p. 243). On the other hand, when he refers to the crossing of the strait by the eastern route, he gives a dynamic description of space found in no other document (below, p. 238). Such details were probably ignored by other authorities (Strabo, Pliny, Eratosthenes etc.), since they may not have regarded them as relevant to scientific knowledge (*epistemê*).

132 See Str. 1, 2, 1; 2, 5, 18; Diod. 3, 38, 2–3.

133 On this "conservative" attitude, see for instance MARCOTTE LXV–LXVI. The ways in which Strabo selected data has been examined by VAN PAASSEN 14–16.

134 Pliny [**22**] expresses his surprise that Juba was unaware of *Berenice epi Dires* (below, p. 216). Consequently, one wonders to which extent Juba was had knowledge of the strait. About Pliny's sources, see DESANGES (a) 86.

135 VAN PAASSEN 2.

136 As BIANCHETTI (b) 158 rightly points out, Arrian does not mention the strait.

137 CASSON 7–9.

138 ARNAUD 27–30.

2.4 An "anonymous" strait

However strange it may seem, the strait between the Arabian Gulf and the Erythraean sea was not given a specific name. Eratosthenes [**2**] and Artemidorus [**15**] refer to the "strait of *Deirê*" or the "strait of the six Islands" (below, p. 216–217), but these expressions are in fact not proper names, nor were they, as far as we know, commonly used. In addition, and most importantly, they were relevant only to the African side of the strait, not to the whole Bab al-Mandab sector. In later sources (Strabo, Pliny, Ptolemy) there are only "neutral expressions" such as "the strait of the Erythraean sea" (τὰ στενὰ ἐν τῇ Ἐρυθρᾷ θαλάσσῃ [**31**]), or "the mouths of the Erythraean sea" (*fauces Rubri maris* [**22**]), which, again, cannot be classified as proper names. The same reasoning applies to Strabo's phrase "Arabian inlet", an expression synonymous with "the strait of the Arabian Gulf".[139] More significantly – because this document certainly reflects the designations employed by the Alexandrian *emporoi* – the author of the *Periplus* refers only to "the strait" (*stena*). In Dionysius' *Periegesis*, a geographical poem depicting the whole *oikoumenê*, which dates back to the mid-2nd century A. D., Bab al-Mandab not only remains anonymous but is not even referred to as a strait.[140]

Unlike the Bosphorus, the Hellespontus and above all the Pillars of Heracles, the strait of Bab al-Mandab was not granted a specific name connected to a particular myth. This fact may seem puzzling when taking into consideration that this strait was not only one of the four "effluents" from the external sea but also an important landmark of the eastern/southern trade routes of the Roman empire. Maybe this spot, of which only the Alexandrian merchants were informed, was too remote from the rest of Mediterranean world to merit its own proper name. The Pillars of Heracles, serving as the gates of the Mediterranean world, were much more important to the Greco-Romans.[141] Perhaps no legend involving a famous mythical character such as Heracles, Io, Helles etc. presented itself as adequate to so distant a place.[142] One could, however, object that Alexander the Great designated the northern part of the Arab-Persian Gulf as "the Sea of *Ikaros*", inspired by the myth of Icarus.[143] But still the king himself had given this name, and I doubt that those who discovered the strait of *Deirê* would have acted the same way.

139 Str. 2, 5, 18 ("All these aforesaid gulfs have narrow inlets from the Exterior Sea, particularly the Arabian inlet and that at the Pillars, whereas the others are not so narrow" [transl. JONES]).

140 Dion. *Perieg.* 43–50: only the columns of Heracles are described (64–68).

141 Plat. *Phaid.* 109c; Diod. 4, 18, 4–5; Plin. *nat.* 3, 3, among many texts.

142 On the contrary, the name "Erythraean <sea>" which was used at least from the 5th century B.C. received a "mythical" explanation in the Hellenistic period, involving a mythical king called Erythras (see e. g. Phot. *Bibl.* 250, 442a–b [= Agatharchides, *GGM* 1, 5]). See, however, VILLENEUVE 161–166, about the *Pon(tus) Hercul(is)* referred to by the Farasan inscription.

143 Arrian. *anab.* 7, 20, 4–5.

3. THE "GRECO-ROMAN BAB AL-MANDAB": A SYNTHETIC REVIEW OF THE GRECO-ROMAN BODY OF KNOWLEDGE

Collecting as thoroughly as possible all the pieces of information scattered among written material, I shall present the strait and the surrounding regions as they were described in Greco-Roman geography. Before surveying what was known to Greco-Romans about places, peoples, fauna and flora, I would, however, like briefly to comment on the genre of the periplus, for the bulk of ancient knowledge was supplied by this particular literary genre.

3.1 The framework of ancient knowledge: the "periplus"

Information relating to places, relief, waters, landscapes, peoples etc. – in a nutshell "geographical information" – could be found in various genres and literary formats: *periodos*, *periegesis*, *khorographia*, *periplous*, historiography etc. With regard to Bab al-Mandab, most data ultimately derives from written reports often referred to as *periplous* (or *paraplous*: "account of a coasting voyage"). Such texts present information about places situated on maritime routes according to the order of a journey along the shoreline. Besides toponyms, distances and directions, the author generally provides information on peoples, fauna, flora etc. In addition, the main features of the hinterland are often commented upon.[144] The Bab al-Mandab region was mostly depicted either in *periploi*, or in texts composed on the structure of a *periplous*: for instance, Strabo draws on Artemidorus' *Geographoumena* which has the format of a *periplous*; Eratosthenes' *Geographika* derives from reports of explorers (*hypomnemata*), some of which were probably written in the form of a periplus. Many words and phrases characterizing this specific literary form can easily be detected in the corpus of texts (*en dexiai apopleousi*[145], *prospleusasi* [**3**], *parakomisthenti*[146], *meta* [**3**], *entos*/*ektos* [**16**], *ephexes* [**16**] etc.).

We must also keep in mind the fact that only a limited number of topics are developed within this framework: the *periplous* is not designed to be an exhaustive or methodical description of the world, nor a description of but a part of it[147], The *periplous* sometimes gives the impression of being a fragmentary or selective report.

More specifically, the "periplus description" focuses primarily on topographical matters (locations, distances, toponyms), places and regions in need of as accurate a position as possible. Concerning the strait of the Erythraean Sea, the main goal of the first explorers was to estimate its position by reckoning the distance between it and *Ptolemais Therôn* (below, p. 241). Besides these fundamental data, a *periplous*, however, usually includes the most noteworthy features of the country

144 About the *periplous*, see GISINGER; MARCOTTE LX–LXIV; about the relationship between *periplous* and *khôrographia*, see MARCOTTE LVII–LIX.

145 Str. 16, 4, 4 ("on the right as one sails" [transl. JONES]).

146 Diod. 3, 39, 4 (= Agatharchides, *GGM* 1, 82 ["as a man coasts along these regions" – transl. OLDFATHER]).

147 See e. g. Arrian. *anab.* 7, 20, 10.

(*idiôma*, *idiôtês*).[148] These generally relate to: a) physical geography (*topoi*, *khôra*, *phainomena*), i. e. rivers, mountains, shapes of the coastline etc.; b) peoples (*ethnê*) and their habits (*nomoi*); c) nature (*phusis*), i. e. animals, plants etc.[149] It is worth adding that marvellous facts (*paradoxon*; *thaumasia*) happened to be taken into account by the authors of *periploi*. Provided that they were based on reliable *opsis* such pieces of information were not discarded as lies or fictions, and belonged to the domain of geographical knowledge.[150]

A strait is formed by two separate landmasses (*êpeiroi* [**6**]) pointing towards each other. Thus, a description based upon the periplus framework was generally divided into two parts, each side appearing in its specific coasting voyage.[151] In the case of a clockwise *periplus* of the whole *oikoumenê* with the shoreline lying on the right (e. g., Pliny the Elder), the Arabian side of the strait is first mentioned. Having reached the bottom of the Red Sea (the Gulf of Suez) the person providing the description turns southwards and then depicts the African side of Bab al-Mandab before he continues his voyage along the rest of the African coast. Strabo's description is, however, organized in a different way: He starts from the bottom of the Red Sea (*Heroônpolis*) and proceeds to the account of the African shoreline in which the "African Bab al-Mandab" is described (= Strabo, 16, 4, 5–13 & [**14**–**16**]). Once this periplus has been completed the author returns to the northern extremity of the Red Sea (say the Gulf of Aqaba) and starts his account with the shoreline on the left, along the Arabian coast, as far as Bab al-Mandab (= Strabo 16, 4, 18–19 & [**17**]).[152] Agatharchides' treatise and the *Periplus of the Erythraean Sea* are arranged this way.

N. B.: owing to its importance, the question of local trade will be treated in a specific section.

3.2 The western side

Exploring the African side of the southern Red Sea, the explorers of Ptolemy II and Ptolemy III reached a cape pointing towards the Arabian side [**2**], which turned out to mark the narrowest part of the Arabian Gulf. They also noticed a small settlement lying here and a monument allegedly built by Sesostris. They assigned

148 See Diod. 3, 38, 5 (= Agatharchides, *GGM* 1, 80): "We shall (…) then sail along (παράπλουν) its two sides past the mainlands, in connection with which we shall describe what is peculiar to them and most deserving of discussion (τὰς ἀξιολογωτάτας κατ᾽ αὐτὰς ἰδιότητας)". See also Str. 1, 1, 13 (idiôtêtas) 1, 2, 23; 16, 4, 22 (idiômatôn); Pol. 3, 58 (idiôtêtas).

149 As Jacob 769 writes, such descriptions were intended to provide "un savoir sur le monde, sur la nature des paysages comme sur les mœurs des habitants."

150 As a consequence this sort of *paradoxon* is different from μυθώδη / πλάσματα ("lies"/"fabrications"/"fictions") which no one should believe (ἄπιστα) and does not belong to ἱστορία: see e. g. Str. 1, 2, 17; 15, 1, 6–7 and 9; Diod. 3, 33, 7 (= Agatharchides, *GGM* 1, 65). The passage in which Strabo describes the plain of the Crau (Str. 4, 1, 7) facilitates an understanding of this original aspect of ancient knowledge.

151 As far as I know, Bab al-Mandab is never described as a single entity. Nonetheless, the Greco-Romans were aware of the strait's unity when considering it in a very general way as the mouth of the Arabian Gulf (e. g. [**26**] and app. 7.3).

152 See also Phot. *Bibl.* 250, 456a (= Agatharchides, *GGM* 1, 80); *P. m. r.* 19.

both this place and the cape with the name *Deirê*, actually a Greek word meaning "neck", or "neck of land projecting a long way out" (Pliny [**22**] gives the Latin translation of this toponym: *cervix*). Either they did not come upon any vernacular name, or they thought that a Greek word would be more convenient for this remarkable spot. The same explorers estimated the distance between *Ptolemais* of the Hunts and *Deirê*, noticing that the shoreline turned slightly eastwards past *Ptolemais* (Eratosthenes drew on such data to devise his map of the *oikoumenê*: below, p. 241–243). The name *Deirê* clearly points to the modern Ras Siyyan, as Desanges rightly states. This long and thin promontory was originally an island until it was linked to the mainland by a coral plateau, so that the Ras Siyyan can really be compared with a "neck".[153] The city of *Deirê* probably lay somewhere between the Ras Siyyan and a nearby place known today as Heirkale.[154] Convincingly, J. Desanges argues that at a later period some Greeks settled in or near *Deirê*, which was then renamed *Berenice* after the name of Ptolemy III's wife [**22**].[155] In the mid-1st century A.D. neither *Berenice* nor *Deirê* are, however, referred to by the *Periplus,* suggesting that this place was of no interest to Greco-Roman *emporoi* at that time, probably because Mediterranean people had long abandoned this settlement. That Ptolemy mentions *Deirê* does not compromise what we learn from the *Periplus*, for this geographer lists all kinds of toponyms regardless of the actual existence of the places referred to.[156]

Sometime after the discovery of the strait at *Deirê* the Ptolemaic explorers noticed, if they did not hear it from natives[157], that the "true" African side of the strait was located further away: "It is not these [sc. the straits at *Deirê*] that are called straits now, but a place farther along on the voyage."[**3**]. There is much to say of this puzzling assertion (see below, p. 240). What is most striking perhaps is the fact that this new location seems to be marked neither by a distinctive physical feature or landmark (such as the characteristic promontory of *Deirê*), nor by a noticeable human settlement. It was merely the final stop (δεῦρο) of a trade traffic (below, p. 228) connecting the African side to the Arabian one (κἀκεῖσε) via six anonymous and mysterious islands. In fact many islands and islets in the waters between *Eumenous limên* (around Assab?) and the open sea were known to the Greeks [**9**; **11; 15**][158]; hence the name "the strait *kata* ("throughout"?) the Six Islands". Regrettably, the distance between *Deirê* and this anonymous spot is missing in the textual evidence. Eratosthenes only informs us that the channel lying between the two opposite land

153 Desanges & Reddé 162, 176–177.

154 Desanges (a) 94–95, 100–101, challenging Müller (b) 760 (*Deirê* = ras Bir; *Arsinoe* = ras Siyyan) and Conti-Rossini 60 (= Rayahta, north of Dumeira).

155 Kortenbeutel 35–36; Desanges (a) 95. However, Desanges & Reddé 178–179 (quoting Rostovtzeff) admit that the name *Berenice* could be given in honour of the daughter of Ptolemy IX *Lathyros* (116–107 and 87–80 B.C.) who married Ptolemy X *Alexandros* I (107–87 B.C.).

156 *Contra*: Kortenbeutel 67, believing that Ptolemy's toponyms referred to places occupied by Romans.

157 The subject of λέγουσι may be the same as the one of κομίζουσι, i. e. the natives involved in the local traffic.

158 Also see Str. 16, 4, 14; *The Red Sea and Gulf of Aden Pilot*, especially chapters 5 and 9.

masses (ἠπείρων) is 200 *stadia* wide, i.e. more than three times wider than the strait at *Deirê*, which in no way aids us in identifying this mysterious point.[159] Artemidorus' text contains nothing [**15**] but the vague idea that the distance between *Deirê* and this unlocated place was not excessive.[160]

These "Six Islands" have been identified as the Sawabi (or Seven Brothers) Islands, six islets lying off the Eritrean coast in the east-west direction. In former times the Ras Siyyan was the seventh of these islets, until it was joined to the landmass by the abovementioned isthmus.[161] This view may, however, be challenged: Firstly the Sawabi and Ras Siyyan are more or less situated at the same latitude. This barely fits with Eratosthenes' report, on the basis of the data gathered by the Ptolemaic *autoptes* (λέγουσι), saying that the so-called strait of the Six Islands lies "farther along the voyage", i.e. somewhere beyond *Deirê* / Ras Siyyan (προσπλεύσασιν ἀπωτέρω) [**3**]. Since Artemidorus' statement is less accurate than that of Eratosthenes it is almost without use here [**15**]. Secondly, the Six Islands are said to "follow one another in a close succession" and to "fill up the channel" (i.e. the interval between the Arabian and African sides [δίαρμα]). A glance at a map will indicate the discrepancy between this account and the actual location of the Sawabi Islands[162], and one wonders how they would make the trip easier by "filling up the channel". Accordingly, I believe that these islets scattered between the two sides of the Red Sea remain unidentified. Whatever their location may be, there is no doubt that in terms of topography and cartography the strait of *Deirê* aroused greater interest to the Greeks and Romans than the passage of the Six Islands, which disappears from the textual evidence after Artemidorus (see below, p. 241).[163]

Artemidorus shows awareness of some places in the area of Bab al-Mandab, which were probably unknown to Eratosthenes and Agatharchides. Since most of them are linked to the hunt for war elephants, they are likely to have been founded in the reign of Ptolemy III and were destined to be abandoned once the hunting for elephant came to an end. North of *Deirê* / *Berenikê* was the Island of Philippos, the Hunting-base of Pythangelos and *Arsinoe* (a city and harbour[164]). Beyond *Deirê*, in the vicinity of which elephants were hunted, were the Hunting-base of *Likhas* and the promontory of Pytholaos with its two remarkable lakes. Given that the "Frankincense bearing country" (*Libanôtophoros*) starts after the promontory of Pytholaos, it seems that these unlocated and unidentified places were a part of the Myrrh-bearing country, whose boundary was *Deirê* [**16**]. Pliny, probably drawing

159 According to Desanges & Reddé 180, "une telle traversée ne peut aboutir, du côté africain, que dans l'anse située à l'ouest du ras Siyyan."

160 Eratosthenes states only that the coast turns toward east and south straight (ἑξῆς) after the strait of the Six Islands, forming a bay (ἐγκολπίζουσι).

161 Müller (b) 760; Desanges (a) 91; Desanges & Reddé 162, 178–180; Desanges (c) 87–88. On the *Stenae Dirae insulae* (Plin. *nat.* 6, 169, actually the *stenae Thyrae*), see Desanges (c) 61–62.

162 This point is accepted by Desanges (a) 91.

163 Note that Agatharchides' informants [**7**; **8**; **10**; **11**] probably referred to the strait of *Deirê.*

164 Str. 16, 4, 14. Kortenbeutel 38 claims that it was founded by Ptolemy *Philopatôr.* Rostovtzeff (a) 305 believes that two *Berenikê* were founded "zur Bewachung *Sabae* und *Deire*", and were "Handelsstationen"). See also Desanges & Reddé 180–181.

on Hellenistic sources, reports the following toponyms: the islands called *Pylae* and *Pseudo-Pylae* (= the Gates and the Pseudo-Gates, maybe the Sawabi Islands), the Island of Diodorus (below, n. 350) and the port of Isis (near *Deirê?*).[165] In addition, there was a number of small indigenous settlements whose vernacular names were either unknown to, or ignored by Artemidorus' informants [**15**].[166] The author of the *Periplus*, who probably did not pay much attention to places that had not been engaged in the Greco-Roman trade, mentions only *Adoulis* – an *emporion* which was to reach its climax in the following centuries – and *Aualitês* (probably Zeila, rather than *Deirê*, Tajura or Assab).[167] Pliny was aware of both toponyms.[168] This topographical review ends with Ptolemy's catalogue (see fig. 5), which is a compilation of ancient (e. g. *Arsinoe*) and more or less recent toponyms (e. g. *Adoulis*, *Aualitês*). Ptolemy does, however, provide several names hitherto not attested (*Mandaïth komê*; *Antiokhou sôlen*), the origin and location of which escape us completely.[169]

The area of the "African Bab al-Mandab" including the hinterland was generally regarded by Hellenistic scholars as a part of *Aithiopia* [**2; 17**]: this term is not easily defined but generally the name *Aithiopia* covered all lands lying south of Egypt as far as the southern Ocean, the Nile commonly being considered to be the boundary between Libya [Africa] and *Aithiopia.* This concept was taken over by most scholars of the imperial period, as attested by Pliny.[170] Similarly, in Ptolemy's *Geography* the whole region of the strait is part of *Aethiopia infra Aegyptum.*[171] This vast country could, however, be divided into smaller units: the inner part of the Bab al-Mandab area was included in what was called the Trôglodytikê [**6**; **20**?; **21**?[172]]. On the basis of some testimonies this country seems to extend beyond the strait of *Deirê* (see **12** and above, n. 46, the dedications offered to Pan by the travellers who safely returned from the country of the Troglodytes). The Ptolemaic explorers, however, would call the region beyond the strait *Smurnophoros* ("myrrh bearing country"), probably because of the presence of trees thought to be myrrh trees [**3**], unless this name originated from a local spice trade taking place beyond *Deirê*. At any rate, this name was still employed by Greek merchants, or hunters, in the following decades [**12**]. Less frequent and accurate was the alternative (?) name

165 Desanges (c) 86–88.

166 See Str. 16, 4, 18: "I am not giving most of the names of the tribes because of their insignificance (ἀδοξίαν) and at the same time because of the oddity of the pronunciations" (transl. Jones).

167 The case of *Aualitês* is discussed by Casson 115–117: although *Deirê* fits the *Periplus'* words, this identification "would leave Avalitês with a very poor harbour indeed"; on the other hand "the *Periplus's* figure of 4800 stades from *Adoulis* to *Aualitês* may seem to favor Zeila". In Ptolemy's geographical handbook *Aualitês* (Ptol. 4, 7, 10) lies far south of *Deirê* (see fig. 5).

168 Plin. *nat.* 6, 172; 174.

169 Müller (b) 760.

170 Plin. *nat.* 6, 173–174.

171 Ptol. 4, 7, 1.

172 Pliny thinks that the *Trogodytice* extends along the entire African side of the Arabian Gulf. See Desanges (a) 83, 90, 96.

Arômatophoros ("Spice bearing country") attested by Artemidorus as well as by the loan agreement published by U. Wilcken.[173]

Whereas the common and ancient name *Aithiopia* remained firmly rooted in Greek and Roman geographical conceptions, the designations of the smaller territorial subdivisions were subject to alterations. It is not easy to explain why such changes in toponymy took place, but this certainly has much to do with the growing Greco-Roman trade activity: merchants and travellers tended to employ the toponyms of their time regardless of ancient geographical *doxa*. In particular, the author of the *Periplus* who, deliberately or not, ignores the name "Troglodytes", calls the coast projecting from *Berenikê* as far as the "Spice Promontory" (*akrôtêrion tôn Aromatôn* = cape Gardafui / Ras Asir) *Barbaria* / *Barbarikê.*[174] In contrast, Pliny relying on reports from unknown sources refers to an *Azanium mare* ("sea of *Azania*") stretching out beyond the strait.[175] Finally, Ptolemy who partly draws on Marinus of Tyrus gives the name *Trôglodytikê* to a wide coastal area (παράλιος χώρα) extending as far as the cape *Elephas* (Ras el-Fil, west of Cape Garfafui).[176] He also removes the *Smyrnophoros* far from its original location in the outer part of the Bab el-Mandab strait.[177]

Far more fascinating than these topographical issues are the issues relating to peoples dwelling in the area of Bab al-Mandab. The 3rd century explorers encountered various tribes which, as was usual in Greek ethnography, were primarily distinguished by their physical appearance and *nomima* ("usages"; "customs"). Fortunately, a reasonable amount of valuable and sometimes lively information has survived.

For obvious reasons coastal sedentary populations were the first to be met by the Greeks. Accordingly, we are quite well informed of the *Ikhthyophagoi* ("Fish Eaters") [**2**; **7; 10; 15; 16**][178], a Greek name pointing to peoples fully dependent on sea products (fish, shellfish, beached whales etc.). The name *Ikhthyophagoi* applied indiscriminately to a number of tribes settled on the African shore of the Red Sea and the Gulf of Aden as well, that is to say inside and outside the strait. They lived in shelters built from every kind of locally available material and grouped in small villages. Such was probably the *polikhnion* ("very small town"[179]) of *Deirê* [**2**], the only one to be noticed by ancient travellers probably because it lay in a conspicuous place. Although the Greeks employed the generic name *Ikhthyophagoi*, they managed to observe the distinctive features of several tribes. As is common in ancient Greek ethnography, these relate to livelihood (*trophê*) and customs. The groups of Fish Eaters settling within the strait were reported to enjoy a comparatively long

173 Wilcken 90; Str. 16, 4, 14.

174 *P. m. r.* 2; 5; 7; 12. See Casson 97–100, 110, 129.

175 Plin. *nat.* 6, 153; 6, 172. See Dihle (b) 562–565; Desanges (c) 75–77.

176 Ptol. 1, 8, 1; 1, 9, 1; 4, 7, 27. Ptolemy refers also to the *Adoulitikos* Gulf (1, 15, 11) and the *Aualitês* Gulf (4, 7, 10), which lies inside the Gulf of Aden (see Müller (b) 759–760; Desanges (a) 97).

177 Ptol. 4, 7, 31.

178 On the *Ikhthyophagoi*, see Longo.

179 Desanges & Reddé 176.

life, due to their simple diet [**7**], and the tribes living beyond the strait (ἐκτὸς τοῦ κόλπου), which seem to have been approached for the first time by Simmias, appeared as extraordinary (the words *thaumasion* / *paradoxon* occur three times in the paragraph 3, 18): they did not need to drink. Whoever has visited the countries of the Red Sea may share the Greeks' amazement. This marvellous fact (*paradoxon*) received a rational explanation: these *Ikhthyophagoi* were supposed to get the minimal amount of fresh water required to survive from the juicy raw fish that they ate.[180] Secondly they could bear any physical and moral pain without complaining [**7**; **10**], as cruel experiments *in vivo* carried out by Simmias himself showed. This conduct of Simmias was presumably an attempt to investigate and understand this *paradoxon*. We are also informed of a tribe called *Khelonophagoi*, i.e. "Turtle Eaters". These lived on the numerous islands lying off the Bab al-Mandab coast in the open sea where they hunted for the sea-turtles that supplied them with flesh and tortoise shell [**11**].[181] On the nearby coast dwelt other Fish Eaters who, due to their low level of expertise, were unable to catch fish and were consequently compelled to feed on beached whales.[182]

Taken together, the tribes who did not depend on sea food were called *Trôg(l)odytes*[183], alternatively Nomads.[184] These peoples were ruled by a "chief" (τύραννος; δυνάστης) and lived "a nomadic life off their flocks".[185] They were confronted with the severe environmental conditions prevailing in this area: for instance they had to "go to war about pasturage" in the dry season.[186] Unlike the *Ikhthyophagoi* forced to dwell on the coastal zone (*paralia*)[187], their livelihood enabled them to occupy the hinterland (*mesogaia* [**15**][188]) but they would occasionally share the coastal zone with the *Fish Eaters* [**13**].[189] Again, unlike the *Ikhthyophagoi*, the Greek travellers seem to have feared the Troglodytes [**12,** and above, n. 46]. Artemidorus, however, mentions "insignificant cities" [**15**], enabling us to infer that a certain number of Troglodytes was committed to agriculture, leading a sedentary life in villages. Two tribes stood out among the various groups settled in the vicinity of the strait, and thus were given a specific name of Greek origin: the *Khreophagoi* ("Flesh Eaters") were undoubtedly a Troglodyte tribe of hunters or breeders. This points to

180 Diod. 3, 18, 2 (= Agatharchides, *GGM* 1, 40).

181 In the parallel version of Diod. 3, 21, 1–5 their islands are said to lie in the ocean near the land (κατὰ τὸν ὠκεανὸν πλησίον τῆς γῆς). Artemidorus (Str. 16, 4, 14) is also aware of the *Khelonophagoi*, but does not clearly say where they dwelt. See Desanges (a) 88.

182 Phot. *Bibl.* 250, 451b (= Agatharchides, *GGM* 1, 48).

183 The "l" is absent from papyri (Burstein 100, n. 1) The *Ikhthyophagoi* who inhabited the African shore of the Red Sea are sometimes considered as *Trôglodytai* (Casson 98).

184 Diod. 3, 32, 1.

185 Diod. 3, 32, 1–3 (= Agatharchides, *GGM* 1, 61); 3, 40, 1 (= Agatharchides, *GGM* 1, 83). Str. 16, 4, 13 (Artemidorus): "The greater part of the people are nomad and those who till the soil (γεωργοῦντες) are few in number" (transl. Jones).

186 Str. 16, 4, 17 (transl. Jones).

187 However a tribe living inside the strait would go every fifth day to the foothills of the mountain in order to get fresh water (Diod. 3, 17, 2–5 [= Agatharchides, *GGM* 1, 37–39]).

188 Also see Str. 16, 4, 4 [Eratosthenes]: "Though there are not many cities on the coast, there are many in the interior that are beautifully settled" (transl. Jones).

189 Tkač (a) 2524–2527.

a carnivorous diet, while other Troglodytes would normally "live off blood and milk which they mix together".[190] According to Artemidorus, they occupied an area inside and beyond *Deirê* and the strait of the Six Islands, probably between Assab and Zeila [**15**; **16**].[191] The Greeks were, however, particularly impressed by the *Koloboi* ("Mutilated glands people"[192]) [**8**; **14**; **15**], most likely another Troglodyte group. No doubt, the sight of men showing a mutilated penis must have struck the Greek observers. Agatharchides claims that they lived inside the strait (= before *Deirê*?), and other evidence proves that the *Koloboi* moved to seek pastures south of *Eumenous limên* (around Assab?) as far as *Deirê*. They probably led a nomadic life, as attested by the anonymous dedicant [**12**] telling us that they could be encountered beyond the strait, somewhere in the *Smyrnophoros*, a fact implicitly confirmed by Artemidorus.[193] This remarkable body of knowledge scarcely improved in later times, apart from a significant shift in the local political organisation reported by the *Periplus*: in the 1st century B.C. a certain Zôskalês turned out to rule "as a king" from south of *Berenikê* as far as the strait (perhaps he had his residence in *Adulis*). Actually, Casson convincingly uses the word "sway" to define Zôskalês' power. He was probably powerful enough to restrict the independence of the above-mentioned tribes, but a general modification of the local tribal organisation seems by no means plausible.[194]

Concerning fauna and flora, as I stated above, the Greeks focused only on the most striking facts. Mangroves (called "olive" groves and "laurel" groves) were probably the most bewildering and fascinating feature of the coastal landscape. Thus Artemidorus [**16**] refers to the groves growing inside the strait (*Deirê*?) and to numerous ones (*pollê*) located beyond. Nowadays mangrove woods survive between Assab and Zeila.[195] Palm trees, the most remarkable "land trees" (mangrove trees were regarded as "sea trees") in so torrid a country, were also reported inside and beyond the strait. Artemidorus' statement that myrrh trees grow beyond *Deirê* ("the first country <after *Deirê*> produces myrrh" [**16**]) seems particularly questionable: firstly, myrrh trees grow in today's Eritrea, north of *Deirê*.[196] Secondly, the identification of myrrh trees and frankincense trees was subject to much confusion in antiquity.[197] As a consequence I severely doubt that Greek observers were able to identify the true myrrh tree (*Commiphora* spp). Conversely, they witnessed people trading aromatics including a special quality of "myrrh" in this area (below, p. 228–229). Hence, a place where myrrh was supplied could be conceived of as an area with myrrh trees, regardless of the botanical determination. As for animals, no

190 Diod. 3, 32, 1.

191 Also see Str. 16, 4, 9: *Kreophagoi* and *Koloboi* could be encountered "above" the harbour of *Antiphilos* (located south of Hawakil Bay by Müller (b) 760).

192 See Burstein 111–112.

193 Str. 16, 4, 14 (past the promontory of *Pytholaos*, in the *Smyrnophoros*, "the people are wholly free from mutilation of the body" – transl. Jones).

194 *P. m. r.* 5. See Casson 109–110.

195 Schneider (a) 364–367. "The coast from ras Jibuti to Zeila (…) is low and swampy, tickly covered with mangrove jungle" (*The Red Sea and Gulf of Aden Pilot* 368).

196 Groom (a) 99.

197 Theophr. *h. plant.* 9, 4, 2–3.

animal species is explicitly linked with the area of the strait[198], except elephants, which were hunted from the shores of the African side of Bab al-Mandab.

3.3 The eastern side

In Greco-Roman geography the Arabian side of the strait was clearly a part of *Arabia Felix / Eudaimôn*. A further look at local topography is the subject of this chapter. When the first Greek explorers reached the Ras Siyyan, they certainly claimed that they had discovered the strait of the Erythraean Sea because of a promontory projecting out from the Arabian side towards the Cape *Deirê*. While Eratosthenes was aware of the Arabian counterpart of *Deirê*, its name seems to be unknown to him. Agatharchides, less informed than Eratosthenes, is of no benefit in this discussion. In contrast, Artemidorus is the first[199] to give the vernacular name of this promontory, *Akila* (*Okêlis*/*Acila*/*Ocilia*) [**14**]. Most scholars identify *Okêlis* with the peninsula of Sheikh Sa'id.[200] Almost no place lying before or beyond *Okêlis* was known to the Greeks: Aden was perhaps referred to by Agatharchides, but the identification of the *Eudaimones* islands (above, n. 35) with this harbour is not firmly established. Thus, in contrast with the African side, the Greeks were poorly informed about its Arabian counterpart, a situation which obviously derives from the lack of direct observation. As stated above, only a few official explorers and *emporoi* sailed along the eastern strait area in the 3rd–2nd centuries B. C. In addition, whereas hunting teams stayed in their African bases for months, no Greek made the Arabian coast his permanent home, except perhaps the colonists from the city of Miletus who allegedly settled in *Ampelome* (above, n. 31).

The growth of Greco-Roman sea traffic from the late 2nd cent. B. C. resulted, however, in an increase in the knowledge of the eastern side of Bab al-Mandab. First of all, the name *Okêlis* appears in the evidence not only as a cape but also as a harbour. It has been generally located it in the north-west part of the peninsula of Sheikh Sa'id.[201] Mediterranean cargo ships would put in there and get water supplies before starting on their voyage to India [**23**; **24**; **28**; **32**[202]]. The author of the *Periplus* does not regard *Okêlis* as an *emporion*, which reinforces the assumption that this port was mainly involved in local trade (below, p. 235). Another toponym appears in Greco-Roman documents of that time, namely *Mouza / Muza*, ranked by the *Periplus* among the "legally limited ports of trade", which has been identified with Mocha / Al Makhā [**27**].[203] Finally, there is *Eudaimôn*/*Adanu*/Aden [**23**; **29**][204],

198 Diod. 3, 35–37 (= Agatharchides, *GGM* 1, 67–78) and Str. 16, 4, 15–16 (= Artemidorus) give a synthetic review of the Ethiopian and Arabian fauna.

199 Dihle (b) 559: "Artemidor als erster".

200 Desanges (a) 92; Dihle (b) 561–562; Casson 157–158; Schiettecatte (a) 235–236.

201 The port of *Okêlis* was probably in a lagoon named Sheikh Sa'id or Khawr Ghurayrah: Desanges & Reddé 180; Dihle (b) 562; Casson 158; Schiettecatte (a) 235–236. Also see von Wissmann (b) 437, 440; Robin (b) 225.

202 See also Plin. *nat.* 6, 104.

203 Also *P. m. r.* 21; 24; Plin. *nat.* 6, 104. See Casson 147–148; Schiettecatte (a) 233–235.

204 Casson 158–160; Desanges (c) 97–98. Ptolemy calls it *Arabias emporion* (see fig. 5). *Adanu* is the form that Pliny drew from Juba: it bears much resemblance with the word *Adaneitês*, "citi-

a place which appeared to some travellers at the last stage of their voyage across the strait (below, p. 239). Both *Mouza* and *Eudaimôn* were integrated into the local trade network. In addition, a certain *portus Laupas* situated between *Camari* (Kamaran Island?) and *Acila* / *Okêlis*, and far less important than the previous places, appears in a list of toponyms coming from Juba [**24**]. The same authority also points out an anonymous promontory, probably in the vicinity of Bab al-Mandab [**25**], as well as several cities (*oppida*) lying somewhere in the region of the strait (*in Rubro litore*): *Merme*, *Marma*, *Corolia*, *Sabbattha*. Almost none of these places can be identified.[205] The myrrh called *Sambracena* was named, Pliny adds, after an unidentified Sabaean city vaguely located near the coastline, perhaps a harbour from which this variety was exported or re-exported.[206] It cannot be established if this place was located in the strait area. Ptolemy's *Geography* (see fig. 5) contains several other original toponyms from either vernacular or Greek origin. The Greek toponyms may have been created by Greco-Roman merchants. Two of them merit special attention, namely *Palindromos akra* (the "cape of the running back again") and *Poseidion akra* (the Cape of Poseidon), south of *Okêlis*. Indeed, the route along the peninsula of Sheikh Sa'id bears some resemblance to a hairpin bend, a feature vividly expresses by the word *Palindromos*. As to the name Cape of Poseidon, it may be linked with the open sea appearing to men after they had passed the strait.[207]

Another significant consequence of the increasing Greco-Roman maritime trade is that, whereas most of the Hellenistic authors (e. g. Artemidorus) were unaware of the islands lying along the Arabian side of Bab al-Mandab, many were reported by later sources. In fact, the route leading to the Arabian side of the strait seems to be dotted with islands. The *Periplus* refers to the *Katakekaumenê nêsos* (Burnt island – Jabal al Ta'ir[208]), after which one comes to *Mouza* and "the very last bay of the left-hand shore of this sea (= the Red Sea)". Pliny/Juba [**23; 24**[209]] points out numerous islands in his presentation of the southern Arabian Red Sea: *insulae Alaea* (possibly a confusion with the *Alalaiou* islands [Dahlak archipelago])[210], *Chelonitis*[211], *Ichthyophagon multae*, *multae Sabaeorum*, *Doricae*, *Pteros*, *Camari*, *Malichu* (the Hanish Islands[212]), not to mention other anonymous islands. No doubt Farasan, an island to which a Roman military detachment was sent in the mid-2nd

zen of Aden". See below, n. 288.

205 Plin. *nat.* 6, 154. See, however, Tkač (b) 1440; von Wissmann (a) 308.

206 Plin. *nat.* 12, 69 (*a civitate regni Sabaeorum mari proxima*).

207 See Groom (b) 69. For the sake of comparison, see Str. 14, 1, 14.

208 *P. m. r.* 20–21. The *K. nêsos* is situated by Ptol. 6, 7, 44 approximately at the latitude of *Mouza*; it is called *Exusta* by Plin. *nat.* 6, 175 (= Juba, *FGrHist* 275, F 35). Casson 147 has identified it as the Jabal al Ta'ir Island (15° 33' N / 40° 50' E), which has "a volcano not yet completely dead".

209 See also Plin. *nat.* 6, 150.

210 Plin. *nat.* 6, 150 (= Juba, *FGrHist* 275, F 33). See *P. m. r.* 4; Diod. 3, 45, 6 (= Agatharchides, *GGM* 1, 96 [an Arabian tribe named *Alilaioi*]). See Müller (a) 184–185; Desanges (c) 84–85.

211 There are Islands of Turtle Eaters on the opposite side (Str. 16, 4, 14 [Artemidorus]; Ptol. 4, 7, 37; Pliny [**24**]).

212 Desanges (c) 95.

century B.C. (above, p. 203), appears in this list under another name.[213] Fewer islands were reported by Ptolemy whose presentation, however, offers two puzzling inaccuracies: the islands *Adanu* (Aden[214]) are wrongly situated in the Arabian Gulf, and the island of Perim (*Diodôrou nêsos*) lying between Ras Siyyan and Sheikh Sa'id is not correctly located (below, p. 244).

Eratosthenes' short account [**1**] contains valuable pieces of information about the ethno-political organisation in the area of Bab al-Mandab, even if it describes the Arabian societies in a very general way. He is aware of four major Arabian states (μέγιστα ἔθνη) occupying the south-west peninsula, namely Ma'in, Saba', Qatabân[215], Hadramawt[216], but only the third one is unambiguously specified as extending as far as the sea. Ma'in is said to lie "on the side [of the Arabian Peninsula] towards the Erythraean Sea [*viz.* the Arabian Gulf[217]]". Quite accurately this reflects the political situation prevailing at the time, since in the early Hellenistic period the *mukarrib* (sovereigns) of Qatabân seem to have prevailed over the Sabaean power.[218] Consequently in the late-4th – early-3rd centuries B.C. Qatabân's rule had probably extended as far as the strait and the coastal plain (Tihamā)[219], although its core and capital (Tamna) remained in the mountains.[220] Nonetheless, the word *diabasis* employed by Eratosthenes implies that this kingdom was to some extent involved in the local sea trade. Agatharchides certainly cannot compare with Eratosthenes' degree of accuracy. Not only is Saba' the only kingdom he refers to (so does Artemidorus[221]), but he wrongly calls its *metropolis* (capital) *Sabas* instead

213 Tkač (b) 1441.

214 Actually the peninsulas of Aden and Little Aden (see Desanges (c) 96). Mela, 3, 80, also locates Aden in the Arabian Gulf: *ab ea parte* (of the Arabian Gulf) *quae introeuntibus sunt Charra* (an unknown place) *et Arabia* (Aden, wrongly repeated?) *et Adanus* (see Desanges (c) 98).

215 The vocalization Qatabân is based on the Greek name *Kattabaneis*; the form Qutbân, quoted by al-Hamdânî, may be the original one. It has been preserved in Pliny's *Gebbanitae* (*Yémen* 98 [A. Avanzini]). According to Pirenne 147–148 (see also von Wissmann [b] 440), the *Kattabaneis* and the *Gebbanitae* were two tribes of the kingdom of Qatabân (however, see below, n. 273).

216 Compare with Theophr. *h. plant.* 9, 4, 2, probably drawing on Anaxicrates: Σαβά, Ἀδραμύτα, Κιτίβαινα, Μαμάλι (= Ma'in? see Tkač (b) 1331–1333). The kingdom of Awsan was apparently unknwown.

217 Berger (a) 299–300, however, thinks that the name *Erythra thalassa* points to the outer ocean, instead of the Arabian Gulf.

218 Robin (a) 52.

219 Tkač (b) 1325; Wissmann (d) 375, 392; Avanzini 90: "De nouvelles études, impliquant une réévaluation des sources classiques, nous portent à croire que, même avant J.–C., les États sudarabiques s'étendaient sur tout le haut plateau et dans les régions côtières: les Sabéens dans la *Tihamā*, les Qatabanites dans la Dathina, vraisemblablement jusqu'à Aden, et même au nord d'Aden, sur la mer Rouge, au temps de leur plus grande splendeur". Also see *Yémen* 94 (Robin); 100 (Avanzini): "Pour le royaume de Qatabân commence alors (sc. 6th cent. B. C.) une longue période de splendeur qui se poursuivra jusqu'au Ier siècle. Son territoire semble s'étendre jusqu'à Bab al-Mandab).

220 Robin (a) 52.

221 Str. 16, 4, 19; 21. Artemidorus, however, is aware of *Mariaba* (Str. 16, 4, 19).

of *Mariaba*.[222] Some scholars argue that Agatharchides used pre-Hellenistic sources reflecting the period when Saba had reached its climax.[223] Alternatively, Agatharchides may have favoured this kingdom, which outshone the rest of south Arabian states in the eyes of the Greek writer.[224] As Ryckmans rightly points out: "La description de l'Arabie par Agatharchide de Cnide (…) après s'être attachée aux détails de la côte de l'Arabie jusqu'à la hauteur du Yémen, substitue à cet endroit, à la description des côtes, une description de Saba' (…) manifestement inspirée de récits fabuleux parvenus par la route terrestre."[225] The phrase "récits fabuleux" naturally refers to the Sabaeans' extraordinary wealth, idleness and luxury, lavishly depicted by Agatharchides. The same author vaguely reports a trading traffic in a place seemingly under Sabaean control and possibly identical to Aden.[226] He also mentions a marvellous fact (*paradoxon*) concerning the coastal part of Saba': "The sea in these parts looks to be white in colour, so that the beholder marvels at the surprising phenomenon and at the same time seeks for its cause."[227]

The documents from the Roman period contain new data and improve the Hellenistic body of knowledge. Pliny quotes Juba's confusing list of south Arabian peoples, among which one can recognise Qatabân (*Catabani*), Saba' (*Sabaei*) – the latter being praised because of their frankincense (*Sabaei Arabum propter tura clarissimi*) – and Hadramawt (*Chatramotitae*).[228] These states are vaguely located before and after Bab al-Mandab (*ad utraque maria porrectis gentibus*[229]). The changing balance of power in this region did not remain unknown to Pliny, the author of the *Periplus* or Ptolemy either. All authors indeed are aware of the growing power of the *Homeritai* [Ḥimyar / Dhu Raydan][230], a kingdom which at that time was competing with Saba' to control the south-west peninsula, while Qatabân's power was vanishing. It is, however, only the *Periplus* that clearly points out the Sabaean-Ḥimyarite rivalry.[231] Their business apparently enabled Alexandrian merchants to obtain a correct understanding of Ḥimyar's administrative organisation, as the *Periplus* reports a local "governor" based in *Sauê* (as-Sawā[232]). This official stood in subordinate relation to the king living in *Saphar*. Finally, by mentioning the road from *Mouza* to the inland cities of *Sauê* (Sawwā) and *Saphar* (Zafār), the author

222 Diod. 3, 47, 4 (= Agatharchides, *GGM* 1, 100). See, however, Pirenne 111–112. *N. B.*: Elsewhere Strabo, drawing on an unknown source, provides a puzzling division of *Arabia Eudaimôn* into five *basileia* (Str. 16, 4, 25).

223 Woelk 233, 260–262 (quoting von Wissmann); Rodinson 59: "On voit que tous les Sudarabiques sont purement et simplement identifiés aux Sabéens. On ne peut savoir jusqu'à quel point cela dérive d'une suprématie réelle de Saba."

224 See Str. 16, 4, 19 (Artemidorus).

225 Ryckmans 86. See Diod. 3, 47, 4 (= Agatharchides, *GGM* 1, 100); 3, 47, 6 (= Agatharchides, *GGM* 1, 102).

226 See above, n. 35. *Contra*: Ryckmans 86: "ce récit (…) ignore (…) l'existence (…) des ports de Qana et Aden."

227 Diod. 3, 47, 8–9 (= Agatharchides, *GGM* 1, 103 – transl. Oldfather).

228 Plin. *nat.* 6, 153–154 (= Juba, *FGrHist* 275, F 33). See Ryckmans 83.

229 Tkač (b) 1439, thinks that *gentibus* refers only to the Sabaeans.

230 Plin. *nat.* 6, 158; 6, 161; *P. m. r.* 23; Ptol. 6, 7, 25.

231 Von Wissmann (d) 311–312; Rodinson 213–214; Robin (a) 52–53.

232 Ryckmans 79; Schiettecatte (a) 234.

informs us that, although Saba' and Ḥimyar were involved in the regional and local sea trade, their real centres did not lie on the coast.[233]

With regards to the coastal area (Tihamā), it was not until Roman times that tribes scattered along this narrow plain, of which some are known from Arabian epigraphy[234], appeared in Greek and Latin sources, even if less attention was paid to such destitute tribes than to the kingdoms surrounding them. The *Periplus* and Pliny [**24**] point to tribes of Fish Eaters (*Ikhthyophagoi*) settling in southern coastal places – certainly including the Bab al-Mandab area – and on Red Sea islands.[235] As for the *Elisaroi*, a people dwelling on the shore between *Adêdos* (Hudayda) and *Okêlis*, they have been identified with the "tribu de al-ʾAšʿar (…) qui s'étendait jusqu'au delà du Bab al-Mandab".[236] Finally, there is a tribe of *Scenitae* Sabaeans (= Sabaeans living in a tent, a clear indication of their nomadic life) situated by Pliny [**24**] at *Ocelis*, whose identity remains obscure. As von Wissmann states, "zur Zeit seiner (sc. Pliny's) Abfassung das Sabäerreich bis zum Bab al-Mandab reichte."[237] At any rate, Pliny's statement offers an interesting parallel to a Ḥimyarite inscription dating from the mid-5th century A.D., in which several kings claim sovereignty over a number of nomadic tribes of the Tihamā.[238]

There is little information concerning climate, fauna, and flora, and almost nothing specifically relating to the area of the strait. In particular, mangroves, which fascinated those who visited the African side of the Arabian Gulf, are ignored by written sources.[239] The few natural facts listed by Eratosthenes concern the hinterland and aim at glorifying a fertile and wealthy country. The region "lying opposite to *Aithiopia*", "watered by summer rains", "sowed twice", having "all kinds of birds" clearly does not point to the barren Tihamā. As for aromatics, the most praised product of south Arabian countries, Eratosthenes provides further details when saying that frankincense was collected in Qatabân, while myrrh was produced "farthest towards the east", in *Khatramôtis* / Hadramawt.[240] These names do, however, hardly apply to the coastal parts of Arabia. Theophrastus, to whom we owe the best description of incense and myrrh trees, claims that Alexander's explorers observed them in the mountains (ἐν τῷ ὄρει), far from the seashore.[241] Agatharchides contrasts with the previous authorities when he claims that two varieties of aromatics respectively named *balsamon* and *kasia*, and a mysterious herb "having a nature peculiar to itself" (ἰδιάζουσαν φύσιν ἔχουσαν [?]) grew along the coast (παράλιον) controlled by the Sabaeans. Agatharchides may have been misled,

233 *P. m. r.* 21–24 (Plin. *nat.* 6, 104, and Ptol. 6, 7, 41–42 are less accurate). See Schiettecatte (a) 276.
234 Gajda 55, 66, 89–90, 195. On the Ṣabir culture which developed in the Tihamā, see Schiettecatte (a) 226–228.
235 *P. m. r.* 20. See Tkač (a) 2528–2529.
236 Ptol. 6, 7, 7; Ryckmans 79–80.
237 Von Wissmann (c) 1341.
238 Gajda 195: "Ils se nommaient ainsi 'roi de Saba', dhū-Raydan, Ḥadramawt et leurs nomades de Ṭwd^{um} et de la Tihāmat'."
239 See, however, Bretzl 102; Schneider (a) 364–367.
240 Str. 16, 4, 4 (= Eratosth. Berger III B 48 – transl. Jones).
241 Theophr. *h. plant.* 9, 4, 4.

thinking that products actually exported (or re-exported) from a Sabaean port were grown in the Sabaean country. On the other hand, he rightly says that the most renowned commodities (frankincense and myrrh) were harvested in the hinterland (μεσόγειον), i. e. the mountains.[242] No noticeable improvement is to be found in later documents. Pliny had much to say about the Arabian spices marketed in Italy but almost nothing about the trees of *Arabia Felix*. It is significant that Pliny expressed disappointment at the *arma Romana* not having improved the description of the incense tree provided by the Greeks.[243]

To conclude, it must be stressed that from the Greco-Roman perspective the Arabian side of Bab al-Mandab served as a contrast to its African counterpart. As regards the western sector, the coastal area (*paralia*) and the hinterland (*mesogaia*) were not perceived as disconnected from each other, while the Arabian *paralia* was overshadowed by the inner region in many respects. In addition, the various tribes (Fish Eaters, Troglodytes, *Koloboi* etc.) scattered on the African side were thought to have a poor livelihood and a weak political organisation (at best they were under the authority of "headmen" [*turannos*]).[244] In contrast, powerful and wealthy kingdoms (*basileia*) flourished in southern Arabia. With their social division into classes (traders, farmers, soldiers etc.), their political organisation and their urban civilization, they appeared to Greco-Romans as structured states.[245]

4. THE LOCAL TRADE NETWORK IN THE AREA OF BAB AL-MANDAB

4.1 Eratosthenes – Agatharchides – Artemidorus

When the first Greek explorers reached the area of Bab al-Mandab, around the mid third century, they noticed a local sea traffic.[246] However poor the extant pieces of evidence are, they give the earliest information about a local network which may have had a long existence.[247] The exchange of goods before this time has been documented only by archaeological remains, e. g. obsidian artefacts. No doubt, this traffic was important and visible enough to attract the attention of Ptolemaic observers, who may have been interested in understanding where valuable commodities

242 Diod. 3, 46, 2–3 (= Agatharchides, *GGM* 1, 97 – transl. Oldfather). Artemidorus repeats Agatharchides' statement (Str. 16, 4, 19).

243 Plin. *nat.* 12, 55.

244 This was to change significantly with the rise of the Aksumite kingdom (Munro-Hay 61–75).

245 Breton 97–114.

246 As far as I know, this topic has not been much investigated: see Delbrueck 21–22, 240–241; Raschke 656, 931, n. 1135–36; Desanges & Reddé 179–180; Casson 19–21; Villeneuve 173–178; Sidebotham (b) 38; Bianchetti (a) 283. Some scholars have overestimated the position of the Mediterranean newcomers, such as Rostovtzeff (b) 747 ("The Ptolemies (…) almost succeeded in monopolizing the African goods for themselves. Only a small part of these wares was still handled by the south Arabian merchants"), or Kortenbeutel 40 ("dadurch, daß die Ptolemäer jetzt das eine Ufer der Enge in der Hand hatten, wird dieser Handel zugunsten der Ptolemäer unterbunden worden sein").

247 De Romanis 98–99; Munro-Hay 61–66; Schiettecatte (a) 225.

imported to the Greek world were coming from. On the other hand, they may have paid little attention to staples of local interest which were likely to have been transported on the same crafts (for instance, foodstuffs or wood[248]).

Some traces of this trade activity can be detected in the half-fantastical narrative of Iamboulos' adventures (above, p. 200): the hero is reported to have been transferred from Arabia to the opposite side by Ethiopian bandits. Considering that between piracy and trade there is often not much difference, this fictional account may echo a local trade activity involving all kinds of people in the strait area. The earliest set of reliable data, however, comes from Eratosthenes [**3**]. The two sides of Bab al-Mandab (*êpeiroi*), he says, were connected by a maritime traffic, apparently only observed from the African side. Light boats (*skhedia* = "raft", "float"[249]) would carry freight from one side to the opposite, not directly but *via* the Six Islands mentioned above (p. 216), which seem to have served as intermediary stops. The natives used small crafts, well suited for short voyages. They put in at moorings which did not require special infrastructure[250], at least on the African side. Wherever the final stop of the "Six Islands crossing" may be,[251] no particular settlement, such as *emporion*, *polis*, *polikhnion*, was noticed by the Greeks in this area. Neither did they mention which African tribes – Troglodytes? *Ikhthyophagoi*? – were involved in this trade exchange. But who were the Arabians coming from the opposite shore? It seems plausible that some people of Qatabân, whose kingdom encompassed the eastern side of the strait (*stena*), would cross the channel (*diabasis* [**1**]) and take part in this trade. Of the freight transported in their crafts, nothing is said by Eratosthenes, but fortunately we learn more from other authorities.

Indeed, Artemidorus [**17**] and Agatharchides provide further information, though the latter neither explicitly mentions the strait[252] nor seems to have a clear

248 For the sake of comparison, see *The Red Sea and Gulf of Aden Pilot* 230: "In the vicinity of ras Siyyan are bushes of considerable extent, where the natives of Perim and the Arabian coast come for wood."

249 Compare with Str. 16, 4, 18 (Nabatean *skhediai*); 16, 4, 20 (Persian *skhediai*).

250 See *P. m. r.* 3: *Ptolemais* of the Hunts "has no harbor and offers refuge only to small crafts (*skaphai*)" (transl. Casson).

251 It has been long assumed that African places located near the strait and called *Sabaïtikon stoma*, *Saba*, *Sabai* (Str. 16, 4, 8; 10 [= Artemidorus]) were ancient "sabaeische Niederlassungen am afrikanische Ufer" (Delbrueck 35; Kortenbeutel 34). Extremely ancient relationships between each side of the Red Sea have been disclosed by archaeological remains (Munro-Hay 61–66). On the so-called Sabaean thalassocracy, see, for instance, Raschke 654–655.

252 Phot. *Bibl.* 250, 459a (= Agatharchides, *GGM* 1, 101, with no similar passage in Diodorus' paraphrase): "All the other <Sabaean> men train for war, work all the land, and voyage from home using large rafts (σχεδίαις). They transport cargoes of various sorts including especially an aromatic plant which grows in the interior (ἐν τῷ πέραν φυόμενον) and is called in the language of the Arabs *larimna* (Str. 16, 4, 19 [*larimnon*]: see Woelk 247; Burstein 167, n. 2). Not a few of the Sabaeans also employ boats made of skins. The tides themselves have instructed them in their use, although they live in luxury." (transl. Burstein). This may point to a "side to side crossing". *Contra* Tkač (b) 1394, referring to Sabaean exportations to India (a statement probably based on the theory of the "Sabaean thalassocracy").

idea of the traffic.[253] That said, both authors emphasize the role of the Sabaeans, as if this nation was in control of these exchanges, differing thus from Eratosthenes in being unaware of the presence of Qatabân. Artemidorus claims that a part of the "masses" (*plethê*) lived off this local trade[254], while Agatharchides vaguely attributes the trading activity to "many Sabeans" (τῶν δὲ Σαβαίων οὐκ ὀλίγοι). According to the former, the Sabaeans imported certain varieties of aromatics, which were not produced in their own country, from the opposite side, certainly because they intended offering for sale a variety of items. The country from which these *arômata* were obtained is called *Aithiopia* by Artemidorus, a name which may point to the Somali coast [**16**].[255]

Arômata and other goods were carried in light boats made of leather (δερματίνοις πλοίοις / δερμ. πορείοις; the case of the σχεδίαι is less certain). Due to lack of details, we cannot determine whether these crafts were made from sewn pieces of leather or were "supported by floats made from animal skin", like the crafts employed by Arabian pirates.[256] At any rate, they were suited for transporting relatively light loads, e. g. spices. Agatharchides concludes with the following puzzling comment: "The tides themselves have instructed (διδασκούσης τῆς ἀναπώτιδος) them (sc. the Sabaeans) in their (= boats) use, although they live in luxury" (transl. S. M. Burstein). Among the Greeks, the Sabaeans had the reputation of being extremely affluent, and were not thought to be simple merchants, hence Agatharchides' surprise. Whatever the value of this final moral inclusion, it indicates that the Arabs were capable of properly using the ebb and flow of the tides and of coping with the dangerous tidal currents in the strait. This interesting remark somehow matches Eratosthenes' statement that the natives sailed from one island to another for security purposes.

4.2 Pliny the Elder

Further information comes from Pliny, whose sources unfortunately cannot be identified – meaning that the historical context escapes us –, and whose statements seem imprecise to the reader on more than one occasion.

Pliny is aware of a local trade network connecting the natives from the coastal parts of the African side of the strait with the hinterland. A certain quantity of myrrh, he says, was conveyed by land and / or by sea from the *Trogodytice* country (= the *Smyrnophoros* country?) to a harbour named "the port of Isis", which probably lay at, or near *Deirê*.[257] No doubt, Pliny refers to a renowned variety of myrrh

253 Agatharchides thought that all aromatic gums and spices exported by the Sabaeans were collected in their own country (Diod. 3, 46, 2–3 [= Agatharchides, *GGM* 1, 97]).

254 On the social organization in the south Arabian kingdoms, see Woelk 245–247; Breton 119–147. See also *Yémen* 102 (Robin): "Le négoce était tenu en piètre estime par les aristocraties guerrières des grandes tribus."

255 The aromatics were also harvested in the hinterland (Str. 16, 4, 14: "the *kinnamômon* is more abundant in the neighbourhood of the places that are deep in the interior" – transl. Jones).

256 Plin. *nat.* 6, 176; Burstein 167. See also *P. m. r.* 27 (leather made crafts at *Kanê*).

257 Plin. *nat.* 6, 174: "And further on, Port of Isis, ten days' row distant from the city of the *Adulitae* (= *Adoulis*), and a centre to which Troglodytic myrrh is brought." (*Isidis portus, decem*

being carried from the harvesting points to an entrepôt located at *Isis portus*. There it was sold to neighbouring peoples, such as the Arabs, which does, however, not exclude the possibility of Greco-Roman merchants purchasing myrrh at *Isis portus*. It must be remembered, at any rate, that this small-scale transport was basically the starting point of the whole trade activity in the Erythraean Sea.

Elsewhere Pliny claimed that the Sabaeans were involved in the local sea traffic bringing them to the opposite side of the sea, where a special kind of myrrh could be obtained: "A kind highly spoken of is also imported from islands, and the *Sabaei* even cross the sea to the Troglodyte country to procure it."[258] (transl. Rackham). The phrase *transitu maris* points to a "side to side crossing" no doubt somewhere in the southern Red Sea and probably in the vicinity of Bab al-Mandab. As far as can be judged from this elliptical text, it is the Troglodytes that loaded myrrh in these unidentified islands (Suqutra?[259]) and shipped it to their homeland, where the Sabaean could purchase it. There this top quality myrrh was transported to a Sabaean port, either for local consumption or more plausibly for re-exporting. As said above, they probably aimed to offer a large variety of aromatics so that Greco-Roman merchants calling at a Sabaean port could get all the items they wanted in the same place. By importing aromatics from other spice bearing regions, the Sabaeans were able to meet the Mediterranean demand.

This variety of commodities was probably made available at the port of *Muza* (Al Mukha), for, according to Pliny, Alexandrian merchants importing spices and aromatics would stop at *Muza*. In contrast, those who desired Indian products would moor at *Ocelis* or *Cane* before they sailed by the south-west monsoon wind: "There is also a third port named *Muza*, which is not called at on the voyage to India, and is only used by merchants trading in frankincense and Arabian perfumes (*quem Indica navigatio non petit, nec nisi turis odorumque Arabicorum mercatores* – transl. Rackham)."[260] The *Periplus* informs us that the *emporoi* heading to Arabia left the Egyptian ports in September, while those going to India had to begin their trip earlier, in July.[261] Even if Pliny speaks of *odores Arabici* not all aromatics grew in Arabia: from the author's point of view, all products exported from *Muza* were regarded as Arabian ones, despite the fact that some were gathered in other countries.

dierum remigio ab oppido Adulitarum distans; in eum Trogodytis myrra confertur – transl. Rackham). Desanges (c) 86–88 avoids converting Pliny's figure into a kilometric distance, and cautiously concludes that this place was situated "around the strait". According to Eratosthenes [**2**], "reproductions of temples of Aegyptian gods" stood in many places along the Red Sea, which may explain the name "Port of Isis".

258 Plin. *nat.* 12, 66 (*convehitur et ex insulis laudata* [sc. *murra*], *petuntque eam ad Trogodytas Sabaei transitu maris*). See also Plin. *nat.* 12, 60 (frankincense from anonymous islands: compare with Theophr. *h. plant.* 9, 4, 10); 6, 155.

259 Groom (a) 116. There is no other island large enough to grow myrrh and incense trees in this area. Neither myrrh nor frankincense are, however, listed in the *P. m. r.* 30 among the goods exported from Suqutra (*Dioskouridês*). See also Pirenne 163 ("Il s'agit évidemment des 'îles sabéennes' de la côte de la mer Rouge dont Pline [**24**] a parlé."). Desanges & Reddé 180 connect this allusive text with the "Six Islands", although these merely seem to be transit points.

260 Plin. *nat.* 6, 104.

261 *P. m. r.* 25; 39; Casson 288–289.

Finally there is the famous narrative dealing with the *cinnamomum* trade[262]:

> *Cinnamomum* (…) grows in Ethiopia, which is linked by intermarriage with the Trogodytes. The latter buy it from their neighbours and convey it over the wide seas in ships that are neither steered by rudders nor propelled by oars or drawn by sails, nor assisted by any device of art (…). Moreover they choose the winter sea about the time of the shortest day, as an east wind is then chiefly blowing. This carries them on a straight course through the bays, and after rounding a cape a west-north-west wind brings them to the harbour of the *Gebbanitae* called *Ocilia*. On this account that is the port most resorted to by these people, and they say that it is almost five years before the traders return home and that many perish on the voyage. In return for their wares they bring back articles of glass and copper, clothing, and buckles, bracelets and necklaces. (transl. Rackham)

No doubt clues are lacking for us to fully understand this text. In particular, the spice named *cinnamomum* is far from being identified. As for the voyage, a duration of five years is simply unbelievable and contradicts the story itself (see in particular the return trip which lasts as long as one winter's time). In short, some data must have been distorted, misunderstood or falsified. Moreover, Pliny cannot help but give a moral judgment – actually a commonplace in his time –, saying that this dangerous voyage was completed in order to fulfil women's desires. This implies that some parts of the story may have been distorted by Pliny himself. In face of such complications, some scholars have rejected this document as spurious, as if it reflects "the commercial lies told to the western merchants."[263]

Despite these difficulties, I believe that Pliny's narrative contains an altered description of an actual local trade circuit involving the Troglodytes.[264] Whatever this mysterious *cinnamomum* may be – either an unidentified but genuine "Ethiopian" spice[265], or a commodity imported from an unknown place in the western Indian Ocean and re-exported from *Aethiopia* – the Troglodytes probably acquired it from a place located on the northern Somalia coast. This view is supported by the following two arguments. First, Pliny claims that this spice was supplied in <a part of> *Aethiopia* "linked by intermarriage with Troglodytes", which means that this area was not too far from the Troglodyte country (it is worth noting that these *Aethiopes* are said to be the neighbours [*conterminis*] of the Troglodytes).[266] Secondly, one must remember that the *Kinnamômophoros* country reported by the Ptolemaic explorers was not far from the *Trogodytice*, and was defined as a part of *Aethiopia* (below, p. 243).[267] Let us turn to Pliny's account in an attempt to suggest a coherent interpretation.

It seems that the Troglodytes purchased (*mercantes*) this *cinnamomum* from their "half-blood" "Ethiopian" neighbours; they reached *Aithiopia* with light

262 Plin. *nat.* 12, 86–88.

263 Warmington 191.

264 Miller 156–159, identifying *cinnamomum* with today's cinnamon, imagined that it was shipped from south-east Asia. De Romanis 110–112, assuming that this *cinnamomum* grew in the Horn of Africa, gives a more convincing commentary to this text. See also Desanges, Stern, Ballet 30.

265 See Raschke 652–655.

266 Only close relationships can lead to intermarriage. Compare with *P. m. r.* 16.

267 See Str. 16, 4, 14 (Artemidorus).

boats (*ratibus*) that, according to Pliny, "are neither steered by rudders (*neque gubernacula regant*) nor propelled by oars (*neque remi impellant*) or drawn by sails (*uel trahant uela*), nor assisted by any device of art (*ratio*)". This description reminds us of the Sabaean rafts pushed by the tidal currents and winds, according to Agatharchides. As far as can be ascertained, the Troglodytes were probably carried along the African seashore by the surface currents and by the north-west winds, which prevail in the southern Red Sea from June to September.[268] They would then cross the strait and reach a place somewhere in northern Somalia. Having loaded some *cinnamomum* they directly (*recto cursu*) sailed back across the Gulf of Aden (*per sinus*). They were aided in this by the north-east or winter (*hibernum; brumam*) monsoon (*euris maxime flantibus*). This detail is in stark contrast to the previous statement that the Troglodyte crafts were not drawn by sails, a fact which makes Pliny's description rather suspect on this point: even if their rafts may have been pushed by surface currents[269], they were no doubt also fitted with some sort of sails. Then the Troglodytes rounded a cape (*promuntarii ambitu*[270], an expression which recalls the *Palindromos akron* referred to by Ptolemy), which is probably to be located near the cape Sheikh Sa'id. Finally, they were carried by a north-west wind (*argestae*) to *Ocilia* (= *Ocelis*).[271] Here, their cargoes of *cinnamomum* were exchanged with the manufactured goods mentioned by Pliny[272], which the Troglodytes then transported back home. This *cinnamomum* was no doubt offered for sale in the *emporium* of *Ocelis* [see **24:** *emporium Acila*], which does not exclude the possibility of a certain quantity being re-exported to *Muza*.

In conclusion, this fascinating text discloses a kind of "triangular" local trade circuit enabling Troglodyte tribes of the southern African Red Sea to obtain manufactured goods in exchange for valuable raw products, in other words by bartering. Pliny's reports that *Ocilia* belonged to the *Gebbanitae* indicate that Qatabân was involved in this trade, which echoes Eratosthenes' statement. Thus, Pliny's source may go back to the time when Qatabân still played a prominent role in this area.[273]

268 See the corresponding charts provided by the National Geospatial-Intelligence Agency (http://msi.nga.mil/NGAPortal/).

269 According to *The Red Sea and Gulf of Aden Pilot* 33, currents run out of the Red Sea from June to September and into from November to April, at the rate of 40 m. per diem during the strength of the monsoons.

270 Compare with Plin. *nat.* 3, 103: *portus Aggasus, promunturium montis Gargani, a Sallentino sive Iapygio CCXXXIIII m. p. ambitu Gargani.*

271 According to Plin. *nat.* 6, 106, the Alexandrian *emporoi* sailing back from India would use the *volturnus* straight to the entrance to the Red Sea, and then (*cum intravere Rubrum mare*) the *africus* or *auster*.

272 Some items may have been imported from Alexandria, or Syria, to *Ocilia*, especially glassware: *P. m. r.* 6–8; Casson 126–127; Desanges, Stern, Ballet 21–61, and especially p. 30–31; de Romanis 111–112.

273 De Romanis 110–111 (quoting Plin. *nat.* 12, 93) rightly points out that the king of *Gebbanitae* had full control of the *cinnamomum* market (also see Breton 64). The question of who the *Gebbanitae* were is, however, debated (see above, n. 215). Beeston (a) 5 thinks they were "a dominant group within Ma'in" (but he does not deal with the present passage). Pirenne 161–166 thinks that this excerpt relates to Qatabân's position in Pliny's time (de Romanis 110 is of

The point is that some African tribes seem to have played an active part in the local trade: Pliny's Troglodytes have almost nothing to do with the "Naturvolk" depicted by Agatharchides in his philosophy oriented presentation. As a matter of fact, it is not by chance that, according to Pliny, they possessed an important port of trade which was destined to have a prominent role in the centuries to come, namely *Adoulis* (*maximum emporium Trogodytarum*).[274]

4.3 The *Periplus maris Erythraei*

In the first part of his narrative (i.e. the voyage from *Myos Hormos* / Quseir al-Qadim to *Rhapta*, near Mombasa) the anonymous author gives a short account of the small port of *Aualitês* [**27**]. Although, according to him, Greco-Roman ships occasionally put in at *Aualitês*, this small harbour seems to have been mostly involved in local trade activities, for light boats – a detail that characterises local trade – stopped here. The author distinguishes two types of crafts: the *skhediai* ("raft"?) and the *skaphê* ("small crafts"?) but, as elsewhere in his work, he does not find it important to describe them and explain their differences.[275] Some of these might be the above-mentioned crafts propelled by tides and currents. At any rate, we learn from this document that *Aualitês* was connected to both *Mouza* and *Okêlis* on the opposite side: "Exports from here, with the transport across to *Okêlis* and *Mouza* on the opposite shore at times carried out by the *Barbaroi* on rafts, are: aromatics [the so-called *cinnamomum* was perhaps included in this general category], a little ivory, tortoise shell, a minimal amount of myrrh but finer than any other." (transl. Casson).

Who were the peoples involved in this local traffic? The author vaguely refers to *Barbaroi* using *skhediai*, and given that he generally refers to people living on the east coast as "Arabs"[276] and states that the "*Barbaroi* who inhabit the place (sc. *Aualitês*) are rather unruly (ἀτακτότεροι)", one may deduce with certainty that the name "*Barbaroi*" points to the inhabitants of *Aualitês*.[277] Thus, the African people of *Aualitês* appear to have played a role in a local trade network. It is, however, unlikely that the Arabs could have been excluded from these activities, and maybe the so-called *skaphê* were operated by them. In conclusion, whatever the true location of *Aualitês* may be (above, n. 167), this place was a centre for the local traffic in the Bab al-Mandab area at the time of the *Periplus*.

Let us now turn our attention to the products listed in the *Periplus*. The following goods are said to be exported to *Mouza* and *Okêlis*: ivory, *arômata*, a special quality of myrrh as well as tortoiseshell. All of these commodities were collected on the African side of Bab al-Mandab, either in the coastal area or in the hinterland. At this stage it is worth emphasizing the presence of two commodities which have

a similar opinion: "commercio del *cinnamomum* o *cinnamum* nel I sec. d.C."). According to Pliny, however, the *Sabaei* were supposed to control *Ocelis* [**24**] (see VON WISSMANN [b] 440; Id. [c] 1341).

274 Plin. *nat.* 6, 173.

275 On these so-called rafts, see CASSON 117–118.

276 *P. m. r.* 20; 25.

277 *P. m. r.* 7 (transl. CASSON) and above, p. 219.

not been referred to in the previous documents, i.e. ivory and tortoiseshell. Ivory was obtained from elephants hunted by the natives, probably in inner Somalia. The Greeks indeed gave up this activity a couple of centuries ago. Thus *Aualitês*, as an ivory entrepôt, was connected to the hinterland and its peoples (*Adoulis* was linked to *Koloê* and Axoum likewise[278]). As for tortoise shell, Agatharchides (above, p. 220) claims that the *Khelonophagoi* were particularly good at hunting sea turtles. Despite the fact that he describes them as a primitive tribe having to fight harsh environmental conditions, it is clear that this allegedly destitute people had managed to take part in the local network by supplying the middlemen of *Aualitês* with this precious material. A similar pattern can be observed in the area of *Adoulis*, which was an entrepôt for tortoise shell brought there by the *Ikhthyophagoi*[279] from the *Alalaiou* Islands (Dahlak Archipelago), who perhaps used very large turtle shells as crafts to reach *Adoulis*.[280]

These commodities were shipped from *Aualitês* to *Mouza* on the Arabian side, confirming that *Mouza* ranked first among the local trade places. Not surprisingly this port is described as particularly busy, full of Arabian shipmasters and sailors trading with the coast on the opposite side (τῇ τοῦ πέραν ἐργασίᾳ) by means of their specific (= local) boats (ἰδίοις ἐξαρτισμοῖς).[281] As a matter of fact, *Mouza* also traded with *Adoulis*, exporting (= re-exporting) the "aforementioned merchandise from *Adoulis* across the water".[282] There was also another African *emporion* called *Malaô* (= Berbera[283]) east of *Aualitês*, from which varieties of *arômata* "were exported to Arabia"[284], which may be an allusion to *Okêlis* and *Mouza* – nonetheless, it cannot be excluded that this refers to *Kanê* (Khor Rori).[285] To what extent *Eudaimôn Arabia* / Aden played a role as a local trade centre is less clear. Supposing that the name Fortunate Islands (*Eudaimones nêsoi*) points to Aden, this place was inserted in the patterns of both regional and local trade in the 3rd century B.C.: not only ships arriving from north-west India and the Arab-Persian Gulf converged on this port, but also numerous "crafts" coming from the whole nearby region (οὐκ ὀλίγας <σχεδίας> […] ἀπὸ […] τῆς σύνεγγυς πάσης).[286] According to the *Periplus*, Greco-Roman captains would, however, moor in this harbour in the mid-1st century A.D., only to be supplied with fresh water[287], as if *Eudaimôn* was disconnected from the Mediterranean sea trade. Still there is no reason to think that Aden was not integrated in the local and regional network, as indicated by the dedication of

278 *P. m. r.* 4 (elephants and rhinoceros were hunted in the hinterland, west of *Axoum*).

279 *P. m. r.* 4 (also Plin. *nat.* 9, 35). See DESANGES (c) 217–218.

280 Str. 16, 4, 14 (these *Khelonophagoi* apparently lived inside the strait). See also Agatharchides [**11**] (*Khelonophagoi* settling beyond the strait); Plin. *nat.* 9, 35.

281 *P. m. r.* 21.

282 *P. m. r.* 24.

283 CASSON 120.

284 *P. m. r.* 8. See CASSON 122–127.

285 *P. m. r.* 27. See SCHIETTECATTE (b) 248, 253–259.

286 Phot. *Bibl.* 250, 459b (= Agatharchides, *GGM* 1, 103).

287 *P. m. r.* 26.

Hermeros, son of Athenion, presenting himself as an Erythraean Sea merchant from Aden [*Adaneitês*], in an inscription dating to A. D. 70.[288]

To sum up, the *Periplus* attests to an important trade activity around the strait of Bab al-Mandab and its vicinity, not only because of the Greco-Roman ships crossing the strait, but also because of the flourishing local trade: *Mouza*, Aden, *Adoulis*, *Malaô*, *Aualitês* were the main spots of a significant local network. Some of these ports were linked to the long distance Greco-Roman network, *Mouza* and *Adoulis* served as entrepôts for the Mediterranean merchants and as terminals for local trade. On the other hand, *Aualitês* certainly received more local crafts than ships belonging to Mediterranaean *emporoi*. The same could be true for *Okêlis* where, according to the *Periplus*, Greco-Roman *emporoi* would not buy and sell. This Arab *kômê* (village) was used as a watering station and a place to moor their boats.[289] Thus, when Rome set up a military detachment in the Farasan Islands in the 2nd century A. D., she judiciously chose a place located near this local trade area, “à peu près à l’entrée nord d’une zone (…) peuplée, contrôlée à l’ouest (…) par le royaume d’Axoum ou le royaume pré-axoumite d’Adoulis, et à l’est, en Arabie, par les royaumes sudarabiques”, as Villeneuve writes.[290]

4.4 Additional observations

This inquiry into local trade inevitably raises the question of to what extent the Bab al-Mandab trade was controlled by local rulers or polities. Unfortunately, only a few documents are available to us when trying to answer this question, most of them being provided by the *Periplus*. Clearly, some rulers were strong enough to set regulations: *Mouza*, for instance, was under the authority of a “tyrant” (τύραννος, that is to say a local king, namely the governor of the region of *Mapharitis*) named *Kholaibos,* who was in turn subordinate to *Kharibael*, the king of Ḥimyar and Saba. Apparently, Arab merchants could do business in *Rhapta* only “through a grant from the king” (παρὰ δὲ τοῦ βασιλέως ὑποφόρον ἔχουσιν οἱ ἀπὸ Μούζα, the word “king” points to *Kharibael* and not *Kholaibos*[291]). In the light of this evidence, one might think that the traffic linking *Mouza* with *Aualitês* was similarly subject to a form of control, at least for tax purposes. For the sake of comparison it must be remembered that according to Pliny (above, n. 273) Qatabân held a sort of monopoly on the cargoes of *cinnamomum* coming to *Ocilia* from the Horn of Africa via the Troglodytes: “The right of controlling the sale of *cinnamomum* is vested solely in the king of the *Gebbanitae*, who opens the market by public proclamation.” (transl. Rackham). On the basis of a somewhat overlooked text we can assume that such a control may have been oppressive. The prices of *cinnamomum*, Pliny writes, were

288 Bernand (b), n° 65 (also see n° 62: *Ada[n]itês emporos*). About Aden’s “renaissance” in the following centuries, see Ryckmans 80–81; Schiettecatte (a) 236–237, 284.

289 *P. m. r.* 25.

290 Villeneuve 155.

291 *P. m. r.* 16; 22–23. See Casson 45; Rodinson 205–206.

"raised to half as much again after the forests (*silvis*)[292] had been burnt, so it is said, by infuriated Barbarians (*ira Barbarorum*)". In this case, Pliny is no doubt thinking of a species growing in the Horn of Africa. These *Barbari* are obviously the inhabitants of *Barbaria* (or *Barbarikê*), a name referring to the north Somalian coast (above, p. 219). Pliny does, however, add that it is not certain whether "this was incendiarism provoked by injustice on the part of those in power (*ob iniquitatem praepotentium* = the local rulers? The middle-men involved in the *cinnamomum* commerce?), or was due to accident."[293] Local authorities also engaged in providing safety to merchants, and this was most likely the duty of those laying claim to a certain trade area. In particular, the *Periplus* informs us that Arabian chiefs and kings fought against the tribes who plundered the shipwrecked sailors in the middle part of the eastern Red Sea.[294] They might have acted the same way in the southern sectors.

As for the African side of the Red Sea, the ruler named Zôskalês (above, p. 221) is reported by the *Periplus* to have controlled an area extending from south of *Berenikê* as far as the strait ("the rest of Barbaria"), including Adulis.[295] Much has been said about this figure whose identity still needs to be established. Be that as it may, he is portrayed as a man "always holding out for getting more" (transl. Casson [τοῦ πλείονος ἐξεχόμενος]). This may explain why he was supplied with items of the finest quality by Greco-Roman *emporoi.* So close a relationship with a number of Alexandrian merchants strongly suggests that he was not indifferent to trade activities (as a matter of fact, *Adoulis* is classified among *emporia nomima* ["legally limited ports of trade"[296]] as is *Mouza* by the *Periplus*). Zôskalês is likely to have been at least somewhat concerned with securing commercial exchanges, since, at the time of the *Periplus*, Adulis was still a risky place to moor[297] as was perhaps also *Aualitês*. Unfortunately we have no knowledge of how, or to what extent the local traffic was regulated by Zôskalês: did the *Ikhthyophagoi* transport tortoiseshell to *Adoulis* on behalf of *Zôskalês*? Did the Troglodytes freely purchase *cinnamomum* from "Ethiopians" and sell it in Arabia? Given the lack of documents there is no point in speculating about these issues, however important they may appear to us.

No one would deny that the development of Greco-Roman sea trade had an impact on the general pattern of commercial exchanges in the western Indian Ocean. For instance, as is well known, the trans-Arabian spice route was less resorted to as more and more Mediterranean seamen ventured to the western Indian Ocean. Likewise, piracy certainly increased in the wake of the increasing quantities of goods being transported.[298] Thus the important alteration in long distance trade

292 It is probably these *cinnamomum* groves that are referred to by Artemidorus in his description of the Horn of Africa: "The *kinnamômon* is more abundant in the neighbourhood of the places that are deep in the interior" (Str. 16, 2, 14 – transl. Jones).

293 Plin. *nat.* 12, 93 (transl. Rackham). These accidental fires are caused by the south winds which "are so hot that they set fire to the forests in summer."

294 *P. m. r.* 20.

295 *P. m. r.* 5.

296 On these ports, see, for instance, Casson 274–276.

297 *P. m. r.* 4.

298 About piracy in the Red Sea, see Schneider (b).

generated by the involvement of Alexandrian merchants may have affected the local trade in the area of Bab al-Mandab, in particular the activity of local markets and the pattern of trade routes. Firstly, we know that several ports absent from early Ptolemaic sources (*Aualitês*, *Mouza*, *Adoulis*) became important places of trade for they supplied the Greco-Roman *emporoi* with local goods. Secondly, by purchasing some commodities and offering others the Greco-Romans inevitably modified the local supply and demand. In particular, the local suppliers will have tried to meet the Mediterranean demand in terms of quantity and / or quality. Fortunately, this is documented by several texts.

Let us consider first the case of tortoiseshell. This material was so highly regarded in the Roman world that the demand necessarily increased from the early-1st century A. D. onwards.[299] For this reason the *Ikhthyophagoi* and the *Khelonophagoi*, the primary producers in the area of Bab al-Mandab, must have become more and more involved in the local network. Interestingly, Pliny mentions a special sort of tortoiseshell called *chelium* (= χέλειον), which seems to have been praised in the Mediterranean world (*eximiae testudinis*)[300]. This commodity was made available to Mediterranean merchants at Adulis, as Pliny elsewhere relates[301]. Next Pliny states that this particular variety was not abundantly supplied (*rarae*): "it (sc. *chelium*) is of tortoise shell of exceptional quality, but it is seldom seen, as the very sharp rocks frighten the Turtle-eater tribe (*Chelonophagi*: see above, n. 181), while the Troglodytes (no doubt settling in the vicinity of *Adoulis*) on whose coasts the turtles swim, worship them as sacred."[302] This statement may imply that despite the probable indirect pressure caused by Mediterranean demand, both tribes refused to engage in dangerous or sacrilegious hunting. The "triangular" *cinnamomum* traffic provides another example: supposing that some of the "manufactured" items offered at *Mouza* and looked for by the Troglodytes were imported from Alexandria, it is possible that they intensified their trade activity in order to purchase more of them.

5. CROSSING THE STRAIT AND MAPPING THE STRAIT: TWO PERCEPTIONS OF SPACE

Despite the meagre amount of surviving evidence, the strait of Bab al-Mandab offers a remarkable case study to those focusing on ancient geography, since two perceptions of space are present in our sources: on the one hand the hodological representation of space[303], a concept thoroughly studied and popularized by P. Janni,

299 Warmington 166–167; Casson 101–102; Burstein 87, n. 1; de Romanis 159–166.
300 Plin. *nat.* 9, 38.
301 Plin. *nat.* 6, 173.
302 Plin. *nat.* 9, 38 (transl. Rackham).
303 "The term 'hodological space' is derived from the Greek word 'hodos', path, way. In contrast to the mathematical concept of space as presented on maps, plans, etc. 'hodological space' is based on the factual topological, physical, social, and psychological conditions a person is faced with on the way from point A to point B, whether in an open landscape or within urban

and the cartographic concept. The hodological depiction of the strait, which reflects the perceptions of those who sailed across it (*diaplous* or *diabasis*), contrast with the more abstract data enabling the mapping (*pinax*) of Bab al-Mandab. As a matter of fact, these two kinds of information may occur simultaneously in most of our evidence.

5.1 Crossing the strait

Significant traces of a hodological perception of space, which encapsulates the traveler's physical and psychological perceptions, appear in Eratosthenes, Agatharchides and Artemidorus' reports. There is nothing suprising about this, as these authorities borrowed most of their topographical information from people who had sailed as far as, and beyond, the strait. The most valuable document relating to this question is, however, the *Periplus of the Erythraean Sea* including a unique description of the crossing (*diaplous*), and Casson may be right when stating that the author's experience "included the African route (…) and the Arabian-Indian route at least down to Cape Comorin".[304] Without doubt, this text offers more "hodological" vocabulary than any other (for a full list of the terms, see app. 7.3). Thus, it is worth beginning with this document, in which three ways to cross Bab al-Mandab are described.

The first of the three accounts concerns the "right-hand" crossing, along the west coast of the Arabian Gulf [**27**] (known today as the "Great Strait"). The author says that past *Adoulis* and the "Obsidian Gulf" (Hauachil Bay) the sealine turns eastwards[305], which means that the two sides of the Arabian Gulf approach each other; in others words, the sea appears to the traveller to be shrinking. Then follows the port of *Aualitês* (probably Zeila; see above, n. 167) at the narrowest section of the passage, a place where Greco-Roman ships and local small boats moor. *Aualitês* also marks the shortest interval (*diaplous*) between the two sides of the Arabian Gulf.

The account of the Arabian route contains a description of the "left-hand" crossing, i.e. along the east shoreline of the Arabian Gulf [**28–29**] (today's "Small Strait"). After a course of 300 *stadia* (about 50 km) past *Mouza* the Arabian and African coast (in the vicinity of *Aualitês*) approach each other. Next the traveller enters an *aulôn* ("channel", or "strait") not very "large" and reminiscent of a double door closing (*apokleiô*), and the two sides of the gulf now appear to lie so close to each other as to provide only a narrow passage. Here lies the *Diodôros* Island (Perim[306]), between the two sides of the strait, which is 60 *stadia* wide. At this point

or architectural conditions." (*Termium Plus, la banque de données terminologiques et linguistiques du gouvernement du Canada*).

304 Casson 8.

305 See Casson 115: "Following the coast, the course from Ptolemais *Therôn* to Adulis would have been roughly south-southeast. From Adulis to Avalitês it would have been southeast, in other words, not due east but merely a little more easterly".

306 Casson 157 (see below, p. 244).

the traveller faces strong currents[307] and a wind coming down from the mountains of the mainland. In this *isthmos* (a word probably pointing to the space between Perim and the Arabian shore) lies *Okêlis*, the first place to drop anchor for "those sailing on (*esô*)".[308] Beyond *Okêlis* the distance between the African and Arabian coasts increases reminiscent of an opening "double door" (in my opinion, *palin* means "contrariwise", a translation seemingly more adequate than that of Casson: ["with the waters again opening out"]). The shoreline trends eastwards, leading gradually (*kata mikron*) to the open sea (*pelagos*). After sailing 1200 *stadia* (about 200 km) one arrives at *Eudaimôn* (Aden), the next place to moor with "its suitable harbours".[309] It seems as if the crossing of the strait actually ended at Aden. This recalls Pliny's words (with Juba as his authority) [**23**] that the "open sea" begins at *Adanu*/Aden. In the hodological perception of space the limit of the "open sea" is not marked by the narrowest interval between the two landmasses but rather by the first place where the traveler can call in having left behind him the dangers of the crossing.

The last way to cross the strait, namely the "side to side" sailing (see the discussion about the local trade activity above, p. 227 sq) is only alluded to [**27**], for Mediterranean merchants would probably avoid this transverse route.[310] Firstly, only light boats could technically manage this difficult crossing: the Mediterranean *ploia* leaving Egyptian harbours headed for the opposite side of the Red Sea more or less beyond *Berenikê*[311] and certainly did not venture into the smalls channels of Bab al-Mandab. Secondly, the Greco-Romans did not take part in the local trade and hence had absolutely no reason to ply between the two sides of the strait. Actually, what is of major interest in this short text is the author perception of space: when defining the smallest width of the strait of Bab al-Mandab the author refers to a sailing "side to side" instead of quoting figures (Αὐαλίτης, καθ' ὃν καὶ στενώτατός ἐστιν ἀπὸ τῆς Ἀραβικῆς εἰς τὸ πέραν διάπλους). Such a crossing only concerns the natives plying with their light crafts (*skhediai*) between a place situated 300 *stadia* from *Mouza* (= *Okêlis*? [**28**][312]) and *Aualitês*. The point is that "shortest" in terms of *diaplous* does not necessarily mean "shortest" in terms of distance.[313] For people facing currents, wind and reefs, the "shortest" crossing was not necessarily a straight course from Sheikh Sa'id to Ras Siyyan, where Africa and Arabia lie nearest to each other as the crow flies, but rather the shortest safe trip.

307 According to Lapidoth 132, the navigation in the Small Strait "is dangerous due to strong and irregular tidal streams".

308 Note that *exô* (or *ek*) generally means "(going) out of the Arabian Gulf" [**13**; **18**; **19**; **21**], and *eis* "into the Gulf" [**19**; Plin. *nat.* 6, 106].

309 Aden is ignored by Plin. *nat.* 6, 104.

310 Casson 157: "Only locals made the crossing".

311 *P. m. r.* 19–20 (also see Juba [**23**]).

312 The strait is situated at a distance of 40 nautical miles (400 *stadia*) from *Mouza* (Casson 147).

313 Compare with the Strait of Gibraltar, which was not crossed at its narrowest part: "A une navigation aléatoire entre la Bétique et la Tingintane traversant le détroit de Gibraltar en son point le plus étroit, les Anciens préféraient une route plus longue mais plus sûre hors des influences dangereuses de celui-ci" (Ponsich 273).

The rest of the evidence (Eratosthenes, Agatharchides, Artemidorus) dates back to the Hellenistic period and precedes the *Periplus*. At that time, Greek ships would skirt the African shore more often than the Arabian side, which explains why the "left hand" crossing is not documented. Although these texts are not as elaborate as the *Periplus*, they offer several "hodological" details. For instance Agatharchides' account [**9**] contains several expressions similar to those employed in the *Periplus*: past the *Tyrkaios* mountains[314], the Arabian Gulf steadily "closes" (*sugkleietai*) and decreases in size (*sunagogê*); "the mainlands" are no longer"out of sight of each other". At this stage it is worth noticing that straits are fundamentally defined by such a "closing" / "opening" sequence.[315] The existence of the strait of the Arabian Gulf was actually not established until the Ptolemaic explorers observed the decreasing/increasing width of the sea. Similarly, the Strait of Hormuz was discovered when Nearchus entering the Arab-Persian Gulf from the "Great Sea" (i. e. the Indian Ocean) observed the shrinking of the passage.[316]

What I, however, believe to be the most striking point is that the *diaplous* mentioned by the *Periplus* was also reported by the Ptolemaic travellers: the so-called "strait of the Six Islands", which is situated beyond the Strait of *Deirê*, is nothing but the sea route used by local people with their *skhedia.* From this alternative perspective, it is the shortest safe crossing which defines the strait, regardless of the actual distance between the two sides. Actually, the strait of the Six Islands is about three times as wide as the strait of *Deirê* [**3**]. This dramatically emphasises the contrast between the hodological perspective, i. e. "the unidimensional view of space as a path"[317] and the two-dimensional space-representation which characterizes maps.

5.2 Mapping the strait

This text in which Eratosthenes refers to the two above-mentioned straits is unique in the sense that two conceptions of space were – probably unintentionally – brought together. On the one hand there is the hodological perception in which the traveller's experience plays a prominent part. On the other hand, when Eratosthenes deals with the strait of *Deirê* (*kata Deirên*) he offers a set of more abstract data which serves as the basis of a geometrical description of the strait, namely points, measurements, and directions.[318] This is, one might say, the two-dimensional or cartographic Bab al-Mandab, which was for the first time set into a "map" (*pinax*) of the *oikoumenê* by this renowned scholar.[319] This geometrical representation of the strait may have remained unchanged for a long time, for only Ptolemy is likely to have modified the

314 This unidentified range is probably situated in the vicinity of Suakin, or Mitswa [Massawa] (Woelk 190; Burstein 133, n. 4).

315 Compare with Str. 2, 5, 19; Diod. 4, 18, 5 (the Pillars of Heracles).

316 Arrian. *anab*. 7, 16, 1–2; *Ind*. 32, 6–9 (also Amm. 23, 6, 10). See Högemann 74.

317 Dan & Geus & Guckelsberger 18.

318 Jacob 114–116.

319 On Eratosthenes' goals, principles and method, see: Jacob 114–115; van Paassen 39–43; Aujac 41–122; Geus 260–288.

mapping of the Bab al-Mandab area (nothing can be said of Agrippa's conception, as there is no surviving fragment relating to the strait in details).

As mentioned above, the explorers serving the interests of the Ptolemies in the 3rd century noticed that the narrowest part of the Arabian Gulf on its western side was marked by the Ras Siyyan (promontory of *Deirê*), a cape rising to 136 metres above the sea level, which could hardly be missed. On the opposite side stood another noticeable promontory with the dominating Djebel Men Ali (229 metres) called *Akila* (*Okêlis*), a name apparently unknown to Eratosthenes but not to Artemidorus. These capes, extending towards one another, appeared to reduce the mouth of the Arabian Gulf to its narrowest[320] (ποιεῖ ἄκρα τὰ στενά [**2**]). In other words these two spots had been regarded by the Ptolemaic explorers as perfect markers when defining the strait. J. Desanges rightly says: "Ces deux repères ont joué en somme le rôle de *Calpê* et *Abila* de part et d'autre du *fretum Gaditanum* (détroit de Gibraltar)."[321] Reports of this supplied Eratosthenes with two topographical points that could be joined by a straight line. This line represented the shortest interval between the two sides of the Arabian Gulf at its mouth. This was the "geometrical strait", unequivocally marking the limit between the Arabian Gulf and the ocean from a cartographic perspective, "die eigentliche Meerenge" as Berger writes[322] (this is still the way we define the Bab al-Mandab strait today). Differently, the strait of the Six Islands – which Berger refers to as "die sogennante Meerenge" – could neither be geometrically defined nor represented as a straight line. Being estimated at 60 *stadia* in width – probably by Timosthenes [**26**] –, the *kata Deirên* strait was thought by some authorities to be as narrow as the Columns of Heracles.[323] There can be little doubt that even if the educated public was aware of the "strait of the Six Islands", Greco-Roman geography in most cases referred to the strait of *Deirê*, for it represented an unambiguous landmark and a clear limit. For instance Artemidorus says that the northern Somali coast extends over an unknown distance, from *Deirê* as far as the *Notou keras* (cape Gardafui [**16bis**]), and ignores the Six Islands. The importance of *Dire* is also present in Pliny's account [**22**].

The importance of the *kata Deirên* strait in the construction of ancient geographical knowledge cannot be underestimated, as the discovery of one of the four inlets of the Outer Ocean was a considerable step. Equally important for improving the Greek geographical knowledge of the time was the set of distances involving this place: 4500 *stadia* were measured from *Ptolemais Therôn* to *Deirê*, while the distance from *Deirê* to the edge of the *Kinnamômophoros* had been estimated at

320 Compare with the discovery of the strait of Hormuz (above, n. 316).

321 Desanges (a) 100. See Ponsich 267.

322 Berger (a) 297. For a modern definition of the Bab al-Mandab strait, see e.g. Lapidoth 130–132, or the *Encyclopaedia Britannica* ("Strait between Arabia (...) and Africa (...) that connects the Red Sea (...) with the Gulf of Aden and the Indian Ocean (...). The strait is 20 miles (32 km) wide and is divided into two channels by Perim Island; the western channel is 16 miles (26 km) across, and the eastern is 2 miles (3 km) wide." See also Warnecke 334–335.

323 According to Str. 2, 5, 19, the Columns of Heracles are 70 *stadia* wide; a different figure (60 *stadia*) is reported elsewhere (7, 3, 6). See Ponsich 259.

5000 *stadia*[324] [**2**; **3**; **4**; **5**; fig. 4]. Although only approximate,[325] these round figures may have met Eratosthenes' needs. As for the length of the opposite side, i. e. the line stretching from the bottom of the Gulf of Aqaba to the promontory facing *Deirê*, Eratosthenes thought that Anaxikratês' (?) figure (14,000 *st.*) was a little excessive.[326] These figures were to be coupled with some indications of directions [**2; 3**] (those relating to the African seashore being more accurate than those concerning the opposite coast): From *Heroônpolis* to *Ptolemais Therôn* the coastline was reported to extend southwards and slightly eastwards (the angle formed by the coastline and the meridian of *Ptolemais* is estimated at 48° by H. Berger[327]); from *Ptolemais* to *Deirê* the shoreline was said to continue southwards, however, turning more eastwards; finally, from *Deirê* to the edge of *Kinnamômophoros*, where the world known to Eratosthenes terminated, and through which an important parallel was established, the sea line was reported to run southwards and eastwards (see also [**5**][328]). In contrast, the whole Arabian coast seems to veer due southwards.[329] Thus, *Deirê* – with the strait – was one of those landmarks which remarkably helped to map the southern part of the inhabited world.

The mapping process includes dividing (*temnein*[330]) the area under consideration into sectors to which a name is given. As regards Eratosthenes, we are not told by our main source, Strabo, how the *sphragis* division (geometric division of the inhabited world determined by natural boundaries[331]) applied to this section of the *oikoumenê*. Nor do we hear of the relevance of the strait of *Deirê* for this matter.[332]

From a general point of view, Eratosthenes regarded the whole area of the strait as a part of Arabia, regarding, as many others, the Nile's right bank as the boundary of both Arabia and Asia – hence the name Arabian Gulf given to the modern Red Sea.[333] The eastern part of the Bab al-Mandab area is included in the part of Arabia called "Blessed" (*Eudaimôn*). Eratosthenes' division of the eastern side rests upon an ethnical basis (*ethnê)*: with Qatabân stretching down to the strait the reader has a useful landmark to roughly locate the three other tribes (kingdoms).[334] As for the African side his sources supplied him with topographical designations. The western coastal part of the Red Sea as far as the strait was known as the *Trôglodytikê.*[335] *Beyond Deirê* lay the *Smyrnophoros* country, next to which was the *Kinnamômo-*

324 The latitude of *Ptolemais* was estimated by astronomical observations (i. e. the duration of the day at the summer solstice). See Berger (a) 295; Desanges (c) 305–307.

325 On this topic see Str. 2, 1, 39; 2, 1, 41; Jacob 207–209.

326 Str. 16, 4, 4. About the alternative figure (12,000 *stadia*) see Berger (a) 294–295; Desanges (c) 39–42.

327 Str. 16, 4, 4. See Berger (a) 297; Geus 278–280.

328 See Berger (a) 297.

329 Desanges (c) 41.

330 On this issue, see Jacob 561–562. See also Str. 2, 1, 30–31.

331 Van Paassen 42–43.

332 On the fourth *sphragis*, see Str. 2, 1, 36; 2, 1, 32; Berger (a) 282–283.

333 See Str. 16, 4, 2; 16, 4, 4. See also Plin. *nat.* 6, 177 (= Juba *FGrHist* 275, F 36); Str. 2, 1, 31–32.

334 The four kingdoms follow each other (see Tkač (b) 1321–3).

335 Str. 16, 4, 4. See also Str. 16, 4, 18, and above, p. 218–219.

phoros.[336] No doubt, however, the country named *Aithiopia* encompassed the Horn of Africa and accordingly also the two latter countries.[337] The case of the *Trôglodytikê* does, thus, demonstrate how difficult it will have been to establish topographical divisions: on the one hand, this region was theoretically included in Arabia, as mentioned above; on the other hand, the same territory may be regarded as a part of *Aithiopia*.[338] Moreover, from an ethnographic perspective most of the Troglodytes could have had physical and cultural features that led them to be characterised them as *Aithiopes* (e.g. the skin colour). In other words, countries might be defined according to at least two different criteria: topography or ethnicity.[339]

Eratosthenes' spatial arrangement was to be slightly improved and/or altered in later times.[340] One may notice that Pliny when referring to *Dire* as a conspicuous place mentions the southeast orientation of the strait [**22**; **26**]. Ptolemy [**30**], Agathemerus [**4**] and Marcianus[341] place the strait at *Deirê,* ignoring the Six Islands, while the former provides a set of new toponyms (above, p. 218). Ptolemy, primarily interested in the location (*thesis*) of places and peoples rather than geometry and figures[342] (see fig. 5) allocated coordinates to the "strait of the peninsula of *Okêlis* and of *Deirê*" [**31**; **32**]. Although this geographer benefited from a greater amount of information than Eratosthenes, there are puzzling statements which must be regarded as inaccuracies or errors. Firstly, the distance between *Okêlis* and *Deirê* is excessive (40' = ~330 *stadia*) [**30**].[343] Secondly, Ptolemy confusingly locates the strait beyond *Okêlis* and before *Deirê*.[344] Thirdly, the opening is oriented in a west-east direction. Finally, the *Koloboi* reported by the Hellenistic observers around Bab al-Mandab are curiously pushed northwards, between *Ptolemais Therôn* and *Adoulis*.[345]

To close this section I would like to examine two issues that are perhaps beyond solution: the first one relates to the width of the strait (see app. 7.4, which offers a collection of figures ranging from 60 *stadia* to 400 *stadia*; I do not take into consideration Agatharchides' "aberrante mesure"[346]). Not only do the numbers vary. In fact they all contradict the actual distances. According to some scholars, this confusion must be explained by the presence of the Perim Island, which actually divides the strait into two channels, namely the great Bab al-Mandab and the small Bab

336 Str. 16, 4, 4.

337 Str. 16, 4, 4; 2, 1, 14; 1, 2, 26. See also Diod. 3, 18, 4 (=Agatharchides, *GGM* 1, 41): the "insensible *Ikhthyophagoi*" are called "Aethiopians".

338 See, for instance, Str. 16, 4, 17: the Troglodytes are not clearly distinguished from the Ethiopians. See Desanges (c) VII–IX.

339 See van Paassen 48.

340 See Geus 286–287.

341 Marcian. Heracl. 1, 11; 1, 15.

342 Jacob 583.

343 Compare with Ptolemy's map of the strait of Gibraltar (Ponsich 266–268, 271–272).

344 About Ptolemy's "inexacte représentation", see Desanges (a) 98–99.

345 Ptol. 4, 7, 7.

346 Desanges (a) 89. Also see Woelk 189; Müller (a) 165. Curiously, the strait is as wide as the *mukhos*, i.e. perhaps the entry of the gulf of Suez, as if it there was a predilection for symmetry. For a full discussion of this topic, see Desanges (c) 39–40, 42.

al-Mandab.[347] Even if this view is accepted as a plausible hypothesis, it does, however, not explain the discrepancy between the most common figure[348] (60 *stadia*) and the actual distance. How could such a short distance be incorrectly measured? If, as I have stated above, the Greco-Romans abstained from crossing the strait from *Deirê* to *Okêlis,* then they may have estimated the distance by visual observations[349], unless they were informed by natives. In both cases the figures were very likely to be rough estimates. Most importantly, it must be remembered that there was not one, but several ways to ply between the two sides. No doubt, this variety of figures – particularly the highest ones, 200 *stadia* and 50 Roman miles – reflects the various routes taken by the natives for trade purposes.

The second question is why the large island of Perim dividing the strait between *Deirê* and *Okêlis* into two unequal parts is so rarely mentioned and badly described. Eratosthenes seems to be unaware of it, as do Agatharchides and Artemidorus. How could such a noticeable spot have been ignored? Considering that the bulk of Hellenistic knowledge originates from observers mostly skirting the African shore and not sailing from *Deirê* to *Okêlis*, this omission could to some extent be understandable. At the time when the Alexandrian merchants would sail down the coast of Arabia on their way to India, Perim could, however, hardly have gone unnoticed. In fact, Pliny reports a certain island of *Diodorus*, which must be Perim, although he wrongly locates it in the Gulf *Aualitu*, *viz.* beyond the strait.[350] In addition, it is surprising that Pliny connects this island to the "right-hand" (i. e. African) periplus, since Perim lies close to the Arabian coast. Ptolemy is even more inaccurate than Pliny. His island of *Diodoros*[351] is mistakenly situated near Adulis.[352] In reality, the only solid information is to be found in the *Periplus* [**28**], which correctly locates the island of *Diodoros* when relating the crossing of the small Bab al-Mandab. Curiously, along with this accurate information we find the common but erroneous figure of 60 *stadia* transmitted, as if the author of the *Periplus* intended to portray himself as a knowledgeable person.

6. IMPERIALISM AND SPACE; POWER AND GLORY: THE RULERS / CONQUERORS AND THE BAB AL-MANDAB STRAIT

Ancient historiography contains several narratives in which the crossing of a strait bears a political significance. For instance, Herodotus tells us that before crossing

347 See DESANGES (a) 89–98 (passim); (c) 92; MÜLLER (b) 760.

348 BERGER (a) 297.

349 See WARNECKE 334. Compare with Arrian. *Ind.* 32, 6 (see HÖGEMANN 74); Str. 16, 3, 2 ("its [= the Arab-Persian Gulf] mouth, he [= Eratosthenes] says, is so narrow that from *Harmozi*, the promontory of Carmania, one can see the promontory at *Macae* in Arabia." – transl. JONES).

350 Plin. *nat.* 6, 174 (*ultra sinus Aualitu, insula Diodori et aliae desertae*). See DESANGES (c) 79, 89, who disagrees with DIHLE. *N.B.*: the island *Citis* referred to by Pliny [**22**] cannot be identified as Perim (DESANGES (c) 71).

351 About this Diodoros, see MÜLLER (b) 787; DESANGES (c) 89.

352 Ptol. 4, 7, 38. According to DESANGES (c) 79, 89, Ptolemy mixed the *Diodorou nêsos* up with the *Didôrou nêsos*. In addition, Ptol. 6, 7, 44 wrongly locates the *Adanu* islands in the Red Sea.

the Bosphorus and reaching Europe Darius the Great erected two steles to commemorate his power: "When he had viewed the Bosporus also, he set up two pillars of white marble by it, engraving on the one in Assyrian and on the other in Greek characters the names of all the nations that were in his army: all the nations subject to him." Alexander symbolically hurled a javelin at land before his ship reached the shore of the Asian Hellespontus, as if the conquest began with the passage across the strait: "Alexander advanced with his army to the Hellespontus and transported it from Europe to Asia. He personally sailed with sixty fighting ships to the Troad, where he flung his spear from the ship and fixed it in the ground, and then leapt ashore himself the first of the Macedonians, signifying that he received Asia from the gods as a spear-won prize."[353]

Crossing a strait is somehow similar to crossing a border and may accordingly convey a political message. As Strabo says in a general discussion relating to the monuments that Heracles and Dionysus left behind in India: "It is not improbable that they who first <visited these regions>, set up boundary marks fashioned by the hand of man, such as altars, towers, and pillars, in the most remarkable situations, to indicate the farthest distance they had reached (and straits, the surrounding mountains, and little islands, are indubitably the most remarkable situations for pointing out the termination or commencement of places)."[354] Does this similarly apply to the Bab al-Mandab strait?

As I have stated in my general introduction, the inhabited world was believed to be an immense island surrounded by the Ocean, from which it received four seas that were regarded as "gulfs". The Bab al-Mandab was one of the four inlets through which the Ocean penetrates the land-masses. Appropriately, the strait, being a marvellous landmark, might enhance the glory of conquerors. Significantly, Strabo, when discussing the controversial question of the identification and location of the Pillars of Heracles, claims that, like the "Pillars of India", the name and celebrity of this boundary of the world had been conferred by conquerors and generals (ἡγεμόνων), not by merchants[355]. It comes as no surprise that several semi-historical conquerors, i. e. the Babylonian Nebuchadnezzar and the Ethiopian Thearkhon (Taharqa), were said to have reached the Pillars of Heracles.[356] Coming back to the Bab al-Mandab strait, it must be remembered that before the discovery of the passage to the open sea by Ptolemaic explorers, Nekôs and Darius were famous for having sailed outside the Arabian Gulf. Could then the Ptolemaic kings, and later the Romans, afford to ignore the role of Bab al-Mandab as a boundary of the inhabited world? In this section I will argue that the surviving evidence supports the view that the *fauces Rubri maris* served the glory of rulers, i. e. they were used for propaganda purposes.

353 Hdt. 4, 87 (transl. Godley); Diod. 17, 17, 2 (transl. Oldfather). See Warnecke 336.

354 Str. 3, 5, 6 (transl. Hamilton).

355 Str. 3, 5, 6. On the "columns" raised by Dionysus in India near the ocean, see e. g. Dion. Per. 623–625.

356 Str. 15, 1, 6 (= Megasthenes, *FGrHist* 715, F 11a).

6.1 The Ptolemaic rulers: Ptolemy II and Ptolemy III

Before examining how the activity of Ptolemy II and Ptolemy III in this part of the world enhanced their glory, it must be remembered that Alexander the Great was a model for his successors. Among his numerous feats, the following are particularly relevant to the issue which I am addressing: firstly, he defeated a famous nation settled at the edge of the world, namely the Indians and their elephants. Secondly, at the end of the Indian expedition he briefly sailed on the "Great Sea" (i. e. the Indian Ocean) surrounding the inhabited world to symbolically but manifestly express that he had reached the edge of the world.[357] Not only was he the first Greek to have achieved this (*prôtos euretês*), but in so doing he proved himself to have outdone Heracles and Dionysus, his mythical predecessors. Finally, he considered subduing a renowned people living in the southern part of the *oikoumenê*, the Arabs.

Similarly, Ptolemy II and Ptolemy III sought the glory of exploring and controlling the most distant countries. For instance the expeditions of Ptolemy II in *Trôglodytikê* and *Aithiopia*[358] – the latter being a name praised by Homer – aimed to present him as the first Greek to extend his power as far as the southern limit of the world. The same rulers also boasted about the successful hunting and taming of African elephants, which was indeed a remarkable achievement. Thus, it is likely that exploring the mouth of the Arabian Gulf may have all the more expanded the prestige of the Ptolemies, since Alexander's men had not actually achieved this.[359] Incidentally, it is worth noticing that Seleukos I and his son Antiochos I gave orders to explore the Caspian Sea with the goal of discovering (through *autopsia*) the fourth inlet of the outer Sea.[360]

Regrettably, no piece of evidence has survived explicitly connecting the crossing of the strait to the esteem of the Ptolemaic kings. We only have several short texts alluding to prestigious royal actions taking place in the area of Bab al-Mandab. However poor this material is, it enables us to reasonably conclude that Ptolemy II and Ptolemy III had realised that reaching this region was beneficial in terms of propaganda.

The name of Ptolemy III appears twice in association with elephant hunting. In an inscription going back to the beginning of his reign, this king proudly claims that he and his father were the first to hunt elephants in *Trôglodytikê* and Ethiopia (= the Nile valley ? the southern part of the Red Sea?) and have them trained for war purposes.[361] In addition, we are informed by Agatharchides of Ptolemy III having founded hunting stations beyond the strait: "The third Ptolemy also, who prided himself (φιλοτιμηθείς) upon the hunting of elephants which are found in this region

357 See e. g. Arrian. *Ind.* 43, 1–10. As van Paassen 265 writes, "Alexander had a geographical picture of the world in mind during his campaign and the problem of the ocean and the boundaries of the world played a great part in it". For the sake of comparison, see Str. 1, 2, 34 on Menelaus in Egypt and Ethiopia.

358 See Desanges (b) 252–279.

359 See, however, above, p. 195.

360 Plin. *nat.* 2, 167; 6, 58.

361 *OGIS* 54.

[περὶ τὴν χώραν ταύτην, *viz.* outside the <Arabian> Gulf], sent one of his friends named Simmias to spy out the land." (transl. OLDFATHER, slightly modified)[362] The word φιλοτιμηθείς in connection with the name of Ptolemy III may preserve some traces of royal propaganda: it seems as if hunting elephants beyond the strait was an achievement more admirable that the previous ones. As a matter of fact, it was during his rule that ships crossed the strait with the purpose of shipping elephants to Egypt. Interestingly, Hipparchus also alludes to the ancient elephant hunting in *Kinnamômophoros* [**18**], which may corroborate the view that this achievement increased the glory of Ptolemy Evergetes.

Besides securing Ptolemy's demand in elephants, explorers and hunters studied this remote part of the world bordered by the outer Sea called the *Kinnamomôphoros*. It is likely that Ptolemy III (if not Ptolemy II: see above, p. 197–199) could claim to have rivalled his glorious predecessor Alexander, as well as the similarly famous Nekôs and Darius by reaching the southern fringe of the *oikoumenê*. In particular, such explorations carried great significance as knowledge of the *oikoumenê* was improved through the endeavour of Ptolemy.[363] In fact, it was the expeditions around Bab al-Mandab that contributed to solving the question of the Nile's summer flood, a major geographical issue. Sometime before the Ptolemies, Alexander and his Friends had explained this phenomenon in a mere theoretical way, by comparing the Nile with Indian rivers overflowing during the rainy season of the summer. Strabo says that thanks to the explorations of Ptolemy (II or III), the phenomenon was explained for the first time not by "conjecture" (στοχασμῷ) but observation (αὐτόπται), since the explorers witnessed the heavy rainfalls falling each summer in "Upper Ethiopia" (τῆς Αἰθιοπίας τῆς ἄνω) and its "most distant mountains" (ἐν τοῖς ἐσχάτοις ὄρεσι). Next Strabo goes on to add: "this fact was particularly clear to those who navigated the Arabian Gulf as far as the *Kinnamomôphoros*, and to those who were sent out to hunt elephants (τοῖς πλέουσι τὸν Ἀράβιον κόλπον μέχρι τῆς κινναμωμοφόρου καὶ τοῖς ἐκπεμπομένοις ἐπὶ τὴν τῶν ἐλεφάντων θήραν) or upon any other business which may have prompted the Ptolemaic kings of Egyt to despatch men thither".[364] Following immediately upon this sentence is the mention of Ptolemy II: "For these kings were concerned with things of this kind (οὗτοι γὰρ ἐφρόντισαν τῶν τοιούτων); and especially the Ptolemy surnamed Philadelphus etc." (transl. Jones). Despite this statement, I do, however, tend to believe that the exploration of northern Somalia should be dated to the rule of Ptolemy III because of the famous Simmias who carried out ethnographic research beyond the strait.

In conclusion, I propose that having pushed his power beyond the African side of Bab al-Mandab, Ptolemy III – that it was Ptolemy II seems uncertain to me – was aware that he had achieved something which no Greek had done before: his ships sailed out of the Arabian Gulf, reached the Outer Ocean and skirted the *Kinnamô-*

362 Diod. 3, 18, 4 (= Agatharchides, *GGM* 1, 41). Compare with Diod. 3, 36, 3 (= Agatharchides, *GGM* 1, 78): Ptolemy II also took pride (φιλοτιμηθείς) in having elephants hunted.

363 Also see Gem. *isag*. 16, 24.

364 Str. 17, 1, 5.

mophoros, a country bearing the name of a mythical spice.[365] Despite the regrettable lack of textual evidence we are at last informed of some achievements of the Ptolemies in the area, namely the quest for war elephants and the explanation of the flooding of the Nile. This may be the remains of Ptolemaic propaganda.

Sesostris is the Greek form of the name of three Egyptian rulers of the 12th Dynasty (Senwosret). In the 5th century B.C. the Greeks were already aware of a sort of "legend of Sesostris". In Herodotus' *Histories* Sesostris (perhaps the third Senwosret: 1872–1853/52 B.C.) is described as a semi-mythical conqueror who, among other achievements, led his armies into *Aithiopia* and undertook a naval expedition in the Erythraean Sea, until he was forced to turn back. This legendary Sesostris may have been a heroic figure invented by some Egyptian priests for ideological purposes in order to challenge the Persian rulers.[366] Be that as it may, the legend of Sesostris was to change during the Ptolemaic period. In Greek Hellenistic sources[367] (or in Latin documents based on Greek sources) the imperial territory of Sesostris tends to coincide with that of Alexander the Great and his successors. In particular, Sesostris is said to have subdued Scythia, Bactria and the countries bordered by the Erythraean Sea as far as India. Diodorus even claims that Sesostris reached the Ocean after he defeated India. This feature clearly echoes Alexander's brief navigation of the "Great Sea" off the mouth of the Indus, probably the acme of his Indian campaign.[368] In the face of such coincidences one can wonder to what extent the Egyptians used the Hellenistic royal ideology and the myth of Alexander to recast Sesostris' legend for either propaganda purposes or because of "nationalist" feelings.[369] Coming back to my topic, this renewed story of Sesostris may plausibly mirror the Ptolemaic propaganda relating to the crossing of the strait.

It is indeed curious that a narrative of Sesostris'conquests comes after the account of Ptolemaic explorations around Bab al-Mandab [**2**]. Having reported the discovery of *Deirê* and the strait Strabo, quoting Eratosthenes, says that Sesostris sailed across the strait and erected a stele, apparently to proclaim that "he was the first to subdue the countries of the Ethiopians and the Troglodytes". Afterwards, crossing the channel he subdued the Arabs and conquered the whole of Asia. Although Strabo's text remains partly obscure, one can assume that the Greek explorers at, or in the vicinity of *Deirê*, found an inscription in hieroglyphic letters.[370] This discovery was later reported to some knowledgeable Egyptians, e.g. priests, who claimed that this proved that the great Sesostris had reached the strait. The point is that *Deirê* marking the end of the Arabian Gulf had been incorporated into the legend of Sesostris. In my opinion this peculiar and novel development of the legend of

365 Hdt. 3, 111: "It [sc. *kinnamômon*] is reported, reasonably enough, to grow in the place where Dionysus were reared" (transl. GODLEY).

366 Hdt. 2, 102–103; 110. See DELBRUECK 20; LLOYD (c) 37–40; MALAISE 252–255, 262–266; OBSOMER 80–113; DESANGES (c) 90–91.

367 Diod. 1, 55–58. See MALAISE n. 366.

368 See e.g. Arrian. *anab*. 6, 19, 4–19.

369 LLOYD (c) 37–40.

370 Plin. *nat*. 6, 174, also reports *stelae* "inscribed with unknown characters" around the strait (see DESANGES (c) 87–88). This has been connected with the Pharaonic voyages to Punt (see e.g. HELCK).

Sesostris reflects the importance attached to the strait by Ptolemy II and/or Ptolemy III. This embellished version of the Pharaoh's expedition was probably invented after the successful Greek exploration. With the crossing of the strait it seems likely that the royal propaganda will have proclaimed the edge of the *oikoumenê* to be under Ptolemaic power, at least symbolically. Some Egyptians, wanting to credit Sesostris with a similar achievement and to outdo the Macedonian kings, put forward the inscription discovered in *Deirê*: not only did it prove that their hero successfully subdued the Arabs, a people who escaped the Ptolemaic domination, but also that he was the first (*prôtos*) to reach the mouth of the Arabian Gulf. This particular aspect of the story of Sesostris is likely to have been devised in order to echo the prestige of the Ptolemies brought about by the successful crossing of Bab al-Mandab.

I stated above that the name *Kinnamômophoros* probably increased the glory of Ptolemy II and/or Ptolemy III. Interestingly, this particular toponym appears in a less known variant of the legend of Sesostris. According to Pliny (probably quoting Juba), Sesostris went (*Sesostris exercitum ducit*) to the cape and port *Mossylites*, at the edge of the Horn of Africa, in the country where the *cinnamomum* grew.[371] Again, Sesostris seems to have been praised for crossing the strait before the Greek kings and travelling as far as they did. Strabo also says that Sesostris led an expedition in the *Kinnamômophoros*, the proof of which could be found in (hieroglyphic) inscriptions: "We may well be surprised (…) by the fact that Sesostris traversed the whole of *Aethiopia* as far as the *Kinnamômophoros*, and that memorials of his expedition, pillars and inscriptions, are to be seen even to this day."[372] Being biased in favour of the Greeks, Strabo does, however, go on to say that the Ptolemies actually outdid Sesostris, as the latter failed to solve the question of the flood of the Nile. At any rate, these few documents no doubt unveil another aspect of the competition between the old Egyptian legend of Sesostris and the Ptolemaic propaganda, in which the strait of the Erythraean Sea and the access to the Outer Ocean will have played an important role.

6.2 Augustus and the Roman *Imperium*

With the annexation of Egypt Augustus to some extent renewed Ptolemy's "southern policy", turning his attention to Arabia, Ethiopia and the Red Sea area. Yet his goals and means were different. Augustus' diplomatic and military action in the Red Sea was guided by a "conception tranquillement œcuménique de l'empire romain", as Nicolet writes.[373] In his late years, Augustus summarized his achievements in the southern part of the *oikoumenê* in his *Res Gestae. Aethiopia* and Arabia are listed among the nations that he subdued: "At my command and under my auspices two armies were led almost at the same time into *Aethiopia* and *Arabia Felix* (*in Aethiopiam et in Arabiam, quae appellatur Eudaemon*); vast enemy forces of both peoples were cut down in battle and many towns captured. *Aethiopia* was penetrated as far as the town of *Nabata*, which adjoins *Meroe*; in Arabia the army (i. e. led

371 Plin. *nat.*, 6, 174.
372 Str. 17, 1, 5.
373 Nicolet 28 (also see Roddaz 262–270).

by Aelius Gallus) advanced into the territory of the Sabaeans to the town of *Mariba* (*in Arabiam usque in fines Sabaeorum processit exercitus ad oppidum Mariba*)."[374] (transl. Brunt & Moore). Augustus' foreign policy indeed focused on two countries of the southern part of the *oikoumenê* famous among the Greco-Romans: *Aethiopia* (*viz*. the kingdom of *Meroe*) and the south Arabian kingdom of Saba. As Strabo [**20**] explains, both were regarded by Rome as neighbours since they were more or less in "contact" with Roman Egypt: Troglodytes and Ethiopians border on Egypt; as for the Arabs, they are located close to Egypt, since the Arabian Gulf, separating them from *Trôglodytikê*, is "extremely narrow".

At first glance neither this text nor the rest of the evidence proves that the *fauces Rubri maris* were of any importance to the *laus imperii* (glory of the Empire). This is not surprising: How could a strait, which had been discovered long ago and was now being traversed by numerous merchants, be of any significance in terms of prestige? A closer look does, however, show how an extraordinary source of glory, so to speak, remained at the disposal of Rome: the Arabian kingdoms in the south-west part of the Arabian Peninsula, which neither Alexander nor the Ptolemies had been able to integrate into their empires. In his *Res Gestae* Augustus, far from mentioning the withdrawal of his army, suggests that the Sabaeans were subdued. It has often been argued that Augustus aimed to control the rich Arabian states for economic purposes[375], but this is consistent with the view that Augustus aimed to overdo the "Alexandrian kings" in terms of prestige by dominating south-west Arabia, an area including the eastern side of Bab al-Mandab.[376] Displaying his forces in this remote area was tantamount to victory, regardless of the outcome of the expedition. By so doing he surpassed the Ptolemies, who never attempted an attack on the Arabian kingdoms, turning most of their attention to the western side of Bab al-Mandab. Similarly, he equalled and even overcame Alexander by subjugating the Arabs, the only people who did not send ambassadors to Alexander[377]: "But at present both they (sc. the Nabataeans and the Sabaeans) and the Syrians are subject to the Romans."[378] No doubt the "Arabian glory" – of which even Augustus' grandson dreamed (above, p. 202) – must have been a source of great political profit. Such must have been the ideological motives lying behind the south Arabian expedition, if there were any.

At any rate, the expeditions led by Aelius Gallus, and maybe Gaius Caesar, changed the balance of power in south-west Arabia and the Bab al-Mandab area[379]:

374 *R. gest. div*. 26, 18–23.

375 Casson 37, n. 71 emphasises the economical aspect of Augustus's plan: "Just as the conquest of Egypt had put in Rome's lap an annual tribute (…) of grain, so the conquest of Arabia would add an annual tribute of precious myrrh and frankincense." Also see e. g. Sidebotham (a) 134, 139–141; Speidel 303–305; *Yémen* (Cuvigny) 69.

376 For the sake of comparison, notice how C. Cornelius Gallus, the first prefect of Egypt, prides himself on leading Roman armies beyond the cataracts (= against *Aethiopia*), claiming – wrongfully – that the (Ptolemaic) kings never attempted such an expedition (Dessau *ILS* 8995).

377 Arrian. *anab*. 7, 19, 6.

378 Str. 16, 4, 21 (transl. Jones).

379 See e. g. Speidel 301–303 (this does not imply a military control of Bab al-Mandab; see, however, Speidel 305). *Contra*: Beeston (b) 12: "Seen from the Sabaean angle, the whole affair

Rome acquired some influence in the eastern part of the Bab al-Mandab area, for in the mid-1st century B.C. the king of Saba' and Dhu Raydan in Saphar had both become "a friend of the emperors, thanks to continuous embassies and gifts".[380] Consequently, the conditions of maritime trading in the Erythraean Sea were necessarily modified. The establishment of direct communications with India in the Ptolemaic period does not imply that Arabian control over the east side of the strait had been weakened, for ships sailing along the Arabian coast may have been forced to stop there (maybe at Aden[381]), unless they headed to India via cape Gardafui and the western side of the strait .[382] At any rate, the Indian trade is likely to have been partly controlled by the Arabs. In contrast, as attested by the following documents, the Roman *imperium* created a serious barrier against the presence of an Arab control in the southern Red Sea. In other words, the "Arabian sea route" was from then on free from real control and pressure. First, as discussed above (p. 203) the (mysterious) sack of Aden enabled a form of free trading, since the Alexandrian merchants were no longer forced to load Indian commodities in this port. Secondly, Strabo, who praises both Aelius Gallus and Augustus for the Arabian expedition [**20**], elsewhere says that the *emporoi* with numerous ships were no longer afraid (ἐθάρρει) of crossing the strait and reach *Aethiopia* or India [**21**]. This means that distant but profitable places of trade in the Indian ocean could be freely accessed.[383] This new situation, which is in great contrast to the weakness of the Ptolemies (τὰ **νῦν** διὰ τοσαύτης ἐπιμελείας οἰκονομούμενα), is obviously linked to Augustus' policy. Roman power had managed to deprive the Arabs of their control over the eastern side of Bab al-Mandab and had provided the Alexandrian merchants with free and unlimited access to the Erythraean products. In this Rome could definitely take great pride.

To some extent, establishing trade with the countries of the Indian Ocean contributed to Rome's prestige. Interestingly, this idea is expressed in two significant texts: The first one is an often quoted excerpt from Aelius Aristides' *Roman Oration*: "No marine rocks and no Chelidonian and Cyanean islands define your empire (…), nor do you rule within fixed boundaries, nor does another prescribe the limit of your power. But the sea is drawn as a kind of belt without distinction through the middle of the inhabited world and your empire. About the sea the continent lie 'vast and vastly spread', ever supplying you with products from those regions. Here is brought from every land and sea all the crops of the seasons and the produce of each land, river, lake, as well as of the arts of the Greeks and barbarians, so that if someone should wish to view all these things, he must either see them by traveling

must have appeared not as a major threat to themselves, but as a minor 'frontier incident.'"

380 *P. m. r.* 23. Casson 37 believes that the diplomatic initiative came from Arabia, for Kharibaêl feared that the emperors repeated Augustus' military attempt. See also Plin. *nat.* 12, 57.

381 Ryckmans 80–81; Desanges (a) 319–321; von Wissmann (d) 312; Dihle (c) 109. The strategic role of *Eudaimôn Arabia* before the Roman expedition is stressed by the *P. m. r.* [**29**]. Note that Plin. *nat.* 6, 100–106 does not mention Aden (*contra*: Warmington 45–46). The Indian castaway who taught Eudoxus the monsoon route (above, p. 200) is likely to have missed Aden (certainly because of a gale).

382 Bianchetti (a) 283–289.

383 Dihle (b) 548–549. Compare with Str. 2, 5, 12.

over the whole world or be in this city. It cannot be otherwise than that there always be here an abundance of all that grows and is manufactured among each people. So many merchant ships arrive there, conveying every kind of goods from every people every hour and every day, so that the city is like a factory common to the whole earth. It is possible to sea so many cargoes from India and even from Arabia Felix, if you wish, that one imagines that for the future the trees are left bare for the people there and that they must come here to beg for their own produce if they need anything."[384] With Pliny we may conclude our inquiry: "For who would not admit that now that intercommunication has been established throughout the world by the majesty of the Roman Empire (*maiestate Romani imperii*), life has been advanced by the interchange of commodities and by partnership in the blessings of peace, and that even things that had previously lain concealed have all now been established in general use?"[385]

7. APPENDICES

7.1 The documents

Eratosthenes [Strabo – Agathemerus]

1. Strabo, 16, 4, 2 [Berger III B 48, p. 290]

… τρίτοι δὲ Κατταβανεῖς καθήκοντες πρὸς **τὰ στενὰ** καὶ **τὴν διάβασιν** τοῦ Ἀραβίου κόλπου, τὸ δὲ βασίλειον αὐτῶν Τάμνα καλεῖται.

… the extreme part of the country above-mentioned (=Arabia Felix) is occupied, third (i. e. after the the *Minaioi* and the *Sabaioi*), by the *Kattabaneis*, whose territory extends down to the straits and the passage across the Arabian Gulf and whose royal seat is called *Tamna* (transl. Jones).

2. Strabo, 16, 4, 4 [Berger III B 48, p. 291]

Ἐντεῦθεν δὲ μέχρι **τῶν στενῶν** ὡς τετρακισχίλιοι καὶ πεντακόσιοι πρὸς τὴν ἕω μᾶλλον. ποιεῖ δὲ **ἄκρα τὰ στενὰ** πρὸς τὴν Αἰθιοπίαν Δειρὴ καλουμένη, καὶ πολίχνιον ὁμώνυμον αὐτῇ· κατοικοῦσι δὲ ἰχθυοφάγοι. καί φασιν ἐνταῦθα στήλην εἶναι Σεσώστριος τοῦ Αἰγυπτίου μηνύουσαν ἱεροῖς γράμμασι **τὴν διάβασιν** αὐτοῦ· φαίνεται γὰρ τὴν Αἰθιοπίδα καὶ τὴν Τρωγλοδυτικὴν πρῶτος καταστρεψάμενος οὗτος, εἶτα **διαβὰς** εἰς τὴν Ἀραβίαν κἀντεῦθεν τὴν Ἀσίαν ἐπελθὼν τὴν σύμπασαν· διὸ δὴ πολλαχοῦ Σεσώστριος χάρακες προσαγορεύονται, καὶ ἀφιδρύματά ἐστιν Αἰγυπτίων θεῶν ἱερῶν. τὰ δὲ κατὰ Δειρὴν **στενὰ συνάγεται** εἰς σταδίους ἑξήκοντα·(*continued n. 3*)

384 Aelius Aristides, *Rhômês egkômion*, 10–12 (transl. Behr). Also see Dion Chrys. *Orat.* 32, 36.
385 Plin. *nat.* 14, 2 (transl. Rackham).

And thence (i. e. from Ptolemaïs of the Hunts), as far as the straits, 4500 *stadia*, in a direction more towards the east. The straits are formed towards *Aethiopia* by a promontory called *Deirê* and by a town bearing the same name, which is inhabited by the *Ichtyophagi*. And here there is a pillar of Sesostris the Aegyptian which tells in hieroglyphics of his passage across the strait[386]; for manifestly he was the first man to subdue the countries of the Aethiopians and the Troglodytes; and he then crossed into Arabia, and thence invaded the whole of Asia; and accordingly, for this reason, there are in many places palisades of Sesostris, as they are called, and reproductions of temples of Egyptian gods. The straits at *Deirê* contract to a width of 60 *stadia* (transl. Jones).

3. Strabo, 16, 4, 4 [Berger III B 48, p. 291]

Οὐ μὴν ταῦτά γε καλεῖται νυνὶ **στενά**, ἀλλὰ προσπλεύσασιν ἀπωτέρω, καθὸ τὸ μέν **δίαρμά** ἐστι **τὸ μεταξὺ** τῶν ἠπείρων διακοσίων που σταδίων, ἓξ δὲ νῆσοι συνεχεῖς ἀλλήλαις **τὸ δίαρμα** ἐκπληροῦσαι στενοὺς τελέως **διάπλους** ἀπολείπουσι, δι' ὧν σχεδίαις τὰ φορτία κομίζουσι δεῦρο κἀκεῖσε, καὶ λέγουσι ταῦτα **στενά**. μετὰ δὲ τὰς νήσους ὁ ἑξῆς πλοῦς ἐστιν ἐγκολπίζουσι[387] παρὰ τὴν σμυρνοφόρον ἐπὶ τὴν μεσημβρίαν ἅμα καὶ τὴν ἕω μέχρι πρὸς τὴν τὸ κιννάμωμον φέρουσαν, ὅσον πεντακισχιλίων σταδίων.

However, it is not these that are called straits now, but a place farther along on the voyage, where the voyage across the gulf between the two continents is about 200 *stadia*, and where are six islands, which follow one another in close succession, fill up the channel, and leave between them very narrow passages; through these merchandise is transported from one continent to the other; and for these the name "straits" is used. After the islands, the next voyage, following the sinuosities of the bays (*sic*), along the myrrh-bearing country in the direction of south and east as far as the cinnamon-bearing country, is about 5000 *stadia* (transl. Jones).

4. Agathemerus, 2, 14, *GGM* 2 p. 475 [Berger III B 49, p. 291–292]

Ὁ δὲ Ἀραβίος κόλπος, στενὸς ὢν καὶ προμήκης, ἄρχεται ἀπὸ Ἡρώων πόλεως, παρὰ τὴν Τρωγλοδυτικὴν ἕως τῆς Πτολεμαΐδος τῆς ἐπὶ θήρας, σταδίων ,θ τὸ μῆκος· [ἐντεῦθεν δὲ ἕως τῶν κατὰ Δειρὴν **στενῶν**] πλοῦς σταδίων,δφ' [Müller δψ' – however, the Latin translation gives 4500] · τὰ δὲ κατὰ Δειρὴν **στενὰ** σταδίων ξ'. ἔνθεν ἑξῆς πλοῦς παρὰ τὴν Ἐρυθρὰν θάλασσαν ἑὼς ὠκεανοῦ σταδίων ,ε · τὸ δὲ λοιπὸν οὐ πλεῖται.

The Arabian Gulf is narrow and fully extends in length. It is 9000 *stadia* long from *Herôônpolis* to *Ptolemais* of the Hunts along the *Trôglodytikê*; the voyage [from there to the strait at *Deirê*] is 4500 *stadia.* The straits at *Deirê* are 60 *stadia* wide. From there the sailing is 5000 *stadia* along the Erythraean sea up to the ocean. No one sailed further.

386 Jones writes: "the gulf".

387 ἐγκολπίζειν (*LSJ*): 1) "to form a bay" 2)" to go into or follow a bay". Desanges (a) 91: "(…) la navigation, quand on sort du golfe (Arabique)."

Eratosthenes and other authorities [Strabo]

5. Strabo, 16, 4, 20

Λέγεται δ' ὑπό τινων τὰ ἀπὸ **τῶν στενῶν** τοῦ Ἀραβίου κόλπου μέχρι τῆς κινναμωμοφόρου τῆς ἐσχάτης πεντακισχιλίων σταδίων, οὐκ εὐκρινῶς, εἴτ' ἐπὶ νότον εἴτ' ἐπὶ τὰς ἀνατολάς.

Some writers[388] say that the distance from the straits of the Arabian Gulf to the extremity of the cinnamon-bearing country is five thousand *stadia*, without distinguishing clearly whether they mean towards the south or towards the east. (transl. JONES).

Agatharchides of Cnidus [Diodorus of Sicily – Photius]

6. Diodorus, 3, 15, 1 [= Agatharchides, *GGM* 1, 31]

Περὶ πρώτων δὲ τῶν Ἰχθυοφάγων ἐροῦμεν τῶν κατοικούντων τὴν παράλιον τὴν ἀπὸ Καρμανίας καὶ Γεδρωσίας ἕως τῶν ἐσχάτων τοῦ μυχοῦ τοῦ κατὰ τὸν Ἀράβιον κόλπον ἱδρυμένου, ὃς εἰς τὴν μεσόγειον ἀνήκων ἄπιστον διάστημα δυσὶν ἠπείροις **περικλείεται** πρὸς **τὸν ἔκπλουν**, τῇ μὲν ὑπὸ τῆς εὐδαίμονος Ἀραβίας, τῇ δ΄ ὑπὸ τῆς Τρωγλοδυτικῆς.

The first people we shall mention are the Ichtyophagi who inhabit the coast which extends from Carmania and Gedrosia to the farthest limits of the recess (= the Gulf of Suez) which is found at the Arabian Gulf, which extends inland an unbelievable distance and is enclosed at his mouth (literally: is enclosed as for the sailing out) by two landmasses, on the one side by Arabia Felix and on the other by the land of Troglodytes (transl. OLDFATHER, slightly modified).

7. Diodorus, 3, 17, 5–18, 1 [= Agatharchides, *GGM* 1, 39–40]

Οἱ μὲν οὖν τὴν παράλιον τὴν **ἐντὸς τῶν στενῶν** κατοικοῦντες οὕτω βιοῦσι, νόσοις μὲν διὰ τὴν ἁπλότητα τῆς τροφῆς σπανίως περιπίπτοντες, ὀλιγοχρονιώτεροι δὲ πολὺ τῶν παρ΄ἡμῖν ὄντες. Τοῖς δὲ τὴν **ἐκτὸς τοῦ κόλπου** παράλιον νεμομένοις πολλῷ τούτων παραδοξότερον εἶναι τὸν βίον συμβέβηκεν, ὡς ἂν ἄδιψον ἐχόντων καὶ ἀπαθῆ τὴν φύσιν.

Now the inhabitants (sc. the Ichthyophagi) of the coast inside the strait lead the kind of life which has been described, and by reason of the simplicity of their food they rarely are subject to attacks of disease, although they are far shorter-lived than the inhabitants of our part of the world. But as for the inhabitants of the coast outside the gulf, we find that their life is far more astonishing than that of the people just described, it being as though their nature never suffers from thirst and is insensible to pain (transl. OLDFATHER, slightly modified).

388 See Str. 16, 4, 19 *in fine*.

8. Diodorus, 3, 32, 4 [= Agatharchides, *GGM* 1, 63]

Τὰ δ᾽αἰδοῖα πάντες οἱ Τρωγοδύται παραπλησίως τοῖς Αἰγυπτίοις περιτέμνονται πλὴν τῶν ἀπὸ τοῦ συμπτώματος ὀνομαζομένων Κολοβῶν· οὗτοι γὰρ μόνοι τὴν **ἐντὸς τῶν στενῶν** νεμόμενοι χώραν ἐκ νηπίου ξυροῖς ἀποτέμνονται πᾶν τὸ τοῖς ἄλλοις μέρος περιτομῆς τυγχάνον.

Moreover all the Troglodytes are circumcised like the Egyptians with the exception of those who, because of what they have experienced, are called *Koloboi*; for only these who live inside the straits have in infancy all that part cut completely off with thee razor which among other peoples merely suffers circumcision (transl. OLDFATHER, modified).

9. Diodorus, 3, 38, 4–5 [= Agatharchides, *GGM* 1, 79]

Ὁ δὲ προσαγορευόμενος Ἀράβιος κόλπος **ἀνεστόμωται** μὲν εἰς τὸν κατὰ μεσημβρίαν κείμενον ὠκεανόν, τῷ μήκει δ᾽ ἐπὶ πολλοὺς πάνυ παρήκων σταδίους τὸν μυχὸν ἔχει περιοριζόμενον ταῖς ἐσχατιαῖς τῆς Ἀραβίας καὶ Τρωγλοδυτικῆς. εὖρος δὲ κατὰ μὲν **τὸ στόμα** καὶ τὸν μυχὸν ὑπάρχει περὶ ἑκκαίδεκα σταδίους, ἀπὸ δὲ Πανόρμου λιμένος πρὸς τὴν ἀντιπέραν ἤπειρον μακρᾶς νεὼς διωγμὸν ἡμερήσιον. τὸ δὲ μέγιστόν ἐστι διάστημα κατὰ τὸ Τύρκαιον ὄρος καὶ Μακαρίαν νῆσον πελαγίαν[389], ὡς ἂν τῶν ἠπείρων οὐχ ὁρωμένων ἀπ᾽ ἀλλήλων. ἀπὸ δὲ τούτου τὸ πλάτος ἀεὶ μᾶλλον **συγκλείεται** καὶ **τὴν συναγωγὴν** ἔχει μέχρι **τοῦ στόματος**. ὁ δὲ παράπλους αὐτοῦ κατὰ πολλοὺς τόπους ἔχει νήσους μακράς, στενοὺς μὲν διαδρόμους ἐχούσας, ῥοῦν δὲ πολὺν καὶ σφοδρόν.

But the Arabian Gulf, as it is called, opens into the ocean wich lies to the south, and its innermost recess, which stretches over a distance of very many *stadia* in length, is enclosed by the farthermost borders of Arabia and the *Trôglodytikê* country. Its width, at the mouth and at the innermost recess, is about sixteen stades, but from the harbour of *Panormos* to the opposite mainland is a day's run for a warship. And its greatest width is at the *Tyrkaion* mountain and *Makaria* (?), an island out at sea, the mainlands there being out of sight of each other. But from this point the width steadily decreases more and more and continually tapers as far as the mouth. And as a man sails along the coast he comes in many places upon long islands with narrow passages between them, where the current runs full and strong (transl. OLDFATHER, slightly modified).

10. Photius, *Bibl.* 250, 450a [= Agatharchides, *GGM* 1, 40]

Ἀλλ᾽οἱ μὲν **ἐντὸς τῶν στενῶν** τὰς οἰκήσεις ἔχοντες ἰχθυοφάγοι οὕτω καταγίνονται · οἱ δὲ τὴν **ἐκτὸς** παραλίαν κεκτημένοι, ἄγρας μὲν ἐνδελεχῶς τοιαύτης εὐποροῦσιν, ὑγροῦ δ᾽ἁπλῶς οὐ δέονται.

389 Μαρίαν δυσπελαγίαν: MÜLLER (a) 165–166; Μαρίαν δυσπελαγίον: BOMMELAER (see BURSTEIN 133, n. 5).

The *Ikhthyophagoi* who dwell inside the strait live in such a way (i. e. they enjoy a simple and healthy life); but those who inhabit the coast ouside <the strait> continually catch an extraordinary quantity of fish and need not water at all.

11. Photius, *Bibl.* 250, 451a [= Agatharchides, *GGM* 1, 47]

Ὅτι **τῶν στενῶν** ἐπέκεινα, φησί, **τῶν συγκλειόντων** τήν τε Ἀραβίαν καὶ τὴν ἀπέναντι χώραν, νῆσοι κεῖνται σποράδες, ταπειναὶ πᾶσαι, μικραὶ τῷ μεγέθει, τὸ πληθος ἀμύθητοι, καρπὸν οὐδένα γεννῶσαι πρὸς τὸν βὶον, οὔτε ἥμερον οὔτε ἄγριον, ἀπέχουσαι μὲν τῆς εἰρημένης ἠπείρου σταδίους ὡς ἑβδομήκοντα, τετραμμέναι δὲ πρὸς τὸ δοκοῦν πέλαγος παρεκτείνειν τὴν Ἰνδικὴν καὶ Γεδρωσίαν.
He says that beyond the strait which is closed by Arabia and the opposite country joining each other there are scattered islands, all low; those are small, unbelievably numerous; they grow neither cultivated nor wild produce which would be useful to live and lie about 70 *stadia* from the above said landmass and face the sea which is supposed to stretch towards India and Gedrosia.[390]

N. B.: the *poros* referred to by Photius, *Bibl.* 250, 460a (= Agath., 108; also Photius, *Bibl.* 250, 441b (= Agath., 2) is unlikely to be the strait (see Schneider [a] 358–361).

Greek inscriptions from the Egyptian Eastern Desert; papyri

12. Bernand (a) n° 164, l. 1–5

Of unknown origin but certainly from the Paneion of Redesiyeh; probably dedicated by a hunter (see euagrôi).[391]
Πανὶ τόδε εὐάγρῳ̣ καὶ ἐπ[ηκό̣]ω̣ι, ὃ̣ς δ̣ιέσωισεν / Τρωγ̣ο̣δυτῶν με [ἐκ] γῆς, πολλὰ παθόντα πόνοις / δισσοῖς̣, Σ[μυρνο]φόρου θ' ἱερᾶς Κ̣ολοβῶν̣ τ̣ε ἀπὸ – – – –[392]/ σώισ̣α̣ς [δὲ (?) ἐν πε]λάγει πλα̣ζ̣ομένους Ἐ̣ρυ̣θ̣ρ[ῷ] etc.

(I dedicate) this to Pan, helper in the hunt, propitious one, for bringing me safely from the land of the Troglodytes, where I suffered many hardships during two enterprises[393] (Bernand: "j'avais été (…) éprouvé par des peines redoublées"), from the holy Myrrh-land and the *Koloboi*. And you saved (us) as we wandered on the [Erythraean ?] sea, by sending a fair wind to our ships when they were drifting around in the sea etc. (transl. Household & Prakken).

390 Compare with Diod. 3, 21, 1: "There are islands in the ocean (ὠκεανόν), which lie near the land etc." (transl. Oldfather). Burstein 85 rightly argues that Agath. refers to Bab al-Mandab (*contra*: Woelk 140–141).

391 First published by Householder & Prakken. According to the editors, this inscription is not later than 200 B. C. See also Bernand (a) 569; Desanges (a) 100 (reign of Ptolemy IV).

392 χώρης? ἀγροῖο? ἄκροιο? The editors referring to Ptolemy prefer the last reading (the cape of the *Koloboi*). Bernand suggests ἔγραψα. For the adjective "holy" see Desanges (a) 100.

393 "This expression probably refers to two stops at different points on the same trip, i. e. at the *Smyrnophoros gê* and among the *Koloboi*." (Household & Prakken 112).

13. Bernand (b) n° 48 (=*OGIS* 69)

Θεοῖς μεγάλοις Σαμοθρᾷξι Ἀπολλώνιος Σωσιβίου Θῆραιος ἡγημὼν τῶν ἔξω τάξεων σωθεὶς ἐγ μεγάλων κινδύνων, ἐκπλεύσας ἐκ τῆς Ἐρυθρᾶς θαλάσσης, εὐχήν.

Apollônios, son of Sôsibios from Thera, officer at disposal[394], (dedicated this monument) to the Great Gods of Samothrace for saving (him) from great dangers as he sailed out from the Erythraean sea, *ex voto.*[395]

13bis. Wilcken p. 90 l. 20

… [τῶν τὴν ἔξω (?) θάλ]ασσαν πλοιζομένων …

(Dêmêtrios, son of Apôllonios), one of those who sail on the outer sea.[396]

Artemidorus [Strabo]

14. Strabo, 16, 4, 5

Φησὶ δ' Ἀρτεμίδωρος **τὸ ἀντικείμενον** ἐκ τῆς Ἀραβίας **ἀκρωτήριον** τῇ Δειρῇ καλεῖσθαι Ἀκίλαν· τοὺς δὲ περὶ τὴν Δειρὴν κολοβοὺς εἶναι τὰς βαλάνους.

Artemidorus says that the promontory on the Arabian side opposite to *Deirê* is called *Akila*; and that the males in the neighbourhood of *Deirê* have their sexual glands mutilated (transl. Jones).

15. Strabo, 16, 4, 13

Μετὰ δὲ Εὐμένους λιμένα μέχρι Δειρῆς καὶ **τῶν** κατὰ τὰς ἓξ νήσους **στενῶν** ἰχθυοφάγοι καὶ κρεοφάγοι κατοικοῦσι καὶ κολοβοὶ μέχρι τῆς μεσογαίας. εἰσὶ δὲ καὶ θῆραι πλείους ἐλεφάντων καὶ πόλεις ἄσημοι καὶ νησία πρὸ τῆς παραλίας.

After the Harbour of Eumenês, as far as *Deirê* and the strait opposite the six islands, the country is inhabited by the *Ikhthyophagoi* and the *Kreophagoi* and the *Koloboi*, who extend as far as the interior. In this region are several hunting-grounds for elephants, and insignificant cities, and islands lying off the coast (transl. Jones).

394 See Bernand (b) 168.

395 Dittenberger 122 suggests that this document dates to the reign of Ptolemy III when an important Ptolemaic garrison was stationned in Thera. He also rightly states that *ekpleusas* means "to sail beyond the strait".

396 Wilcken 97: "Am wahrscheinlichsten ist mir, daß mit der *exô thalassa* hier das Rote Meer gemeint ist, aber auch darüber hinaus das Meer bis zur Somaliküste (Punt) und zur Südküste Arabiens, denn da (…) sehe ich keinen Grund, weshalb man hier etwa die Enge von Bab el Mandeb als Endgrenze für diesen Begriff nehmen sollte." I personaly tend to believe that *exô thalassa* applies only to the sea outside the strait rather than both the Red Sea and the Gulf of Aden.

16. Strabo, 16, 4, 14

Πᾶσα δ' ἡ παραλία φοίνικάς τε ἔχει καὶ ἐλαιῶνας καὶ δαφνῶνας, οὐχ ἡ **ἐντὸς τῶν στενῶν** μόνον ἀλλὰ καὶ τῆς **ἐκτὸς** πολλή (...). ἀπὸ δὲ τῆς Δειρῆς ἡ ἐφεξῆς ἐστιν ἀρωματοφόρος, πρώτη μὲν ἡ τὴν σμύρναν φέρουσα, καὶ αὕτη μὲν ἰχθυοφάγων καὶ κρεοφάγων.

And the whole of the coast has palm-trees, olive groves, and laurel groves, not only the part inside the straits, but also most of the part outside. (...). The next country after *Deirê* produces aromatics, the first that produces myrrrh (this country belongs to the *Ichthyophagoi* and *Kreophagoi*) (transl. Jones).

16bis. Strabo, 16, 4, 15

(...) κατὰ τὴν γνώριμον παραλίαν τὴν ἀπὸ Δειρῆς μέχρι Νότου κέρως, τὸ δὲ διάστημα οὐ γνώριμον.

(*the pillars and altars of Pytholaos, Likhas, Pythangelos* etc. *lie*) along the known coast extending from *Deirê* as far as the *Notou keras* (= cape Gardafui), but the distance is unknown (transl. Jones).

17. Strabo, 16, 4, 19

Τὰ δὲ πλήθη τὰ μὲν γεωργεῖ τὰ δ' ἐμπορεύεται τὰ ἀρώματα τά τε ἐπιχώρια καὶ τὰ ἀπὸ τῆς Αἰθιοπίας, **πλέοντες** ἐπ' αὐτὰ **διὰ τῶν στενῶν** δερματίνοις πλοίοις.

But the masses (*sc.* of Sabaeans) engage partly in farming and partly in the traffic in aromatics, both the local kinds and those from *Aethiopia*; to get the latter they sail across the straits in leathern boats (transl. Jones).

Hipparchus [Strabo]

18. Strabo, 2, 5, 35

Τῷ δὲ λεχθέντι μεσημβρινῷ παράλληλός πως παράκειται ἕωθεν ὁ Ἀράβιος κόλπος· τούτου δ' ἔκβασις εἰς τὸ ἔξω πέλαγος ἡ Κινναμωμοφόρος ἐστίν, ἐφ' ἧς ἡ τῶν ἐλεφάντων γέγονε θήρα τὸ παλαιόν.

The Arabian Gulf lies eastward parallel to the said meridian (i. e. of *Syenes*). Its egress into the exterior ocean is [in the same latitude as] the *Kinnamômophoros*, the place where anciently they used to hunt the elephants (transl. Hamilton).

Strabo

19. Strabo, 2, 3, 5

Τίς γὰρ ἡ πιθανότης πρῶτον μὲν τῆς κατὰ τὸν Ἰνδὸν περιπετείας; Ὁ γὰρ Ἀράβιος κόλπος ποταμοῦ δίκην στενός ἐστι καὶ μακρὸς [πεντακισχιλίους] ἐπὶ μυρίοις που σταδίους μέχρι **τοῦ στόματος**, καὶ τούτου **στενοῦ** παντάπασιν ὄντος· οὐκ εἰκὸς δ' οὔτ' ἔξω που τὸν πλοῦν ἔχοντας εἰς τὸν κόλπον παρωσθῆναι τοὺς Ἰνδοὺς κατὰ πλάνην (**τὰ** γὰρ **στενὰ ἀπὸ τοῦ στόματος** δηλώσειν ἔμελλε τὴν πλάνην), οὔτ' εἰς τὸν κόλπον ἐπίτηδες **καταχθεῖσιν** ἔτι πλάνης ἦν πρόφασις καὶ ἀνέμων ἀστάτων.

For, in the first place, what plausibility is there in the "strange mischance" which the Indian tells about? Why, the Arabian Gulf is like a river in its narrowness, and it is about 15000 *stadia* (~ 2400 km) long up to its mouth, which, in its turn, is extremely narrow; and so it is not likely that the Indians who were voyaging outside this gulf were pushed out of their course into it by mistake (for its narrowness at its mouth would have shown their mistake), nor, if they sailed into the gulf on purpose, did they any longer have the excuse that they mistook their course or encountered inconstant (transl. JONES).

20. Strabo, 16, 4, 22

Τοῦτον δ' ἔπεμψεν ὁ Σεβαστὸς Καῖσαρ διαπειρασόμενον τῶν ἐθνῶν καὶ τῶν τόπων τούτων (*viz.* the Arabians) τε καὶ τῶν Αἰθιοπικῶν, ὁρῶν τήν τε Τρωγλοδυτικὴν τὴν προσεχῆ τῇ Αἰγύπτῳ γειτονεύουσαν τούτοις (*viz.* the Ethiopians), καὶ τὸν Ἀράβιον κόλπον στενὸν ὄντα τελέως τὸν διείργοντα ἀπὸ τῶν Τρωγλοδυτῶν τοὺς Ἄραβας· προσοικειοῦσθαι δὴ διενοήθη τούτους ἢ καταστρέφεσθαι· ἦν δέ τι καὶ τὸ πολυχρημάτους ἀκούειν ἐκ παντὸς χρόνου, πρὸς ἄργυρον καὶ χρυσὸν τὰ ἀρώματα διατιθεμένους καὶ τὴν πολυτελεστάτην λιθείαν, ἀναλίσκοντας τῶν λαμβανομένων τοῖς ἔξω μηδέν· ἢ γὰρ φίλοις ἤλπιζε πλουσίοις χρήσεσθαι ἢ ἐχθρῶν κρατήσειν πλουσίων.

He (*viz.* Aelius Gallus) was sent by Augustus Caesar to explore the tribes and the places, not only in Arabia, but also in *Aethiopia*, since Caesar saw that the *Trôglodytikê* which adjoins Aegypt neighbours upon *Aethiopia*[397], and also that the Arabian Gulf, which separates the Arabians from the Troglodytes, is extremely narrow. Accordingly he conceived the purpose of winning the Arabians over to himself or of subjugating them. Another consideration was the report, which had prevailed from all time, that they were very wealthy and that they sold aromatics and the most valuable stones for gold and silver, but never expended with outsiders any part of what they received in exchange; for he expected either to deal with wealthy friends or to master wealthy enemies (transl. JONES, modified).

21. Strabo, 17, 1, 13

Ὅπου οὖν ὁ κάκιστα καὶ ῥαθυμότατα τὴν βασιλείαν διοικῶν τοσαῦτα προσωδεύετο, τί χρὴ νομίσαι τὰ νῦν διὰ τοσαύτης ἐπιμελείας οἰκονομούμενα καὶ τῶν Ἰνδικῶν ἐμποριῶν καὶ τῶν Τρωγλοδυτικῶν ἐπηυξημένων ἐπὶ τοσοῦτον; πρότερον μέν γε οὐδ' εἴκοσι πλοῖα ἐθάρρει τὸν Ἀράβιον κόλπον διαπερᾶν ὥστε **ἔξω τῶν στενῶν ὑπερκύπτειν**, νῦν δὲ καὶ στόλοι μεγάλοι στέλλονται μέχρι τῆς Ἰνδικῆς καὶ τῶν ἄκρων τῶν Αἰθιοπικῶν, ἐξ ὧν ὁ πολυτιμότατος κομίζεται φόρτος εἰς τὴν Αἴγυπτον (...)

397 See DESANGES (b) 308, n. 6.

If, then, the man who administered the kingdom in the worst and most careless way obtained so large a revenue, what should one think of the present revenues, which are managed with so much diligence, and when the commerce with the Indians and the Troglodytes has been increased to so great an extent? In earlier times, at least, not so many as twenty vessels would dare to traverse the Arabian Gulf far enough to get a peep outside the straits, but at the present time even large fleets are despatched as far as India and the extremities of Aethiopia, from which the most valuable cargoes are brought to Aegypt (…) (transl. JONES).

See also Str. 2, 5, 12.

Juba of Mauretania; Pliny the Elder

22. Plin. *nat.* 6, 170 [= Juba, *FGrHist* 275, F 34]

Juba, qui videtur diligentissime persecutus haec, omisit in eo tractu (…) et tertiam (sc. Berenicen) quae epi Dires, insignem loco: est enim sita in cervice longe procurrente, ubi **fauces** Rubri maris VII·D p. ab Arabia distant. Insula ibi Citis, topazum ferens et ipsa. (text edited by DESANGES).

Juba, who appears to have investigated these matters extremely carefully, has omitted to mention in this district (…) and a third called *Berenice on the Neck,* which is remarkable for its situation, being placed on a neck of land projecting a long way out, where the straits at the mouth of the Red Sea separate Africa (sic) from Arabia by a space of only 7 ½ miles.[398] Here is the island of *Citis*, which itself also produces topaz. (transl. Rackham).

23. Plin, *nat.* 6, 175 [= Juba, *FGrHist* 275, F 35]

Cujus (sc. Juba) tota sententia hoc in loco subtrahenda non est. a promunturio Indorum quod vocetur Lepte Acra, ab aliis Drepanum, proponit recto cursu praeter Exustam ad Malichu insulas $|\overline{XV}|$ p. esse, inde ad locum quem vocant Sceneos[399] $|\overline{CCXXV}|$ p., inde ad insulas Adanu $\overline{CL}$; sic fieri ad apertum mare$|\overline{XVIII}|\cdot\overline{LXXV}$ p. (text edited by DESANGES)

He (sc. Juba) puts forward the view that the distance from the cape of the Indians called in Greek the Narrow Head, and by others the Sickle (= Ras Banas, near *Berenikê*)], in a straight course past Burnt Island to the Islands *Malichu* is 1500 miles, from there to the place called *Sceneos* 225 miles, and on from there to the *Adanu* Islands 150 miles making 1875 miles to the open sea. (transl. RACKHAM, slightly modified).

398 Literally: "where the strait (…) is 7 ½ miles from Arabia".

399 DESANGES (c) 95: "Le *locus* appelé *Sceneos* semble identifiable (…): il s'agirait d'*Ocelis*".

24. Plin. *nat.* 6, 151 [Juba, *FGrHist* 275, F 33]

Insulae Chelonitis, Ichthyophagon multae, Odanda deserta, Basa, multae Sabaeorum. flumina Thanar, Amnum, insulae Doricae, fontes Daulotos, Dora, insulae Pteros, Labatanis, Coboris, Sambrachate et oppidum eodem nomine in continente. a meridie insulae multae, maxima Camari, flumen Musecros, portus Laupas, Scenitae Sabaei, insulae multae, emporium eorum Acila, ex quo in Indiam navigatur.

Chelonitis islands and a number of islands of the Fish-eaters, the uninhabited *Odanda*, *Basa*, a number of islands belonging to the *Sabaeans*. The rivers *Thanar* and *Amnum*, the Islands *Doricae*, the *Daulotos* and *Dora* springs, the islands of *Pteros*, *Labatanis*, *Coboris* and *Sambrachate* with the town of the same name on the Mainland. Many islands to the southward, the largest of which is *Camari*, the river *Musecros*, Port *Laupas*; the *Scenitae*[400] *Sabaeans*, many islands and their *emporium*, *Acila*, from where one sails for India (transl. RACKHAM, modified).

25. Plin. nat. 6, 154 [Juba, *FGrHist* 275, F 33]

Promuntorium a quo ad continentem Trogodytarum $\overline{\text{L}}$.

Then a cape[401] the distance between which and the mainland in the Trogodytae's territory is 50 miles. (transl. RACKHAM).

26. Plin. *nat.* 6, 163–164

Nunc reliquam oram Arabiae contrariam persequemur. Timosthenes totum sinum quadridui navigatione in longitudinem taxavit, bidui in latitudinem, **angustias** $\overline{\text{VII}}$·D p.; Eratosthenes ab **ostio** $|\overline{\text{XII}}|$ in quamque partem; Artemidorus Arabiae latere $|\overline{\text{XVII}}|$· L, Trogodytico vero $|\overline{\text{XI}}|$·$\overline{\text{XXXVII}}$·D p. Ptolomaida usque; Agrippa $|\overline{\text{XVII}}|$·$\overline{\text{XXXII}}$ sine differentia laterum. Plerique latitudinem $\overline{\text{CCCCLXXV}}$ prodiderunt, **fauces**que hiberno orienti obversas alii $\overline{\text{III}}$, alii $\overline{\text{VII}}$, alii $\overline{\text{XII}}$ patere (text edited by DESANGES).

We will now follow along the rest of the the coast lying opposite to Arabia. Timosthenes estimated the length of the whole gulf at four days' sail, the breadth at two, and the width of the strait as 7½ miles; Eratosthenes makes the length of the coast on either side from the mouth of the gulf 1200 miles; Artemidorus gives the length of the coast on the Arabian side as 1750 miles and on the side of the Trogodytae country as far as Ptolemais 1137½ miles; Agrippa says that there is no difference between the two sides, and gives the length of each as 1732 miles. Most authorities give the breadth as 475 miles, and the mouth of the gulf facing south-east [Rackham: south-west] some make 4 miles wide, others 7 and others 12 (transl. RACKHAM).

400 *Scenitae* ("living in tents") generally applies to the nomadic Arabs of the northern Arabian Peninsula (e. g. Str. 16, 3, 1) and does not fit the Sabaeans *stricto sensu*.

401 This cape is situated in the region of the *Larendani*, *Catabani* and *Gebbanitae*, somewhere beyond *Nagia* and *Thomna*.

Periplus maris Erythraei

27. *P. m. r.* 7

Ἤδη [δὲ] ἐπ' ἀνατολὴν ὁ Ἀραβικὸς κόλπος διατείνει καὶ κατὰ τὸν Αὐαλίτην μάλιστα **στενοῦται**. Μετὰ δὲ σταδίους ὡσεὶ τετρακισχιλίους, κατὰ τὴν αὐτὴν ἤπειρον εἰς ἀνατολὴν πλεόντων, ἔστιν ἄλλα ἐμπόρια Βαρβαρικά, τὰ πέραν λεγόμενα (…) Πρῶτος μὲν ὁ λεγόμενος Αὐαλίτης, καθ' ὃν καὶ **στενώτατός** ἐστιν **ἀπὸ τῆς Ἀραβικῆς εἰς τὸ πέραν διάπλους**. Κατὰ τοῦτον τὸν τόπον μικρὸν ἐμπόριόν ἐστιν ὁ Αὐαλίτης, σχεδίαις καὶ σκάφαις εἰς τὸ αὐτὸ προσερχομένων (…) Φέρεται δ' ἐξ αὐτῆς, ποτὲ καὶ τῶν βαρβάρων ἐπὶ σχεδίαις **διαφερόντων** εἰς τὴν ἀντικρὺς Ὄκηλιν καὶ Μούζα, ἀρώματα καὶ ἐλέφας ὀλίγος καὶ χελώνη καὶ σμύρνα ἐλαχίστη, διαφέρουσα δὲ τῆς ἄλλης.

By now (sc. after *Adulis*) the Arabian Gulf trends eastward and at *Aualitês* is it at its narrowest. After about 4000 stades (from *Adulis*: Casson 115) on an eastward heading along the same coast, come the rest of the ports of trade of the *Barbaroi*, those called "far-side" (…). The first is called *Aualitês;* at it the crossing from Arabia to the other side is the shortest. At this place there is a small port of trade, namely *Aualitês*, where rafts and small crafts put in (*list of imported goods*). Exports from here, with the transport across to *Okêlis* and *Mouza* on the opposite shore at times carried out by the *Barbaroi* on rafts, are: aromatics, a little ivory, tortoise shell, a minimal amount of myrrh but finer than any other. (transl. Casson)

28. *P. m. r.* 25

Μετὰ δὲ ταύτην ὡσεὶ τριακοσίους παραπλεύσαντες σταδίους, ἤδη **συνερχομένης** τε τῆς Ἀραβικῆς ἠπείρου καὶ τῆς πέραν κατὰ τὸν Αὐαλίτην Βαρβαρικῆς χώρας, **αὐλὼν**[402] **ἐστιν οὐ μακρὸς**, ὁ **συνάγων** καὶ **εἰς στενὸν ἀποκλείων τὸ πέλαγος**, οὗ **τὸν μεταξὺ πόρον** ἑξήκοντα σταδίων **μεσολαβεῖ** νῆσος ἡ Διοδώρου· διὸ καὶ **ῥοώδης**, καταπνεόμενος ἀπὸ τῶν παρακειμένων ὀρῶν, ἐστὶν **ὁ κατ' αὐτὴν διάπλους**. Κατὰ **τοῦτον τὸν ἰσθμὸν** παραθαλάσσιός ἐστιν Ἀράβων κώμη τῆς αὐτῆς τυραννίδος Ὄκηλις, οὐχ οὕτως ἐμπόριον ὡς ὅρμος καὶ ὕδρευμα καὶ πρώτη καταγωγὴ τοῖς ἔσω διαίρουσι.

About a 300-stade sail past this port (viz. *Mouza*), the Arabian mainland and the country of *Barbaria* across the water in the vicinity of *Aualitês* converge to form a strait, not very long, that contracts the water and close them into a narrow passage; here, in the middle of the channel, 60 stades wide, stands *Diodoros* Island. For this reason, and because a wind blows down from the mountain that lies alongside, the sail through along the island meets strong currents. Along this strait is *Okêlis*, an Arab village on the coast that belongs to the same province; it is not so much a port of trade as a harbor, a watering station, and the first place to put in for those sailing on (transl. Casson).

402 *LSJ*: "hollow between hills or banks, defile, glen" (Hdt. 7, 128; 129"); "channel, trench" (Hdt. 2, 100, 127, Xen. *anab.* 2, 3, 10); "strait", (Μαιωτικός Aesch. *Pr.* 731; πόντιαι α. sea-straits, channels, Soph. *Tr.* 100 [lyr.]).

29. <u>*P. m. r.* 26</u>

Μετὰ δὲ Ὄκηλιν, **ἀνοιγομένης** πάλιν **τῆς θαλάσσης** εἰς ἀνατολὴν καὶ κατὰ μικρὸν **εἰς πέλαγος ἀποφαινομένης**, ἀπὸ σταδίων ὡς χιλίων διακοσίων ἐστὶν Εὐδαίμων Ἀραβία, κώμη παραθαλάσσιος, βασιλείας τῆς αὐτῆς Χαριβαὴλ, τοὺς ὅρμους μὲν ἐπιτηδείους καὶ ὑδρεύματα γλυκύτερα [καὶ] κρείσσονα τῆς Ὀκήλεως ἔχουσα, ἤδη δὲ ἐν ἀρχῇ κόλπου κειμένη τῷ τὴν χώραν ὑποφεύγειν. Εὐδαίμων δ' ἐπεκλήθη, πρότερον οὖσα πόλις, ὅτε, μήπω ἀπὸ τῆς Ἰνδικῆς εἰς τὴν Αἴγυπτον ἐρχομένων μηδὲ ἀπὸ [τῆς] Αἰγύπτου τολμώντων **εἰς τοὺς ἔσω τόπους διαίρειν**, ἀλλ' ἄχρι ταύτης παραγινομένων, τοὺς παρ' ἀμφοτέρων φόρτους ἀπεδέχετο, ὥσπερ Ἀλεξάνδρεια καὶ τῶν ἔξωθεν καὶ τῶν ἀπὸ τῆς Αἰγύπτου φερομένων ἀποδέχεται. Νῦν δὲ οὐ πρὸ πολλοῦ τῶν ἡμετέρων χρόνων Καῖσαρ αὐτὴν κατεστρέψατο.

Beyond *Okêlis*, with the water again opening out towards the east and little by little revealed to be open sea, about 1200 stades distant is *Eudaimôn Arabia*, a village on the coast belonging to the same kingdom, Charibaêl's. It has suitable harbors and sources of water much sweeter than at *Okêlis*. It stands at the beginning of a gulf formed by the receding of the shore. *Eudaimôn Arabia*, a full-fledged city in earlier days, was called *Eudaimôn* when, since vessels from India did not go on to Egypt and those from Egypt did not dare sail to the places further on but came only this far, it used to receive the cargoes of both, just as Alexandria receives cargoes from overseas as well from Egypt. And now, not long before our time, Caesar sacked it (transl. CASSON).

Ptolemy

30. <u>Ptolemy, *Geographia*, 1, 15, 11</u>

(…) Πτολεμαΐδος δὲ καὶ τοῦ Ἀδουλιτικοῦ κόλπου <ἀναλικωτέρα ἐστι> **τὰ στενὰ** τὰ κατὰ Ὄκηλιν τὴν χερσόνησον καὶ Δείρην σταδίοις τρισχιλίοις πεντακοσίοις

The straits on *Okêlis* peninsula and *Deirê* are 3500 stadia east of Ptolemaïs and the Adoulitikos Gulf.

31. <u>Ptolemy, *Geographia*, 4, 7, 9–10</u>

Μετὰ **τὰ στενὰ** ἐν τῇ Ἐρυθρᾷ θαλάσσῃ, Δείρη πόλις ἐν **ἄκρᾳ**, οδ° L′ – ια°. Εἶτα ἐν τῷ Αὐαλίτῃ κόλπῳ, Αὐαλίτης ἐμπόριον, οδ° – γ° γιβ′ · Μαλαὼ (ἢ Μάλεως) ἐμπόριον, οε° – ς° L′.

Past the strait of the Erythraean sea *Deirê*, a town located on a cape: 74° 30′ – 11°. Beyond, in the Gulf *Aualitês*, the port of trade of *Aualitês*: 74° – 8° 35′; the port of trade of *Malaô*: 75° – 6° 30′.

32. <u>Ptolemy, *Geographia* 6, 7, 7–8</u>

Ὄκηλις ἐμπόριον, οε° – ιβ° · Παλίνδρομος **ἄκρα**, οδ° L′ – ια° γο′· Ἐρυθρᾶς θαλάσσης μετὰ **τὰ στενὰ**, Ποσείδιον ἄκρον … οε° – ια L′·

The port of trade of *Okêlis*: 75° 12°; the cape *Palindromos* ("going back again"): 74° 30′ 11° 40′. Beyond the strait of the Erythraean sea, the cape *Poseidion*: 75° 11° 30′.

7.2 Maps

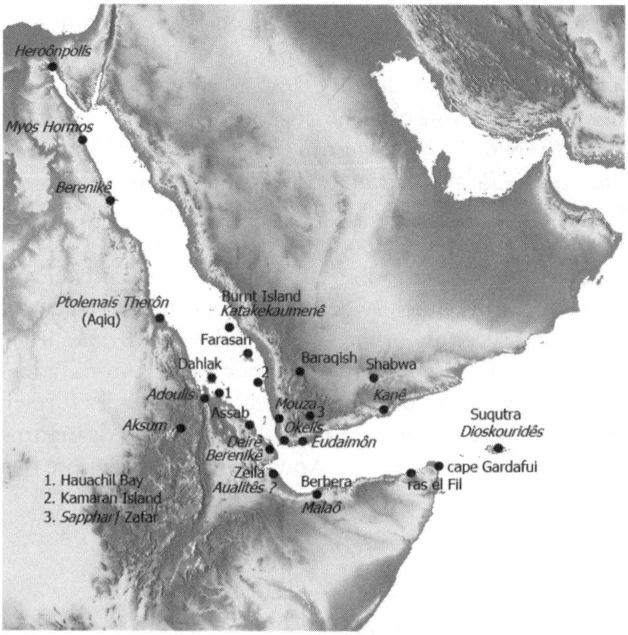

Fig. 1: the Arabian Gulf and the Horn of Africa

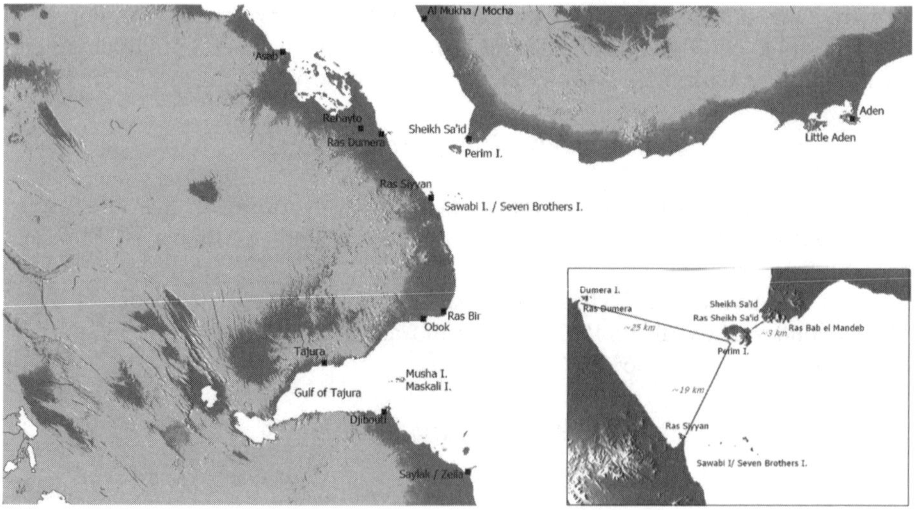

Fig. 2: the Bab al-Mandab area

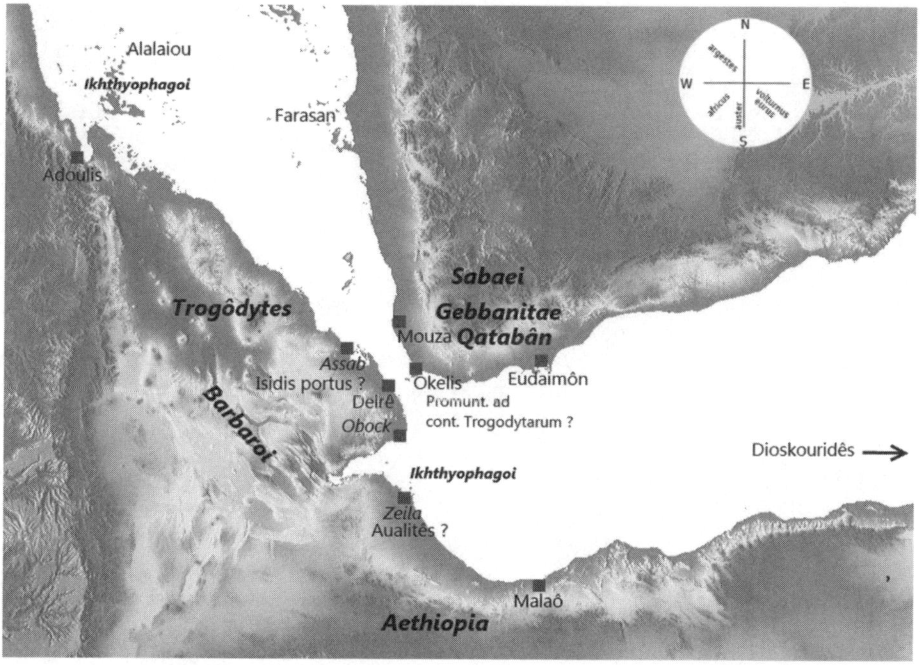

Fig. 3: the local trade network

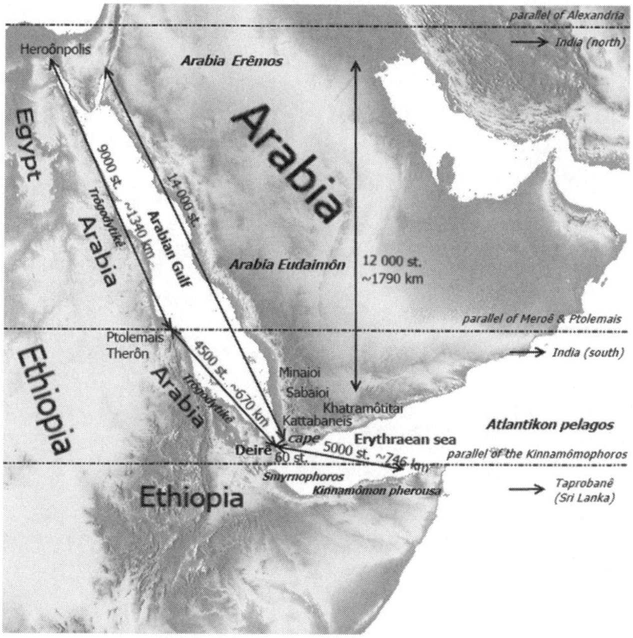

Fig. 4: Deirê in Eratosthenes' map

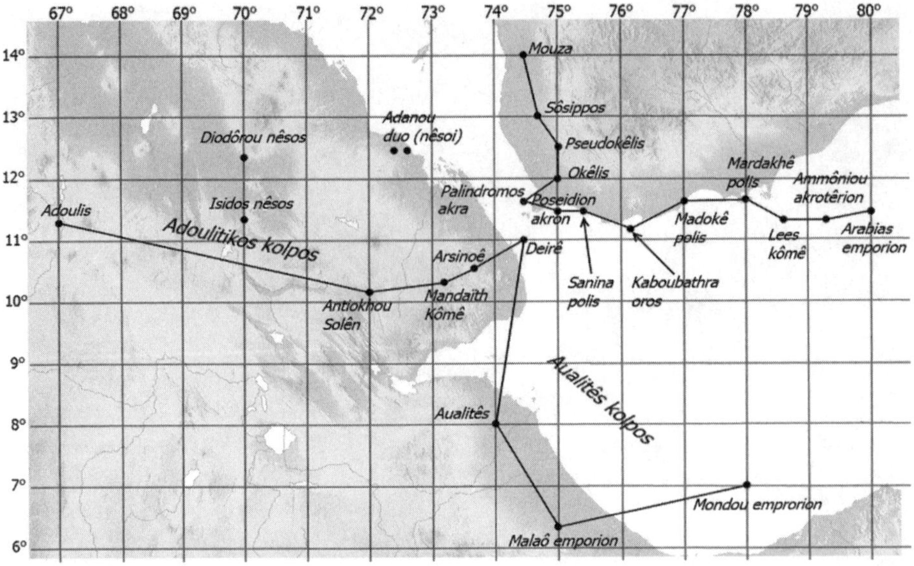

Fig. 5: Deirê in Ptolemy's map. (N. B.: There is no correspondence between Ptolemy's coordinates and the scale of the map in the background.)

7.3 Vocabulary relating to the strait (synoptic table)

	stena and other words relating to the strait – landmarks	inside / beyond – interval	widening / narrowing	crossing – sailing
Eratosthenes	*stena – akra*	*diarma to metaxu tôn êpeirôn–*	*sunagetai*	*diabasis / diabas – diaplous*
Agatharchides	*stena – stoma*	*entos (tôn stenôn) – ektos (tou kolpou)*	*sunagôgê – perikleietai – anestomatai – sugkleietai*	*ekploun*
Artemidorus	*stena – to antikeimenon akroterion*	*entos tôn stenôn – ektos (tôn stenôn)*		*pleontes dia*
Strabo	*stena – stoma – stena apo tou stomatos*			*huperkuptein exô tôn stenôn – kataktheisin eis ton kolpon*
Pliny the Elder	*fauces – angustiae – ostium*			
Periplus m. r.	*isthmon – ton metaxu poron – mesolabei (nêsos) – aulôn*		*stenoutai – sunerkhomenês –sunagôn (to pelagos) – apokleiôn eis stenon (to pelagos) – anoigomenês tês thalassês – apophainomenês (eis pelagos)*	*diaplous (stenotatê.) – diapherontôn (eis tên antikrus) – eis tous eisô topous diairein – diaplous – eis tous esô topous diairein*
Ptolemy	*stena – akra / akron*			

7.4 The width of the strait (synoptic table)

Ancient figures (original figures are printed in bold)				
Eratosthenes	[**2**; **4**] (Timosthenes)	~11 km	**60 st.**	7 ½ R.m.[403]
	[**3**]	~37 km	**200 st.**	25 R.m.
Agatharchides	[**9**]	~3 km	**16 st.**	2 R.m.
Pliny the Elder	[**26**]	~6 km	32 st.	**4 R.** m.
	[**26**]	~10.3 km	56 st.	**7 R.** m.
	[**26**] (Timosthenes)	~11 km	60 st.	**7 ½ R.** m.
	[**12**]	~17.7 km	96 st.	**12 R.** m.
	[**25**] (Juba?)	~74 km	400 st.	**50 R.** m.
Periplus maris Erythraei	[**28**]	~11 km	**60 st.**	7 ½ R.m.
Actual figures				
The Great Strait (between the African coast and Perim)		~18 km	~97 st.	~12 R.m.
The Small Strait (between Perim and the Arabian coast)		~3 km	~16 st.	~2 R.m.
The Strait		~23 km	~124 st.	~15 ½ R.m.

8. BIBLIOGRAPHY

Altheim, F., *Die Araber in der alten Welt*, vol. 1, Berlin, 1964.

Amigues, S., "L'expédition d'Anaxicrate en Arabie occidentale", *Topoi (Orient-Occident)* 6.2 (1996), 671–677.

Avanzini, A., "Le commerce des aromates et les États de l'Arabie méridionale préislamique", *Res orientales* 11 (1998) [*Parfums d'orient*], 85–92.

Arnaud, P., "Le *Periplus maris Erythraei*. Une œuvre de compilation aux préoccupations géographiques", *Topoi (Orient-Occident)*, Supplém. 11 (2012), 27–61.

Aujac, G., *Ératosthène de Cyrène, le pionnier de la géographie*, Paris, 2001.

Beeston, A. F. L. (a), "Pliny's *Gebbanitae*", *Proceedings of the 5th Seminar for Arabian Studies*, vol. 2, London, 1972, 4–8.

Beeston, A. F. L (b), "Some Observations on Greek and Latin Data Relating to South Arabia", *BSOAS* 42 (1979), 7–12.

Berger, H. (a), *Die geographischen Fragmente des Eratosthenes*, Leipzig, 1880.

Berger, H. (b), *Geschichte der wissenschaftlichen Erdkunde der Griechen*, Leipzig, 1887.

403 According to Desanges's conversion ratio (Desanges (c) 39).

Bernand, A. (a), *Inscriptions métriques de l'Égypte gréco-romaine*, Paris, 1969.

Bernand, A. (b), *Les portes du désert: recueil des inscriptions grecques d'Antinooupolis, Tentyris, Koptos, Apollonopolis Parva et Apollonopolis Magna*, Paris, 1984.

Bernand, A. (c), *Le Paneion d'El-Kanaïs: les inscriptions grecques*, Leiden, 1972.

Bianchetti, S. (a), "Die Seerouten nach Indien in hellenistischer und in römischer Zeit", E. Olshausen & H. Sonnabend (eds.), *Zu Wasser und zu Land. Verkehrswege in der antiken Welt*, Stuttgart, 2002, 283–292.

Bianchetti, S. (b), "La ‚scoperta' della Penisola Arabica nell'età di Alessandro Magno", *Geographia Antiqua* 18 (2009), 153–163.

Borca, F., "*In orbem intrare*: l'Oceano, il Mediterraneo e le colonne d'Ercole", M. Khanoussi, P. Ruggeri & C. Vismara (eds.), *L'Africa romana* 14, Rome, 2002, vol. 1, 123–128.

Bowersock, G.W. (a), "A Report on Arabia Provincia", *JRS* 61 (1971), 219–242.

Bowersock, G.W. (b), *Roman Arabia*, Cambridge (Ma) – London, 1983.

Braunert, H., "Die heilige Insel des Euhemeros in der Diodor-Überlieferung", *RhM* 108 (1965), 255–268.

Breton, J.-Fr., *L'Arabie heureuse au temps de la reine de Saba': VIIIe–Ier siècles av. J.-C.*, Paris, 1998.

Bretzl, H., *Botanische Forschungen des Alexanderzuges*, Leipzig, 1903.

Briant, P., *Histoire de l'empire perse de Cyrus à Alexandre*, Paris, 1996.

Burstein, S.M., Agatharchides of Cnidus, *On the Erythraean Sea*, transl. and comm. by S.M. Burstein, London, 1989.

Casson, L., The *Periplus Maris Erythraei*, Princeton, 1989.

Cohen, G., *The Hellenistic Settlements in Syria, the Red Sea Basin and North Africa*, Berkeley – Los Angeles – London, 2006.

Conti-Rossini, C., *Storia d'Etiopia*, Bergama, 1928.

Dan, A.; Geus, K.; Guckelsberger, K., "What is Common Sense Geography? Some Preliminary Thoughts from a Historical Perspective", Geus K. & Thiering M. (eds.), *Features of Common Sense Geography. Implicit knowledge structures in ancient geographical texts*, Vienna – Berlin, 2014, 17–38.

Delbrueck, R., "Südasiatische Seefahrt im Altertum", *BJ* (1955–56), 8–58, 229–308.

Desanges, J. (a), "Le littoral africain du Bab el-Mandeb d'après les sources grecques et latines", *Annales d'Éthiopie* 11 (1978), 83–101.

Desanges, J. (b), *Recherches sur l'activité des Méditerranéens aux confins de l'Afrique (VIe siècle avant J.-C.–IVe siècle après J.-C.)*, Rome, 1978.

Desanges, J. (c), Pline l'Ancien, *Histoire naturelle* (livre VI, 4e partie: l'Asie africaine sauf l'Égypte; les dimensions et les climats du monde habité), édition, traduction et commentaire de J. Desanges, Paris, 2008.

Desanges, J. & Reddé, M., "La côte africaine du Bab el-Mandeb dans l'Antiquité", Berger C., Clerc G., Grimal N. (eds.), *Hommages à Jean Leclant*, vol. 3, *Études isiaques*, Cairo, 1994, 161–194.

Desanges, J.; Stern, E.M.; Ballet, P., *Sur les routes antiques de l'Azanie et de l'Inde. Le fonds Révoil du Musée de l'Homme (Heis et Damo, en Somalie)*, Paris, 1993.

Dihle, A. (a), *Umstrittene Daten. Untersuchungen zum Auftreten des Griechen am Roten Meer*, Cologne, 1965.

Dihle, A. (b), "Die entdeckungsgeschichtlichen Voraussetzungen des Indienshandels der römischen Kaiserzeit", *ANRW* 2. 9. 2 (1978), 546–580.

Dihle, A. (c), "Der Seeweg nach Indien", V. Poeschl & H. Petersmann (eds.), *Antike und Orient. Gesammelte Aufsätze*, Heidelberg, 1984, 109–117.

Fraser, P.M., *Ptolemaic Alexandria*, vol.1 (text), Oxford, 1972; vol. 2 (notes), Oxford, 1972.

Gajda, I., *Le royaume de Ḥimyar à l'époque monothéiste. L'histoire de l'Arabie du Sud ancienne de la fin du IVe siècle de l'ère chrétienne jusqu'à l'avènement de l'islam*, Paris, 2009.

Geus, K., *Eratosthenes von Kyrene. Studien zur hellenistischen Kultur- und Wissenschaftsgeschichte*, Munich, 2002.

Gorman, R., "Οἱ περί τινα in Strabo", *ZPE* 136 (2001), 201–213.

Gisinger, F., *RE* 19.1 (1937), s. v. “Periplus 2”, c. 841–850.

Groom, N. (a), *Frankincense and Myrrh: a Study of the Arabian Incense Trade*, London, 1981.

Groom, N. (b), “Eastern Arabia in Ptolemy's Map”, *Proceedings of the Seminar for Arabian Studies* (vol. 16) [Proceedings of the 19th S. f. A. S. held at Oxford on 30th July – 1st August 1985], London, 1986, 65–75.

Helck, W., *RE* Suppl. 9 (1962), s. v. “Deirê”, c. 24.

Householder, Jr F. W. & Prakken, D. W., “A Ptolemaic graffito in New York”, *TAPhA* 76 (1945), 108–116.

Huss, W., *Ägypten in hellenistischer Zeit: 332–30 v. Chr.*, Munich, 2001.

Miller, J. I., *The Spice Trade of the Roman Empire (29 B. C. to A. D. 641)*, Oxford, 1969.

Jacob, Chr., *Géographie et culture en Grèce ancienne: essai de lecture de la* Description de la terre habitée *de Denys d'Alexandrie*, Lille, 1987 (microfiche).

Jacoby, F., *RE* 6.1 (1907), s. v. “Euemeros 3”, c. 953–972.

Janni, P., *La mappa e il periplo. Cartografia e spazio odologico*, Rome, 1984.

Kortenbeutel, H., *Der ägyptische Süd- und Osthandel in der Politik der Ptolemäer und römischen Kaiser*, Berlin, 1931.

Kroll, W., *RE* 9.1 (1914), s. v. “Iambulos”, c. 681–683.

Lapidoth-Eschelbacher, R., *The Red Sea and the Gulf of Aden*, The Hague-Boston-London, 1982.

Lloyd, A. B. (a), Herodotus. Book 2 (vol. 1: Introduction), Leiden, 1975.

Lloyd, A. B. (b), Herodotus. Book 2 (vol. 2: Commentary 1–98) Leiden, 1976.

Lloyd, A. B. (c), “Nationalist Propaganda in Ptolemaic Egypt”, *Historia* 31.1 (1982), 33–55.

Longo, O., “I mangiatori di pesci: regime alimentare e quadro culturale”, *MD* 18 (1987), 9–55.

MacDonald, M., “Arabi, Arabia e Greci. Forme di contatto e percezione”, S. Settis (ed.), *I Greci. Storia, cultura, arte, società*, vol. 3.1, *Greci oltre la Grecia*, Turin, 2001, 230–266.

Malaise, M., “Sésostris, Pharaon de légende et d'histoire”, *CE* 41 (1966), 244–271.

Marcotte, D., *Les Géographes grecs* (vol. 1), *Introduction générale*; Ps.-Scymnos, *Circuit de la terre*, texte établi et traduit par D. Marcotte, Paris, 2000.

Müller, K. (a), *Geographi Graeci minores*, vol. 1, Paris, 1853–1861.

Müller, K. (b), *Claudii Ptolemaei geographia*, Paris, 1883.

Munro-Hay, S., *Aksum. An African Civilisation of Late Antiquity*, Edinburgh, 1991.

Nicolet Cl., *L'inventaire du monde. Géographie et politique aux origines de l'Empire romain*, Paris, 1988.

Otto, W. & Bengtson, H., *Zur Geschichte des Niederganges des Ptolemäerreiches: ein Beitrag zur Regierungszeit des 8. und des 9. Ptolemäers*, Munich, 1938.

Paassen, Ch. van, *The Classical Tradition of Geography*, Groningen, 1957.

Peremans, W., “Diodore et Agatharchide de Cnide, *Historia* 16 (1967), 432–455.

Pfister, Fr., “Das Alexander-Archiv und die hellenistisch-römische Wissenschaft”, *Historia* 10.1 (1961), 30–67.

Pirenne, J., *Le royaume sud-arabe de Qatabân et sa datation d'après l'archéologie et les sources classiques*, Louvain, 1961.

Ponsich, M., “La navigation antique dans le détroit de Gibraltar”, R. Chevallier (éd.), *Littérature gréco-romaine et géographie historique. Mélanges offerts à R. Dion*, *Caesarodonum* 9bis, Paris, 1974, 256–273.

Potts, D. T., *The Arabian Gulf in Antiquity* (vol. 2). *From Alexander the Great to the Coming of Islam*, Oxford, 1990.

Radt, S., *Strabons Geographika (vol. 8). Buch XIV–XVII. Kommentar,* Göttingen, 2009.

Raschke, M. G., “New Studies in Roman Trade with the East”, *ANRW* 2.9.2, Berlin – New York, 1978, 604–1378.

The Red Sea and Gulf of Aden Pilot, Great Britain Hydrographic Service, 5th edition, London, 1900.

Rathbone, D., “Koptos the emporion. Economy and society. I–III A. D.”, *Topoi. Orient-Occident* Suppl. 3 (2002), 179–198.

Robin, Chr. (a), “Cités, royaumes et empires de l'Arabie avant l'Islam”, *Revue du monde musulman et de la Méditerranée* 61 (1991), 45–54.

ROBIN, Chr. (b), "La Tihama yéménite avant l'Islam: notes d'histoire et de géographie historique", *Arabian Archaeology and Epigraphy* 6 (1995), 222–235.

RODINSON, M., École pratique des hautes études. 4e section, Sciences historiques et philologiques. Annuaire 1975–1976 (1976), 201–224.

ROMANIS, F. de, *Cassia, cinnamomo, ossidiana. Uomini e merci tra Oceano Indiano e Mediterraneo*, Rome, 1996.

RODDAZ, J.-M., "Auguste et les confins", *L'Africa romana*, 15.1 (*Ai confini dell'Impero: contatti, scambi, conflitti*), Roma, 2004, 261–276.

ROSTOVTZEFF, M. I. (a), "Zur Geschichte des Ost- und Südhandels im ptolemaisch-römischen Ägypten", *APF* 4 (1908), 298–315.

ROSTOVTZEFF, M. I. (b), "Foreign commerce of Ptolemaic Egypt", *Journal of Economic and Business History* 4 (1931–32), 728–769.

RYCKMANS, J., "Petits royaumes sud-arabes d'après les auteurs classiques", *Muséon* 70 (1957), 75–96.

SALLES, J.-Fr., "Découvertes du golfe Arabo-Persique aux époques grecque et romaine", *REA* 94.1–2 (1992), 79–97.

SCHIWEK, H., "Der Persische Golf als Schiffahrts- und Seehandelsroute in Achämenidischer Zeit und in der Zeit Alexanders der Grossen", *BJ* 162 (1962), 4–97.

SCHNEIDER, P. (a), "La connaissance des mangroves tropicales dans l'Antiquité (éléments complémentaires)", *Topoi (Orient-Occident)* 17 (2011) 353–402.

SCHNEIDER, P. (b), "Before the Somali Threat: Piracy in the Ancient Indian Ocean", *The Journal of the Hakluyt Society* (2014).

SCULLARD, H. H., *The Elephants in the Greek and Roman World*, London, 1974.

SCHIETTECATTE, J. (a), *D'Aden à Zafar. Villes d'Arabie du Sud préislamique*, Paris, 2011.

SCHIETTECATTE, J. (b), "L'Arabie du Sud et la mer, du 3e siècle av. au 6e siècle apr. J.-C.", *Topoi (Orient-Occident)*, Supplém. 11 (2012), 237–273.

SCHMITTHENNER, W., *RE* Suppl. 14 (1974), s. v. "Anaxikrates 9", c. 44–47.

SELAND, E. H., "The 'Periplus' Report of a Roman Attack on Aden: an Unintended Result of Successful Propaganda?", *SO* 80 (2005) 60–67.

SIDEBOTHAM, S. E. (a), *Roman Economic Policy in the Erythra Thalassa 30 B. C. – A. D. 217*, Leiden, 1986.

SIDEBOTHAM, S. E. (b), *Berenike and the Ancient Spice Route*, Berkeley, 2011.

SPEIDEL, M. A., "Außerhalb des Reiches? Zu neuen lateinischen Inschriften aus Saudi-Arabien und zur Ausdehnung der romischen Herrschaft am Roten Meer", *ZPE* 163 (2007), 296–306.

TARN, W. W. (a), "Ptolemy II", *JEA* 14 (1928), 246–260.

TARN, W. W. (b), "Ptolemy II and Arabia", *JEA* 15 (1929), 9–25.

THIEL, J. H., *Eudoxus of Cyzicus. A Chapter in the History of the Sea-route to India and the Routes round the Cape in Ancient Times*, Groningen, 1966 (engl. translation).

TKAČ, J. (a), *RE* 9.1 (1914), s. v. "Ichthyophagi", c. 2524–2531.

TKAČ, J. (b), *RE* 1.A.2 (1920), s. v. "Saba", c. 1298–1515.

VILLENEUVE, Fr., PHILLIPS, C., FACEY, W., "Une inscription latine de l'archipel Farasan (sud de la mer Rouge) et son contexte archéologique et historique", *Arabia: revue de sabéologie* 2 (2004), 141–192.

WARMINGTON, E. H., *The Commerce between the Roman Empire and India*, Cambridge, 1928.

WARNECKE, H., s. v. "Meerenge", Sonnabend H. (ed.), *Mensch und Landschaft in der Antike. Lexicon der Historischen Geographie*, Stuttgart, 1999, 334–336.

WELLESLEY, K., "The Fable of a Roman Attack on Aden", *PP* 39 (1954), 401–405.

WILCKEN, U., "Punt-Fahrten in der Ptolemäerzeit", *ZÄS* 60 (1925), 86–102.

WINIARCZYK, M., "Das Werk des Jambulos. Forschungsgeschichte (1550–1988) und Interpretationsversuch", *RhM* 140.2 (1997), 126–153.

WISSMANN, H. von (a), "De mari Erythraeo", Herrmann-Lautensach Festschrift (Stuttgarter geographische Studien Band 69), Stuttgart, 1957, 289–324.

WISSMANN, H. von (b), "Himyar, Ancient History", *Muséon* 77 (1964), 429–497.

Wissmann, H. von (c), *RE* Suppl. 11 (1968), s. v. “Zangenae”, c. 1337–1348.

Wissmann, H. von (d), “Die Geschichte des Sabäerreichs und der Feldzug des Aelius Gallus”, *ANRW* 2. 9.1, Berlin – New York, 1976, 308–544.

Woelk. D., *Agatharchides von Knidos: über das rote Meer. Ubersetzung und Kommentar*, Bamberg, 1966.

Yémen. Au pays de la reine de Saba (exposition présentée à l’Institut du monde arabe [27 oct. 1997–28 fév. 1998]), Paris, 1997.

Young, G. K., *Rome’s Eastern Trade. International Commerce and Imperial Policy. 31 B. C.–A. D. 305*, London – New York, 2001.

GLOBALISIERUNG IN DER ANTIKE? SEEGESTÜTZTER FERNHANDEL ZWISCHEN ROM UND INDIEN

Monika Schuol

> Der Begriff der Globalisierung ist ein Neuzugang im terminologischen Arsenal der Historiker. Erst seit 1993 kann man von einer weiten Verbreitung in öffentlichen Diskussionen sprechen; danach aber nahm seine Diffusion epidemische Züge an. Noch später hat Globalisierung als analytischer Begriff in die Geschichtswissenschaft Einzug gehalten.[1]

Diese Aussage trifft zweifellos für die Neuere Geschichte und die Zeitgeschichte zu, nicht jedoch für die Alte Geschichte, die sich mit dem Blick über die Grenzen des Römischen Reiches schwer tut und die ökonomischen und sozialen Zusammenhänge größerer Reichweiten ausschließlich aus der Perspektive des antiken Mittelmeerraumes betrachtet.

I. FRAGESTELLUNG UND VORGEHENSWEISE

Im Mittelpunkt der folgenden Ausführungen steht die Frage, inwiefern das Funktionsmodell der Globalisierung das althistorische Methodenspektrum insbesondere im Hinblick auf die antike Wirtschaftsgeschichte bereichern kann. Ich werde versuchen, „aus der Perspektive von Globalisierung einen neuen Blick auf die Vergangenheit zu werfen"[2] und insbesondere die Etablierung seegestützter antiker Fernhandelsbeziehungen zwischen Rom und Indien in den ersten drei nachchristlichen Jahrhunderten als grenzüberschreitende Vernetzungs- und Austauschprozesse betrachten. Das Ziel ist es, den heuristischen Nutzen des Funktionsmodells der Globalisierung für die althistorische Forschung auszuloten.

Vorangestellt sei eine knappe Definition des Begriffes „Globalisierung" (Punkt II). Die weiteren Gliederungspunkte der folgenden Ausführungen sind Roms Indienhandel im Spiegel der antiken Überlieferung (Punkt III), die Einordnung der Antike in die Globalisierungsgeschichte (Punkt IV), die „globale" Vernetzung in Roms Indienhandel (Punkt V) und eine Zusammenfassung (Punkt VI).

II. DEFINITION „GLOBALISIERUNG"

Die einfachste und wohl anerkannteste Definition von Globalisierung ist Expansion von Interaktionsradien, Verdichtung, Stabilisierung und institutionelle Unter-

1 Conrad 2013, 146.
2 Osterhammel/Petersson 2012, 10.

mauerung großräumiger Netzwerke und wechselseitige Beeinflussung entfernter Regionen. Dieser Prozess wird oft als „Raum-Zeit-Kompression“ (*time-space-compression*), Verdichtung von Raum und Zeit, umschrieben[3]. Der Begriff „Verdichtung“ bezeichnet bei David Harvey jene prozessualen geschichtlichen Schnittpunkte, welche die „objektiven“ Eigenschaften von Raum und Zeit revolutioniert und damit die menschlichen Vorstellungen von „unserer“ Welt radikal verändert haben. „Verdichtung“ meint bei ihm also die Folgen der technischen Innovationen im Transport- und Kommunikationsbereich, dass nämlich Waren, Personen, Kapital und Informationen schneller, häufiger, kostengünstiger und in größerem Umfang rund um den Globus verkehrten, die Relevanz räumlicher Distanzen also zurückgedrängt wird. Die Überwindung räumlicher Schranken durch technischen Fortschritt lasse in der menschlichen Wahrnehmung die Welt schrumpfen, Raum und Zeit immer enger zusammen fallen. Die für diese „raumzeitliche Implosion“[4], die scheinbare Verdichtung der Welt zu einem „global village“, verantwortlichen neuen Transport- und Kommunikationstechnologien werden daher als „time-space-shrinking-technologies“[5] bezeichnet.

III. ROMS INDIENHANDEL IM SPIEGEL DER ANTIKEN ÜBERLIEFERUNG

Ausgehend von dieser Definition von „Globalisierung“ widme ich mich nun zwei „Klassikern“ zum Fernhandel zwischen Rom und Indien, nämlich Plinius' Berichten über die Ausnutzung des Monsuns für die Überfahrt von der südarabischen Küste nach Indien (Plin. *nat.* 6, 100 f.):

> … diuque ita navigatum est, donec conpendia invenit mercator lucroque India admota est: quippe omnibus annis navigatur, sagittariorum cohortibus inpositis; etenim piratae maxime infestabant.
>
> … und lange Zeit ist man so gesegelt (sc. vom Kap Syagros nach Zigeros), bis ein Kaufmann noch weitere Abkürzungen fand und so durch seine Gewinnsucht Indien (uns) noch näher gebracht worden ist: Man segelt nämlich jetzt das ganze Jahr hindurch, wobei man Kohorten von Bogenschützen (auf den Schiffen) mitnimmt, denn Seeräuber machten (die Fahrt) sehr gefährlich. … (Plin. *nat.* 6, 101)

Plinius liefert (aus römischer Perspektive) einen aufschlussreichen Bericht über die politische Grenzen überspringende Erschließung transozeanischer Handelsrouten, die schrittweise Intensivierung, Effektivierung und Verstetigung der Fernhandels-

3 Zu dem Begriff „Raum-Zeit-Kompression“ vgl. Harvey 1989, 240; Robertson 1990, 15–31; Thompson 1995; Werlen 2014, 54–66, hier 59–62. – Eine Zusammenfassung des Harvey'schen Konzepts von „Verdichtung“ bietet Neugebauer 2007.

4 Werlen 2000, 23; vgl. auch Lippuner 2005, 29–39; Lippuner 2008, 365–392, hier 377; Kessler 2009, 28–79, hier 30–32 (kritische Auseinandersetzung mit der Benutzung des Begriffs „global village“, der der weiterhin bestehenden Bedeutung räumlicher Distanz nicht gerecht werde); Werlen 2014, 59–62.

5 Dicken 2011, 81–97; zusammenfassend Giese/Mossig/Schröder 2011, 76–80.

aktivitäten zwischen dem Mittelmeerraum und Indien[6]. Mit der Nutzung des Monsuns hatte sich, von der südarabischen Küste oder der Nordostspitze Somalias aus gerechnet, der Seeweg zu den bedeutenden westindischen Hafenplätzen auf drei bis vier Wochen verkürzt. Strabo (2, 5, 12 [118]) weiß zu berichten, dass die westlichen Händler mit „ganzen Flotten“ agierten und „dass regelmäßig nicht weniger als 120 Schiffe“ nach Indien gesegelt seien, „während früher unter den Ptolemäerkönigen nur ganz wenige die Fahrt und den Handel mit den indischen Waren gewagt hätten.“ Die „Welt“ war kleiner, Entferntes miteinander verknüpft geworden. Raum und Zeit büßten ihre trennende Dimension immer mehr ein; für die Händler Ägyptens und des östlichen Mittelmeerraumes hatte der Indische Ozean, in der modernen Forschung bisweilen als die „Wiege der Globalisierung“ (*cradle of globalization*[7]) tituliert, seine Wirkung als naturräumliche Interaktionsbarriere verloren und war zum „central hub“[8] und „nucleus of ancient world trade“[9] geworden. An anderer Stelle moniert Plinius (*nat.* 12, 84; vgl. auch 13, 15):

> … minimaque computatione miliens centena milia sestertium annis omnibus India et Seres et paeninsula illa imperio nostro adimunt: tanti nobis deliciae et feminae constant!
> … und nach der niedrigsten Schätzung rauben Indien, China und die (arabische) Halbinsel dem Römischen Reich Jahr für Jahr 100 Millionen Sesterzen. Soviel kosten uns Luxus und Frauen!

Er beklagt ein Ausbluten des Imperiums, eine Flucht des Geldes aus Rom und spitzt seine Schreckensvision noch einmal zu mit der Diagnostizierung einer extrem defizitären Handelsbilanz. Plinius inszeniert die erschreckende Größenordnung von 100 Millionen Sesterzen regelrecht, um mit diesem ungeheuren Ausmaß der Luxus-Dekadenz in Rom die fatalen Auswirkungen der Fernhandelsgüter auf die römische Gesellschaft plastisch werden zu lassen. Für ihn ist diese Riesensumme ein jährlich wiederkehrender Skandal[10].

Was Plinius hier nachzeichnet, ist – pauschal formuliert – die Beschleunigung und Dynamisierung von Gütertransport über große Distanzen hinweg, aber auch die Rückkopplungseffekte zwischen den wirtschaftlichen Austauschbeziehungen und der italischen Gesellschaft – alles Prozesse, die man in der jüngeren wirtschaftswissenschaftlichen und soziologischen Forschung und auch in der Geschichte der Neuzeit mit dem Begriff „Globalisierung“ erfasst.

6 Schuol 2000, 96–99. 417. 428–430; Bianchetti 2002, 280–292; Ruffing 2002, 360–378, hier 370 mit Anm. 58; Whittaker 2004, 163–180; McLaughlin 2010, 40–42.
7 Wilkinson o. J.; Moorthy/Jamal 2009, 9; vgl. auch Conrad 2013, 209; Wiarda 2014, 88.
8 Seland 2013, 373.
9 Seland 2008, 70.
10 Zu den Plinius-Belegen vgl. z. B. Young 2001, 202; Murphy 2004, 96–105. – Zum Luxus im antiken Rom und der Luxuskritik in der Römischen Republik und in der Frühen Kaiserzeit vgl. Baltrusch 1988; Wagner-Hasel 2002, 325–253; Weeber 2003, 7. 105–114; Kloft 2007, 65–75; Weeber 2007, 3–15; Baltrusch 2010, 711–738, hier 718–724.

IV. PHASEN DER GLOBALISIERUNG: GLOBALISIERUNG IN DER ANTIKE?

Kann man, wenn man das Harvey'sche Konzept der „Verdichtung von Raum und Zeit“ zu Grunde legt, in der Antike von Globalisierung sprechen? Waren nicht am Anfang die Homogenität kleinerer, begrenzter Räume und zu viele „Raumteiler“, d.h. naturräumliche und kulturell-institutionelle Interaktionsbarrieren, die einer gegenseitigen Beeinflussung und Vernetzung dieser vielen kleinen Lebenswelten entgegenstanden? Ist die Kausalkette: großräumige Verflechtungsprozesse – Dynamisierung der Wirtschaft – Beschleunigung des Privat- und Arbeitslebens – eine vom „Tempo-Virus“[11] infizierte Gesellschaft nicht vielmehr Ausdruck des neuzeitlichen Lebensgefühls? Anders gefragt: Galt die Devise *time is money!* bereits in der Antike?

Das Nachdenken über die Frage „Wann begann die Globalisierung?“ ist nicht neu. „Globalisierung“ in der Antike thematisiert meines Wissens erstmals Malitz, der unter „Globalisierung“ allerdings im Wesentlichen die wichtigsten Facetten der Romanisierung neu eroberter Territorien subsumiert, nicht jedoch expandierende Interaktionsradien und „transnationale“ Netzwerke mit ihren von Reziprozität geprägten, gesellschaftlichen und wirtschaftlichen Folgen als Kennzeichen von Globalisierungsprozessen; nicht diskutiert wird, inwiefern die Globalisierung der altertumswissenschaftlichen Forschung heuristische Hilfestellungen bieten kann und in der Antike von „Globalisierung“ gesprochen werden kann[12]. Im Zusammenhang mit den antiken Fernhandelsbeziehungen im Indischen Ozean wird der Begriff „Globalisation“ von Seland gebraucht: Die Kernprozesse von „Globalisierung“ werden nur grob umrissen; die Frage nach der Anwendbarkeit des Begriffs „Globalisierung“ auf die Antike bleibt offen – war die antike Welt „globalised, or should we perhaps say 'oikumenised'?“ –, und die Auswirkungen von Fernhandel auf politische und soziale Strukturen der Herkunfts- und Zielregionen von Menschen, ihrem Wissen und Warenströmen finden keine Berücksichtigung[13]. Stark auf den römischen Machtbereich und den Mittelmeerraum fokussierend, aber offenbar nicht als Auswirkung von Fernhandelsbeziehungen auf die wirtschaftlich-sozialen Verhältnisse im Römischen Reich wird „Globalisation“ jüngst in dem Sammelband „Globalisation and the Roman World“ diskutiert[14].

In der archäologischen, anthropologischen, wirtschaftshistorischen und soziologischen Forschung wird durchaus auf weitgespannte Fernhandelsbeziehungen, überregionale Marktintegration sowie auf ausgedehnte Interaktionsräume zwischen China, Indien, Mesopotamien und dem Mittelmeerraum bereits in der vorchristlichen Zeit verwiesen. Derzeit neigen die meisten Historiker dazu, die Anfänge der modernen Globalisierung im frühen 16. Jahrhundert zu verorten, denn die Erschließung globaler Räume hätte überhaupt erst die Etablierung großräumiger (transkon-

11 Zu dem Begriff „Tempo-Virus“ vgl. Borscheid 2004.

12 Malitz 2000, 37–52.

13 Seland 2008, 67–79, Zitat S. 76.

14 Vgl. z. B. Morley 2015, 49–68; Pitts 2015, 69–98. Der Sammelband war mir bislang nicht zugänglich.

tinentaler) Netzwerke politischer, ökonomischer und kultureller Akteure ermöglicht. Politik-, Wirtschafts- und Sozialwissenschaftler konzentrieren sich hingegen auf das 19. und 20. Jahrhundert und charakterisieren den Zeitraum 1840–1914 als erste Phase der modernen Globalisierung[15].

Auf diesem Zeitstrahl der Globalisierungsgeschichte, der lediglich als ungefähre Orientierung dient, wird die Antike der präglobalen Epoche zugeschlagen, zu deren Kennzeichen auf der Haben-Seite großräumige Herrschaftsgebilde mit Weltherrschaftsanspruch, Expansion von Interaktionsradien und Ansätze zur Netzwerkverdichtung zählten. Auf der Soll-Seite hingegen stehen (wenn man die westeuropäische Perspektive zum Maßstab erhebt) fehlende politische Stabilität, defizitäre geographische Kenntnisse, mangelndes technisches *know-how* (etwa navigatorische Fähigkeiten) und die Beschränkung des Fernhandels auf Luxuswaren; und im Hinblick auf die Netzwerke werden separierte Interaktionsräume (z. B. die Beschränkung auf Eurasien/Afrika) und ein Zuwenig an Intensität, mangelnde Engmaschigkeit der Netzwerke und fehlende Dauerhaftigkeit der überregionalen Verflechtungen geltend gemacht, um bereits in der Vormoderne von Globalisierung sprechen zu können[16].

Für diese Einordnung der Vormoderne in den Globalisierungsprozess sprechen zweifellos die bereits erwähnten Innovationen im Verkehrswesen und der interkontinentalen Telegraphie, konkretisierbar etwa an den strukturellen Wandlungsprozessen der Medien, die in der sogenannten „massenmedialen Sattelzeit“[17] am Ende des 19. Jahrhunderts einen neuen, stark beschleunigten und weltweiten Kommunikationsraum generierten. Mit der Gründung von Nachrichtenagenturen und der Etablierung von Korrespondentennetzen veränderten sich die Strukturen der Presse von Grund auf. Die Telegraphie (in einem freilich etwas hinkenden Vergleich gelegentlich als „das viktorianische Internet“ bezeichnet[18]) und Telefon sorgten zudem für eine fast lichtschnelle Nachrichtenübermittlung im Vergleich zum physisch-gebundenen Transport. Die Verlegung der ersten Transatlantikkabel (1857/58, 1866) zwischen London und New York als Auftakt der telegraphischen Umrundung der Erde (erreicht 1902!) transformierte regionale maritime Räume zu globalen medialen Landschaften, die eine zentrale Voraussetzung waren für die Entstehung eines globalen Bewusstseins, transnationaler Öffentlichkeiten und einer Totalvernetzung in Echtzeit[19].

15 Zur Periodisierung von Globalisierung vgl. z. B. Schäfer 2003, 75–88; Bayly 2004; Fässler 2007, 32 f. 46–51 (zusammenfassend). 52–175 (ausführliche Darstellung der Globalisierungsphasen); Bayly 2007, 81–108; Eckert 2012, 11–18.

16 Zur Charakterisierung der präglobalen Epoche vgl. z. B. Fässler 2007, 52–59. – Zur Problematik dieser gängigen Periodisierung von Globalisierung anhand eines auf europäischer Geschichtsschreibung beruhenden Stufenmodells und Forderungen nach unterschiedlichen weltregionalen Perspektiven vgl. z. B. Komlosy 2011, 50–66.

17 Zu dem Begriff „massenmediale Sattelzeit“ für den Zeitraum zwischen 1880–1960, in dem sich das Verhältnis zwischen Medien und Gesellschaft grundlegend verändert habe, vgl. Knoch/Morat 2003, 9–33, hier 20.

18 Standage 1999; vgl. dazu auch die kritische Stellungnahme von Gugerli 1999, 75.

19 Zur Bedeutung der Transatlantik-Seekabel in der Kommunikationstechnologie vgl. z. B. Holtorf 2001, 86–97; Holtorf 2005, 157–178; Osterhammel 2009, 1012–1027; Buschauer 2010,

V. ROMS INDIENHANDEL: „GLOBALE“ VERNETZUNG, INTERAKTION UND TRANSFORMATION

Definiert man diese neuzeitliche „transport- und kommunikationstechnische Doppelrevolution“[20] als unabdingbare Voraussetzung einer weltweiten Vernetzung und zum Motor der Globalisierung, kann man freilich in der Antike nicht von Globalisierung sprechen: Für die Fahrt durch das Rote Meer bis an die Nordostspitze Somalias (Kap Guardafui) und von dort über den Indischen Ozean nach Indien werden jeweils 30 bis 40 Tage Reisezeit veranschlagt[21], während das weltumspannende Telegraphenkabelnetz in der zweiten Hälfte des „langen“ 19. Jahrhunderts die Übermittlungszeit von Nachrichten auf Minuten oder allenfalls wenige Stunden verkürzte und das Zusammenwirken neuer Verkehrsmittel (Eisenbahn, Dampfschiff) Verstetigung, enorme Beschleunigung und Berechenbarkeit des Gütertransports bewirkte.

In der (Vor-)Geschichte von Globalisierung beobachtbar ist jedoch sowohl die Integration von Individuen oder Kleingruppen in weitreichende Interaktionszusammenhänge als auch deren Partizipation an verschiedenen überlappenden, aber nicht deckungsgleichen sozialen Beziehungsgeflechten; diese Akteure können deshalb auch nicht kleingerechnet werden als „Teile“ einer räumlich umgrenzten Gemeinschaft[22]. Dass lokal begrenzte Themen über den engeren räumlichen Horizont ihres Schauplatzes hinausführen und auch eine globalgeschichtliche Dimension haben können, das Meer in der Antike also keineswegs ein geschichtsloser Raum oder eine „naturgegebene Grenze des Raumes menschlicher Unternehmungen“[23] ist, soll am Rom der frühen Kaiserzeit und der Intensivierung des seegestützten Indienhandels aufzeigt werden: Die politischen, sozialen und ökonomischen Umbrüche in der späten Römischen Republik ließen im Mittelmeerraum einen rasant expandierenden, finanzkräftigen Absatzmarkt für exotische und hochpreisige Güter entstehen, der wiederum der Motor war für die Vergrößerung des Aktionsradius der mittelmeerischen Händler. Neben Plinius die ausführlichste Beschreibung des großräumigen antiken Warenaustauschs bietet der *Periplus maris Erythraei*, ein griechisches Seefahrerhandbuch aus der Zeit zwischen 40 und 70 n. Chr.: Beschrieben wird der römische Süd- und Osthandel in voller Blüte, die Handelsschifffahrt unter Ausnutzung des Monsuns zwischen Ägypten, Südarabien, Ostafrika, Indien und dem Golfgebiet. Die Routenbeschreibungen beinhalten Angaben zu den wichtigsten Stationen, der dortigen Bevölkerung und den politischen Verhältnissen und natürlich die für die Händler wichtigsten Informationen über die günstigsten Reisezeiten, die Reisedauer, die Warenumschlagplätze und die Handelsgüter selbst[24].

111–120; Holtorf 2013.

20 Torp 2008, 41–59, hier 53.

21 Plin. *nat.* 6, 104. – Zu den Reisezeiten nach Indien vgl. z. B. Casson 1980, 21–36, hier 28–35; Hourani 1995, 26; Schuol 2000, 429; McLaughlin 2010, 40–42. 69 mit Anm. 83.

22 Osterhammel/Petersson 2012, 20.

23 Blumenberg 1993, 10; vgl. auch Braudel 2006, 37–60.

24 Eine Edition des Textes mit Übersetzung und ausführlichem Kommentar bietet Casson 1989. – Warmington 1974, 53; Dihle 1978, 546–580, hier 558; Raschke 1978, 604–1361, hier 645.

Jenseits des Bāb al-Mandab den Kaufleuten aus dem Mittelmeerraum bekannte Warenumschlagplätze waren z. B. in Südarabien Kane, an der somalischen Nordküste die als „far-side ports“ (πέραν ἐμπορία) bezeichneten Hafenplätze (*Periplus* §§ 8–13)[25], an der afrikanischen Ostküste der Haupthafen Rhapta im Süden Azanias (*Periplus* §§ 16–18; Ptol. 4, 7, 11 f.), an der westindischen Küste Barygaza und Muziris sowie an der ostindischen Küste Poduke (Arikamedu). Die Bedeutung dieser Orte für den Fernhandel spiegelt sich auch im archäologischen Befund wider[26].

Man wird davon auszugehen haben, dass die spätestens seit dem dritten vorchristlichen Jahrtausend genutzten Kommunikationslinien im Indischen Ozean weitgehend in der Hand ostafrikanischer, südarabischer und indischer Händler blieben, die sich mit der Konstruktion ihrer Schiffe und der Organisation des Warentransportes insgesamt den klimatischen und hydrologischen Verhältnissen bestens angepasst hatten und Konkurrenz von ihren wichtigen Handelsplätzen fernhalten wollten[27]. Letztgenannte handelspolitische Maßnahmen schreibt der *Periplus* den ḥaḍramitischen Königen in der Mitte des 1. Jh. n. Chr. zu (§§ 16. 31), die offenbar einen von Südarabien über Soqoṭrā bis nach Azania reichenden Sperrriegel etablierten, um die Einschaltung arabischer Mittelsmänner in den Fernhandel zwischen Indien und dem Mittelmeerraum zu erzwingen[28].

Aufbau, Verdichtung und zunehmende Bedeutung von (bezogen auf die antike Oikumene) „weltweiter“ Vernetzung in der Antike, die Kernprozesse von Globalisierung, lassen sich am Beispiel von Roms Indienhandel gut nachweisen[29]. In der frühen Kaiserzeit nachweisbar ist die wachsende Präsenz mittelmeerischer Händler in den seegestützten Handelsnetzwerken im Verflechtungsraum Mittelmeer/Rotes Meer/Indischer Ozean. Wohl vor allem Geschäftsleute ägyptischer oder syrischer Provenienz waren es, die sich als neue Akteure in den transozeanischen Warenaustausch außerhalb des Mittelmeerraumes einklinkten und damit in Fernhandelsnetzwerke des Indischen Ozeans vorstießen, die schon über viele Jahrhunderte oder Jahrtausende hinweg bestanden, bevor europäische Akteure daran teilhatten. Wenn mittelmeerische Händler auf afrikanische, arabische, vorderasiatische und indische Kaufleute treffen und das ägyptische Alexandria mit seinen Rotmeer-Hafenplätzen

660; Hourani 1995, 23–36; Ray 2003, 52; Crone 2004, 35; Biedermann 2006, 30–38; McLaughlin 2010, 79; Seland 2010, passim; Seland 2014, 367–402; vgl. auch Marcotte 2011 und die Beiträge in Boussac/Salles/Yon 2012.

25 Zu den „far-side ports“, den aus der Perspektive römisch-ägyptischer Händler jenseits des Bāb al-Mandab an den Küsten des Golfs von Aden gelegenen Handelsplätzen, vgl. Casson 1989, 115. 120–123; McLaughlin 2010, 67–69.

26 Chittick 1976, 117–133; Chittick 1980, 364–366; Smith/Wright 1988, 115–141; Meyer 1992, 58; Stern 1993, 21–61; Avanzini 2008; Tomber 2008, 93–108. 117–144; Salles/Sedov 2010; Sidebotham 2011.

27 Zu dem Bestreben der arabischen und indischen Händler, die Seewege nach Indien vor den „westlichen“ Kaufleuten zu verheimlichen, vgl. Rougé 1988, 59–74.

28 Vgl. z. B. Biedermann 2006, 31; McLaughlin 2010, 79. – Ein vergleichbares Handelsmonopol und vielleicht sogar Blockaden durch südarabische Händler an der ostafrikanischen Küste, um fremde Kaufleute von den dortigen Hafenplätzen fernzuhalten, vermutet Salles 1988, 75–102, hier 96; Salles 1993, 493–523, hier 506–508; Salles 1996, 251–267; demgegenüber skeptisch äußert sich Sheriff 2010, 142.

29 Zu antiken Handelsnetzwerken im Indischen Ozean vgl. zuletzt Seland 2013, 373–389.

und südarabische Häfen wie Kane oder das indische Muziris als bedeutende Knotenpunkte an den Schnittstellen mehrerer Wirtschaftsräume (arabische Halbinsel, Mittelmeerregion, Ostafrika und Indien) in den Aktionsradius römischer Händler integriert werden, setzt das Prozesse interkultureller Interaktion in Gang, die politische, soziale, ökonomische und kulturelle Auswirkungen auf alle beteiligten Akteure hatten: Entlang der Fernhandelsrouten im gesamten Verflechtungsraum Mittelmeer/Rotes Meer/Indischer Ozean bilden sich stetig wachsende Knotenpunkte (Kommandozentralen) heraus, die Ströme von Menschen, Waren und Informationen bündeln und kanalisieren.

Im Kontext dieser „weltweiten" Verflechtungsprozesse und der neuen Konkurrenzkonstellationen zu verorten ist die Generierung von Wissen über fremde Wirtschaftssysteme, Absatzmärkte und Einkaufsmöglichkeiten, wie sie für die mittelmeerischen Händler im *Periplus Maris Erythraei* dokumentiert ist. Ebenfalls der Gewinnung von Marktanteilen und der Aufrechterhaltung gewonnener zentraler *Netzwerkpositionen* geschuldet sind technische Innovationen, etwa die außerordentlich stabile Konstruktion der mittelmeerischen Schiffe und ihre Ausstattung mit immer größerer Ladekapazität, um sehr viel größere Warenmengen transportieren zu können als die arabische und indische Konkurrenz; mit diesen „Ozeanriesen", die auch die Bewunderung der indischen Handelspartner fanden, konnte man unter Ausnutzung des Monsuns nahezu das ganze Jahr die überseeischen Märkte anlaufen und erheblich kürzere Reisezeiten erzielen[30]. Die Ausrüstung von Schiffen in dieser Größenordnung und der Transport eines breiteren Warensortiments rückte für die mittelmeerischen Kaufleute überhaupt erst durch das Rechtsinstitut des Seedarlehens (*fenus nauticum*) in den Bereich der finanziellen Möglichkeiten, brachte ihnen gegenüber den indigenen Händlern weitere Wettbewerbsvorteile ein und versetzte sie in die Lage, im Raum des Indischen Ozeans die Potenziale rapide expandierender Handlungsräume auszuschöpfen[31].

Die Einbürgerung der Monsunschifffahrt in der frühen Kaiserzeit veränderte nicht nur die Akteurs-und Wettbewerbskonstellationen in den Fernhandelsnetzwerken zwischen dem Mittelmeerraum und Indien, sondern zeitigte auch tiefgreifende innergesellschaftliche sozio-ökonomische Veränderungen: Plinius bringt die wirtschaftliche Prosperität Roms in den ersten drei nachchristlichen Jahrhunderten in Verbindung mit der Expansion von Handlungsradien im Fernhandel und zeigt – modern gesprochen – *spill-over-effects* auf, dass nämlich die Erschließung immer weiterer Räume durch zunehmend dichtere Netzwerke Konsumgewohnheiten veränderte, also in Prozessen der Reziprozität und Reflexivität auf die in die überregionalen (transozeanischen) Interaktionsprozesse einbezogenen Gesellschaften zu-

30 Höckmann 1985, 52–74; Casson 1991a, 167. 174. 208; Casson 1991b, 8–12. – Zu den Belegen in der tamilischen Sangam-Literatur für die Präsenz westlicher Händler (*Yavanas*) in Südindien vgl. Kulke 2005, 36; Kulke/Rothermund 2010, 137; McLaughlin 2010, 52–56 u. ö.

31 Harrauer/Sijpesteijn 1985, 124–155; Thür 1987, 229–245; Thür 1988, 229–233; Casson 1994, 195–206; Rathbone 2000, 39–50; Young 2001, 42–48; Drexhage/Konen/Ruffing 2002, 137 f. 153–156; Morelli 2011, 199–233; Romanis 2012, 75–101; Romanis 2014a, 73–89; Romanis 2014b; Romanis 2014c, im Druck.

rückwirkte. Diese Rückkopplungseffekte waren es dann auch, die an der römischen Luxus-Kritik ablesbare Gegenreaktionen auslösten.

Der Rahmen für das Zusammenwachsen der großen Wirtschaftsräume zwischen dem Mittelmeer und dem indischen Subkontinent ist nicht nur durch die wirtschaftliche Prosperität und politische Stabilität des Römischen Reiches abgesteckt, sondern auch durch die mächtigen politischen Gebilde im südarabischen Raum (Saba, Himyar, Ḥaḍramawt) und Indien (die Reiche der Kuschanas, Guptas und Shatavahanas und im „fernen Süden" die Stammesfürstentümer der Colas, Pandyas und Ceras[32]), die ihre Entstehung, ihre Konsolidierung und ihren Aufstieg ebenfalls dem florierenden Seehandel der ersten drei nachchristlichen Jahrhunderte verdanken. Die „Tempo-Macher" in dieser von Plinius (*nat.* 6, 100 f.) als Beschleunigungsphase von Roms Indienhandel definierten Zeit sind aber zweifellos die soeben erwähnten Neuerungen im mittelmeerischen Schiffbau und die Finanzkraft der westlichen Händler.

VI. ZUSAMMENFASSUNG

Kernprozesse von Globalisierung – überregionale (transozeanische) Austauschbeziehungen mit den entsprechenden Rückkopplungseffekten, Expansion von Aktions- und Interaktionsradien, und Erosion räumlicher Interaktionsbarrieren – lassen sich bereits für die Antike insbesondere am Beispiel von Roms seegestützten Indienhandel zweifelsfrei nachweisen.

Globalisierung als neue Perspektive der althistorischen Forschung lässt sich also durchaus legitimieren, verlangt uns AlthistorikerInnen aber viel ab:

1) Den Blick in den großen Raum außerhalb der griechisch-römisch dominierten Welt, der politisch, wirtschaftlich und kulturell mit der antiken Mittelmeerwelt auf vielen Ebenen in stetigem Austausch stand,
2) Kompetenzen in mehreren Disziplinen, Kulturen und Regionen,
3) Offenheit für die Forschungsergebnisse der Nachbardisziplinen,
4) Aufbau fächerübergreifender Kooperationen[33].

Dennoch ist der Begriff „Globalisierung" für die althistorische Forschung insbesondere dann attraktiv, wenn es um die Analyse von Prozessen der Verdichtung von Netzwerken auf unterschiedlichen Ebenen und Austauschbeziehungen über politische, regionale und kulturelle Grenzen hinweg geht. Die Globalisierung als heuristisches Modell bereichert das althistorische Methodenspektrum insbesondere im Bereich der antiken Wirtschaftsgeschichte: Die Nachzeichnung und Kontextualisierung der großräumigen antiken Vernetzungsprozesse auf der Folie der Globali-

32 Zur Geschichte Indiens in den drei nachchristlichen Jahrhunderten vgl. Kulke/Rothermund 2010, 100–138, zu den Handelsverbindungen zwischen Indien und dem Mittelmeerraum 135–138; vgl. auch Kulke 2005, 25–42, zum Seehandel zwischen den Mittelmeer-Anrainern und Indien 35–37.

33 Für einen Paradigmenwechsel in der althistorischen Beschäftigung mit dem Vorderen Orient vgl. Wiesehöfer 2007, 595–616; Wiesehöfer 2013, 485–507, hier 493. 495

sierung bietet unverzichtbare Grundlagen für ein tieferes, über die bloße Kenntnis des Güteraustausches zwischen Rom und Indien hinausgehendes Verständnis der kaiserzeitlichen Wirtschafts- und Sozialgeschichte. Sozioökonomische Auf- und Abstiegsszenarien und daraus resultierende, mitunter auch größere Räume der antiken Welt umfassende wirtschaftliche und gesellschaftliche Transformationsprozesse erscheinen im Kontext der Globalisierungsgeschichte in neuem Licht. In Überwindung von Eurozentrismus und der ausschließlichen Fokussierung auf den antiken Mittelmeerraum lässt sich der römische Indienhandel auf diese Weise in den umfassenderen Kontext transozeanischer und transkontinentaler Interaktion situieren. Die Entwicklungen im Mittelmeerraum können kontextualisiert und einer Neuinterpretation unterzogen werden.

Das Funktionsmodell der (ökonomischen und kulturellen) Globalisierung als Perspektive lenkt also den Blick auf eine multizentrische antike Welt (die eben nicht nur aus dem Mittelmeerraum besteht), auf Relationen zwischen Akteuren unterschiedlichster geographischer Verortung und die Einbindung europäischer Entwicklungen in sehr viel größere Interaktionszusammenhänge. Die Globalisierung als heuristisches Modell liefert neue Interpretationsansätze für transozeanische Vernetzungsprozesse bereits in der Antike und trägt bei zu einer methodisch-analytisch notwendigen Weiterentwicklung der althistorischen Forschung über den Mittelmeerraum hinaus.

VII. LITERATURVERZEICHNIS

ASSKAMP et al. 2007 = Aßkamp, Rudolf / Brouwer, Marijke / Christiansen, Jörn / Kenzler, Herwig / Wamser, Ludwig (Hrsg.), Luxus und Dekadenz. Römisches Leben am Golf von Neapel, Mainz 2007.

AVANZINI 2008 = Avanzini, Alessandra (Hrsg.), A Port in Arabia between Rome and the Indian Ocean (3rd C. B. C–5rd C. A. D.), Roma 2008 (= Khor Rori Report 2).

BALTRUSCH 1988 = Baltrusch, Ernst, *Regimen morum*. Die Reglementierung des Privatlebens der Senatoren und Ritter in der römischen Republik und frühen Kaiserzeit, München 1988 (= Vestigia 41).

BALTRUSCH 2010 = Baltrusch, Ernst, in: Reallexikon für Antike und Christentum 23 (2010) 711–738 s. v. Luxus II (Luxuskritik).

BAYLY 2004 = Bayly, Christopher A., The Birth of the Modern World, 1780–1914. Global Connections and Comparisons, Malden 2004.

BAYLY 2007 = Bayly, Christopher A., „Archaische“ und „moderne“ Globalisierung in Eurasien und Afrika, ca. 1750–1850, in: Conrad, Sebastian / Eckert, Andreas / Freytag, Ulrike (Hrsg.), Globalgeschichte. Theorien, Ansätze, Themen, Frankfurt/Main 2007, 81–108.

BIANCHETTI 2002 = Bianchetti, Serena, Seerouten nach Indien in hellenistischer und römischer Zeit, in: Olshausen/Sonnabend 2002, 280–292.

BIEDERMANN 2006 = Biedermann, Zoltán, Soqotra. Geschichte einer christlichen Insel im Indischen Ozean vom Altertum bis zur frühen Neuzeit, Wiesbaden 2006 (= Maritime Asia 17).

BLUMENBERG 1993 = Blumenberg, Hans, Schiffbruch mit Zuschauer. Paradigma einer Daseinsmetapher, Frankfurt am Main 1993.

BORSCHEID 2004 = Borscheid, Peter, Das Tempo-Virus. Eine Kulturgeschichte der Beschleunigung, Frankfurt am Main 2004.

BOUSSAC/SALLES/YON 2012 = Boussac, Marie-Françoise / Salles, Jean-François / Yon, Jean-Baptiste (Hrsg.), Autour du Périple de la mer Érythrée, Lyon 2012 (= Topoi, Supplément 11).

BRAUDEL 2006 = Braudel, Fernand, Das Meer, in: Ders. / Duby, Georges / Aymard, Maurice (Hrsg.), Die Welt des Mittelmeeres, Frankfurt am Main 2006, 37–60.
BUSCHAUER 2010 = Buschauer, Regine, Mobile Räume. Medien- und diskursgeschichtliche Studien zur Tele-Kommunikation, Bielefeld 2010.
CASSON 1980 = Casson, Lionel, Rome's Trade with India, in: TAPhA 110 (1980) 21–36.
CASSON 1989 = Casson, Lionel, The Periplus Maris Erythraei. Text with Introduction, Translation, and Commentary, Princeton 1989.
CASSON 1991a = Casson, Lionel, The Ancient Mariners. Seafarers and Sea Fighters of the Mediterranean in Ancient Times, Princeton ^{2}1991.
CASSON 1991b = Casson, Lionel, Ancient Naval Technology and the Route to India, in: Begley, Vimala / Puma, Richard D. de (Hrsg.), Rome and India. The Ancient Sea Trade, Madison 1991 (= Wisconsin Studies in Classics), 8–12.
CASSON 1994 = Casson, Lionel, New Light on Maritime Loans: P. Vindob. G 40822, in: Zeitschrift für Papyrologie und Epigraphik 89 (1994) 195–206.
CHITTICK 1976 = Chittick, Neville, An Archaeological Reconnaissance in the Horn: The British Somali-Expedition 1975, in: Azania 11 (1976) 117–133.
CHITTICK 1980 = Chittick, Neville, Pre-Islamic Trade and Ports of the Horn, in: Leakey, Richard E. / Ogot, Bethwell A. (Hrsg.), Proceedings of the 8th Panafrican Congress of Prehistory and Quaternary Studies: Nairobi, 5 to 10 September 1977, Nairobi 1980, 364–366.
CONRAD 2013 = Conrad, Sebastian, Globalgeschichte. Eine Einführung, München 2013.
CRONE 2004 = Crone, Patricia, Meccan Trade and the Rise of Islam, Princeton 2004.
DICKEN 2011 = Dicken, Paul, Global Swift. Mapping the Changing Contours of the World Economy, London/Thousand Oaks//New Delhi ^{6}2011.
DIHLE 1978 = Dihle, Albrecht, Die entdeckungsgeschichtlichen Voraussetzungen des Indienhandels der römischen Kaiserzeit, in: ANRW II 9/2 (1978) 546–580.
DREXHAGE/KONEN/RUFFING 2002 = Drexhage, Hans-Joachim / Konen, Heinrich / Ruffing, Kai, Die Wirtschaft des Römischen Reiches (1.–3. Jahrhundert). Eine Einführung, Berlin 2002 (= Studienbücher Geschichte und Kultur der Alten Welt).
ECKERT 2012 = Eckert, Andreas, Globalisierung, in: Boer, Pim den et al. (Hrsg.), Europäische Erinnerungsorte, Band 3: Europa und die Welt, München 2012, 11–18.
FÄSSLER 2007 = Fäßler, Peter E., Globalisierung, Köln 2007.
GIESE/MOSSIG/SCHRÖDER 2011 = Giese, Ernst / Mossig, Ivo / Schröder, Heike, Globalisierung der Wirtschaft, Paderborn 2011.
GUGERLI 1999 = Gugerli, David, Gab es ein viktorianisches Internet? Süffige Anekdotensammlung zur Geschichte der Telegrafie, in: Neue Zürcher Zeitung Nr. 234, 8. Oktober 1999, 75.
HARRAUER/SIJPESTEIJN 1985 = Harrauer, Herrmann / Sijpesteijn, Pieter J., Ein neues Dokument zu Roms Indienhandel (P. Vindob. G 40822), in: Anzeiger der Österreichischen Akademie der Wissenschaften, Phil.-Hist. Kl. 122 (1985) 124–155.
HARVEY 1989 = Harvey, David, The Condition of Postmodernity. An Enquiry into the Origins of Cultural Change, Oxford 1989.
HÖCKMANN 1985 = Höckmann, Olaf, Antike Seefahrt, München 1985 (= Beck's Archäologische Bibliothek).
HOLTORF 2001 = Holtorf, Christian, Der erste Draht zur Neuen Welt – Eine Archäologie der Zukunft, in: Matejovski, Dirk / Kamper, Dietmar / Weniger, Gerd-Christian (Hrsg.), Mythos Neandertal. Ursprung und Zeitenwende, Frankfurt am Main/New York 2001, 86–97.
HOLTORF 2005 = Holtorf, Christian, Die Modernisierung des nordatlantischen Raumes. Cyrus Field, Taliaferro Shaffner und das submarine Telegraphennetz von 1858, in: Geppert, Alexander C. T. / Jensen, Uffa / Weinhold, Jörn (Hrsg.), Ortsgespräche. Raum und Kommunikation im 19. und 20. Jahrhundert, Bielefeld 2005, 157–178.
HOLTORF 2013 = Holtorf, Christian, Der erste Draht zur neuen Welt. Die Verlegung des ersten transatlantischen Telegrafenkabels, Göttingen 2013.
HOURANI 1995 = Hourani, George F., Arab Seafaring in the Indian Ocean in Ancient and Early Medieval Times, Princeton ^{2}1995.

KESSLER 2009 = Kessler, Johannes, Der Mythos vom globalen Dorf: Zur räumlichen Differenzierung des Globalisierungsniveaus, in: Ders. / Steiner, Christian (Hrsg.), Facetten der Globalisierung. Zwischen Ökonomie, Politik und Kultur, Wiesbaden 2009, 28–79.

KLOFT 2007 = Kloft, Hans, Luxuria. Begriff und Sache in der römischen Kaiserzeit, in: Aßkamp et al. 2007, 65–75.

KNOCH/MORAT 2003 = Knoch, Habbo / Morat, Daniel, Medienbilder und Gesellschaftsbilder 1880–1960, in: Dies. (Hrsg.), Kommunikation als Beobachtung. Medienwandel und Gesellschaftsbilder zwischen 1880–1960, München 2003, 9–33.

KOMLOSY 2011 = Komlosy, Andrea, Globalgeschichte. Methoden und Theorien, Wien 2011.

KULKE 2005 = Kulke, Hermann, Indische Geschichte bis 1750, München 2005.

KULKE/ROTHERMUND 2010 = Kulke, Hermann / Rothermund, Dietmar, Geschichte Indiens. Von der Induskultur bis heute, München ^{2}2010.

LIPPUNER 2005 = Lippuner, Roland, Raum, Systeme, Praktiken. Zum Verhältnis von Alltag, Wissenschaft und Geographie, Stuttgart 2005 (= Sozialgeographische Bibliothek 2).

LIPPUNER 2008 = Lippuner, Roland, Körper, Raum und mediale Repräsentation, in: Döring, Jörg / Thielmann, Tristan (Hrsg.), Spatial Turn. Das Raumparadigma in den Kultur- und Sozialwissenschaften, Bielefeld 2008, 365–392.

MALITZ 2000 = Malitz, Jürgen, Globalisierung? Einheitlichkeit und Vielfalt des Imperium Romanum, in: Schreiber, Waltraut (Hrsg.), Vom Imperium Romanum zum Global Village. „Globalisierungen" im Spiegel der Geschichte, Neuried 2000 (= Eichstätter Kontaktstudium zum Geschichtsunterricht 1), 37–52.

MARCOTTE 2011 = Marcotte, Didier, Géopolitique de l'océan Indien au début de l'Empire, in: Geographia antiqua 20 (2011) 13–24.

MCLAUGHLIN 2010 = McLaughlin, Raoul, Rome and the Distant East. Trade Routes to the Ancient Lands of Arabia, India and China, London/New York 2010.

MEYER 1992 = Meyer, Carol, Glass from Quseir al-Qadim and the Indian Ocean Trade, Chicago 1992 (= Studies in Ancient Oriental Civilizations 53).

MOORTHY/JAMAL 2009 = Moorthy, Shanti / Jamal, Ashraf (Hrsg.), Indian Ocean Studies: Cultural, Social, and Political Perspectives, New York 2009 (= Routledge Indian Ocean Series).

MORELLI 2011 = Morelli, Federico, Dal Mar Rosso a Alessandria: Il *verso* (ma anche il *recto*) del 'papiro di Muziris' (SB XVIII 13167), in: Tyche 26 (2011) 199–233.

MORLEY 2015 = Morley, Neville, Globalisation and the Roman Economy, in: Pitts/Versluys 2015, 49–68.

MURPHY 2004 = Murphy, Trevor, Pliny the Elder's Natural History. The Empire in the Encyclopedia, Oxford 2004.

NEUGEBAUER 2007 = Neugebauer, Christian: http://www.european-spaces.eu/2007/03/15/harvey-david-1990-zeit-und-raum-im-projekt-der-aufklaerung/ (zuletzt aufgerufen am 3.1.2015).

OLSHAUSEN/SONNABEND 2002 = Olshausen, Eckart / Sonnabend, Holger (Hrsg.), Zu Wasser und zu Land. Verkehrswege in der antiken Welt. Kolloquien zur Historischen Geographie des Altertums 7, 1999, Stuttgart 2002 (= Geographica Historica 17).

OSTERHAMMEL 2009 = Osterhammel, Jürgen, Die Verwandlung der Welt. Eine Geschichte des 19. Jahrhunderts, München 2009.

OSTERHAMMEL/PETERSSON 2012 = Osterhammel, Jürgen / Petersson, Niels P., Geschichte der Globalisierung. Dimensionen, Prozesse, Epochen, München ^{5}2012.

PITTS 2015 = Pitts, Martin, Globalisation, Circulation and Mass Consumption in the Roman World, in: Pitts/Versluys 2015, 69–98.

PITTS/VERSLUYS 2015 = Pitts, Martin / Versluys, Miguel J. (Hrsg.), Globalisation and the Roman World. World History, Connectivity and Material Culture, New York 2015.

RASCHKE 1978 = Raschke, Manfred G., New Studies in Roman Commerce with the East, in: ANRW II 9/2 (1978) 604–1361.

RATHBONE 2000 = Rathbone, Dominic W., The "Muziris" Papyrus (SB XVIII 13167). Financing Roman Trade with India in: Bulletin Sociéte Archéologique d'Alexandrie 46 (2000) 39–50.

Ray 2003 = Ray, Himanshu P., The Archaeology of Seafaring in Ancient South Asia, Cambridge 2003.
Robertson 1990 = Robertson, Roland, Mapping the Global Condition: Globalization as the Central Concept. In: Featherstone, Mike (Hrsg.), Global Culture, London 1990, 15–31.
Romanis 2012 = Romanis, Federico de, Playing Sudoku on the Verso of the 'Muziris Papyrus': Pepper, Malabathron and Tortoise Shell in the Cargo of Hermapollon, in: Journal of Ancient Indian History 27 (2012) 75–101.
Romanis 2014a = Romanis, Federico de, Time to repay a Maritime Loan: A Note on SB III 7169 and SB XVIII 13167 Recto, in: Sileno 40 (2014) 73–89.
Romanis 2014b = Romanis, Federico de, Ivory from Muziris, in: ISAW Papers 8 (2014), http://dlib.nyu.edu/awdl/isaw/isaw-papers/8/ (zuletzt aufgerufen am 3.1.2015).
Romanis 2014c = Romanis, Federico de, Structural Aspects of a Commercial Enterprise to Muziris (on SB XVIII 13167 again), in: Topoi Suppl. 13 (2014), im Druck.
Rougé 1988 = Rougé, Jean, La navigation en mer Érythrée dans l'Antiquité, in: Salles, Jean-François (Hrsg.), L'Arabie et ses mers bordières. Itinéraires et voisinages, Lyon 1988 (= Travaux de la Maison de l'Orient 16), 59–74.
Ruffing 2002 = Ruffing, Kai, Wege in den Osten. Die Routen des römischen Süd- und Osthandels (1. bis 2. Jahrhundert n. Chr.), in: Olshausen/Sonnabend 2002, 360–378.
Salles 1988 = Salles, Jean-François, La circumnavigation de l'Arabie dans l'Antiquité, in: Ders. (Hrsg.), L'Arabie et ses mers bordières. Itinéraires et voisinages, Lyon 1988 (= Travaux de la Maison de l'Orient 16), 75–102.
Salles 1993 = Salles, Jean-François, The Periplus of the Erythraean Sea and the Arab-Persian Gulf, in: TOPOI 3/2 (1993) 493–523, hier 506–508 = Ders., in: Boussac, Marie-Françoise / Salles, Jean-François (Hrsg.), Athens, Aden, Arikamedu. Essays on the Interrelations between India, Arabia, and the Eastern Mediterranean, New Delhi ²1995 (ND New Delhi 2005), 115–146.
Salles 1996 = Salles, Jean-François, Achaemenid and Hellenistic Trade in the Indian Ocean, in: Reade, Julian (Hrsg.), The Indian Ocean in Antiquity, London 1996, 251–267.
Salles/Sedov 2010 = Salles, Jean-François / Sedov, Alexander V., Qānī': Le Port Antique du Ḥaḍramawt entre la Méditerranée, L'Afrique et l'Inde. Fouilles Russes 1972, 1985–1989, 1991, 1993–1994, Turnhout 2010 (= Indicopleustoi 6).
Schäfer 2003 = Schäfer, Wolf, The New Global History. Toward a Narrative for Pangea Two, in: Erwägen – Wissen – Ethik 14 (2003) 75–88.
Schuol 2000 = Schuol, Monika, Die Charakene. Ein mesopotamisches Königreich in hellenistisch-parthischer Zeit, Stuttgart 2000 (= Oriens et Occidens 1).
Seland 2008 = Seland, Eivind H., The Indian Ocean and the Globalisation of the Ancient World, in: West and East 7 (2008) 67–79.
Seland 2010 = Seland, Eivind H., Ports and Political Power in the Periplus. Complex Societies and Maritime Trade on the Indian Ocean in the First Century AD, Oxford 2010 (= BAR International Series 2102/Society for Arabian Studies Monographs 9).
Seland 2013 = Seland, Eivind H., Network and Social Cohesion in Ancient Indian Ocean Trade: Geography, Ethnicity, Religion, in: Journal of Global History 8 (2013) 373–389.
Seland 2014 = Seland, Eivind H., Archaeology of Trade in the Western Indian Ocean, 300 BC–AD 700, in: Journal of Archaeological Research 22 (2014) 367–402.
Sheriff 2010 = Sheriff, Abdul, Dhow Cultures and the Indian Ocean: Cosmopolitanism, Commerce and Islam before the Coming of the Europeans, New York 2010.
Sidebotham 2011 = Sidebotham, Steven E., Berenike and the Ancient Maritime Spice Route, Berkeley / Los Angeles 2011.
Smith/Wright 1988 = Smith, Matthew C. / Wright, Henry T., The Ceramics from Ras Hafun in Somalia. Notes on a Classical Maritime Site, in: Archaeological Research in Africa 23/1 (1988) 115–141.
Standage 1999 = Standage, Tom, Das viktorianische Internet. Die erstaunliche Geschichte des Telegraphen und der Online-Pioniere des 19. Jahrhunderts, St. Gallen / Zürich 1999.

Stern 1993 = Stern, Eva M., The Glass from Heis, in: Desanges, Jehan / Stern, Eva M. / Ballet, Pascale (Hrsg.), Sur les routes antiques de l'Azanie et de l'Inde. Le Fonds Révoil du Musée de l'Homme (Heïs et Damo, en Somalie), Paris 1993 (= Mémoires de l'Académie de l'Inscriptions et Belles-Lettres, Nouvelle Série 13), 21–61.

Thompson 1995 = Thompson, John B., The Media and Modernity, Cambridge 1995.

Thür 1987 = Thür, Gerhard, Hypotheken-Urkunde eines Seedarlehens für eine Reise nach Muziris und Apographe für die Tetarte von Alexandreia (P. Vindob. G 40822), in: Tyche 2 (1987) 229–245.

Thür 1988 = Thür, Gerhard, Zum κατὰ Μουζεῖριν Seedarlehen: P. Vindob. G 40822, in: Tyche 3 (1988) 229–233.

Tomber 2008 = Tomber, Roberta, Indo-Roman Trade. From Pots to Pepper, London 2008 (= Duckworth Debates in Archaeology).

Torp 2008 = Torp, Cornelius, Weltgesellschaft und Weltereignis. Bemerkungen aus historischer Perspektive, in: Nacke, Stefan / Unkelbach, René / Werron, Tobias (Hrsg.), Weltereignis. Theoretische und empirische Perspektiven, Wiesbaden 2008, 41–59.

Wagner-Hasel 2002 = Wagner-Hasel, Beate, Verschwendung und Politik in Rom. Überlegungen zur politischen Semantik des Luxuskonsums in der späten Republik und frühen Kaiserzeit, in: Historische Anthropologie 10 (2002) 325–253.

Warmington 1974 = Warmington, Eric H., The Commerce between the Roman Empire and India, London ²1974.

Weeber 2003 = Weeber, Karl-Wilhelm, Die Schwelgerei, das süße Gift … Luxus im alten Rom, Darmstadt 2003.

Weeber 2007 = Weeber, Karl-Wilhelm, Luxuria, das „süße Gift“, in: Aßkamp et al. 2007, 3–15.

Werlen 2014 = Werlen, Benno, Globalisierung, in: Lossau, Julia / Freytag, Tim / Lippuner, Roland (Hrsg.), Schlüsselbegriffe der Kultur- und Sozialgeographie, Stuttgart 2014, 54–66.

Werlen 2000 = Werlen, Benno, Sozialgeographie. Eine Einführung, Bern 2000.

Whittaker 2004 = Whittaker, Charles R., Indian Trade within the Roman Imperial Network, in: Ders., Rome and its Frontiers. The Dynamics of Empire, London / New York 2004, 163–180 = Le commerce romain avec L'Inde et la prise de décision économique, in: Topoi 10 (2000) 267–288.

Wiarda 2014 = Wiarda, Howard J., On the Boundaries: When International Relations, Comparative Politics, and Foreign Policy meet, Lamham 2014.

Wiesehöfer 2007 = Wiesehöfer, Josef, Alte Geschichte und Alter Orient, oder: Ein Plädoyer für Universalgeschichte, in: Rollinger, Robert / Luther, Andreas / Wiesehöfer, Josef (Hrsg.), Getrennte Wege? Kommunikation, Raum und Wahrnehmung in der Alten Welt, Frankfurt am Main 2007 (= Oikumene 2), 595–616.

Wiesehöfer 2013 = Wiesehöfer, Josef, Anstelle eines Nachwortes: Methodische und rezeptionsgeschichtliche Überlegungen zu ‚Orient-Okzident-Beziehungen‘ in der Antike, in: Zenzen, Nicolas / Hölscher, Tonio / Trampedach, Kai (Hrsg.), Aneignung und Abgrenzung. Wechselnde Perspektiven auf die Antithese von ‘Ost’ und ‘West’ in der griechischen Antike, Heidelberg 2013 (= Oikumene 10), 485–507.

Wilkinson o. J. = Wilkinson, J. T., Indian Ocean: Cradle of Globalization. Scholar Voices, in: http://www.accd.edu/sac/history/keller/IndianO/Wilkin.html und http://ccat.sas.upenn.edu/indianocean/index.html (zuletzt aufgerufen am 3.1.2015).

Young 2001 = Young, Gary K., Rome's Eastern Trade. International Commerce and Imperial Policy, 31 BC–AD 305, London 2001.

BY THE WATERS OF PHAZIMON – A NOTE ON THE RESEARCH HISTORY OF *STUDIA PONTICA III* 24

Søren Lund Sørensen

Introduction

A well-known Greek inscription from Havza in Turkey (Samsun province) contains a dedication to the god Asklepios by Titus Plankis Peiso, who had been cured of some unknown ailment. Although Johann Heinrich Mordtmann (1852–1932) had published the inscription already in 1884,[1] its rediscovery by the renowned Scottish scholar of Asia Minor John George Clark Anderson (1870–1952) and the subsequent publication in *Studia Pontica III* brought the inscription to the attention of a wider audience.

The inscription (fig. 1) is built into the outside wall of the *imaret* in Havza and is visible even today. Havza lies over the site of 'the hot springs of the Phazimonitai' mentioned by Strabon.[2] The text is carved inside a circular field of a stone measuring 60×33 cm. Only the right half of the stone is preserved, leaving scope for several restorations. The text has been revised and improved since its publication in *Studia Pontica III*,[3] but it is the editions by Anderson and Franz Cumont (1868–1947) that are the focus of this paper.

Text

In *Studia Pontica III* 24 the restored text reads as follows:

ἀγαθῇ
τ]ύχῃ·
[Τίτος(?) Πλά]νκις Πείσων
[πορευ]θεὶς τὴν κατα-
[ντικρὺ κ]ρήνην, σεμνῶς
[καὶ κοσμί]ως πλήσας τὸν χρό-
[νον δύο(?) ἐτ]ῶν, εὐχαριστῶ, τῷ
[σωτῆρι θε]ῷ Ἀσκληπιῷ κὲ ταῖς
[Νύμφαις, εὐ]χαριστῶ δὲ καὶ ἀ-
[ρχιάτρῳ Σωσ(?)]άνδρῳ καὶ τοῖα π-
[άθη καθαί]ροντι καὶ τῷ π-
[αναρίστῳ Σ]ειλίῳ Λεπίδῳ
[διοικοῦντ]ι ἄνακτος Ἀσκληπ-

1 Mordtmann 1884, 48 no. 12.
2 Strab. 12.3.38 (560C): θερμὰ ὕδατα τῶν Φαζημονιτῶν.
3 SEG 13 526, Bean 1953, 172–175 no. 11; Olshausen, Biller 1984, 83; Merkelbach, Stauber 2001, 357–359 no. 11/04/01.

[ιοῦ λουτ]ρόν. Πείσων ἐγὼ
[σωθεὶς στείχ]ω ζαθέην ἐπὶ πά-
[τρην ἐν κώμ]ῃ πλήσας γερα-
[ρᾷ δισσὸ]ν λυκάβαντα.
[ἤδη δ'(?)] ἡμετέρες[4]
[ὄλβο]ς ζώοιτο
[γεν]έθλις

Good luck!
I, Titus Plankis Peiso, who came straight down to the spring and spent two years there in a solemn and orderly manner, give thanks to the saviour god Asklepios and the Nymphs. I also give thanks to the official doctor Sosandros, a curer of such illnesses, and to the best of all, Seilius Lepidus, the manager of the hot baths of lord Asklepios. Having been cured, I Peiso go (back) to my sacred fatherland having spent two years in this honourable village. May our family now be blessed![5]

From the many square brackets in the Greek text it is obvious that most of the conclusions drawn from the text are based on restorations. We are not sure of the name of the dedicator, nor that of the official doctor, if he is indeed an official doctor (ἀρχιατρός), to mention only a few of the critical restorations. Other restorations are more plausible, such as Νύμφαις, based on the frequent coupling of nymphs with Asklepios in other inscriptions from Havza.[6]

In the notes accompanying the inscription, Cumont stressed that most of the restorations were the work of Anderson, but Cumont had also made a copy and taken a photo of the inscription.[7] In the archives of the Sackler Library in Oxford are kept the personal notebooks of Anderson, among which two are of special importance for the region of Pontos. In these two notebooks, which have been given the numbers 3 and 7, several of the inscriptions (as well as milestones) that make up the two fascicles of *Studia Pontica III*, of which the last still awaits publication, are recorded. In addition to these inscriptions, Anderson took many notes regarding the roads traversed, the days and hours spent in one place, etc. Many of these notes were incorporated in *Studia Pontica I*.[8]

In notebook 7 there are several notes pertaining to *Studia Pontica III* 24: a reconstruction of the text with notes (fig. 2), a photo of the inscription on the back of which is a reconstruction of the text, a drawing of the inscription with a reconstruction and a letter from a college fellow to Anderson dated June 5, 1903, which must have been inserted into the notebook after this date. On the back of this letter Anderson has made notes on the inscription. From Anderson's notebooks and *Studia Pontica I* we can gather that he saw the inscription sometime between August 8 and 10, 1899.[9]

4 The stone clearly reads ἡμετέρες. According to the editors of SP III, p. 38: 'Ἡμετέρες doit représenter ἡμετέραις et [γεν]έθλις ου γενέθλοις ου γενέθλιος.' Cf. Wright 1895, 59. γενέθλις is not attested in Greek literature, but the present author has, unlike the editors of SP III 24, taken the word to be similar to γενέθλη, meaning 'family.'

5 Author's translation.

6 SP III 25–27.

7 SP III, p. 37–38.

8 Many thanks are due to Dr Graham Piddock (Sackler Library, University of Oxford) for assisting me in my work on Anderson' notebooks.

9 SP I, 79–83.

Fig. 1: Studia Pontica 24 built into the wall of the imaret, Havza (author's photo).

The various documents pertaining to *Studia Pontica III* 24 in notebook 7 offer us a fascinating glimpse into the working methods of Anderson. Since the inscription from Havza is inscribed in a circular frame, Anderson made a drawing of the legible part of the stone and mirrored this part onto the illegible part thereby providing him with the maximum extent of the left part of the frame. Before settling for the reading printed in *Studia Pontica III* 24, in which we must expect that Cumont had a share, Anderson experimented with several variant readings. Many of these readings, which are mentioned in the notes to *Studia Pontica III* 24, are based on parallels found in Georg Kaibel's collection of epigrams.[10] There are, however, additional restorations which, taken together, provide a different version of the text from that printed in *Studia Pontica III* 24. Although Anderson, as well as most of us, will have been unenthusiastic about having his personal notes and scribbles published, the restorations he proposes are very interesting, and his provisional edition deserves to be printed here:

10 Kaibel 1878.

Text

Ἀγαθῇ
τ]ύχῃ·
[Τίτος(?) Πλά]νκις Πείσων
[εἰσαφιχ]θεὶς τὴν κατὰ
[Φαζίμονα[11] κ]ρήνην, σεμνῶς
[καὶ κοσμί]ως[12] πλήσας τὸν χρό-
[νον ἐνιαυτ]ῶν, εὐχαριστῶ τῷ
[ἀνάκτι[13]] Ἀσκληπιῷ κὲ ταῖς
[Νύμφαις, ε]ὐχαριστῶ δὲ καὶ Ἀ-
[πόλλωνι[14]—]ΑΝΔΡΩ καὶ τοῖα π-
[άθη—]ροντι καὶ τῷ π-
[αναρίστῳ Α]ἰλίῳ Λεπίδῳ
[τῷ ἱερε]ῖ ἄνακτος Ἀσκληπ-
[ιοῦ λουτ]ρόν.[15] Πείσων ἐγὼ
[σωθεὶς ἵκ]ω ζαθέην ἐπὶ πά-
[τρην —]ΙΗ πλήσας γερα-
[ῖως χρόνο]ν[16] λυκάβαντα.
[εἰ παρ'] ἡμετέρες
[ὄλβο]ς ζώοιτο
[γεν]έθλις.

Good luck! I, Titus Plankis Peiso, having arrived at the spring by Phazimon, spent the years here in a solemn and orderly manner. I give thanks to lord Asklepios and the Nymphs, but I also give thanks to Apollon … such illnesses … and to the best of all Aelius Lepidus, priest of the hot baths of lord Asklepios. I Peiso, having been healed, have come to my sacred fatherland after having fulfilled a period of time here with dignity. May our family be blessed!

Commentary

Overall, the meaning of the text is still the same, but the names are different, and no official doctor is attested. On a completely different level, however, is line 5, where Anderson among other things proposed to read Φαζίμονα, i.e. Phazimon. There does not seem to have been sufficient space for this conjecture, and among Anderson's notes we also find the restoration that found its way into *Studia Pontica III* 2. Still, what reasons should we think made Anderson come up with the conjecture Φαζίμονα? When describing Havza in *Studia Pontica I* Anderson mentions Strabon 12.3.38 (560C), where the Pontic historian describes the hot springs of the Phazimonitai. These lay within the Phazimonitis, i.e. the territory of Phazimon, near which Pompey the Great in 64 BC founded a Greek city under the name Neapolis, later Neoklaudiopolis-Andrapa. At the same time, Pompey also changed the name of Phazimonitis to Neapolitis. Although the identification of Havza with the

11 Anderson also proposes: κατ' ἀντικρύ/κατ' Ἀσκληπιεῖον/Νεάπολιν.
12 Anderson also proposes: ἀμεμπτῶς.
13 Anderson proposes: θεῷ σωτῆρι.
14 SP III, p. 38.
15 Anderson proposes: ἱε]ρόν.
16 Anderson also proposes: ἑκατό]ν.

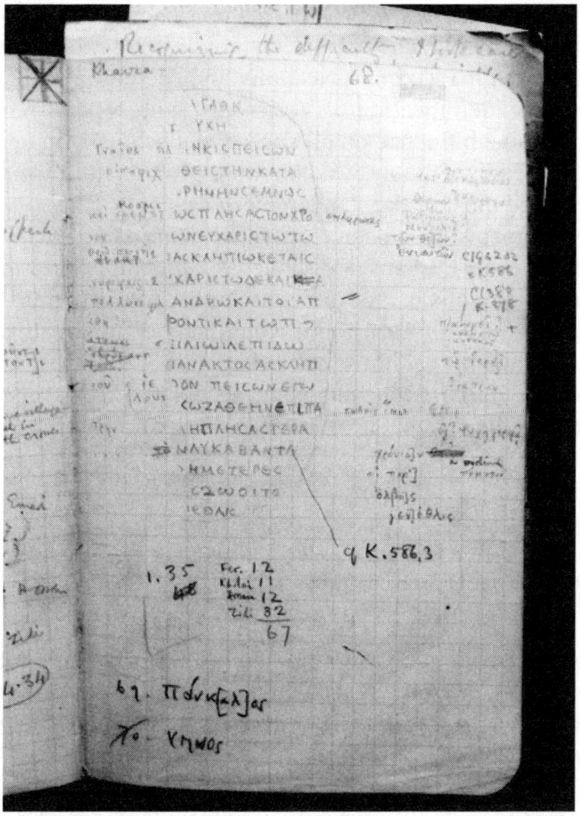

Fig. 2: Notebook 7: Reconstruction of the text with notes (author's photo. By permission of the Sackler Library, University of Oxford).

hot springs of the Phazimonitai had seemed natural on the basis of the hot springs that are still in use there today, the modern counterpart of the Pompeian foundation Neapolis-Neoklaudiopolis-Andrapa was only agreed upon in the years of Anderson's and Cumont's travels in Pontos.[17] Traversing Pontos in 1899 Anderson and John Arthur Ruskin Munro (1864–1944) were looking for Neoklaudiopolis-Andrapa, which they assumed, on the basis of Heinrich Kiepert (1818–1899), was identical to İskilip.[18] Secondly, Anderson supposed that Phazimon-Neapolis was to be found in the town of Vezirköprü. In İskilip Anderson and Munro, however, found no traces of Neoklaudiopolis. In April Anderson found an inscription in Vezirköpru mentioning 'the council and people of Neoklaudiopolis,' thereby identifying Neoklaudiopolis with Vezirköprü.[19] One year later Cumont similarly came to Vezirköprü and was shown an unpublished inscription containing an oath of allegiance

17 Bekker-Nielsen 2013, 11–14.

18 Kiepert 1884, 57; Anderson 1900, 152; SP I, 4–5; 91–92; Bekker–Nielsen 2013, 10.

19 Anderson 1900, 152 no. 1 = SP III 67 = IGR 3 139.

taken to the emperor Augustus in 3 BC. This text mentions the Phazimonitai and contains the syllable λιν, which Cumont restored as [Νεάπο]λιν, enabling him to identify Vezirköprü with Phazimon-Neapolis and thereby also with Neoklaudiopolis-Andrapa.[20] The identification of Vezirköprü with Neapolis and Neoklaudiopolis has had a massive influence on all subsequent scholarship on Pontos, but the inscription discovered by Cumont was the more prominent of the two documents, and it quickly entered most of the standard corpora.

No further documents have been discovered mentioning Neapolis, but in 1990 Gerhard Kahl and Eckart Olshausen discovered yet another inscription containing the name of Neoklaudiopolis in the Kurşunlu Camii in Vezirköprü.[21] The identification of Neoklaudiopolis with Vezirköprü would, one could argue, have been made anyway, and putting it a bit pessimistically, Anderson's discovery can be dispensed with, unlike that of Cumont. In the published account of his travels, *Studia Pontica I*, Anderson clearly indicated that he was not looking for Neoklaudiopolis but rather Neapolis:

> 'For our own part, we were so confident about the correctness of this identification that we eagerly scanned each new inscription in support of it; and when at last we found one (no. 67) with the promising formula ἡ βουλὴ καὶ ὁ δῆμος followed by Νε..., we rushed to the conclusion Νεαπολειτῶν and let our hopes be dashed!'[22]

The quotation from *Studia Pontica I* can hardly be interpreted in any other way than that Anderson was disappointed at finding an inscription mentioning Neoklaudiopolis and not Neapolis, even though this helped identify the two cities as one and the same. Likewise, we may guess that Anderson was somewhat dissatisfied at not having come upon the inscription containing the imperial oath, when he visited Vezirköprü.

When Anderson identified Vezirköprü with Neoklaudiopolis-Andrapa, he assumed that Havza would be the site of Phazimon-Neapolis.[23] This is perhaps where we should look for a reason for the restoration τὴν κατὰ [Φαζίμονα κ]ρήνην. Not only would the reference to Phazimon have been in accord with Strabon's reference to Havza as θερμὰ ὕδατα τῶν Φαζημονιτῶν, it would also have provided Anderson with an identification of Phazimon-Neapolis with Havza. In addition, from the notes made by Anderson we learn that he did in fact also propose to read Νεάπολιν. Whichever reading, it would have made this inscription more prominent, on a par with Cumont's famous discovery.

20 Cumont 1900; Cumont 1901 = SP III 66 = OGIS 532 = IGR 3 137 = ILS 8781.

21 Kahl 1995; Kahl, Olshausen 1991.

22 SP I, 91.

23 SP I, 92: 'Surely, then, Neoclaudiopolis with its native title Andrapa is a different place from Neapolis with its native name Phazimon, and the latter must be placed elsewhere?' Similarly, Munro 1901, 61: 'The frequented hot baths of Khavsa are obviously the θερμὰ ὕδατα τῶν Φαζημωνιτῶν noticed by Strabo. The expression seems at first sight to separate them from the town, but Phazemon, and therefore Neapolis, now appears to have been actually at the springs.'

Conclusion

As mentioned above, there is not sufficient space for the restoration Φαζίμονα, and Anderson realized this himself, as can be gathered from his notebooks. Eventually, Anderson and Cumont, who had also seen and copied the inscription in Havza, settled upon the reading found in *Studia Pontica III* 24, and the springs of Phazimon were abandoned. Nonetheless, the reconstructions proposed by Anderson in his notebooks attest to the years around the turn of the last century when a consensus view as to the location of Neapolis-Neoklaudiopolis-Andrapa was slowly emerging, and when every inscription was carefully scanned to see whether it provided useful information about the geography of the Pompeian settlements in northern Anatolia.

Abbreviations

IGR 3: Inscriptiones Graecae ad res Romanas pertinentes. Tomus tertius edendum curavit R. Cagnat auxiliante G. Lafaye, Paris 1906.
ILS: Inscriptiones Latinae selectae. Volumen II, pars II edidit H. Dessau, Berlin 1906.
OGIS: Orientis Graeci inscriptiones selectae. Volumen alterum edidit W. Dittenberger, Leipzig 1905.
SEG: Supplementum epigraphicum Graecum edited by A. Chaniotis, T. Corsten, N. Papazarkadas, R.A. Tybout, Leiden.
SP I: J.G.C. Anderson, A journey of exploration in Pontus, Bruxelles 1903 (Studia Pontica I).
SP III: Recueil des inscriptions grecques et latines du Pont et de l'Arménie publiées par J.G.C. Anderson, F. Cumont, H. Grégoire (fascicule I), Bruxelles 1910 (Studia Pontica III).

Bibliography

Anderson 1900: J.G.C. Anderson, Pontica, in: Journal of Hellenic Studies 20, 151–158.
Bean 1953: G.E. Bean, Inscriptions from Pontus, in: Belleten 17, 167–178.
Bekker-Nielsen 2013: T. Bekker-Nielsen, 350 years of research on Neoklaudiopolis (Vezirköprü), in: Orbis Terrarum 11, 3–31.
Cumont 1900: F. Cumont, Inscription grecque de Vézir-Keupru dans l'ancienne Paphlagonie (Asie Mineure), in: Comptes-rendus de l'Académie des Inscriptions et des Belles-Lettres 44, 687–691.
Cumont 1901: F. Cumont, Un serment de fidélité à l'empereur Auguste, in: Revue des etudes grecques 14, 26–45.
Kaibel 1878: G. Kaibel, Epigrammata Graeca ex lapidibus conlecta, Berlin.
Kahl 1995: G. Kahl, Pontica I: Eine Inschrift aus Vezirköprü, in: Orbis Terrarum 1, 109–120.
Kahl, Olshausen 1991: G. Kahl – E. Olshausen, Bericht über die epigraphische und numismatische Landesaufname im Samsun ili 1990, in: Araştırma Sonuçları Toplantısı 1991, 611–616.
Kiepert 1984: H. Kiepert, Gegenbemerkungen zu der Abhandlung von Hrn. G. Hirschfeld über die Lage von Tavium, in: Sitzungsberichte der Berliner Akademie der Wissenschaften 1884, 47–57.
Merkelbach, Stauber 2001: R. Merkelbach – J. Stauber, Steinepigramme aus dem griechischen Osten. Band 2. Die Nordküste Kleinasiens (Marmarameer und Pontos), München.
Mordtmann 1884: J.H. Mordtmann, Επιγραφαί της επαρχίας του Πόντου, in: Ο εν Κωνσταντινούπολει Ελληνικός Φιλολογικός Σύλλογος 15 Παράρτημα, 44–49.
Munro 1901: J.A.R. Munro, Roads in Pontus, royal and Roman, in: Journal of Hellenic Studies 21, 52–66.

Olshausen, Biller 1984: E. Olshausen – J. Biller, Historisch-geographische Aspekte der Geschichte des Pontischen und Armenischen Reiches, Wiesbaden.
Wright 1895: J. H. Wright, A votive tablet to Artemis Anaïtis and Mên Tiamu in the Boston Museum of Fine Arts, in: Harvard Studies in Classical Philology 6, 55–74.

THE LOCATION OF *NOVAESIUM:* A NEW INTERPRETATION OF PTOLEMY'S COORDINATES[1]

Irina Tupikova, Klaus Geus

ABSTRACT

We propose a new explanation for the misplacement of *Novaesium*/Neuss, which Ptolemy listed among the German cities on the east bank of the Rhine in his *Geography*. Our approach is based on two arguments: First, Ptolemy's main error originated in a wrong calculation of Earth's circumference (probably taken over from Posidonius); second, Ptolemy found no information about the position of *Novaesium* in relation to *Colonia Agrippinensis* / Cologne in this source material. We suggest that Ptolemy placed *Novaesium* on his map through linkage with places in Gaul along the old route connecting *Burdigala*/Bordeaux with *Germania magna*. Hence, after recalculating the coordinates to Earth's size measured by Eratosthenes, the position of *Novaesium* matches very well the modern one of Neuss.

INTRODUCTION

The location of the city *Novaesium* in Ptolemy's *Geography* poses a serious problem. Ptolemy listed *Novaesium*, called Νουαίσιον or Ναυαίσιον in his *Geography* (2.11.29), under *Germania magna* in his catalogue – and hence on the right side of the Rhine. Also, according to his coordinates, *Novaesium* lies in the *Abnoba mountains*, which is considered commonly as the Schwarzwald. Modern scholars identify, with some notable exceptions,[2] ancient *Novaesium* with modern Neuss, despite the fact that it lies neither on the right side of the Rhine nor in the Schwarzwald.

This identification seems historically reasonable and can nowadays be called the *opinio communis*. Among the arguments, the linguistic similarity of the two names;[3] Tacitus' many references to *Novaesium* as a *castrum* in *Germania inferior*,[4] and archaeological findings near Neuss from Roman times.[5] Last but not least, the

1 We should thank Anca Dan (Paris), Hartmut Blum (Tübingen), and Søren Lund Sørensen (Berlin) for their valuable comments.

2 Mannert 1820, 463–4; Ukert 1843, 444; Patzig 1917, 29; Mehlis 1918, 110; Steche 1937, 161; contra, e. g., Hansen 1991, 578.

3 Krahe 1954, 126 (contemplating a personal name).

4 Tac. *hist*. 4.26.3, 33.3, 35.1. 2, 57.3, 62.1, 70.5, 77.2, 79.4; 5.22.1; cf. Itin. Ant. 255 (*Nevensio*), 370 (*Novaesiae*); Amm. 18.2.4; survey of the ancient sources in Franke 1936.

5 Chantraine et al. 1984.

discovery of a legionary camp by Constantin Koenen (1854–1929) in the suburbs of modern Neuss strongly suggests that the city the Romans knew by the name *Novaesium* is nothing else than Neuss.[6]

But how to explain Ptolemy's gross error? Previous attempts have attributed *Novaesium*'s misplacement to the scanty knowledge of the Romans about the northern part of the *oikumene* in general, and Ptolemy's limited knowledge of *Germania magna* in particular. But given Ptolemy's huge amount of data for this region, such an answer is hardly satisfying. Being one of oldest and most important Roman military posts in the early Principate,[7] *Novaesium* must, in all probability, have been known to Ptolemy.

THE PROBLEM OUTLINED

Let us take a closer look at the problem.
The Ptolemaic coordinates of *Novaesium* are (*geogr*. 2.11.29):
latitude 51° 10' N, longitude 31°.[8]
The modern coordinates of Neuss are:
latitude 51° 12' N, longitude 6° 41' E.
We compare Ptolemy's location of Novaesium with that of Colonia Agrippinensis[9]/Cologne, the capital of Germania Inferior (geogr. 2.9.15). His coordinates for Cologne are:
latitude 51° 30' N, longitude 27° 40'.
The modern coordinates of Cologne are:
latitude 50° 56' N, longitude 6° 57' E.

By comparing these numbers, we see that the real longitudinal difference between Cologne and Neuss is about 16', whereas it amounts to the huge 3° 20' in Ptolemy's *Geography*.

Before we proceed to discuss the Ptolemaic positions, we should first underline the following point: Whereas Ptolemy gives the latitudes measured from the equator exactly as in our actual geographical coordinate system, the longitudes are measured from his "prime" meridian, the *Insulae Fortunatae* (i. e. the Canaries). Therefore, for comparing the modern geographical positions with their Ptolemaic counterparts, one needs to know the position of Ptolemy's prime meridian relative to the modern Greenwich meridian. We will discuss this problem later in this text. At this stage, it suffices to say that *Colonia Agrippinensis* / Cologne lies at its proper meridian. The relative positions of the other significant locations in the region,

6 See, e. g., STADT NEUSS 1989.

7 See, e. g., FORBIGER 1877, 179, n. 70; CAPELLE 1937, 449; SCHÖNING 1962, 131; STÜCKELBERGER/GRASSHOFF 2006, 233, n. 253. The article "Novaesium" by WIEGELS 2000 in "Der Neue Pauly" does not even bother to mention alternative identifications.

8 The X recension has 31° 30' which puts *Novaesium* even more to the east.

9 Throughout the text, we use the names as indicated in Ptolemy's *Geography* (ed. STÜCKELBERGER/GRASSHOFF 2006).

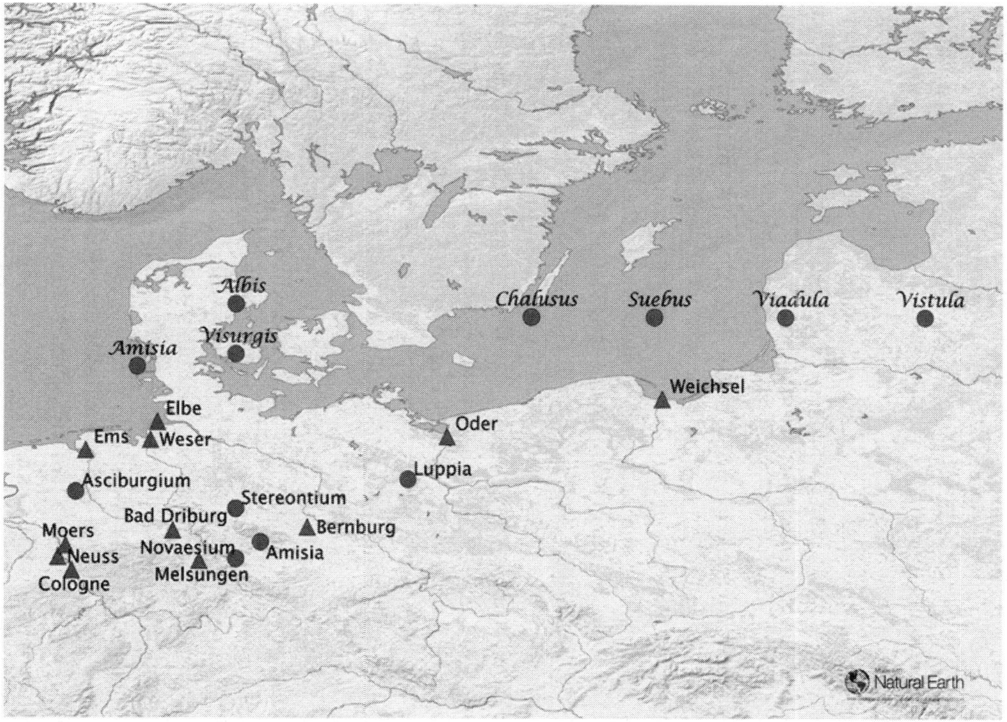

Fig. 1: Real positions (triangles) of Novaesium/Neuss, Colonia Agrippinensis / Cologne, and some selected locations in Germania magna in comparison with their Ptolemaic positions (grey circles). The position of the "Greenwich" meridian is defined by the identification of the position of Cologne with the meridian of Colonia Agrippinensis.

including the mouths of the rivers along the sea coastline, are defined accordingly relative to Cologne (see Fig. 1).

Being accustomed to the usual longitudinal distortions of Ptolemy's maps in other regions of the *oikumene*, nearly all modern historians have accepted the latitudinal precision of *Novaesium* while ignoring, or rather complaining about its misplacement at the "wrong" (i. e. eastern) side of river Rhine in Ptolemy's *Geography* as well as its eastern (instead of the almost northern) position in relation to Cologne.

Since Ptolemy made no conscious effort to link *Novaesium* with Cologne, the capital of *Germania inferior*, the case can be made that he found no reliable information about a connection between *Novaesium* and *Colonia Agrippinensis*. This would mean that he was either drawing on a source from a time when Neuss was not playing a major role in this region anymore (e. g. after the sack of *Novaesium* during the revolt of the Batavi and the subsequent dislocation of *legio VI*) or – since his ordering of cities in *Germania magna* according to climate is a unique feature of his *Geography* – that he was unable to match his different sources here.

In an alternative view, ancient *Novaesium* has nothing to do with modern Neuss. *Novaesium* might be a city in Ptolemy's catalogue which happens to have the same latitude as modern Neuss. Scholars of the 19th and early 20th century have guessed that *Novaesium* is "nordöstlich von Marburg, Schloß Nienhus an der Mönne bei Neheim" (Ukert 1843, 444; followed by Mehlis 1918, 110) or near Warburg at the Diemel (Steche 1937, 161) or at the Fulda banks (Mannert 1820, 463–4). This theory has not found much resonance with later researchers. But recently, a DFG research group on Ptolemy's *Geography* at the TU Berlin enunciated that by geodetical rectification of the Ptolemaic coordinates, *Novaesium* has to be identified with Nausis (a quarter of Spangenberg) or Melsungen in the Schwalm-Eder-Kreiss, Hessen), some 250 km off Neuss towards the east.[10] Since this approach claims to solve an age-old riddle once and for all with mathematical accuracy, it deserves a fuller evaluation.

The method of this group assumes that the distortion of Ptolemy's map can be "rectified". The basic approach can be described as such: First find some "transformation coefficients" which minimize the deviations of the coordinates for the correctly identified locations from their real positions within a certain region; then apply these coefficients to the coordinates of the other – yet unidentified – places. Last, put these coordinates on a modern map and try to make sense out of these "positions" from a historical perspective.

This procedure is quite simple and basically sound. But there is a two-fold problem with it: First, we have only a handful of surely identified locations in *Germania magna* (as the example of Neuss itself shows) which renders the whole mathematical algorithm quite fuzzy. In fact, such an approach can provide convincing results only for well-known regions with many identified places. For lesser-known regions with few identified locations this procedure has tight limits. In general, the fewer places are identified within a region, the more problematic is the result.

10 See the discussion of Kleineberg et al. 2010, 52. On p. 30 they name only Melsungen.

Second, the method ignores the fact that links or contacts (i. e. "roads" or "routes") existed not only between neighbouring cities but also to other places, which were more remote or even lying outside the regional maps (e. g. to important centres of the ancient world like Rome, Alexandria, or Marseille). This is not surprising since a geographer – as Ptolemy himself stated in his introduction – has to give an *overall* and *continuous* picture of the world.[11] Such global mapping introduces locations connected with other regions (with their own systematic errors) into the local maps and thus spoils the "transformation coefficients" for a respective region; the coordinates of these locations cannot be "rectified" within the same algorithm. What is even worse, the statistical methods may produce wrong results for such locations.

The case of Neuss seems to be a neat example of this *caveat*: The Ptolemaic coordinates in *Germania magna* show a significant northern digression; *Novaesium* identified as Neuss, to the contrary, lies at its proper latitude, thus supporting the impression that it was not linked to the other locations in this region (see Fig. 1).

RECALCULATION OF PTOLEMAIC POSITIONS

Nevertheless, we deem a mathematical approach combined with an interpretation of both the ancient sources and the numerical results from a historical perspective as a principally sound method. But we have to get rid of the statistical procedure which eliminates the "fingerprints" of Ptolemy's mapping method.

Our own approach is based on the assumption that many distortions of Ptolemy's maps are generated by his adoption of the wrong circumference of the earth, i. e. the 180,000 stades proposed by Posidonius and other ancient geographers instead of the 252,000 stades calculated by Eratosthenes.

The length of a stadion is highly debatable, but what is clear is that the Posidonian value is in every case too small: even with a length of a stadion of 185m, 180,000 would give 33,300 km for the circumference of the Earth whereas the actual value is about 40,000 km. Most scholars tend to think that this is not a big issue, but this is simply wrong. Ptolemy should recalculate his distance data in spherical coordinates and here the adopted circumference of the Earth is of principal importance.

Ptolemy converted terrestrial data ("distances") gained at Earth's surface in metrical units to angular units according to the rule "500 stadia equals 1°". If the circumference of the Earth does not equal 180,000 stadia (whatever the length of the Greek *stadion* was), the transformation rule should be different. On the "Eratosthenic" Earth , e. g., 1° would be equivalent to 700 stadia. That means that a measured distance of 700 stades would be represented as 1° in longitude or latitude at the "Eratosthenic" map and as 700:500 = 1.4° at the Ptolemaic map. In the same way, every distance measured in stades would be represented by Ptolemy on

11 Ptol. *geogr.* 1.1.2: Τῆς δὲ γεωγραφίας ἴδιόν ἐστι τὸ μίαν τε καὶ συνεχῆ δεικνύναι τὴν ἐγνωσμένην γῆν.

Earth's surface with an arc of the big circle which will be with the factor 1.4 bigger than that of Eratosthenes. This simple idea can explain the "overextension" of the Ptolemaic *oikumene* along the east-west direction. Since Ptolemy adhered, as far as possible, to astronomical data, especially to latitudes of important cities, he had to adjust the distances expressed in a far too big angular measure (degrees) to the known latitudinal circles of these cities thus placing them farther to the east or to the west as they should actually lie.

As a result of such an erroneous mapping, the Ptolemaic maps extend in general in every direction relative to each position he choose as the starting point of adjusting the known distances. Hence, the position of the modern Greenwich meridian relative to the Ptolemaic "prime" or "zero" meridian on Ptolemy's maps is not fixed. The longitudinal difference between the Greenwich meridian and the *Insulae Fortunatae* which should be calculated to compare the Ptolemaic positions with their modern counterparts must rather be determined *locally* relative to the selected reference point.

The main assumption of our algorithm is that the Ptolemaic distances were measured (or rather estimated) on a sphere whose circumference was approximately equal to 252,000 units and transferred to a sphere with the circumference of 180,000 units. Although our recalculations *do not use any metrical value of the stadion*, the impressive coincidence of the recalculated coordinates with their actual counterparts hints towards very good estimation of the circumference of the Earth by Eratosthenes.[12] Consequently, the estimation of the "Eratosthenic" stadion as 157.5m (see HULTSCH 1882, 61) seems to be very reasonable. This result is also confirmed by some recent statistical investigations (RUSSO 2012; GUCKELSBERGER 2013).

Our programs[13] allow us first to retrieve the original data, that is, the terrestrial measurements and directions – most probably, found in *itineraria* and *periploi* – from the Ptolemaic coordinates in stades, then to recalculate these data in order to gain the longitudes and latitudes Ptolemy *would have obtained* with the "Eratosthenic" size of the Earth. Finally, we can identify the ancient places listed in his catalogue with modern ones.[14]

We would like to stress that in contrast to the statistical geodetical method of KLEINEBERG et al. 2010, where the Ptolemaic positions were "improved" relative to selected identified locations to minimize the deviations, all the underlying errors and uncertainties of Ptolemy's data are, however, retained when employing our algorithm, where the original Ptolemaic coordinates are merely "transferred" to the surface of the "Eratosthenic" Earth. To make this point clear: our operation is not a *rectification* but a *recalculation*. Thus we do not sieve out "secondary" errors like scribal errors, wrong identifications by Ptolemy, inaccurate measurements etc.[15]

12 The other possible scenarios are discussed in TUPIKOVA/SCHEMMEL/GEUS 2014.

13 Our programs are written in MAPLE.

14 See TUPIKOVA/GEUS 2013 and TUPIKOVA 2014.

15 It is clear that these "secondary" errors leave valuable "fingerprints" for identifying the sources used by Ptolemy. For an example, see TUPIKOVA/GEUS 2013.

Looking for an appropriate reference point in the vicinity of *Novaesium*, we made *Colonia Agrippinensis* / Cologne our first choice. Thus, the longitude of the *Insulae Fortunatae* is given at our maps as the modern longitude of Cologne, 6° 57', minus the Ptolemaic longitude of Cologne, 27° 40', that is, 20° 43' W. Then the Ptolemaic longitude of *Novaesium*, e. g., is given in the modern system as 10° 17' (i. e. 31° – 20° 43'). To recalculate a position, we have constructed a spherical triangle with vertices at the chosen reference point (in this case: Cologne), at the North Pole and at the position of a location itself. The sides of such triangle are the arcs of the great circles at Earth's surface, representing the shortest distance between the vertices. Their values were firstly gained in angular value from the Ptolemaic coordinates and after that, recalculated to the angular value at the surface of the "Eratosthenic" Earth.

The directions and the lengths of the routes were (i) recalculated from the Ptolemaic coordinates for every pairing of *Colonia Agrippinensis* with a location, (ii) transferred to the actual position of Cologne (because they were in fact known relative to the true position of Cologne), and after that (iii) recalculated to the spherical geographical coordinates for a sphere with the circumference of 252,000 units.

The results of our recalculation are given in Fig. 2.[16] The Ptolemaic positions of some selected locations adjusted to the modern map in this way are given in Fig. 1. To demonstrate the precision of our algorithm, we have also recalculated some other places in *Germania magna* and displayed the results on the same map. Since most identifications in this area are hotly debated, we have limited us here to the positions of the rivers' mouths and to three cities – *Asciburgium* (modern Moers-Asberg),[17] *Amisia* (identified as Geismar at Fulda),[18] *Luppia* (identified as Bernburg an der Saale)[19] and *Stereontium* (identified as Bad Driburg)[20] – in the interior of *Germania magna.*

One can see that after our recalculation[21] to the "Eratosthenic" size of the Earth with Cologne as a reference point, the positions of the mouths of the German rivers "slide" towards their proper places at the East Sea coastline. Especially impressive is the precision of coordinates of *Vistula*/Weichsel's mouth. The bigger digression in the positions of the mouths of the other rivers can be explained by their linking with another reference point in *Germania magna*.[22]

16 Due to the character of Ptolemy's original data – e. g., known distance and/or known latitudes and/or known direction –, our programs use two sets of mathematical solutions for this region, which are discussed in Tupikova/Geus 2013 as Case 1 and Case 2.

17 See, e. g., Schöning 1962, 117–118.

18 Kleineberg et al. 2010, 46–47.

19 Kleineberg et al. 2010, 48. For a different view, see, e. g., Schöning 1962, 124: Ruppin.

20 Kleineberg et al. 2010, 46.

21 For mathematical details see Tupikova/Geus 2013.

22 On a side note: After the recalculation of Ptolemy's coordinates we have found the results for the North and East Sea coastlines (also in the east of the river *Vistula*/Weichsel) very impressive, too. The veracity and accuracy of the Roman knowledge about these regions outside the *Imperium Romanum* is simply amazing and must probably be explained with Roman military intelligence in the time of Augustus and Tiberius. See Geus/Tupikova 2013.

Also, the recalculated position of *Asciburgium* fits remarkably well with that of modern Moers (Asberg), its undisputed identification. In addition, the recalculated coordinates for *Luppia* coincides with Bernburg and those of *Stereontium* with Bad Driburg,[23] both already proposed as probable identifications by other scholars. But the identification of *Amisia* with Geismar[24] still seems questionable.[25]

In addition, Figure 2 shows our recalculated position of Neuss in comparison with the geographical position of the proposal for *Novaesium* = Melsungen. In contrast to the other cases, this identification digresses considerably from the results of our recalculation. Also, other proposals for the identification of *Novaesium*, e. g., Nausis bei Spangenberg (Patzig 1917, 29), Neheim (Mehlis 1918, 95), Warburg (Steche 1937, 161) – do not match the recalculated position of *Novaesium*; the deviations here are even larger.

In other words: all attempts to identify *Novaesium* with a place on the right side of the Rhine rivers go seriously astray.[26] So, how to explain these ambiguous results?

NEW ARGUMENTS FOR AN OLD IDENTIFICATION

The results of our previous investigation[27] show that due to an erroneously adopted size of the Earth, places with approximately known latitudes were placed by Ptolemy at their (approximately) proper latitudinal circles. But such places do necessarily exhibit an aberration of longitudinal difference relative to the reference point of mapping. This seems to be the case here. We suppose that the latitude of Neuss was well-known but its erroneous longitudinal distance (compared to Cologne) was due to the usage of *another* reference point in Ptolemaic mapping procedure. And by adjusting two local maps constructed relative to different reference points distortions are bound to happen.

To make this point clear: due to the different sources Ptolemy had at his disposal for drafting his maps, one should rather expect "layers" of different "maps"[28]

23 The coordinates of *Asciburgium*, *Stereontium* and *Novaesium* were obtained with the formulae of Case 1, the coordinates of *Luppia* and the mouths of the rivers of the East sea with the ones of Case 2 (see Tupikova 2014).

24 This was suggested by Kleineberg et al. 2010, 47. For a different view, see also Steche 1937, 161: "Der Ort muß in der Gegend der heutigen Städtchen Rietberg, Wiedenbrück oder Rheda gelegen haben"; Schöning 1962, 120: Hameln.

25 Amisia is surely Bad Ems. We will prove this with our mathematical approach in a forthcoming article.

26 The results of our recalculation do not only confirm the majority of the proposed identifications (within a tolerance interval), but also the amazingly high precision of the Ptolemaic distances data and of Eratosthenes' measurement of the Earth.

27 Tupikova/Geus 2013.

28 By "map" we do not necessarily mean a *graphical* representation of an area. Such maps, especially for certain regions, were rare in antiquity. Rather we mean a pattern or cluster of locations measured relatively to each other and/or to a common reference point. Thus, it can be said that ancient geographical texts like *itineraria* or *periploi* produce "maps".

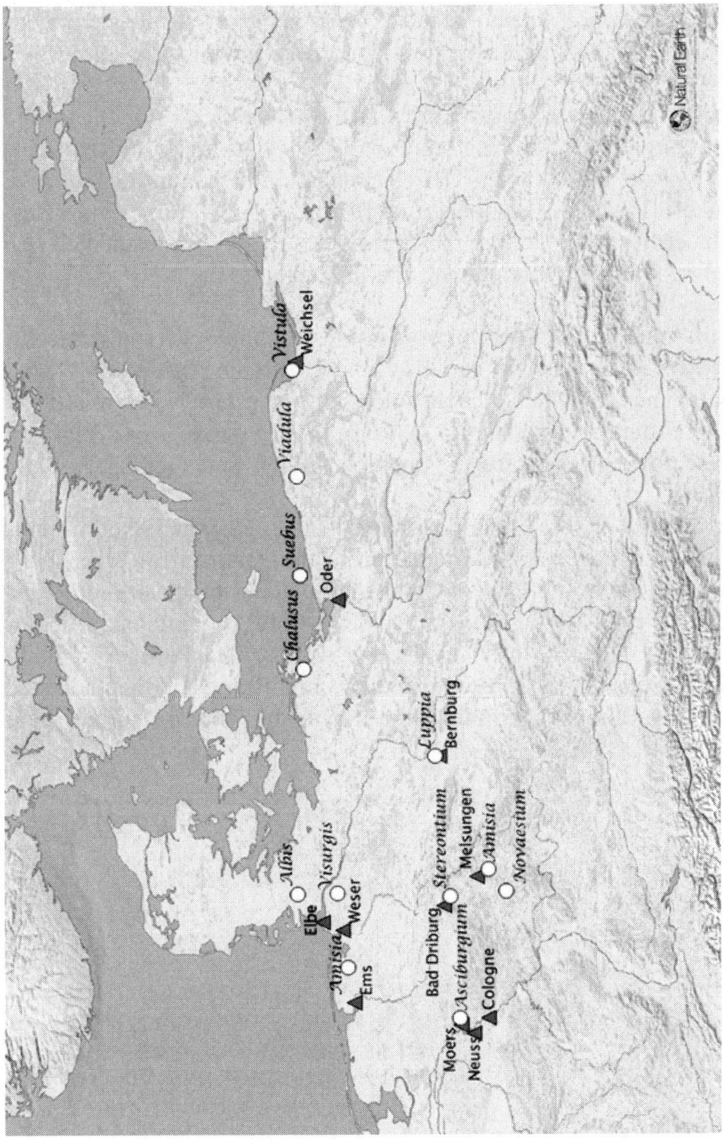

Fig. 2: Real positions (triangles) of Novaesium/Neuss, Colonia Agrippinensis/Cologne, and some selected locations in Germania magna in comparison with the recalculated Ptolemaic positions (white circles). The position of the "Greenwich" meridian is defined by the identification of the position of Cologne with the meridian of Colonia Agrippinensis.

(each connected with a selected reference point) than a single map. As shown, all these maps are adjusted differently to the Greenwich meridian. In fact, Fig. 1 is an overlay of at least two local maps. The crucial point in identifying locations in our procedure is the proper adjustment of the different recalculated Ptolemaic maps.

While it is obvious that most locations of *Germania magna* belong to a map with Cologne as reference point, the position of Neuss is clearly mapped from another source. Our quest for this reference point is motivated by historical consideration.

Our starting point was the well-known fact that *Novaesium* definitely lies at the crossroads of important routes from the south and the west. Since the legionary camp was founded by Roman troops c. 16 BC, a connection with important cities in Gaul seems very likely. By linking the position of *Novaesium* with *Burdigala*/Bordeaux we notice that, after recalculating the coordinates to the "Eratosthenic" size of the Earth, the position of the Ptolemaic *Novaesium* matches the modern position of Neuss pretty well.

Along this main route, at least two other locations, *Cenabum*/Orleans and *Augusta Viromanduorum* / Saint-Quentin, are also linked with the position of *Novaesium*; our recalculation has shown that the distances to *Novaesium*/Neuss relative to these cities was even better known (see Fig. 3). *Augusta Viromanduorum* seemingly served as a reference point for mapping the position of the Roman legionary camp at *Lacus Flevum*.[29]

As we have shown elsewhere,[30] the position of *Burdigala* was linked by Ptolemy directly with Rome. After our recalculation to the Eratosthenic size of the Earth relative to the capital of the Roman Empire, it matches the position of Bordeaux remarkable well. Fig 3 shows the Ptolemaic position of Burgidala in comparison with its recalculated position.[31] Hence we assume that Ptolemy initially mapped the position of *Burdigala*/Bordeaux in relation to Rome and then adjusted *Novaesium*/Neuss to the same map with the help of available *itineraria* along these places.

29 See Geus/Tupikova 2013 for details. We placed the camp at the mouth of Vlie which probably flowed into the sea through what is now the Vliestrom channel (before the formation of the Zuidersea).

30 Tupikova/Geus 2013.

31 The results of recalculations were obtained with the formulae of Case1.

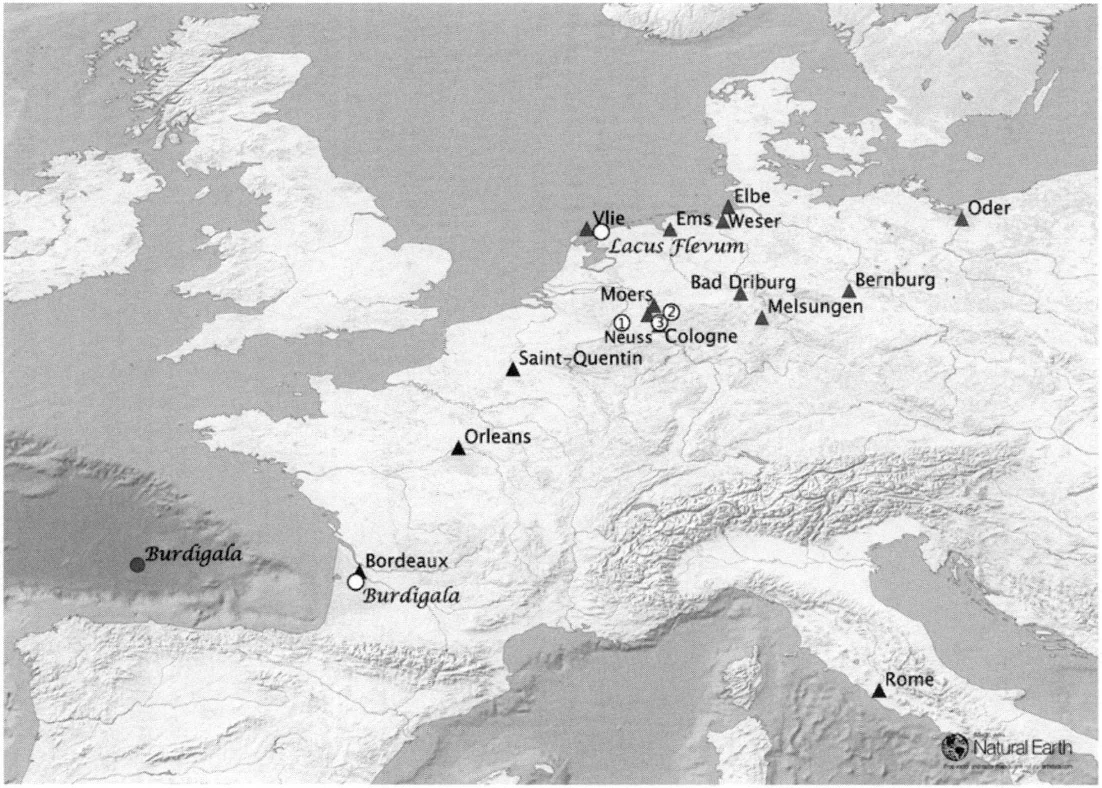

Fig. 3: Real position of Novaesium/Neuss in comparison with its position recalculated relative to Burdigala/Bordeaux (white circle numbered 1), Cenabum/Orleans (white circle numbered 2), and Augusta Viromanduorum/Saint-Quentin (white circle numbered 3). The Ptolemaic position of Burdigala relative to Rome (grey circle) is given in comparison with its recalculated position (white circle); the position of Lacus Flevum/Vlie is given in comparison with its position recalculated relative to Augusta Viromanduorum (white circle).

Another interesting side effect of such a mapping procedure is worth mentioning here. It concerns the fact that the Ptolemaic position of *Novaesium* adapted to the position of the Greenwich meridian defined relative to Rome attains 51° 10' N, 6° 50' E, which matches the longitudinal distance to the real coordinates of Neuss (51° 12' N, 6° 41' E) within 17 km![32] This is surprising, since Ptolemy firstly, had no reliable means to measure longitudinal distances,[33] and secondly, as already mentioned, mapped the places onto an Earth which is "too small". One would expect there to be a longitudinal aberration of Neuss. This coincidence can only be understood as a result of a special sequence of mapping procedures applied by Ptolemy at a sphere of a wrong size. Fig. 4 explains this result.

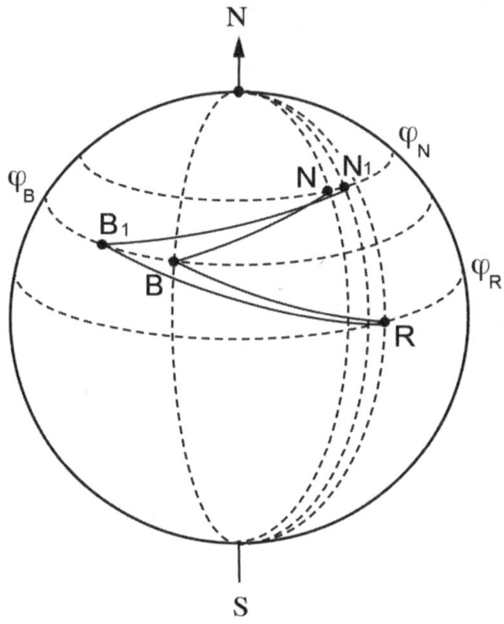

Fig. 4: Schematical illustration of the mapping of the position of Neuss (N) relative to Bordeaux (B) which itself is mapped relative to Rome (R).

Since the latitudes of Rome (φ_R), Bordeaux (φ_B) and Neuss (φ_N) seem to have been well-known in antiquity, and the measured distances from Rome to *Burdigala* (RB) and from *Burdigala* to *Novaesium* (BN) attain on the „smaller" Ptolemaic Earth

32 Up to our knowledge, this has not been noticed earlier. The reason may lie in the common adjusting of the Ptolemaic longitudes either to *Insulae Fortunatae* or to Alexandria.

33 The only possibility to determine longitudinal distances in antiquity was via the observation of (lunar) eclipses. But even in this case, without synchronized clocks and agreed conditions of observations, the precision of coordinates, which Ptolemy exhibits in his *Geography* (i. e. 5'), could not be achieved. The only evidence about using eclipses in ancient geographical mapping concerns the famous solar solstice in *Arbela* during Alexander's campaign. No evidence exists for *Germania magna*.

bigger angular values (great arcs RB_1 and B_1N_1, respectively), Ptolemy was forced to put *Burdigala* onto his map west of its true position, i. e. at B_1 (to attain the same latitudinal circle with a bigger arc), and then to match the longitudinal circle of *Novaesium* with an arc of the great circle B_1N_1 (which is bigger than the "true" arc BN). Due to this special configuration, the resulting Ptolemaic position of *Novaesium* referred to the meridian of Rome differs not much from its modern position – the errors of the bidirectional tracking have been compensated. This consideration proves again that *Novaesium* belongs not to the "Cologne map" but rather to the "Rome map".

CONCLUSION

Statistical and/or geodetical methods of "improving" the Ptolemaic coordinates should be applied very carefully: they imply that Ptolemy used only or at least mainly *local* information during his mapping procedure. Although this might be true in many cases, Ptolemy also made a conscious effort to link remote places. This causes a kind of "overmapping" in many cases, where Ptolemy had different data for a region at his disposal and needed to combine and adjust this data. Such a procedure necessarily involves in a map different sets of locations with different error patterns, e. g. in longitudes. Consequently, modern statistical "improvements" or "rectifications" of Ptolemy's mixed coordinates like that of Kleineberg et al.[34] will either classify locations as "unidentified" because of the huge aberration of their coordinates or – what is even more dangerous – provide a wrong solution by applying deformation factors which were gained on the basis of a local map. The case of Neuss seems to be a neat example of such an "overmapping".

We are able to explain the misplacement of *Novaesium*, surely to be identified with modern Neuss, on Ptolemy's map from another mathematical perspective. While converting terrestrial distances to angular coordinates, Ptolemy adopted the wrong value for the circumference of the earth (180,000 instead of 250,000 stades). This procedure prompted cities, connected in *itineraries*, to shift on his map. The case of *Novaesium* is clearly an example of this phenomenon. By linking it with cities in *Galliae* via an ancient road, Ptolemy had to put *Novaesium* on the right side of the Rhine and list it among the cities of *Germania magna*. Our programs now provide a digital tool for identifying the dozens of hitherto unknown cities in Ptolemy's description of *Germania magna*.

34 A similar approach was recently proposed by Rinner 2013.

REFERENCES

BERNECKER, Annemarie: Die Feldzüge des Tiberius und die Darstellung der unterworfenen Gebiete in der "Geographie des Ptolemaeus". Bonn: Rudolf Habelt, 1989.

BERTHELOT, André: La Germanie d'après Ptolémée: Configuration général et orographie. In: Revue des études anciennes 37 (1935), 34–44.

BILIŃSKI, Bronisław: Le vie dell'ambra, la Vistola e le carte geografiche di Tolemeo. In: Archaeologia Polona 7 (1964), 135–59.

CAPELLE, Wilhelm (ed.): Das alte Germanien: Die Nachrichten der griechischen und römischen Schriftsteller. Jena: Eugen Diederichs, 1937.

CHANTRAINE, Heinrich; GECHTER, Michael; HORN, Heinz Günter; KNÖRZER, Karl-Heinz; MÜLLER, Gustav; RÜGER, Christoph B.; TAUCH, Max: Das römische Neuss. Stuttgart: Konrad Theiss, 1984.

FORBIGER, Albert: Handbuch der alten Geographie. Aus den Quellen bearbeitet. Dritter und letzter Band. Europa. 2. umgearb. u. vielfach verb. Aufl. Leipzig: Mayer und Wigand, 1877. Repr. Graz: Akademische Druck- u. Verlagsanstalt 1966.

FRANKE, Alfred: Novaesium. In: RE 33: Nereiden–Numantia. Stuttgart: Druckenmüller, 1936, 1131–34.

GEUS, Klaus; TUPIKOVA, Irina: Von der Rheinmündung in den Finnischen Golf … Neue Ergebnisse zur Weltkarte des Ptolemaios, zur Kenntnis der Ostsee im Altertum und zur Flottenexpedition des Tiberius im Jahre 5 n. Chr. In: Geographia Antiqua 22 (2013), 125–43.

GNIRS, Anton: Das östliche Germanien und seine Verkehrswege in der Darstellung des Ptolemaeus: Ein Beitrag zur alten Geographie von Germanien; Mit einer Karte. Prag: Rohliček und Sievers, 1898 (Prager Studien aus dem Gebiete der Geschichtswissenschaft; H. 3).

GREWE, Klaus: Alle Wege führen nach Rom: Römerstraßen im Rheinland und anderswo. In: Landschaftsverband Rheinland / Rheinisches Amt für Bodendenkmalpflege (ed.): Alle Wege führen nach Rom: Internationales Römerstraßenkolloquium Bonn, Pulheim Brauweiler: Rhein Eifel Mosel Verlag, 2004, 9–42.

GUCKELSBERGER, Kurt: Private communication on the longitudinal values of Ptolemy's coordinates. 2013.

HANSEN, Günther Christian: Klaudios Ptolemaios (Kommentar, Geographie). In: Herrmann, Joachim (ed.): Griechische und lateinische Quellen zur Frühgeschichte Mitteleuropas bis zur Mitte des 1. Jahrtausends u.Z.. Dritter Teil: Von Tacitus bis Ausonius (2. bis 4. Jh. u. Z.). Berlin: Akademie, 1991 (Schriften und Quellen der alten Welt; 37, 3), 553–87.

HOLZ, Georg: Beiträge zur deutschen Altertumskunde. I. Heft: Über die germanische Völkertafel des Ptolemaeus. Mit einer Tabelle. Halle a. S.: Max Niemeyer, 1894.

HULTSCH, Friedrich: Griechische und römische Metrologie, 2. Aufl., Berlin: Weidmann, 1882 (Repr. Graz 1971).

KLEINEBERG, Andreas; MARX, Christian; KNOBLOCH, Eberhard; LELGEMANN, Dieter: Germania und die Insel Thule: Die Entschlüsselung von Ptolemaios "Atlas der Oikumene". Darmstadt: Wissenschaftliche Buchgesellschaft, 2010.

KLEINEBERG, Andreas, MARX, Christian; LELGEMANN, Dieter: Europa in der Geographie des Ptolemaios; Die Entschlüsselung des "Atlas der Oikumene"; Zwischen Orkney, Gibraltar und den Dinariden. Darmstadt: WBG (Wissenschaftliche Buchgesellschaft), 2012.

KRAHE, Hans: Sprache und Vorzeit: Europäische Vorgeschichte nach dem Zeugnis der Sprache. Heidelberg: Quelle & Meyer, 1954.

MANNERT, Konrad: Geographie der Griechen und Römer. Dritter Theil: Germania, Rhaetia, Noricum, Pannonia. 2., ganz umgearb. Aufl. Leipzig: Hahn'sche Verlags-Buchhandlung, 1820.

MARX, Christian; KLEINEBERG, Andreas: Die Geographie des Ptolemaios: Geographike Hyphegesis Buch 3: Europa zwischen Newa, Don und Mittelmeer. Berlin: epuli, 2012.

MEHLIS, Christian: Cl. Ptolemaeus und der Teutoburger Wald. In: Mitteilungen der Geographischen Gesellschaft in München 13, 1 (1918), 57–125.

Patzig, Hermann: Die Städte Großgermaniens bei Ptolemäus und die heut entsprechenden Orte. Dortmund: Fr. Wilh. Ruhfus, 1917.

Rathmann, Michael: Die Reichsstraßen der Germania Inferior. In: Bonner Jahrbücher 204 (2004), 1–45.

Rathmann, Michael: Die Reichsstraßen der Germania Inferior – ein Überblick. In: Landschaftsverband Rheinland / Rheinisches Amt für Bodendenkmalpflege (ed.): Alle Wege führen nach Rom: Internationales Römerstraßenkolloquium Bonn, Pulheim Brauweiler: Rhein Eifel Mosel Verlag, 2004, 225–34.

Rinner, Elisabeth: Zur Genese der Ortskoordinaten Kleinasiens in der Geographie des Klaudios Ptolemaios. Bern: Bern Studies in the History and Philosophy of Science, 2013.

Russo, Lucio: Ptolemy's Longitudes and Eratosthenes' Measurement of the Earth's Circumference. In: Mathematics and Mechanics of Complex Systems 1, 1 (2012), 67–79.

Sadowski, J. N. von: Die Handelsstrassen der Griechen und Römer durch das Flussgebiet der Oder, Weichsel, des Dniepr und Niemen an die Gestade des Baltischen Meeres. Übers. v. Albin Kohn. Amsterdam: Meridian Publishing Co. 1963. Urspr. Jena 1877.

Schöning, A[rtur]: Germanien in der Geographie des Ptolemäus. Mit einer Fehlerkritik. Detmold: Ernst Schnelle, 1962.

Stadt Neuss (ed.): Novaesium: Neuss zur Römerzeit. Neuss: Kulturamt der Stadt Neuss, 1989 (Schriftenreihe der Volkshochschule Neuss; H 4).

Steche, Theodor: Altgermanien im Erdkundebuch des Claudius Ptolemäus. Leipzig: Kurt Kabitzsch, 1937.

Stückelberger, Alfred; Grasshoff, Gerd (eds.): Klaudios Ptolemaios. Handbuch der Geographie, Griechisch-Deutsch, 2 vols. Basel: Schwabe, 2006.

Tupikova, Irina: Ptolemy's circumference of the earth. Berlin: Max Planck Institute for the History of Science, 2014 (TOPOI – Towards of Historical Epistemology of Space = Preprint; 464).

Tupikova, Irina; Geus, Klaus: The Black Sea in Ptolemy's *Geography* and (Ps.-)Arrian's *Periplus Ponti Euxini:* Explaining some differences and errors. In: Dan, Anca; Gehrke, Hans-Joachim; Podossinov, Alexandr (eds.): Arrian's Black Sea [in print].

Tupikova, Irina; Schemmel, Matthias; Geus, Klaus: Travelling along the Silk Road: A new interpretation of Ptolemy's coordinates. Berlin: Max-Planck-Institut für Wissenschaftsgeschichte, 2014 (MPIWG Preprints; 465).

Ukert, F. A.: Germania nach den Ansichten der Griechen und Römer dargestellt. Weimar: Verlag des Geographischen Instituts, 1843.

Wiegels, Rainer: Novaesium. In: Der Neue Pauly 8 (2000), 1018–20.

Sonja Nadolny

Die severischen Kaiserfrauen

PALINGENESIA – BAND 104

DIE AUTORIN
Sonja Nadolny studierte Alte Geschichte, Musikwissenschaft und Deutsche Philologie an der TU Berlin und an der Universität Bergen (Norwegen). Promotion bei Prof. Matthäus Heil über die Kaiserfrauen der severischen Dynastie. Forschungsschwerpunkt: frühe römische Kaiserzeit, Geschlechtergeschichte

Die antiken Historiker betonten wiederholt die außergewöhnliche Machtstellung der severischen Kaiserfrauen Julia Domna, Julia Maesa, Julia Soaemias und Julia Mamaea: Sie hätten Herrschaft und Reichspolitik der Jahre 193 bis 235 n. Chr. in erheblichem Maße bestimmt. Angesichts dieser Einschätzung verwundert es, dass die Severerinnen in den zahlreichen Publikationen über römische Kaiserfrauen noch immer unterrepräsentiert sind und auch die Frage nach den Gründen für ihre herausragende Stellung nur selten gestellt wurde.
Zum Ausgangspunkt ihrer Untersuchung nimmt Sonja Nadolny die öffentliche Rolle der Kaiserfrauen, die in einer großen Anzahl numismatischer und epigraphischer Quellen zutage tritt. Diese bezeugen nicht nur ihre intensive politische Einbeziehung in die kaiserliche Selbstdarstellung, welche das Kaiserhaus immer wieder als Familienherrschaft präsentierte, sondern auch ihre besondere Rolle für die Dynastiebildung und eine allgemeine Akzeptanz- und Verehrungsbereitschaft der römischen Öffentlichkeit gegenüber den Kaiserfrauen. Ihre herausragende Stellung ist dabei nicht als historischer Zufall anzusehen, sondern als logische Folge der zunehmenden Autokratisierung römischer Herrschaft.

2016
257 Seiten mit 10 Abbildungen, 7 Graphiken und 3 Übersichtenn
978-3-515-11311-3 GEBUNDEN
978-3-515-11315-1 E-BOOK

Franz Steiner Verlag

Hier bestellen:
www.steiner-verlag.de

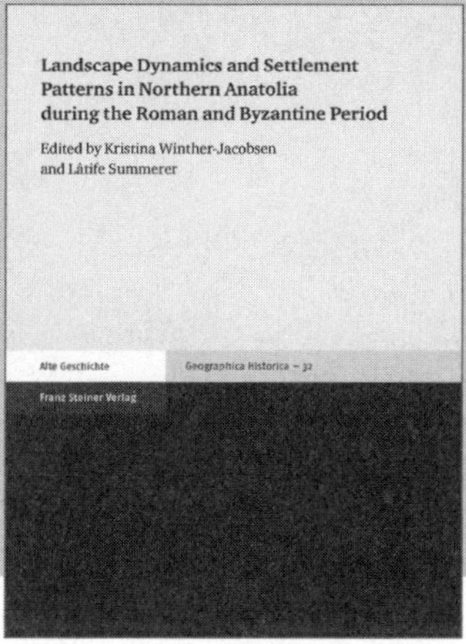

Kristina Winther-Jacobsen / Lâtife Summerer (ed)

Landscape Dynamics and Settlement Patterns in Northern Anatolia during the Roman and Byzantine Period

GEOGRAPHICA HISTORICA – VOL. 32

CONTRIBUTORS
Philip Bes, Owen Doonan, Peri Johnson, Kristina Winther-Jacobsen, Peter Bikoulis / Hugh Elton / John Haldon / Jim Newhard, Max Ritter, Celal Özdemir, Pavol Hnila, Vera Sauer, Julia Koch, Muzaffer Doğanbaş, İlkay İvgin, Baran Aydin / Laura Buccino / Lâtife Summerer, Luisa Musso, Esra Keskin, Tønnes Bekker-Nielsen / Rainer M. Czichon, Christian Marek, Vera Sauer

After his victory over Mithradates VI the Roman general Pompey founded a number of cities in order to shape the newly founded Roman province in the inland of the conquered former Pontic kingdom, in the southern Black Sea region. This sparked the beginning of an intense process of urban and rural development peaking in the 2nd century AD and continuing until the Byzantine period, a level of intensity never accomplished in northern Anatolia until today. The reorganization of space through the development of new urban centers affected the whole region and transformed the territory. This volume reviews current knowledge regarding these new founded Roman cities in relation to their territories, necropoleis and sanctuaries. It consists of 18 articles, which explore dynamics in settlement patterns, architecture, urban and mortuary spaces, monetary circulation and epigraphic habit. Some articles present the results of recent field research, others review little known material ripe for new interpretations, while new archaeological data is provided by the reports of rescue excavations carried out by local museums.

354 pages with 61 col. and
79 b/w photos, 9 col. and
40 b/w illustrations, 8 tables
978-3-515-11214-7 PB.
978-3-515-11216-1 E-BOOK

Franz Steiner Verlag

Hier bestellen:
www.steiner-verlag.de

Wolfgang Havener

Imperator Augustus

Die diskursive Konstituierung der militärischen *persona* des ersten römischen *princeps*

STUDIES IN ANCIENT MONARCHIES – BAND 4

DER AUTOR
Wolfgang Havener ist Akademischer Rat auf Zeit am Seminar für Alte Geschichte und Epigraphik der Ruprecht-Karls-Universität Heidelberg. Seine Forschungsschwerpunkte liegen u.a. im Bereich Kaiser, Militär und Eliten im römischen Prinzipat.

Die völlige Kontrolle über den militärischen Apparat war ein Grundpfeiler der Macht des ersten römischen *princeps*. Von Beginn seiner Herrschaft an nahm Augustus militärischen Erfolg und das daraus erwachsende politische und soziale Prestige exklusiv für sich in Anspruch. Den Mitgliedern der senatorischen Elite wurde damit eine zentrale Ressource im Kampf um Ehren, Ämter und Ansehen genommen.
Wolfgang Havener zeigt, wie es dem *princeps* gelang, sein Monopol im militärischen Bereich zu etablieren und es in eine Form zu überführen, die für die Senatoren akzeptabel war. Wie entwarf Augustus seine Rolle als ultimativer Sieger und oberster Feldherr? Wie wurde diese Rolle von den Angehörigen der senatorischen Oberschicht wahrgenommen? Welche Phasen im Kommunikationsprozess zwischen ihnen und Augustus lassen sich nachzeichnen? Zur Beantwortung dieser Fragen analysiert der Autor zentrale Themenfelder: Augustus' Handeln während der Bürgerkriege; die öffentliche Präsentation seines Sieges über Antonius und Kleopatra; die Darstellung des Verhältnisses von Krieg, Sieg und Frieden im augusteischen Prinzipat; und schließlich den Umgang des *princeps* mit dem Siegesritual des Triumphs.

2016
424 Seiten
978-3-515-11220-8 KART.
978-3-515-11221-5 E-BOOK

Franz Steiner Verlag

Hier bestellen:
www.steiner-verlag.de